Treatise on Heavy-Ion Science

Volume 6

Astrophysics, Chemistry, and Condensed Matter

Treatise on Heavy-Ion Science

EDITED BY D. ALLAN BROMLEY

Volume 1: Elastic and Quasi-Elastic Phenomena

Volume 2: Fusion and Quasi-Fusion Phenomena

Volume 3: Compound System Phenomena

Volume 4: Extreme Nuclear States

Volume 5: High-Energy Atomic Physics

Volume 6: Astrophysics, Chemistry, and Condensed Matter

Volume 7: Instrumentation and Techniques

Treatise on Heavy-Ion Science

Volume 6
Astrophysics, Chemistry, and Condensed Matter

EDITED BY
D. ALLAN BROMLEY
Henry Ford II Professor of Physics
Yale University
New Haven, Connecticut

PLENUM PRESS • NEW YORK AND LONDON

Library of Congress Cataloging in Publication Data

Main entry under title:

Treatise on heavy-ion science.

Includes bibliographies and index.
Contents: v. 1. Elastic and quasi-elastic phenomena— —v. 6. Astrophysics, chemistry, and condensed matter.
1. Heavy ions—Collected works. I. Bromley, D. Allan (David Allan), 1926–
QC702.7.H42T74 1984 539.7 84-8384
ISBN 0-306-41786-3 (v. 6)

A Division of Plenum Publishing Corporation
233 Spring Street, New York, N.Y. 10013

Printed in the United States of America

For Pat, Lynn, and David

Preface to Treatise on Heavy-Ion Science

After a long gestation period, heavy-ion physics and chemistry is now, worldwide, the most rapidly growing area of nuclear science, and the concepts, techniques, and instrumentation of this heavy-ion work are finding ever-widening application in other areas of science and technology. Although there remain broad regions at higher energies, with heavier projectiles, and at higher excitations and angular momenta where heavy ions still provide gateways into the totally unknown, intensive studies over the past two decades have provided a sound framework of understanding of many of the salient features of interactions induced by these new heavier projectiles and a basis for coherent planning of future studies.

It thus seemed appropriate, at this point in the history of the field, to pull together in one place and in as coherent a fashion as possible, an overview of what has been accomplished and some enlightened speculation about where we go next. It is my hope that these volumes will provide a definitive reference for those working in this and neighboring fields—both students and professional scientists; beyond that I would also hope that they will make accessible to a much wider audience in other sciences and technologies some of the richness of heavy-ion research, and perhaps help to stimulate the transfer of techniques and concepts that I have already mentioned.

I have been singularly fortunate in planning these volumes in being able to persuade internationally recognized authorities to write in their areas of special expertise and interest, and also fortunate that I have not had to restrict them to any artificial or externally imposed procrustean restrictions on the scope or length of their chapters. I have asked each author to include sufficient background to make the chapter accessible to students and to the nonspecialist, to provide a broad selection of illustrations, and to feel free to extrapolate and to speculate about future directions.

In inviting contributions to these volumes I have made arbitrary decisions concerning both topics and contributors, and I make no claim to completeness. Indeed, a few chapters that I would have liked to include do not appear because of illness or other reasons which prevented their authors from completing them.

I should like to take this opportunity to thank all the authors represented for taking time in already full schedules to share with a wider audience their special experience and expertise in heavy-ion science. As was inevitable from the outset in a multiple-author venture of this scope—with over 65 different authors—my original scheduling and deadlines proved unrealistic. To all those authors who responded to them and produced manuscripts on or before the original deadline—in many cases, I am aware, at substantial personal cost—my most sincere thanks and appreciation. To them, too, however, go my apologies for the delay in bringing their work into print. I have delayed publication for over a year so that I might include a number of very important chapters without which the work would have been obviously incomplete.

Volumes 1–4 of the *Treatise on Heavy-Ion Science* are devoted to aspects of heavy-ion nuclear science, beginning with an overview of the historical development of the science and some of its simpler interactions—elastic and quasi-elastic, fusion and quasi-fusion phenomena—and moving from them to compound system phenomena and to much more complex and less well understood phenomena involving very heavy nuclear species and very high energies. Volume 5 is devoted to high-energy atomic physics, an entirely new field of science first made accessible by the availability of a broad range of heavy-ion beams, and one still very much in its infancy. Volume 6 considers the impact of heavy-ion studies on other sciences including astrophysics, chemistry, surface physics, condensed matter physics, materials science, and heavy-ion-induced fusion power. Volume 7, the concluding volume of this treatise, is devoted to some of the instrumentation peculiar to heavy-ion science and its applications.

Special thanks go to Ellis Rosenberg and Bennett K. Ragnauth of Plenum Press with whom it has been a pleasure to work on these volumes, and to Mary Anne Schulz for all her help in producing them. And I would also acknowledge my indebtedness to the Alexander von Humboldt Stiftung for a Humboldt Award that I was privileged to hold during part of the time these volumes were in preparation.

New Haven, Connecticut D. Allan Bromley

Contributors to Volume 6

Richard C. Arnold, Max Planck Institut für Quantenoptik, Garching, West Germany

Charles A. Barnes, W. K. Kellogg Radiation Laboratory, California Institute of Technology, Pasadena, California 91125

Jochen P. Biersack, Hahn–Meitner Institute for Nuclear Research, Berlin, West Germany

Fulvio Cacace, University of Rome, 00100 Rome, Italy

Sheldon Datz, Physics Division, Oak Ridge National Laboratory, P.O. Box X, Oak Ridge, Tennessee 37831

Colin A. English, Metallurgy Division, Atomic Energy Research Establishment, Harwell, Didcot, Oxfordshire OX11 0RA, England

Donald S. Gemmell, Physics Division, Argonne National Laboratory, Argonne, Illinois 60439

Peter K. Haff, Division of Physics, Mathematics, and Astronomy, California Institute of Technology, Pasadena, California 91125

Michael L. Jenkins, Department of Metallurgy and Science of Materials, University of Oxford, Parks Road, Oxford, England

William A. Lanford, Department of Physics, State University of New York/ Albany, 1400 Washington Avenue, Albany, New York 12222

Charles D. Moak, Physics Division, Oak Ridge National Laboratory, P.O. Box X, Oak Ridge, Tennessee 37831

John M. Poate, AT&T Bell Laboratories, Murray Hill, New Jersey 07974

Stephen Trentalange, W. K. Kellogg Radiation Laboratory, California Institute of Technology, Pasadena, California 91125

Zeev Vager, Department of Nuclear Physics, The Weizmann Institute of Science, Rehovoth, Israel

Shiu-chin Wu, W. K. Kellogg Radiation Laboratory, California Institute of Technology, Pasadena, California 91125

James F. Ziegler, Watson IBM Research Laboratories, Yorktown Heights, New York 10598

Preface to Volume 6

Although still a very young field, the techniques, instrumentation, and concepts of heavy-ion science are finding broad application in other areas of science. In this sixth volume of this treatise on heavy-ion science, I include representative examples of such application to ten different areas of science. Barnes *et al.* discuss the role of heavy-ion reactions in nuclear astrophysics and in nucleogenesis and stellar energy generation in particular. Cacace considers their role in hot atom chemistry—so-called because the reactant species are not in their normal atomic or molecular ground states but rather in excited configurations reflecting the passage of energetic ions. Ziegler and Biersack review the essential data on the stopping range of heavy ions in matter—information of critical importance in the understanding of heavy-ion phenomena in almost all fields. Poate discusses ion implantation, a technique of burgeoning importance in the fabrication of ultra large-scale integrated electronic circuits, in ion–beam mixing to produce entirely new alloys and products and in ionic treatment of surfaces to obtain new wear, corrosion and other resistance, often in relatively inexpensive substrates. The fact that rapidly moving ions, when properly oriented and directed, can pass directly through lattice channels in crystalline targets with anomalously low energy loss has been recognized as providing a powerful probe for study of the interatomic forces and potentials in solids. Reflecting the very large local electrostatic and magnetic fields encountered by the ions within such a lattice, there may well be substantial practical applications. Datz and Moak review these questions in detail. During their passage through matter, highly charged ionic species have been shown to have dramatic effects on the electron gas in that matter; such phenomena as wakes and Mach shock waves have been clearly identified as have the Coulomb explosions that occur when a molecular ionic bond is broken and the fragments separate under their mutual electrostatic repulsion. Gemmell and Vager, pioneers in this work, review its present status. Haff reports on the modification of surfaces, terrestial and extraterrestial, under ionic bombardment and sputtering. This is yet another process that although very poorly understood, as yet, from any fundamental viewpoint, has already found widespread technological application. English and Jenkins consider heavy-ion-induced damage in solids; this is of major importance in the mechanical hardening of materials through the introduction of dislocations, in the simulation of long term damage under neutron bombardment in both fission and fusion reactors, and as a further probe for the structure of solids. There is already a growing history of utilization of heavy ions as analytic probes in Rutherford backscattering, induced X radiation, and the like; Lanford considers all these analytic techniques including those in which he has done major work e.g., the use of ^{15}N beams in quantitative, nondestructive hydrogen profiling in solids. Finally, Arnold considers one of the possible longer range applications of heavy-ion beams in the induction of inertial confinement fusion. Here the heavy ions have the great advantages of tight coupling to the fuel pellet, hence efficient energy delivery, and well understood beam production and handling techniques. And as emphasized above, this volume represents only a sampling of the application of heavy ions to other sciences and to technology. I regret, for example, that it was not possible to include a chapter here on the medical and clinical uses of heavy-ion beams. Successful use of GeV/A beams of silicon, from the Berkeley Bevalac, in the treatment of otherwise intractable pancreatic cancer is only a single example of such use.

New Haven, Connecticut D. Allan Bromley

Contents

2. Heavy Ions in Hot Atom Chemistry

Fulvio Cacace

3. The Stopping and Range of Ions in Matter

James F. Ziegler and Jochen P. Biersack

4. Ion Implantation

John M. Poate

5. Heavy-Ion Channeling

Sheldon Datz and Charles D. Moak

6. The Electronic Polarization Induced in Solids Traversed by Fast Ions

Donald S. Gemmell and Zeev Vager

9. Analysis with Heavy Ions

William A. Lanford

10. Heavy-Ion-Induced Fusion Power

Richard C. Arnold

Contents of Volumes 1–5, 7

Volume 1: Elastic and Quasi-Elastic Phenomena

Volume 2: Fusion and Quasi-Fusion Phenomena

Volume 3: Compound System Phenomena

Volume 4: Extreme Nuclear States

Volume 5: High-Energy Atomic Physics

Volume 7: Instrumentation and Techniques

1

CHARLES A. BARNES (top) is Professor of Physics and Co-Principal Investigator at the W. K. Kellogg Radiation Laboratory of the California Institute of Technology. He received a B.A. degree in 1943 from McMaster University, and an M.A. in 1944 from the University of Toronto; in 1950 he received his Ph.D. degree from Cambridge University. From 1944 to 1946 he was a Research Physicist with the joint British–Canadian Atomic Energy Project in Montreal and Chalk River, Canada. From 1950 to 1953, and during 1955–1956, he was an instructor and Assistant Professor at the University of British Columbia. From 1953 to 1955, and since 1956, he has been affiliated with Caltech as Research Fellow, Senior Research Fellow, Associate Professor and, since 1962, Professor of Physics. During 1962–1963, he was an NSF Senior Fellow at the Niels Bohr Institute and, during 1973–1974, a NORDITA Guest Professor at the same Institute.

S. TRENTALANGE (bottom left) is currently an Assistant Research Physicist at the University of California, Los Angeles. He received his Ph.D. degree in Physics from Caltech in 1983. His undergraduate training was at the Rensselaer Polytechnic Institute, from which he received a B.S. degree in 1976. As an undergraduate, he participated in experimental studies at the Bates Linear Accelerator Laboratory at MIT.

S.-C. WU (bottom right) has been an Associate Professor of Physics at the National Tsing-Hua University in Taiwan since 1982. She received a B.S. degree in Physics from Tsing-Hua University in 1974, and a Ph.D. degree in Physics from Caltech in 1978. She was a Research Fellow during 1978 at Rutgers University, and from 1979 to 1981 she was a Research Fellow at Caltech.

Heavy-Ion Reactions in Nuclear Astrophysics

C. A. BARNES, S. TRENTALANGE, AND S.-C. WU

1. Introduction

Because of the enormous ranges of temperature and density that exist in various places throughout the Universe, it is not hard to imagine situations where the kinetic energies should be high enough for heavy ions to react with one another. There are many dramatic photographs from astronomical observatories around the world showing galaxies undergoing titanic explosions which must involve temperatures high enough for Coulomb barrier penetration by heavy ions. On a smaller scale, but still with enormous energy releases, supernovas occur at a rate in the neighborhood of one per galaxy per 50 years. Very high temperatures are reached in their cores during the explosions. Supernovas are now presumed to leave neutron stars or black holes as remnants, at least part of the time; in the enormous gravitational fields around such remnants, very high temperatures must also exist.

However, the Universe is fundamentally different from the nuclear physics laboratory in several crucial respects; perhaps the most important of these is that matter is generally in local thermal equilibrium, or close to it, in astrophysical contexts. Thus, in addition to the ion energies being distributed according to a Maxwell–Boltzmann distribution, there will be an intense bath of thermal electromagnetic radiation in equilibrium (or close to equilibrium) with the matter. The high-energy tail of this photon distribution can photodisintegrate nuclei as rapidly as they are built if the temperature rises above about 5×10^9 K, even though kT is only roughly 0.5 MeV at this

C. A. BARNES, S. TRENTALANGE, AND S.-C. WU • W. K. Kellogg Radiation Laboratory, California Institute of Technology, Pasadena, California 91125. Work supported in part by the National Science Foundation PHY79-23638.

temperature. Thus, heavy-ion nuclear reactions as we usually study them in the laboratory are mainly precluded by photodisintegration at temperatures far lower than necessary for heavy ions to react significantly. Nucleosynthesis does occur in stars at temperatures up to a few times 10^9 Kelvins during supernovae because the nuclei present are not in thermal equilibrium; in equilibrium the abundances would be determined only by binding energies and partition functions. As nucleosynthesis starts with the lightest nuclei enormously more abundant than the heavier nuclei, there will still be a net nucleosynthetic flow towards heavier nuclei, even at temperatures above the 5×10^9 K range.

Heavy ions exist with small abundances in cosmic radiation. However, since hydrogen and helium are both enormously more abundant in interstellar space than the total of all the elements heavier than helium, the most likely fate for cosmic ray heavy ions is to be disintegrated or "spalled" to lighter nuclei by collisions with hydrogen or helium.

We conclude that only the heavy-ion reactions between carbon and oxygen nuclei within stellar interiors occur with sufficient frequency to be important in astrophysics, because of the lower temperatures at which these "light" heavy ions can react, and because carbon and oxygen are likely to be by far the most abundant low-Z nuclei present after helium burning.

The next section of this paper describes briefly our present understanding of the scenario for stellar nucleosynthesis, in which carbon burning and oxygen burning play a significant role. Section 3 discusses the laboratory techniques usually employed for measurement of heavy-ion fusion cross sections at very low energies. Section 4 reviews the sub-Coulomb-barrier cross-section data for the $^{12}C + {}^{12}C$, $^{12}C + {}^{16}O$, and $^{16}O + {}^{16}O$ reactions, and Section 5 discusses the models commonly employed to describe these reactions, and to extrapolate their excitation functions to energies below those for which measurements have proved feasible. Section 6 summarizes the present astrophysical status of the heavy-ion work.

2. Nuclear Astrophysics

2.1. The Primeval Big Bang

The initial phase of our current scenario for the history of the Universe is the "hot big bang." There is abundant evidence that the Universe is expanding, and the hypothesis that the Universe started from a high-temperature, high-density singularity provides a convincing explanation for most of the observed general features of the Universe, including the isotropic black-body radiation characterized by a temperature of approximately 3 K (Pe 65, Di 65). This radiation is interpreted as the relic of the black-body radiation from an earlier epoch during which the radiation was in thermal equilibrium with charged matter, now greatly red-shifted by the universal expansion.

When the temperature of the Universe dropped below about 10^{10} K, nucleosynthesis began, starting with the mix of protons and neutrons provided by thermal equilibrium at the temperature, T_w, at which the weak interaction processes coupling the neutron and proton became too slow to maintain thermal equilibrium. Thus the ratio of the initial mass fractions of neutrons and protons is given by

$$X_n/X_p = \exp[-(M_n - M_n)\, c^2/kT_w] \tag{1}$$

As this ratio is approximately 0.1, and all the neutrons would rapidly capture protons to form ^{4}He via the nuclear reaction sequence,

$$n + {}^1\text{H} \rightarrow {}^2\text{D} + \gamma \tag{2}$$

$${}^1\text{H} + {}^2\text{D} \rightarrow {}^3\text{He} + \gamma \tag{3}$$

$${}^3\text{He} + {}^3\text{He} \rightarrow {}^4\text{He} + 2\,{}^1\text{H} \tag{4}$$

the mass fraction of ^{4}He from the big bang is expected to be about 0.2. The most recent calculations of nucleosynthesis in the big bang yield a ^{4}He mass fraction of approximately 0.22, in agreement with the minimum mass fractions of ^{4}He derived from measurements of the galaxies with the lowest abundance of the elements heavier than helium (Le 79, Fr 80).

Of special interest from the point of view of later nucleosynthesis is the current view that most of the ^{3}He and D in the Universe, as well as most of the ^{4}He, are the result of nuclear reactions during the expansion of the primeval fireball. It appears also that much of the ^{7}Li, a rare nuclide in the Universe, comes from big-bang nucleosynthesis, but that the amounts of all other nuclides produced in the big bang are negligible compared with the production in later nucleosynthesis (Wa 67, Fo 67, Wa 73), mainly because of the difficulty of bridging the "mass gaps" at $A = 5$ and $A = 8$. For ^{5}He and ^{5}Li, the lifetimes for nucleon emission are about 10^{-21} s, and for ^{8}Be, about 10^{-16} s for breakup into α particles. (There seems to be no reasonable way to make the much longer-lived β-active nuclei ^{8}He, ^{8}Li, ^{8}B, and ^{8}C). To a small degree, the mass gaps can be bridged by reactions such as $^3\text{He} + {}^4\text{He} \rightarrow {}^7\text{Be} + \gamma$, $^7\text{Be} + {}^4\text{He} \rightarrow {}^{11}\text{C} + \gamma$, and $3\ {}^4\text{He} \rightarrow {}^{12}\text{C} + \gamma$. The temperatures and densities required are simply not available in the big bang by the time ^{3}He and ^{4}He are made in sufficient quantities, although these reactions are important later in stellar nucleosynthesis, at the appropriate densities and temperatures.

2.2. Some Mathematical Preliminaries

Nucleosynthesis within stars has been reviewed many times more fully than the topic can be dealt with here. (See, for example, Fo 67b, Cl 68, Fo 75, Tr 75, Ro 78, Ba 80, Ba 82.) Nevertheless we will present a brief outline of the subject to illustrate the role of heavy-ion reactions.

During the formation and early phases of the evolution of stars, there cannot be large concentrations of neutrons. If the density is low, as during contraction of the protostellar gas cloud, any neutrons present undergo β-decay. When the density becomes slightly higher, they will be captured on the abundant hydrogen present. The nuclear evolution of the star is thus largely a question of charged-particle-induced nuclear reactions, starting with matter that is largely hydrogen and helium-4, "salted" only lightly with traces, $\lesssim 10^{-3}$ by number, of heavier nuclei if the stars are second or later generation objects.

The earliest reactions in the sequence of stellar burning stages are nonresonant, and it is convenient to factor their cross sections,

$$\sigma(E) = [S(E)/E] \exp(-2\pi\eta) = [S(E)/E] \exp(-2\pi Z_1 Z_2 e^2/\hbar v) \qquad (5)$$

where we expect $S(E)$ to be only weakly dependent on the kinetic energy, E. The other factor $[1/E] \exp[-2\pi\eta]$ is approximately the s-wave point-nucleus Coulomb barrier penetration probability. Thus $\sigma(E)$ varies approximately as $E^{-1} \exp(-bE^{-1/2})$, which yields a very steep energy dependence at low energies. The rate for the nuclear reaction

$$1 + 2 \rightarrow 3 + 4 \qquad (6)$$

in a thermal equilibrium distribution is given by weighting the cross section by the Maxwell–Boltzmann velocity distribution. After converting the independent variable in the integral to kinetic energy, the thermally averaged reaction rate is

$$\langle \sigma v \rangle = [8/\pi M (kT)^3]^{1/2} \int_0^\infty \sigma(E) \exp(-E/kT)\, E\, dE \qquad (7)$$

where M is the reduced mass.

The integrand in this expression has the form of a somewhat skewed peak with its maximum at energy E_0, usually called the Gamow peak, and this fact provides a simple method for approximating the integral. Expressing $S(E)$ as a Maclaurin series leads to (Ba 66, Fo 67b)

$$\langle \sigma v \rangle = (2/M)^{1/2} [\Delta E_0/(kT)^{3/2}]\, S_{\text{eff}} \exp(-\tau) \qquad (8)$$

where

$$E_0 = [\pi\alpha Z_1 Z_2 kT(Mc^2/2)^{1/2}]^{2/3} \tag{9}$$

$$\Delta E_0 = 4[E_0 kT/3]^{1/2} \tag{10}$$

$$S_{\text{eff}} = S(0)\{1 + (5/12\tau) + [E_0 + (35kT/36)][S'(0)/S(0)] + \cdots\} \tag{11}$$

and

$$\tau = 3E_0/kT \tag{12}$$

ΔE_0 is the full width of the Gamov peak at $1/e$ of the maximum.

For nuclear reactions in which the cross section exhibits narrow resonances, the cross section may conveniently be written in the Breit–Wigner form,

$$\sigma(E) = \pi\lambda\!\!\!^{-2}\omega\Gamma_{12}\Gamma_{34}/[(E - E_R)^2 + \Gamma^2/4] \tag{13}$$

If $\Gamma \ll \Delta E_0$, the resonance acts as a δ function, and

$$\langle\sigma v\rangle = [2\pi\hbar^2/MkT]^{3/2}[\omega\gamma/\hbar]\exp[-E_R/kT] \tag{14}$$

In these expressions

$$\omega = (2J_R + 1)/(2I_1 + 1)(2I_2 + 1) \tag{15}$$

and

$$\gamma = \Gamma_{12}\Gamma_{34}/\Gamma \tag{16}$$

The total reaction rate for the resonant case is the sum of terms like that in equation (14).

2.3. Stellar Burning Processes

The strong dependence of $\sigma(E)$ on energy, and the dependence on Z_1 and Z_2, as displayed in equation (5), lead to the grouping of the nuclear reactions in stars into successive processes or stages. These processes are quasistatic (hydrostatic) because the tendency of the star to contract under the action of gravity is balanced by the thermal pressure. Heat is supplied by the nuclear reactions at the rate needed to balance the outward energy flow in the star. It is easily seen that this situation is self-stabilizing. If the energy production is too small to balance the outflow, the temperature and hence the pressure fall. The star will then contract a bit, raising the density and converting some gravitational energy to heat; the temperature thus rises and the rate of nuclear energy generation increases until it balances the outflowing energy. After a nuclear fuel is exhausted, the star contracts sufficiently to ignite the nuclear fuel with the next higher Coulomb barrier.

The order of these burning stages is hydrogen, helium, carbon, neon, oxygen, and silicon burning. A newly formed star evolves from core hydrogen burning to shell burning of hydrogen around the resulting helium core. When the helium core becomes massive and hot enough, helium burning will begin at the center. This kind of succession continues so that a massive, evolved star may have a core burning one fuel, while other fuels are burning in shells at various radial distances outwards from the center.

The hydrogen-burning phase takes most of the life of the star, and the subsequent stages become progressively shorter mainly because the luminosity (energy output) of the star increases as the star evolves from one stage to the next. The increased luminosity during helium burning is produced by a great increase in the size of the star; it becomes a red giant by the time helium burning begins. The high luminosity for the later stages of stellar evolution results from neutrino emission directly from the burning region of the star, which bypasses the slow transport of energy to the surface of the star via photon emission and absorption. For the $25M_{\odot}$ (25 solar mass) star studied by Weaver *et al.* (We 78), the various hydrostatic (presupernova) stages are summarized in Table 1. At the conclusion of these hydrostatic stages of stellar evolution, nuclear energy production has run its course. Thermal pressure can no longer resist gravity and the core of the star must inevitably collapse. Although the details of the reversal of this collapse are still not well understood, there is a bounce, possibly caused by the hard core of the nucleon–nucleon force and the resulting stiffening of the equation of state; a strong shock wave propagates outward through the mantle and envelope of the star. The energy in the shock, and the shock-triggered nuclear energy release, are enough to disperse very nearly the whole star. Whether or not a remnant is left depends on the fine details of the bounce.

How rapidly and how far a star evolves through the sequence of

Table 1. Summary of Burning Stages in a 25 $M_{\odot}$ Star

Burning stage	Approximate temperature (10^9 K)	Principal products	Energy production (erg/g)	Approximate duration	Principal luminosity
Hydrogen	0.02	^{4}He, ^{14}N	8×10^{18}	7×10^6 yr	Photons
Helium	0.2	^{12}C, ^{16}O, ^{22}Ne	7×10^{17}	5×10^5 yr	Photons
Carbon	0.8	^{20}Ne, ^{23}Na, ^{24}Mg	5×10^{17}	6×10^2 yr	Neutrinos
Neon	1.5	^{16}O, ^{24}Mg, ^{28}Si	1×10^{17}	1 yr	Neutrinos
Oxygen	2.0	^{28}Si, ^{32}S	5×10^{17}	180 d	Neutrinos
Silicon	3.5	^{54}Fe, ^{56}Ni, ^{52}Cr	3×10^{17}	1 d	Neutrinos

burning stages listed above depends on the stellar mass. A still poorly known quantity in stellar evolution is the rate and total amount of mass loss. This may be extremely large for very massive stars; stars more massive than about $60M_{\odot}$ are not seen, either because they do not form at all or because they lose much of their mass quickly. In addition to continuous stellar winds, during the double-shell burning stage of stellar evolution, i.e., with simultaneous helium-burning and hydrogen-burning shells, instabilities occur which may cause a large, pulsed mass loss, even from relatively small stars, in the form of planetary nebulae—outward flowing shells of gas.

For stars with initial masses somewhat less than a solar mass, the hydrogen-burning lifetime becomes longer than the age of the Universe. For stellar masses in the approximate range 0.8–$6M_{\odot}$, it is believed that mass loss during the life of the star, especially in the double-shell-burning phase, will reduce the mass of the star to less than the Chandrasekhar mass ($\sim 1.4M_{\odot}$), in which case the star will become a white dwarf consisting mainly of helium and products of helium burning. In such a star, nuclear energy production has ceased and electron-degeneracy pressure supports the star. White dwarfs must eventually cool to oblivion unless they accrete more matter, and ignite nuclear burning again.

2.4. Carbon, Neon, and Oxygen Burning

It appears that there may be a narrow range of initial masses from ~ 6 to $\sim 8M_{\odot}$ in which the star may complete core helium burning and ignite carbon burning under a condition of strong electron degeneracy. Although there is still considerable uncertainty about the fate of such a star, most investigations lead to the conclusion that the temperature will rise to such a high value before thermal pressure can exceed the pressure of the degenerate electrons, that sufficient nuclear energy will have been released to completely disperse the star. One of the problems with this hypothesis is that much of the star—a significant fraction of a solar mass, at least—is expected to be turned into iron, and there may be a problem of overproduction of iron by such objects. These stars are called carbon-detonation supernovas, although it is not clear whether this class of star exists.

For stars more massive than $\sim 8M_{\odot}$, hydrostatic carbon burning can begin in the core; when carbon is exhausted there it will continue to burn outwards in a thin shell. The products of helium burning are principally ^{12}C and ^{16}O, in amounts determined by the relative rates of the $3\ ^{4}He \rightarrow\ ^{12}C$ and $^{12}C(\alpha, \gamma)^{16}O$ reactions. With the measured rates available at this time, it appears that comparable amounts of ^{12}C and ^{16}O will be produced for a wide range of stellar masses, the ratio of ^{16}O to ^{12}C increasing with stellar mass. The reaction $^{16}O(\alpha, \gamma)\ ^{20}Ne$ does not play a significant role in helium burning except for the most massive stars known. However, there will be

^{20}Ne in the star, formed in previous generations of stars, and the burning process following hydrostatic carbon burning is called neon burning.

Neon burning consists of the photo-alpha disintegration of neon by the high-energy tail of the Planck spectrum, and the addition of the released α particles to the other nuclei present. Thus we have mainly

$$\gamma + {}^{20}\mathrm{Ne} \rightarrow {}^{16}\mathrm{O} + {}^{4}\mathrm{He} \tag{17}$$

$$ {}^{4}\mathrm{He} + {}^{20}\mathrm{Ne} \rightarrow {}^{24}\mathrm{Mg} + \gamma \tag{18}$$

$$ {}^{4}\mathrm{He} + {}^{24}\mathrm{Mg} \rightarrow {}^{28}\mathrm{Si} + \gamma, \qquad \text{etc.} \tag{19}$$

Although the first of these is endoergic, with a rather low threshold, there is a net energy release in neon burning because of the higher energy released in the other reactions. For stars with initial masses up to at least $15M_{\odot}$, the sequence of neon- and oxygen-burning stages is extremely complicated (Sp 80), and there may be many flashes of both neon and oxygen burning. Neon burning may even begin quite far from the center of the core, because significant temperature inversions can develop as a result of the very high neutrino emission rates from the highly compressed electron-degenerate core. For more massive stars, core neon burning begins in a more orderly fashion, and is followed by a short-lived stage of hydrostatic oxygen burning.

While carbon burning directly produces ^{20}Ne, ^{23}Na, and ^{23}Mg, the alpha particles, protons, and the (small abundance of) neutrons produced lead to a spread of nuclei up to or slightly beyond ^{28}Si. Oxygen burning directly produces mainly ^{28}Si, ^{31}P, and ^{31}S. The alpha particles, protons, and neutrons released cause the buildup of nuclei up to the region of scandium, with ^{28}Si, ^{32}S, ^{36}Ar, and ^{40}Ca the most abundant products.

2.5. Silicon Burning and Supernovas

The final hydrostatic process, silicon burning, consists of a very large number of different proton-, neutron-, and alpha-induced reactions, at peak temperatures up to about $3.5 \times 10^9\,K$, or even higher. Most of the reactions are in quasiequilibrium, i.e., forward and reverse reactions occur at nearly equal rates. The alpha particles, protons, and neutrons arise mainly from photodisintegration of the most abundant nuclides, ^{28}Si and ^{24}Mg. Hence there is a "mass flow" up to and slightly beyond the well-known abundance peak in the neighborhood of iron. Beyond this region, the energy gained by adding nucleons to nuclei is less than that needed to photodisintegrate the abundant nuclei in and below the iron region. This effectively terminates nuclear energy production as well as charged-particle-induced nucleosynthesis.

After the cessation of nuclear energy production, and the onset of the strongly endoergic photodisintegration of iron down to helium and nucleons, the stellar core collapses, the bounce alluded to earlier occurs, and the resultant outgoing shock wave briefly raises the temperature of all of the outer layers of the star. The shock-wave-triggered nuclear reactions that result are referred to as explosive nuclear burning. At this time, supernova models are not able to give final answers to the questions of how much of each nuclide is produced in the presupernova evolution, and how much in the explosive burning during the supernova shock wave. Great progress has, however, been made recently in supernova models (see, for example, We 78, We 80, Wo 82). In these calculations it has been possible to include, for the first time, both a large enough nuclear network to account for most of the nuclear energy production and the dynamical evolution of the presupernova. For the supernova phase, a simple parametrization of the core bounce and subsequent shock wave, which agreed with the dynamical supernova calculations of Wilson (Wi 80), was employed. From these studies it seems that the ejected nuclides up to ^{28}Si are largely unaltered by the supernova shock wave, but that explosive oxygen burning significantly enhances nuclides such as sulfur, argon, and calcium. It now appears that detailed calculations of the type described here for a $25M_{\odot}$ supernova may have to be made for a large range of supernova masses before firm conclusions can be drawn about the sites and processes responsible for all the nuclides through the iron abundance peak.

It may be interesting to note that the observational data are also improving rapidly. Large enhancements of sulfur, argon, and calcium have been seen in the optical studies made of certain "filaments" in the expanding 300-year-old supernova remnant in Cas A (Ch 78, Ch 79b; see also Ki 80). Holt *et al.* (Ho 79) have also observed enhancements of these elements in Cas A x-ray spectra from the Einstein satellite. The x-ray spectrum of the remnant of Tycho Brahe's supernova of 1572 also shows strong enhancements of silicon, sulfur, argon, and calcium (Be 80). These are just the expected products of a zone that has been evolved through earlier stages and then modified by explosive oxygen burning before being projected, more or less unmixed, into interstellar space.

2.6. ^{12}C + ^{12}C, ^{12}C + ^{16}O, and ^{16}O + ^{16}O in Explosive Oxygen Burning

The predominant overall effect of oxygen burning is to produce one alpha particle per ^{28}Si,

$$^{16}\mathrm{O} + {}^{16}\mathrm{O} \rightarrow {}^{28}\mathrm{Si} + {}^{4}\mathrm{He} \tag{20}$$

At the high temperatures prevailing during oxygen burning, some oxygen will be photodisintegrated, in spite of the 7-MeV separation energy,

$$^{16}\mathrm{O} + \gamma \rightarrow {}^{12}\mathrm{C} + {}^{4}\mathrm{He} \tag{21}$$

and the carbon may react with oxygen, because of the low abundance of carbon,

$$^{12}\mathrm{C} + {}^{16}\mathrm{O} \rightarrow {}^{24}\mathrm{Mg} + {}^{4}\mathrm{He} \tag{22}$$

After one of these alpha particles is captured by the ^{24}Mg, the net result of the reaction is

$$^{16}\mathrm{O} + {}^{16}\mathrm{O} \rightarrow {}^{24}\mathrm{Mg} + 2\,{}^{4}\mathrm{He} \rightarrow {}^{28}\mathrm{Si} + {}^{4}\mathrm{He} \tag{23}$$

which is the same as equation (20).

If, on the other hand, the ^{12}C produced by the photodisintegration of the ^{16}O reacts with carbon, the alpha-particle production will be higher because the ^{20}Ne produced will very likely be photodisintegrated at the elevated temperatures of oxygen burning. Thus we have the more complicated chain of reactions,

$$2\,{}^{16}\mathrm{O} + 2\gamma \rightarrow 2\,{}^{12}\mathrm{C} + 2\,{}^{4}\mathrm{He} \tag{24}$$

$$^{12}\mathrm{C} + {}^{12}\mathrm{C} \rightarrow {}^{20}\mathrm{Ne} + {}^{4}\mathrm{He} \tag{25}$$

$$^{20}\mathrm{Ne} + \gamma \rightarrow {}^{16}\mathrm{O} + {}^{4}\mathrm{He} \tag{26}$$

which is stoichiometrically equivalent to

$$^{16}\mathrm{O} + {}^{16}\mathrm{O} \rightarrow {}^{16}\mathrm{O} + 4\,{}^{4}\mathrm{He} \tag{27}$$

The calculations of Woosley *et al.* (Wo 71) have shown that ^{16}O + ^{16}O and ^{16}O + ^{12}C do not produce enough alpha particles per ^{28}Si to build the observed abundances up to $A \approx 42$. The last chain above can easily augment the number of alpha particles. It thus appears that the relative rates of ^{16}O + ^{16}O, ^{12}C + ^{16}O, and ^{12}C + ^{12}C at supernova shock temperatures are all necessary in order to understand the nucleosynthetic results of explosive oxygen burning (Wo 71).

3. Measurement Techniques for Sub-Coulomb-Barrier Heavy-ion Reactions

In this section, we discuss the methods most commonly used to measure heavy ion reaction cross sections in the astrophysically interesting, sub-Coulomb-barrier energy range. For most purposes, such as the energy production rate in stars, or the general course of nucleosynthesis, it is only the total fusion reaction cross section that is needed. The reason for this is that the alpha particles, protons, and neutrons liberated in the reactions are readily captured by the fusion residues (the heavy resultant nuclei), at the temperatures at which the heavy ions will react. Thus, for example, the reactions

$$^{12}C + {}^{12}C \rightarrow {}^{20}Ne + {}^{4}He \tag{28}$$

$$\rightarrow {}^{23}Na + {}^{1}H \tag{29}$$

$$\rightarrow {}^{23}Mg + n \tag{30}$$

lead quickly to

$$^{12}C + {}^{12}C \rightarrow {}^{24}Mg + Q \tag{31}$$

as far as energy production and the general flow of nucleosynthesis is concerned. Partial reaction cross sections such as (28), (29), and (30) may have their value, however, as input for the massive nucleosynthetic codes in which the flow of nuclear reactions involving hundreds of nuclei is followed. The nucleosynthesis is frequently calculated by using simplified reactions, of the type of (31), until the stellar dynamical evolution is determined, and then analyzing, in more detail, what happens to the alpha particles, protons, and neutrons produced, in order to determine the nucleosynthetic yield of nuclides of both low and high abundance.

For hydrostatic burning of carbon and oxygen, the cross sections are often needed at energies well below those for which measurements can currently be made in the laboratory. In this case, it is necessary to make the best measurement possible, and then to use a nuclear model to extrapolate the cross sections to the energies needed for the astrophysical situation.

Aside from the very low cross sections encountered in the sub-Coulomb-barrier energy range, which make the measurements difficult, there is another experimental problem for some of the measuring techniques. It will frequently be the case that, after the compound nucleus emits one particle—an alpha particle, proton, or neutron—the residual nucleus will still have a high enough energy to emit a second particle. This means that, aside from having to keep track of many different reaction products for each interacting pair of nuclei, those methods that depend on counting alpha particles, protons, and neutrons may have severe double-counting problems, and it is the total number of reactions that is of primary interest. The Q-value diagrams of the $^{12}C + {}^{12}C$, $^{12}C + {}^{16}O$, and $^{16}O + {}^{16}O$ reactions are shown in Figures 1a, 1b, and 1c, which illustrate the very large number of open channels.

The most frequently used techniques for studying these heavy-ion reactions near and below the Coulomb barrier are (i) detection of the residual radioactivities, (ii) detection of the fusion residues, (iii) detection of the emitted particles, (iv) detection of the γ rays emitted by the fusion residues, and (v) elastic scattering measurements. Each of these methods has its individual strengths and weaknesses.

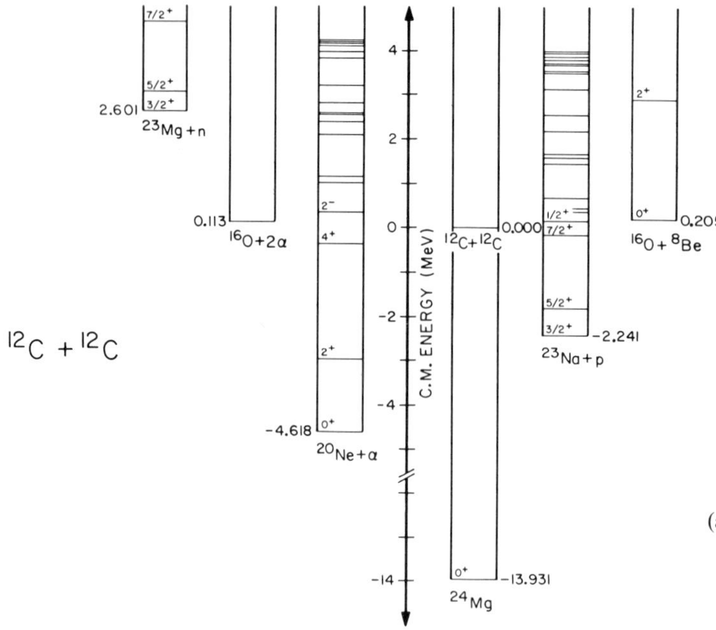

¹²C + ¹²C
C.M. ENERGY (MeV)
2.601
²³Mg + n
0.113
¹⁶O+2α
-4.618
²⁰Ne+α
¹²C+¹²C
0.000
²⁴Mg
-13.931
²³Na+p
-2.241
0.205
¹⁶O+⁸Be

(a)

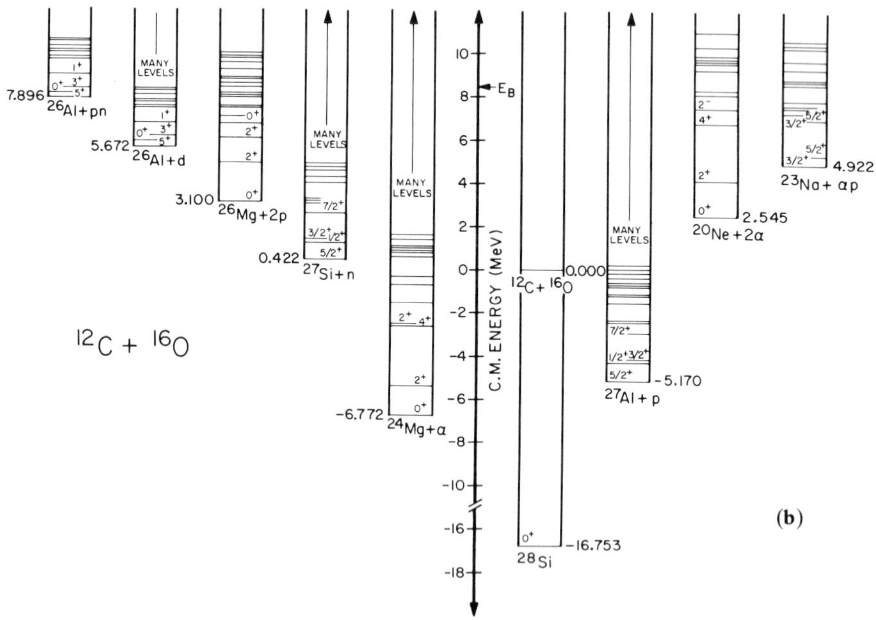

¹²C + ¹⁶O
C.M. ENERGY (MeV)
E_B
7.896
²⁶Al+pn
MANY LEVELS
5.672
²⁶Al+d
3.100
²⁶Mg+2p
0.422
²⁷Si+n
-6.772
²⁴Mg+α
¹²C+¹⁶O
0.000
²⁸Si
-16.753
²⁷Al+p
-5.170
2.545
²⁰Ne+2α
4.922
²³Na+αp

(b)

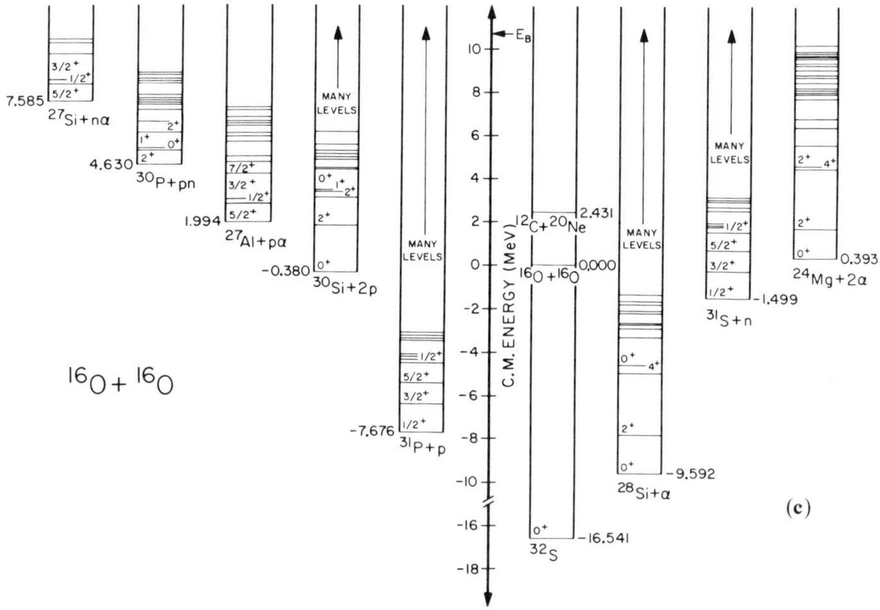

Figure 1. Q-value diagrams for (a) $^{12}C + ^{12}C$, (b) $^{12}C + ^{16}O$, and (c) $^{16}O + ^{16}O$. Only the lower-energy excited states are shown in each residual nucleus.

3.1. The Residual Radioactivity Method

The fusion of a pair of heavy ions, followed by the emission of a neutron or proton, often leads to the production of a radioactive residual nucleus. This means that background can usually be reduced by pulsing the beam at an appropriate rate, and counting positrons, annihilation radiation, electrons, or γ rays characteristic of the daughter nucleus, while the beam is off. Because of the low background resulting from counting with the beam off, the excitation functions of reactions leading to radioactive residual nuclei can often be followed to lower energies than is the case for reactions leaving stable residual nuclei.

Of course, there are usually still background problems. The most serious of these is the production of radioactive nuclides from contaminants in the target (or, occasionally, in the beam), which cannot be distinguished from the activity of interest, because of similarities in lifetime and/or end-point energy. If there are γ rays characteristic of the daughter product, the use of Ge(Li) or hyperpure Ge γ-ray detectors will usually produce a positive identification of the desired reaction.

An even more troublesome background is sometimes encountered—production of a desired radioactive residual nucleus by the interaction of the beam particles with a contaminant in the target. The best solution to this is, of course, to reduce the abundance of the offending contaminant to the lowest practicable level. Sometimes the difference in excitation function, usually reflecting a different Coulomb barrier or a different reaction mechanism such as Coulomb excitation, may be used to separate the contaminant reaction from the desired reaction.

The residual radioactivity method gives no information about the excited states populated by alpha-particle, proton, or neutron evaporation. Although this information is both interesting and important for nuclear physics purposes, it plays no role in astrophysical situations. We have found the radioactivity technique useful to confirm or check other methods of determining the cross sections leading to radioactive residual nuclei. In the fusion of carbon and oxygen nuclei, the radioactivities occur mainly after neutron evaporation, or neutron and proton or alpha evaporation, leaving positron emitters such as ^{22}Na, ^{23}Mg, ^{26}Al, ^{27}Si, ^{30}P, and ^{31}S.

3.2. Detection of Fusion Residues

The various fusion residues, i.e., the heavy residual nuclei, may be detected and identified with an appropriate selection of flight time, $\Delta E/\Delta x$ measurement, energy measurement, and momentum measurement (if a magnet is employed). Typical two-dimensional $\Delta E/\Delta x$ vs. E plots are shown in (Ey 76) or (Ko 79b), for example.

Fusion residue measurements, of course, require that measurements be made at many angles, in order to obtain angle-integrated cross sections. If the angular distributions are of interest for nuclear physics purposes—reaction mechanism, or spin and parity determination for resonances, for example—the effort may be quite worthwhile. From the astrophysical point of view, angular distributions are of little interest. For purely kinematic reasons, fusion residue angular distributions are forward peaked. The rapid variation in energy with angle, $dE/d\theta$, means that small intervals in θ must be employed, to obtain good energy resolution, and may impose some technical problems in detecting and identifying the nuclei because of the large range of laboratory energies encountered.

The fusion residue technique has been most successful at bombarding energies far above the Coulomb barrier. At such high energies, the fusion cross sections are relatively large, the elastic scattering cross sections are quite moderate, and the energy of the fusion residues is relatively high. As the bombarding energy is lowered below the Coulomb barrier, the fusion cross sections drop roughly exponentially, the elastic scattering cross sections (mainly Rutherford scattering) climb as $1/E^2$, and the fusion residues are of low energy and are much harder to detect and identify. It is clear that it will be a major experimental challenge to extend this cross-section measurement method to low energies, although the method is, in principle, the least ambiguous to interpret. We will argue later that a major effort should be made to extend the technique to low energies because of discrepancies among the results obtained with other methods.

3.3. Detection of the Emitted Particles

The light particles—alpha particles, protons, and neutrons—evaporated from the compound nucleus may be detected and identified in counter telescopes, usually consisting simply of a very thin solid-state or gas-filled $\Delta E/\Delta x$ detector and a residual energy detector for the charged particles, or a "long-counter" or time-of-flight system for the neutrons. As for the fusion residues, angular distributions must be measured in order to get the needed angle-integrated cross sections. When the elastically scattered target or beam particles are too numerous to be handled by the detectors or their associated electronic circuits, they must be "ranged out" with absorbing foils. This means that a portion of the emitted charged-particle spectrum is lost, corresponding to the high excitation energy region of the residual nuclei. This is sometimes a serious limitation of the technique, although it is usually possible to tailor the foil thickness at each angle, i.e., use the minimum possible foil thickness, so that the theoretically calculated lost part of the spectrum does not cause a serious error.

The spectrum of the emitted charged particles normally consists of well-defined higher-energy groups, corresponding to the ground and low-lying states of the residual nuclei, and a continuum corresponding to unresolved higher-density excited states of the residual nuclei. Because large fractions of the protons or alpha particles are in such unresolved continua, there is no way to tell whether these particles come from single-particle evaporation, or the evaporation of two or more particles; in the latter case, double counting of fusion events is possible. Coincidence measurements can throw some light on this question, but such measurements take a prohibitive amount of time at low energies, because of the very low yields. In the absence of coincidence information, one has to resort to a theoretical calculation (see Section 3.4), or be satisfied with upper and lower limits for certain residual nuclei.

3.4. Gamma-Ray Measurements

The fusion residues are usually produced with considerable excitation energy. When further particle emission is energetically impossible, or strongly hindered by Coulomb and angular momentum barriers, the nuclei decay to the ground state by gamma-ray emission, usually by multiple-step cascades. Nevertheless, most of the cascades are usually "funneled" through one, or a few, low-lying, high-J states.

This suggests immediately two quite different techniques for measuring fusion cross sections via the γ rays in the residual nuclei. (Gamma-ray emission directly from the highly-excited compound nucleus has a low probability compared with particle emission.) Čujec and Barnes (Cu 76) have shown that two large NaI(Tℓ) scintillation counters, close to the target, have a detection efficiency (for noncoincident counts) that is roughly independent of the number of steps in a cascade. Thus, a measurement of the total number of counts as a function of beam energy gives the excitation function relatively quickly, because of the high detection efficiency. The large solid angle subtended by the detectors in the appropriate geometry means that angular distribution effects are unimportant. Of course, particle evaporations from the compound nucleus, or from highly excited residual nuclei, that lead directly to the ground states of the final nuclei are not detected by the γ-ray method. Statistical model calculations indicate that such events constitute a small fraction (usually $\lesssim 10\%$) of all particle evaporations at low beam energies, and that they can be corrected for with acceptable accuracy. For higher energies in the laboratory, more frequent sequential multiple-particle evaporations are expected to make the ground-state particle emissions a more serious correction.

We have already noted that high detection efficiency is the principal advantage of the NaI γ-detection technique. The principal disadvantage is that the pulse-height spectrum obtained with this technique shows few resolved γ-ray peaks, and it is necessary to sum over a largely featureless pulse-height distribution. The experimenter is therefore vulnerable to γ rays from contaminant reactions in the target; even low abundance contaminants with lower Z than the target nuclei can dominate the yield at low energies and limit the energy range of the experiment. Also, because of the poor energy resolution of the detector, little or no information is obtained about the relative yields of the various possible fusion residues. This information is needed only for the most detailed nucleosynthesis calculations, however.

More recently, Ge(Li) or hyperpure Ge γ-ray detectors have been employed to overcome the problems caused by the poor resolution of the NaI scintillators, in spite of their lower intrinsic detection efficiencies. In the Ge detector technique, one usually detects one or more characteristic γ rays from the low-lying states of each of the fusion residues. This largely eliminates the troubles caused by contaminants, although Coulomb excitation of low-lying contaminant states sometimes causes problems when the contaminant is also one of the expected fusion residues. To some extent, the quite different excitation functions help with this problem, at least down to the energies where the Coulomb excitation, with its flatter excitation function, dominates the yield. The fact that ^{12}C and ^{16}O are both common contaminants, as well as the principal targets for the present study, may cause special problems. The same residual nuclei may be produced in reactions with both target nuclei. This causes trouble especially when the contaminant nucleus is the one with lower Z. Thus, because of the lower Coulomb barrier for ^{12}C, a small amount of carbon contamination may produce a significant yield of some residual nucleus when ^{16}O is the target. In most cases, the way around this problem is to determine the contaminant yield by looking at residual nuclei that come from the contaminant alone, and then subtracting off the contaminant contributions.

As for the NaI detector work, to improve detection efficiency, the detector is usually placed as close to the target as possible. This helps to average over unknown γ-ray angular distributions, although these are somewhat reduced already by the extensive cascading. For most nuclei, the γ-ray decay schemes of the low-lying states of the residual nuclei are already well known, and this helps in interpreting the observed γ-ray peaks in the detector pulse height spectrum.

The γ-ray yield from the low-lying states is not a direct measure of the fusion cross section, but must be related to it by models which determine the populations of the excited states and the cascades among them. In many cases, the excitation energies are so far above the thresholds for particle emission to the excited states of the various residual nuclei that the transmission coefficients are essentially unity. In such cases, the populations of the final states are just proportional to $(2J + 1)$, where J is the angular momentum of the final state. For the states for which the γ-branching ratios are known, it is then easy to relate the γ intensities to the transition probabilities. A more sophisticated analysis (which is often necessary) requires a Hauser–Feshbach statistical model calculation, instead of the simple $(2J + 1)$ assumption, especially for the situations where successive particle emissions occur. This is just the situation when the energy available for emitting a particular kind of particle is not large compared with the threshold energy. The calculations involve assumptions about energy level densities, because excitations above those for which the states have been well studied occur in many reactions.

From the statistical model calculation and the known efficiency of the detector, one can derive what are called summing and branching corrections. The summing correction takes account of the fact that a particular cascade may escape detection, not only when no γ ray in the cascade interacts with the detector, but when two or more γ rays from the same cascade interact. This has the result of removing a γ ray from the pulse-height peak corresponding to the full γ ray energy, and putting the event somewhere else in the spectrum, usually somewhere in the continuum pulse height distribution arising from Compton events in the detector. The branching correction gives the probability that a given cascade will pass through the state which emits the γ ray in question.

As has been discussed above, the excellent energy resolution of the solid-state γ detectors is their principal advantage. Their principal disadvantages are that their detection efficiency is lower than that of scintillators, and that one must rely on a model calculation to relate the observed γ rays to the total number of fusion events leading to each residual nucleus. Calculations of this kind are described in (Wu 78) and (Da 76), in more detail than can be provided here.

3.5. The Elastic Scattering Method

By invoking unitarity, it is possible, at least in principle, to extract the total reaction cross section from measurements of the elastic scattering. This sounds particularly appealing because the measurement of the relatively intense elastic scattering is usually much easier than the measurement of the very small sub-Coulomb-barrier reaction cross sections. The total amplitude for elastic scattering is the sum of the Rutherford (or Mott, in the case of identical nuclei) amplitude, f_{Coul}, and a nuclear compound-elastic amplitude f_N,

$$f = f_{\text{Coul}} + f_N \tag{32}$$

Although the angle-integrated cross section for Coulomb scattering is infinite, the optical theorem leads to the finite expression for the reaction cross section (Tr 80, Ho 65),

$$\begin{aligned}\sigma_R = 2\pi \int_{\theta_0}^{\pi} [(d\sigma_{\text{Coul}}/d\Omega) - (d\sigma_{\text{el}}/d\Omega)] \sin\theta \, d\theta - 4\pi\, |f_N(0)|^2 \sin^2(\theta_0/2) \\ + 4\pi\lambda\!\!\!^{-} \operatorname{Im}\{f_N(0) \exp[2i\eta \ln \sin(\theta_0/2) - 2i\sigma_0]\}\end{aligned} \tag{33}$$

where $d\sigma_{\text{Coul}}/d\Omega$ is the Rutherford (or Mott) differential cross section, $d\sigma_{\text{el}}/d\Omega$ is the measured elastic differential cross section, $f_N(0)$ is the non-Coulomb finite-range elastic scattering amplitude evaluated at 0^0, η is the Sommerfeld parameter $Z_1Z_2e^2/\hbar v$, σ_0 is the Coulomb phase shift for $l = 0$, and θ_0 is a scattering angle satisfying $\theta_0 < 1/l_{\text{grazing}}$ (Tr 80). In the case of Mott scattering, the integral should be cut off at $\pi/2$. It has been shown that the last two terms on the right-hand side can be neglected for heavy-ion scattering because $f_N(0)$ is small as a result of both η and the absorption being large (Wo 76). Basically, this is because the high Coulomb barrier makes the transmission coefficients for fusion very small, and the cross section for compound elastic scattering involves the square of this small transmission coefficient.

Without the last two terms, the reaction cross section becomes

$$\sigma_R = 2\pi \int_{\theta_0}^{\pi} [(d\sigma_{\text{Coul}}/d\Omega) - (d\sigma_{\text{el}}/d\Omega)] \sin\theta \, d\theta \tag{34}$$

for nonidentical nuclei, and for Mott scattering,

$$\sigma_R = 2\pi \int_{\theta_0}^{\pi/2} [(d\sigma_{\text{Coul}}/d\Omega) - (d\sigma_{\text{el}}/d\Omega)] \sin\theta \, d\theta \tag{35}$$

As the elastic scattering cross section rises when the energy is lowered, unlike the reaction cross section, it might appear that the elastic scattering method would be a promising method for determining reaction cross sections at very low energies. Although the reaction cross sections that have been derived in this way for $^{12}C + {}^{12}C$ are somewhat larger than those obtained in other measurements, the fusion resonances derived by the elastic scattering method agree well with those found in direct measurements of the reaction cross sections, down to $E_{\text{c.m.}} = 5.5$ MeV, the lower limit of the scattering measurements (Tr 80). To extend reaction cross-section determinations by the elastic scattering method to very low center-of-mass energies appears difficult. The problem is clearly that of subtracting two very large cross sections to obtain a very small one. Even determining $(d\sigma_{\text{Coul}}/d\Omega)$ from theory with sufficient accuracy requires knowing the beam energy and the beam energy loss in the target with very high accuracy.

Another feature of this method which could give serious problems in determining fusion cross sections in general is that neutron or alpha-particle transfer processes might dominate the reaction cross section for favorable Q values (Sw 74, Cu 79), and the elastic scattering method simply measures the flux lost from the elastic channel, whatever the reaction mechanism. For the three heavy-ion reactions of special interest here, neither nucleon nor alpha-particle transfers compete with fusion (compound nucleus) reactions (Ko 70, We 74), because their Q values are not optimal (Sw 74).

3.6. The Effects of Resonances

As a general note, we point out that for nonresonant nuclear reactions in general, and heavy-ion reactions in particular, the excitation functions at low energies are steep, nonlinear functions of the center-of-mass energy. This means that determining the effective beam energy is not completely trivial when finite thickness targets are employed. A technique called the "*e*-folding" method has been outlined by Mak *et al.* (Ma 74), and many other techniques have also been devised to handle the problem, usually requiring some assumption about the functional form of the energy dependence of the cross section, e.g., that it varies with energy as in equation (5), or in some other simple way.

In the presence of strong resonances, the problem is considerably more complicated. The experimenter must guess the form of the true cross-section curve and then integrate over the target thickness to get the effective energy for each measurement. This procedure must be iterated until the curve of cross section vs. effective energy, integrated over the target thickness, looks like the measured excitation function.

4. Experimental Data

Partial and total fusion cross sections, obtained by the methods discussed above, will be presented in this section of the paper for the reactions resulting from the bombardment of ^{12}C by ^{12}C, ^{12}C by ^{16}O, and ^{16}O by ^{16}O, at sub-Coulomb-barrier energies. The selection of these three reaction pairs, and the low-energy region, from the very large literature on light heavy ions, is partly to limit the length of the present paper, but mainly because only these three interaction pairs clearly play a major role in astrophysics, and this role is at subbarrier energies.

In order to reduce the enormous energy dependence of the cross section for the sub-Coulomb-barrier region, the cross sections are usually converted to S factors by equation (5):

$$S(E) = \sigma(E)E \exp(2\pi\eta) = \sigma(E)E \exp[2\pi Z_1 Z_2 e^2/\hbar v] \tag{36}$$

This is a model-independent way to display the cross section, and divides out the s-wave, point-nucleus barrier penetrability. For the three reactions, $2\pi\eta$ takes the values

$$\begin{aligned} 2\pi\eta &= 87.26/\sqrt{E} \quad &\text{for} \quad {}^{12}\text{C} + {}^{12}\text{C} \quad &(37) \\ &= 124.37/\sqrt{E} \quad &\text{for} \quad {}^{12}\text{C} + {}^{16}\text{O} \quad &(38) \\ &= 179.09/\sqrt{E} \quad &\text{for} \quad {}^{16}\text{O} + {}^{16}\text{O} \quad &(39) \end{aligned}$$

It will be clear from the figures to be presented later that converting from $\sigma(E)$ to $S(E)$ "overcompensates" for the energy dependence caused by Coulomb barrier penetration, since it converts a steeply rising function of E to a (less steep) decreasing function of E. The main reason for this is that the interacting nuclei have finite sizes. Sometimes, experimenters have converted their measured cross sections to a more nearly constant form by dividing out the calculated Coulomb penetrabilities. Unfortunately, this is no longer a model-independent way of displaying the cross section data, as it requires specifying a radius for the nuclei, and also a diffuseness if diffuse-edged nuclei are assumed. In a similar manner, a function $\tilde{S}(E)$ has sometimes been used, where the simplest finite-size correction has been included. Thus

$$\sigma(E) = [\tilde{S}(E)/E] \exp(-2\pi\eta) \exp(-gE) \quad (40)$$

where g is given theoretically (Fo 67b) by

$$g = 0.122(AR^3/Z_1Z_2)^{1/2} \text{ MeV}^{-1} \quad (41)$$

In this expression, $A = A_1A_2/(A_1 + A_2)$ is the reduced atomic weight. Usually, g is treated as a parameter to be adjusted to give as nearly a constant $\tilde{S}(E)$ as possible, and differs considerably from the theoretical value. From (40),

$$\tilde{S}(E) = E\sigma(E) \exp(2\pi\eta) \exp(gE) \quad (42)$$

In addition to specifying the value of g derived from the data, R must also be specified, as was required above.

We shall restrict our usage to the originally defined $S(E)$, except in Figure 7, to preserve the model independence of this manner of presenting cross sections. Confusion has occasionally arisen in the past in interpreting published data because of the various forms of S employed by different authors. In Section 5 of this paper, we employ $S(E)$ as the plotted form. Throughout Section 4, we display partial and total cross sections.

4.1. The $^{12}C + {}^{12}C$ Reactions

The $^{12}C + {}^{12}C$ data discussed in this paper are summarized in Table 2. The partial cross sections for the various reactions that occur when ^{12}C bombards ^{12}C are shown in Figures 2–6, for the data from Patterson *et al.* (Pa 69), Mazarakis and Stephens (Ma 73), High and Čujec (Hi 77), Becker (Be 78), and Dayras *et al.* (Da 77). The agreement between the various data sets is far from perfect. Some of the disagreement in the excitation functions is caused by differences in target thickness. However, a careful comparison of the data suggests that there are also substantial energy shifts between the data of different authors.

Even a small error in beam energy can very strongly distort the shape of a curve of $S(E)$ vs. E, or $\tilde{S}(E)$ vs. E. In Figure 7, the data of Mazarakis and Stephens (Ma 73) are shown as $\tilde{S}(E)$ vs. E. The points joined by the solid curve are the data as published by these authors. The points joined by the dashed curve were obtained by (arbitrarily) shifting the beam energy upwards by 50 keV, and the points joined by the dotted curve were obtained by shifting the beam energy upwards by 100 keV. Such a shift reduces the differences between the data of (Ma 73) and the data of the other authors. At

Table 2. Summary of $^{12}C + ^{12}C$ Measurements

$E_{c.m.}$(MeV)	Method employed	Measured quantities	Reference
3.23–8.75	Light particles	Total and partial fusion σ's	(Pa 69)
2.45–4.91	Light particles	Total and partial fusion σ's	(Ma 73)
3.68–7.49	Gammas, NaI	Total fusion σ	(Sp 74)
3.89–7.98	Solid-state detector	Elastic σ at 90 deg	(Sp 74)
3.54–4.94	Gammas, Ge(Li)	Neutron σ	(Da 77)
4.0–8.0	Light particles	Energy and J, alpha resonances	(Ga 77)
2.45–6.15	Gammas, Ge(Li)	Alpha and proton σ's	(Ke 77, Ke 80)
2.46–5.88	Gammas, Ge(Li)	Alpha, proton, and total fusion σ's	(Hi 77)
6.5–12.0	Solid-state detector	Alpha σ, energy and J of resonances	(Vo 77)
7.4–31.2	Fusion residues	Total and partial fusion σ's	(Ko 79b)
5.9–10.0	Gammas, Ge(Li)	Alpha and proton σ's, energy and J of resonances	(Er 80)
5.5–12.8	Elastic scattering	Total fusion σ	(Tr 80)
2.8–6.3	Light particles	Alpha and proton σ's	(Be 78)

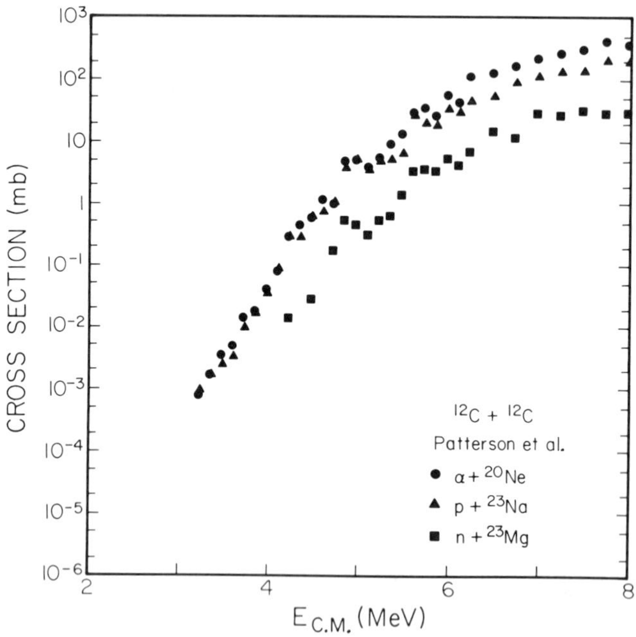

Figure 2. Partial cross sections for the $^{12}C + ^{12}C$ reactions from Patterson *et al.* (Pa 69).

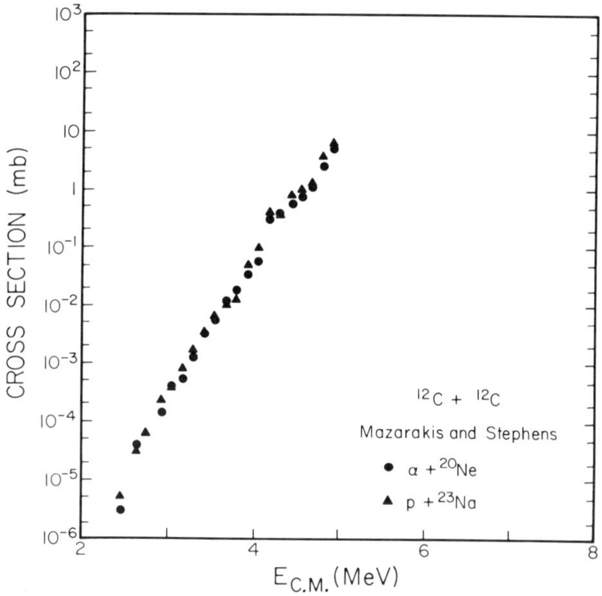

Figure 3. Partial cross sections for the $^{12}C + ^{12}C$ reactions from Mazarakis and Stephens (Ma 73).

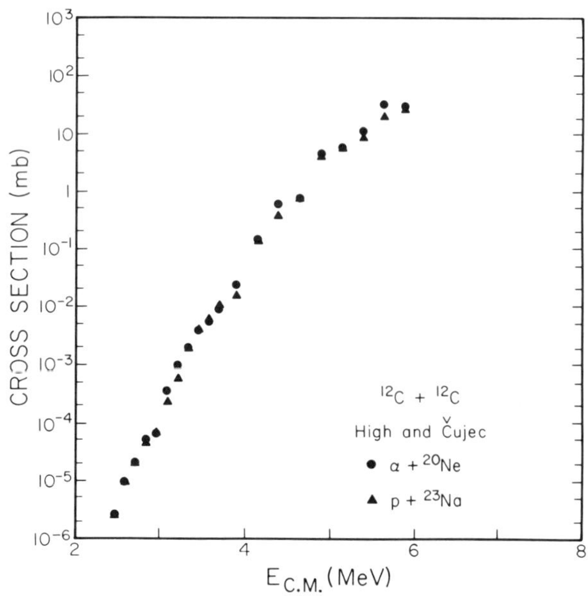

Figure 4. Partial cross sections for the $^{12}C + ^{12}C$ reactions from High and Čujec (Hi 77).

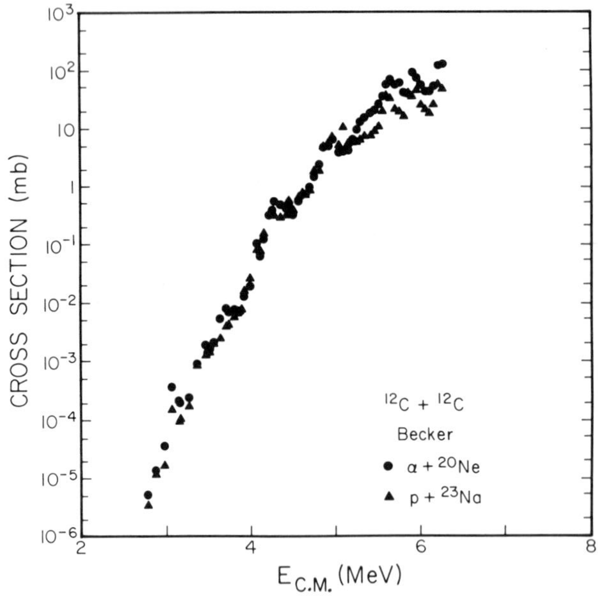

Figure 5. Partial cross sections for the $^{12}C + {}^{12}C$ reactions from Becker (Be 78).

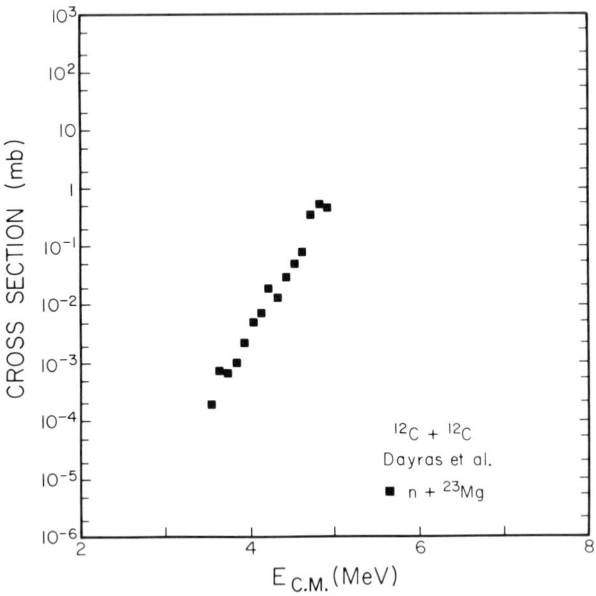

Figure 6. Partial cross section for the neutron channel in the $^{12}C + {}^{12}C$ reactions from Dayras *et al.* (Da 77).

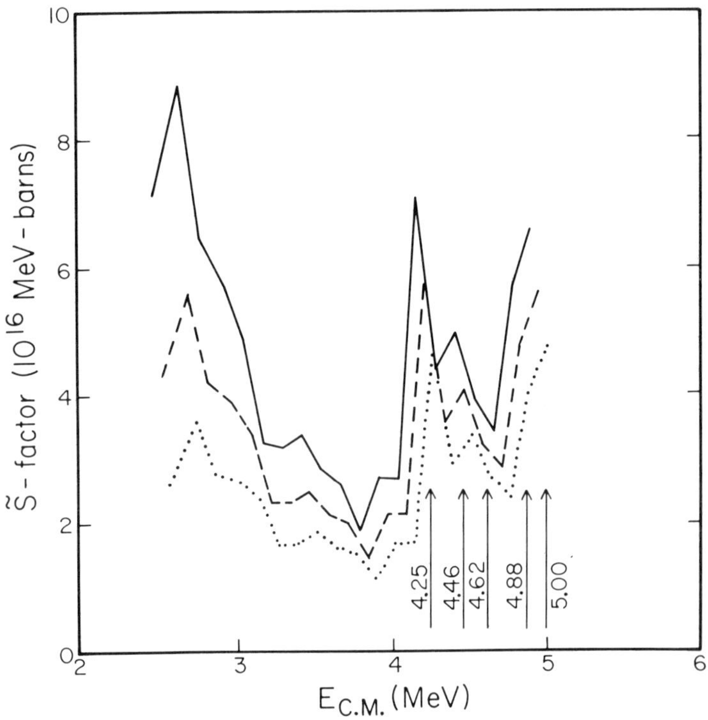

Figure 7. Effect of beam energy uncertainty on $\tilde{S}(E)$. The solid curve connects the data points of Mazarakis and Stephens (Ma 73), displayed as $\tilde{S}(E)$ vs. E. The dashed and dotted curves connect the data shifted upwards by 50 and 100 keV, respectively. The arrows mark the positions of the known resonances, reported in Galster *et al.* (Ga 77).

a center-of-mass energy of 2.5 MeV, a shift of only 100 keV in the beam energy assigned to a data point changes $\tilde{S}$ by a factor of 3.

The total $^{12}C + {}^{12}C$ reaction cross sections from various sources are shown in Figure 8. Here, the energy shifts between the various data sets are quite visible, and there seem to be systematic differences in the shapes of the cross-section energy dependences as well. Possibly the higher cross sections of (Pa 69) at high energies are due to double counting of reactions with sequential (or simultaneous) emission of two particles (Vo 77). At very low energies, the disagreement is likely to be caused by errors in the energy scales and, to a lesser extent, the effects of target thickness. As discussed above, even small errors in the energy scale at very low energies are just as damaging as careless cross-section measurements. The cross sections of Treu *et al.* (Tr 80), obtained by the elastic scattering method, are higher than the other results, for reasons which are not known. The fusion residue measurements, available only at the highest energies, are lower than those of

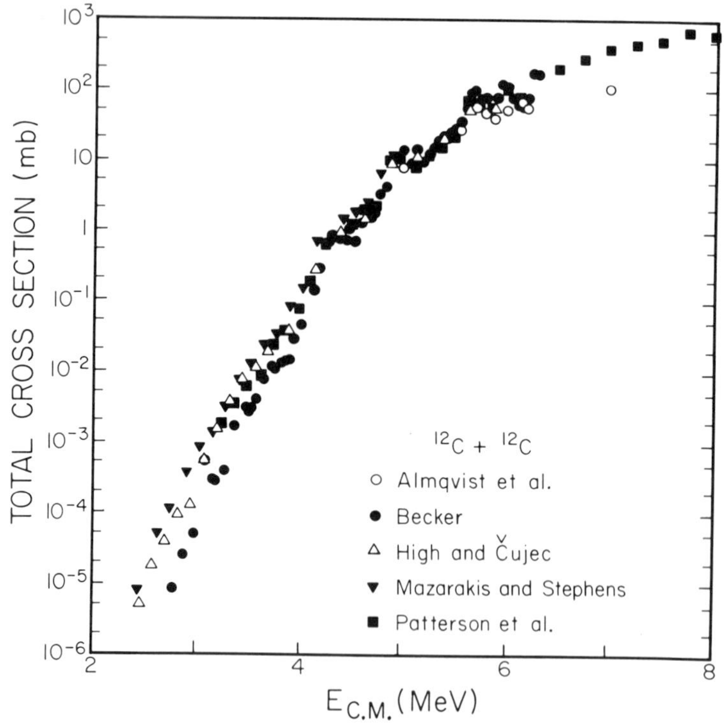

Figure 8. Total reaction cross section for $^{12}C + {}^{12}C$ from Patterson *et al.* (Pa 69), Mazarakis and Stephens (Ma 73), High and Čujec (Hi 77), Becker (Be 78), and Almqvist *et al.* (Al 64).

most of the other methods in the region where they overlap, possibly because the energies of the fusion residues are low and they are hard to detect and identify.

4.2. The $^{12}C + {}^{16}O$ Reactions

Far fewer data exist for the $^{12}C + {}^{16}O$ reactions at sub-Coulomb-barrier energies, as shown in Table 3. There are proton and alpha-particle data from Patterson *et al.* (Pa 71), and data from Ge(Li) detector studies by Christensen and Switkowski (Ch 77a) on the alpha, proton, two-proton, proton + alpha, proton + neutron, and two-alpha channels. The later measurements cover a wider range of energies but agree fairly well with the older measurements, where they overlap.

The partial cross-section measurements are shown in Figures 9 and 10. There are additional measurements by Čujec and Barnes (Cu 76) on the total cross sections by the total γ-ray (NaI detector) method, and these are compared with the other two measurements in Figure 11. As can be seen,

Table 3. Summary of $^{12}C + ^{16}O$ Measurements

$E_{c.m.}$(MeV)	Method employed	Measured quantities	Reference
4.54–8.53	Solid-state detector	Alpha, proton, and total fusion σ's	(Pa 71)
5.47–10.20	Solid-state detector	Elastic scattering σ at 90, 116, 125 deg	(Sp 74)
6.99–13.99	Fusion residues	Partial and total fusion σ's	(Ey 76)
4.00–11.78	Gammas, NaI	Total fusion σ	(Cu 76)
3.90–11.93	Gammas, Ge(Li)	Partial and total fusion σ's	(Ch 77a)
12.9–27.0	Fusion residues	partial and total fusion σ's	(Ko 79b)

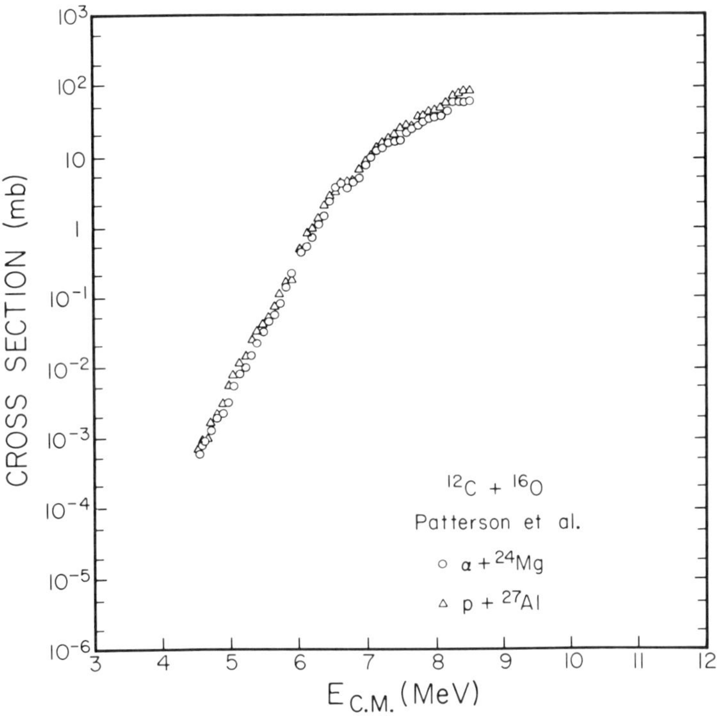

Figure 9. Partial cross sections for the $^{12}C + ^{16}O$ reactions from Patterson *et al.* (Pa 71).

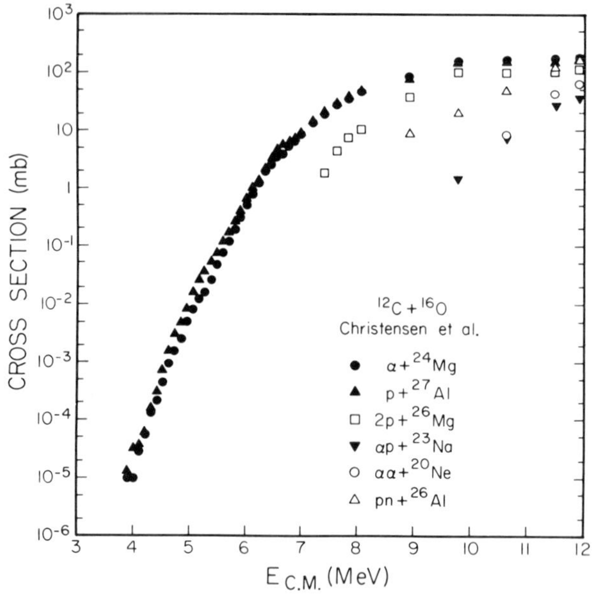

Figure 10. Partial cross sections for the $^{12}C + ^{16}O$ reactions from Christensen and Switkowski (Ch 77a).

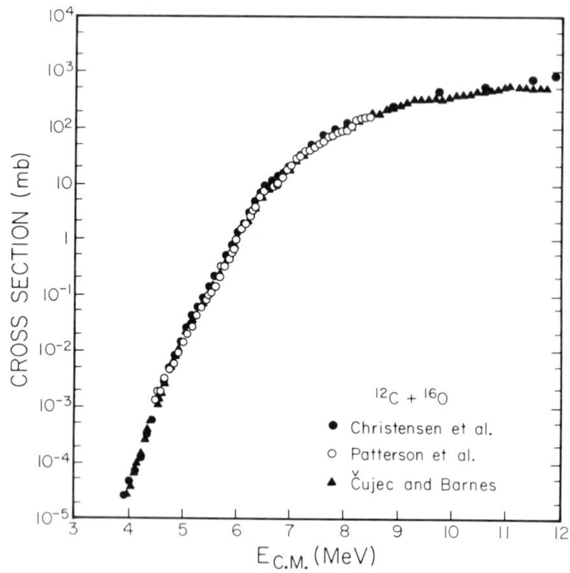

Figure 11. Total reaction cross section for $^{12}C + ^{16}O$ from Patterson *et al.* (Pa 71), Čujec and Barnes (Cu 76), and Christensen and Switkowski (Ch 77a). The cross sections of (Pa 71) and (Ch 77a) have been multiplied by 1.1 to account for unobserved transitions.

there is some resonance structure at low energies, but far less than for $^{12}C + {}^{12}C$. There are some small differences between the Ge(Li) and NaI measurements at low energies. The NaI measurements have not been corrected for the unobserved ground-state transitions, which might be more important, as a fraction of the total, at low energies. The resonance structure is not sufficient to cause serious difficulties in fitting the energy-averaged excitation functions.

4.3. The $^{16}O + {}^{16}O$ Reactions

There have been many measurements of the $^{16}O + {}^{16}O$ reactions in the past few years. Unfortunately, there is considerable disagreement between the measurements made with various techniques, as will be clear shortly. The sub-Coulomb-barrier data are summarized in Table 4.

Figures 12, 13, and 14 show the partial fusion cross sections of Spinka and Winkler (Sp 74), Wu (Wu 78), and of Hulke *et al.* (Hu 80). There are some differences in the partial cross sections of the three sets of measurements.

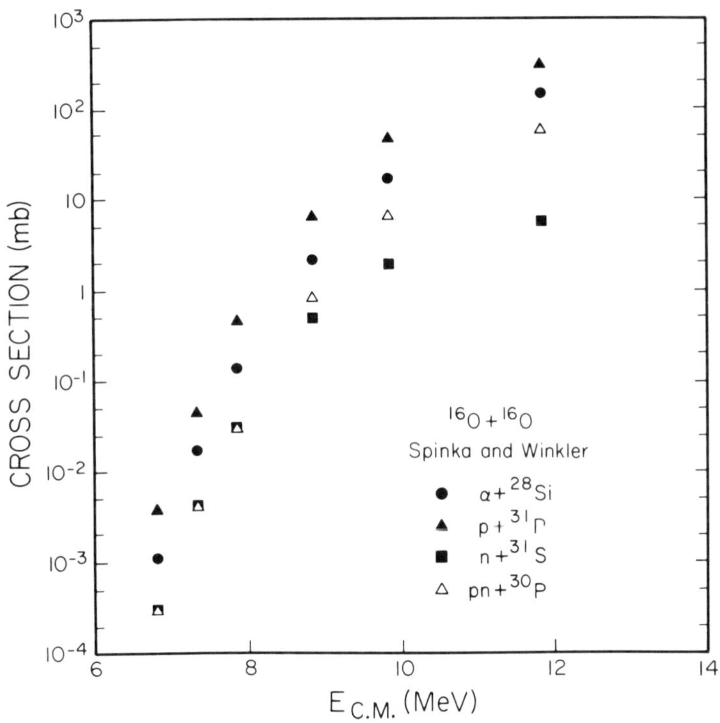

Figure 12. Partial cross sections for the $^{16}O + {}^{16}O$ reactions from Spinka and Winkler (Sp 74).

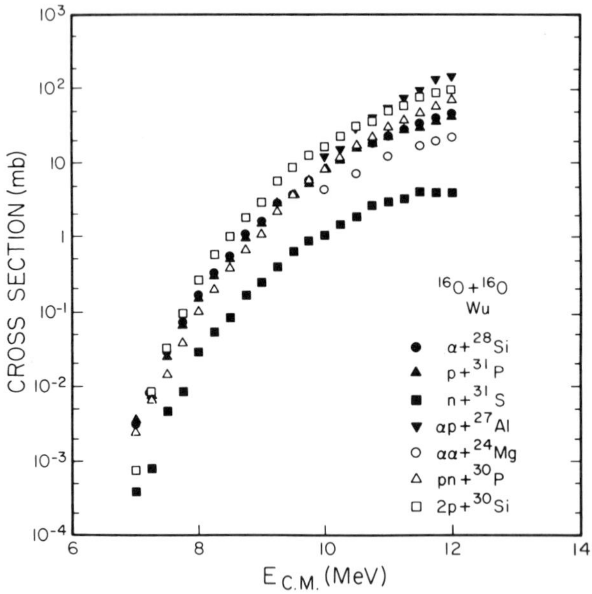

Figure 13. Partial cross sections for the $^{16}O + ^{16}O$ reactions from Wu (Wu 78).

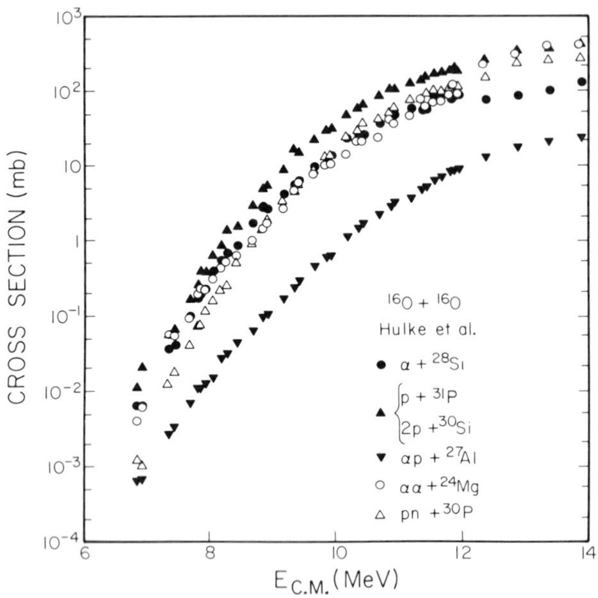

Figure 14. Partial cross sections for the $^{16}O + ^{16}O$ reactions from Hulke *et al.* (Hu 80).

The total fusion cross sections of the various authors are compared in Figure 15, and the total cross sections at $E_{c.m.} = 7$ MeV and 12 MeV are listed in Table 4. From the figure and the table it can be seen that there is fairly good agreement between the particle detection measurements and relative γ-ray measurements of Spinka and Winkler (Sp 74), the Ge(Li) detector γ-ray measurements of Wu and Barnes (Wu 78), and the Ge(Li) detector γ-ray measurements of Hulke, Rolfs, and Trautvetter (Hu 80). The γ-ray measurements of Cheng *et al.* (Ch 79a) and Kolata *et al.* (Ko 79a), on the other hand, give cross sections only about one-half as large. It is significant that the latter two γ-ray measurements are based only on the yields of the detected γ rays. No correction has been made for unobserved transitions, or for summing effects which cause the loss of events. The measurements of Wu would agree approximately with the results of Cheng *et al.* and Kolata *et al.* if no corrections were made for summing and branching, or for unobserved transitions such as ground states.

The measurements at higher energies of Tserruya *et al.* (Ts 78) and

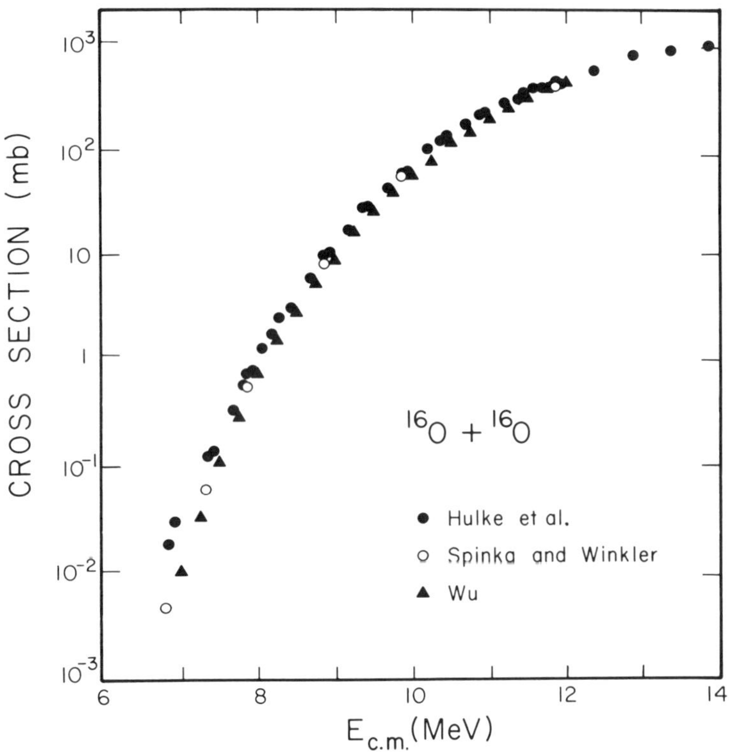

Figure 15. Total reaction cross section for $^{16}O + {}^{16}O$ from Spinka and Winkler (Sp 74), Wu (Wu 78), and Hulke *et al.* (Hu 80).

Table 4. Summary of $^{16}O + ^{16}O$ Measurements

$E_{c.m.}$(MeV)	Method employed	Measured quantities	Fusion cross sections 7 MeV	Fusion cross sections 12 MeV	Reference
7.0–15.0	Solid-state detector	Elastic scattering σ and angular distributions	—	—	(Br 61)
10–36	Solid-state detector	Elastic scattering σ and angular distributions	—	—	(Ma 69)
6.80–11.85	Light particles and activity	Alpha, proton, neutron, and total fusion σ's	13 μb	410 mb[a]	(Sp 74)
7.0–11.83	Gammas, NaI	Total fusion σ	25 μb	380 mb	(Sp 74)
7.34–14.37	Solid-state detector	Elastic scattering σ at 90 deg.	—	—	(Sp 74)
12.0–40.0	Gammas, Ge(Li)	Partial and total fusion σ's		160 mb	(Ko 77) (Ko 79a)
13.5–33.0	Fusion residues	Partial and total fusion σ's		~200 mb[a]	(Ts 78)
9–30	Gammas, Ge(Li)	Partial and total fusion σ's		150 mb	(Ch 79a)
14.9–36.0	Fusion residues	Partial and total fusion σ's		~125 mb[a]	(Ko 79b)
6.83–13.87	Gammas, Ge(Li)	Partial and total fusion σ's	40 μb	480 mb	(Hu 80)
7.0–12.0	Gammas, Ge(Li)	Partial and total fusion σ's	10 μb	438 mb	(Wu 78)
6.89–12.92	Solid-state detector	Elastic scattering angular distribution	—	—	(Wu 84

[a] Extrapolated

Kovar *et al.* (Ko 79b), by detection of the fusion residues, may be extrapolated down to 12 MeV c.m., by assuming that the trend of the low-energy portion of the excitation functions of these papers may be continued to lower energies. This yields cross sections of roughly 200 and 125 mb, respectively. It is disturbing that the fusion residue measurements also give (extrapolated) cross sections considerably lower than those reported in the first three papers. Perhaps this is because we are extrapolating from the lowest quoted cross sections, at 13.5 and 14.9 MeV, respectively, where it may be possible that the technique employed has been extended beyond the region where it is most reliable.

This kind of strong disagreement certainly leaves one with considerable worry as to what the best cross-section values really are. In the next section, we shall see that there are also problems in obtaining $^{16}O + ^{16}O$ fits that are consistent with both the reaction and elastic scattering data.

5. Reaction Model Calculations

5.1. The Shapes of $S(E)$ Curves

As many astrophysical calculations need cross sections for energies well below those for which measurements have been possible, it is necessary to fit the measurements with credible models which can be extrapolated to lower energies.

From comparison of the measured excitation functions for many of the light heavy-ion reaction cross sections that have been studied, it can be seen that the shapes of the excitation functions for the various reactions are very different, and we do not have a convincing explanation of this difference. Figure 16 shows in schematic form the $S(E)$ vs. E plots for several heavy-ion reactions in which one of the reacting partners is ^{12}C. Among the explanations that have been suggested for the large differences in the shapes of the subbarrier $S(E)$ curves are differences in compound nucleus level density, and differences in microscopic structure, such as the number of valence nucleons outside of a closed sunshell. The $S(E)$ curves of the reactions $^{12}C + ^{12}C$, $^{12}C + ^{13}C$, and $^{13}C + ^{13}C$ also differ strongly in shape from one another, aside from the resonance structure seen in $^{12}C + ^{12}C$ (Da 82). Both of the explanations just mentioned may play a role here.

Among the three reactions chosen for special examination in the present paper, there are also differences in the shape of the $S(E)$ curves, which are not well understood.

5.2. Transfer Reactions

When Q values are appropriate and spectroscopic factors are not too small, transfer reactions may play a significant role in heavy-ion reactions, especially at higher energies; it is quite common to find significant transfer reaction cross sections at higher energies.

At very low energies it is also possible for transfer reaction cross sections to be large compared with fusion cross sections. Switkowski *et al.* (Sw 74) have derived an expression for the optimum Q value for transfer reactions, valid in the low-energy region where the Coulomb barrier plays a major role. These authors have examined experimentally several cases where neutron transfer approximately satisfies the criterion of having an optimal Q value, and have shown that transfer reactions may, in favorable cases, completely dominate the reaction cross section at low energies.

A strong case has been made by Čujec *et al.* (Cu 79) that alpha transfer plays a dominant role in the low-energy region of the reaction

$$^{9}Be + ^{16}O \rightarrow ^{5}He + ^{20}Ne^{*} \tag{43}$$

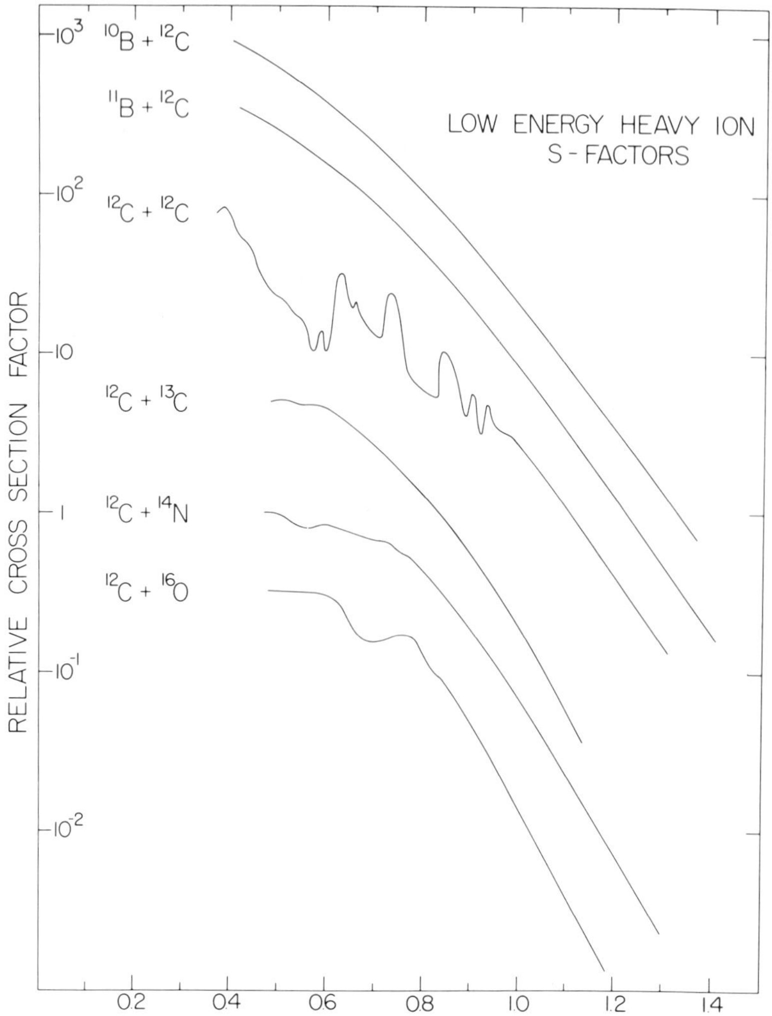

Figure 16. $S(E)$ for heavy-ion reactions involving ^{12}C as one of the reacting pair, illustrating the differences in shape of the S-factor curves observed for different reacting pairs. The abscissa is the ratio of E_{cm} to the Coulomb Barrier energy.

because of a near optimal Q value, and because ^{9}Be has a substantial alpha-particle spectroscopic factor. This work also showed a near optimal Q-value neutron transfer in the reaction

$$^9\text{Be} + {}^{16}\text{O} \rightarrow {}^8\text{Be} + {}^{17}\text{O}^* \tag{44}$$

The possibility that such subbarrier dominant transfer reactions might be important for the reaction pairs $^{12}C + {}^{12}C$, $^{12}C + {}^{16}O$, and $^{16}O + {}^{16}O$

must be considered. Because of the $N = Z =$ even structure of these nuclei, the Q values for nucleon transfers are strongly negative, ranging from -10 to -14 MeV; such transfers are therefore energetically impossible or strongly hindered at sub-Coulomb-barrier energies. It is less obvious that alpha transfer can be ignored as a contributor to the total reaction cross section, and as a source of alpha particles. For the reactions

$$^{12}\mathrm{C} + {}^{12}\mathrm{C} \rightarrow {}^{8}\mathrm{Be} + {}^{16}\mathrm{O} \tag{45}$$

$$^{12}\mathrm{C} + {}^{16}\mathrm{O} \rightarrow {}^{8}\mathrm{Be} + {}^{20}\mathrm{Ne} \tag{46}$$

$$^{16}\mathrm{O} + {}^{16}\mathrm{O} \rightarrow {}^{12}\mathrm{C} + {}^{20}\mathrm{Ne} \tag{47}$$

the Q values are -0.205, -2.637, and -2.432 MeV, respectively. The large negative Q value of the latter two reactions makes low-energy alpha transfer for these reacting pairs very unlikely. Standard alpha-transfer calculations, normalized to the ^{20}Ne production rate near or above the Coulomb barrier (certainly an overestimate of the alpha-transfer cross section), show that the alpha-transfer cross section is completely negligible compared with the fusion cross section for these reactions at low energies.

The Q value for the first of the three reactions above is less negative, but still considerably below the optimal Q value (which is positive). A subbarrier transfer calculation has been carried out by Kozlovsky (Ko 70) for this case, and the transfer cross section has been shown to be negligible compared with the fusion cross section at astrophysically interesting energies.

5.3. Optical Model Fits for Heavy-Ion Reactions

Optical models have proved to be very useful for describing both transfer and fusion reactions with light-ion projectiles, and it is entirely natural that such models would be employed for heavy-ion reactions as well. The most common optical model potentials have a Woods–Saxon radial dependence for the real potential, and either a Woods–Saxon potential or the derivative of a Woods–Saxon potential, for the imaginary potential which gives rise to the reaction cross section (absorption from the incident wave). The imaginary derivative potential has been used to emphasize absorption near the surface of the nucleus. Without considering refinements such as spin–orbit potentials, choosing Woods–Saxon form factors for both real and imaginary potentials means that there are still six adjustable parameters that can be varied to fit experimental transfer, fusion, or elastic scattering data. In this case,

$$V(r) = U(r) + iW(r) + V_{\mathrm{Coul}}(r) + V_{\mathrm{cent}}(r) \tag{48}$$

where

$$U(r) = -U_0\{1 + \exp[(r - R_r)/a_r]\}^{-1} \tag{49}$$

and

$$W(r) = -W_0\{1 + \exp[(r - R_i)/a_i]\}^{-1} \tag{50}$$

In equations (48), (49), and (50), the subscripts r and i label the real and imaginary potentials, respectively. Sometimes U_0 and/or W_0 are taken to be energy dependent; a fairly common prescription is to keep U_0 independent of energy, and take $W_0 = a + bE$, i.e., a linear increase in W with energy. The Coulomb potential is frequently approximated as that of a sphere of radius, R_c, filled with a uniform volume-charge-density, with total charge Ze. For a separation of the two ions greater than the sum of the Coulomb radii, the repulsive Coulomb potential is just that of two positive point charges; it approaches a finite value as $r \to 0$. The centrifugal potential takes the usual form,

$$V_{\text{cent}} = l(l+1)\,\hbar^2/2\mu r^2 \tag{51}$$

where l is the orbital angular momentum and μ is the reduced mass of the colliding ions.

An elegant method of determining the real and imaginary potentials is to derive them from a folding model, which should include complete antisymmetrization of the nucleon wave functions, although this greatly complicates the derivation. In the simplest versions, one starts with the nucleon–nucleon potential and first derives a nucleon–nucleus potential by integrating over all the nucleons in one of the ions. Then this potential is integrated over all of the nucleons in the other ion. Much more sophisticated treatments of folding models have been developed, but we omit the details for reasons of length.

If one considers the lamentable lack of agreement among the data of the various experiments reported above, it is immediately clear that an unsophisticated theoretical analysis is sufficient; there are simply too many parameters to be determined from the rather discrepant data to justify a sophisticated treatment. To reduce the number of parameters, we may choose the real and imaginary potential form factors to be the same, and also take $R_c = R_r$. This still leaves four parameters U_0, W_0, the radius, R, and the diffuseness, a, to be adjusted. Commonly R is taken as $r_0\ (A_1^{1/3} + A_2^{1/3})$, where r_0 is called the radius parameter, and A_1 and A_2 are the numbers of nucleons in the two ions. Similarly, $R_c = r_c(A_1^{1/3} + A_2^{1/3})$.

A "standard" strong-absorption optical potential that has been used quite often [see, for example, (Mi 73), or (St 76)], takes $U_0 = 50$ MeV, $W_0 = 10$ MeV, $r_0 = 1.26$ fm, and $a = 0.4$ fm. With this potential, there are no shape resonances because of the strong absorption. An optical potential this simple cannot accommodate the very different shapes of the excitation functions of the three reactions. However, if one relaxes some of the simplifying conditions that we have imposed on the optical potential, for example, allowing U_0, r_0, and a to take different values for each of the three reactions, fairly good fits to the $^{12}C + ^{16}O$ and $^{16}O + ^{16}O$ data, and to the $^{12}C + ^{12}C$ data averaged over the observed intermediate structure resonances, can be found. Even better fits can be obtained if the real and imaginary parameters are separately varied, of course, and a great number of fits have been made by various authors. As is well known, optical potentials have both continuous and discrete ambiguities, i.e., the parameters can be varied in certain restricted ways without changing the predicted cross sections, and this adds to the variety of potentials that have been proposed.

As examples, Figures 17 and 18 show optical model fits to the $^{12}C + ^{12}C$ fusion data of (Al 64), (Pa 69), (Ma 73), (Hi 77), and (Be 78),

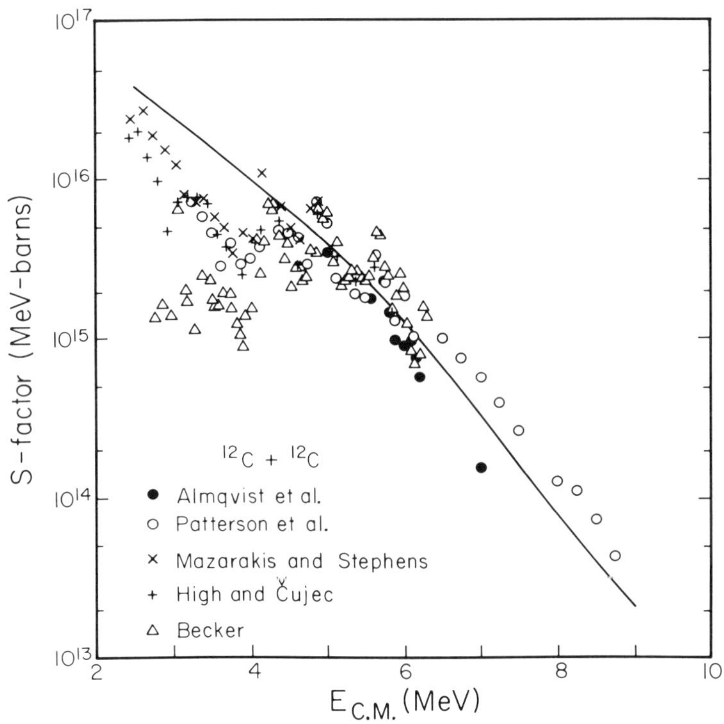

Figure 17. Typical strong-absorption optical model fit to the $^{12}C + ^{12}C$ $S(E)$ data of Almqvist *et al.* (Al 64), Patterson *et al.* (Pa 69), Mazarakis and Stephens (Ma 73), High and Čujec (Hi 77), and Becker (Be 78), with potential parameters $U_0 = 50$ MeV, $W_0 = 10$ MeV, $r_0 = 1.26$ fm, and $a = 0.4$ fm (St 76).

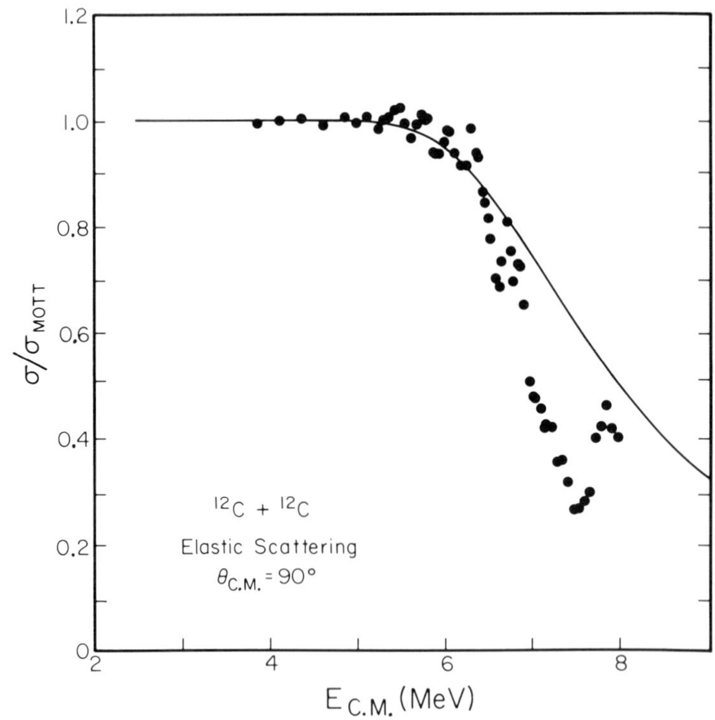

Figure 18. Fit to the 90° $^{12}C + {}^{12}C$ elastic scattering data of Spinka and Winkler (Sp 74) with the potential parameters used in Figure 17 (St 76).

and the 90° $^{12}C + {}^{12}C$ elastic scattering data of (Sp 74), with the potential parameters listed above: $U_0 = 50$ MeV, $W_0 = 10$ MeV, $r_0 = 1.26$ fm, and $a = 0.4$ fm (St 76). As noted, these fits are far from perfect, but describe the data within about a factor of 2, over most of the range of energies, except for the strong resonances. This particular fusion fit is decidedly poorer at low energies, and has somewhat too steep a slope. These faults could be improved if desired by varying more parameters.

Figures 19 and 20 show the somewhat better optical model fits for the $^{12}C + {}^{16}O$ fusion data of Patterson *et al.* (Pa 71), Čujec and Barnes (Cu 76), and Christensen *et al.* (Ch 77a), and the elastic scattering data of Spinka and Winkler (Sp 74), obtained with the same potential parameters as listed in the previous paragraph. As in the $^{12}C + {}^{12}C$ case, the strong absorption potential chosen here cannot produce the resonance structure present in the $^{12}C + {}^{16}O$ reaction at very low energies.

As applications of a quite different potential, Figures 21 and 22 show fits to the $^{16}O + {}^{16}O$ data of (Sp 74) and (Wu 78), with the Gobbi potential

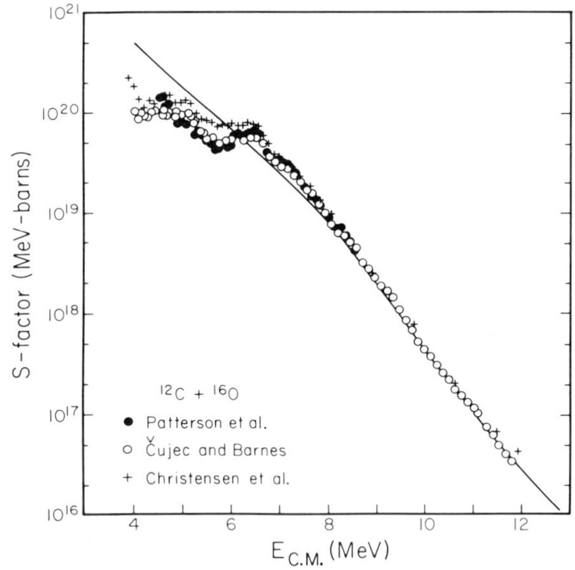

Figure 19. Strong absorption optical model fit to the $^{12}C + ^{16}O$ $S(E)$ data of Patterson *et al.* (Pa 71), Čujec and Barnes (Cu 76), and Christensen and Switkowski (Ch 77a) with the same potential parameters as used in Figure 17.

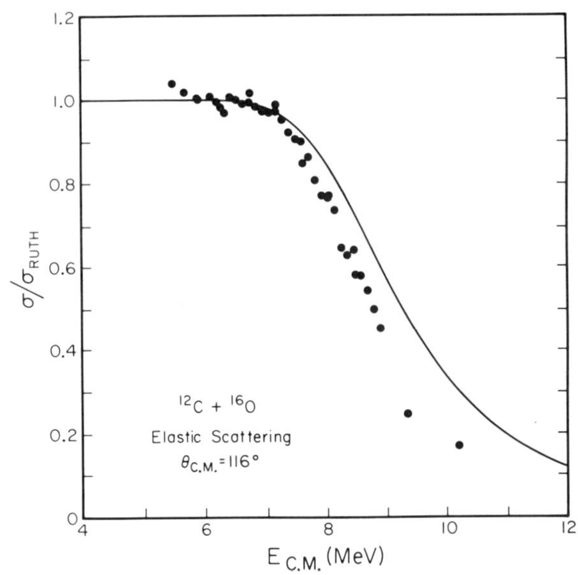

Figure 20. Fit to the 116° $^{12}C + ^{16}O$ elastic scattering data of Spinka and Winkler (Sp 74) with the same potential as used in Figure 19.

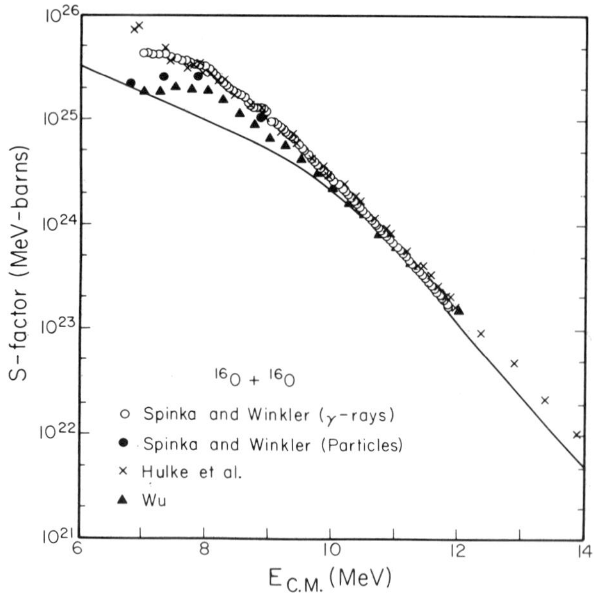

Figure 21. Optical model fit to the $^{16}O + ^{16}O$ $S(E)$ data of Spinka and Winkler (Sp 74) and Wu (Wu 78) with the Gobbi potential (Go 73).

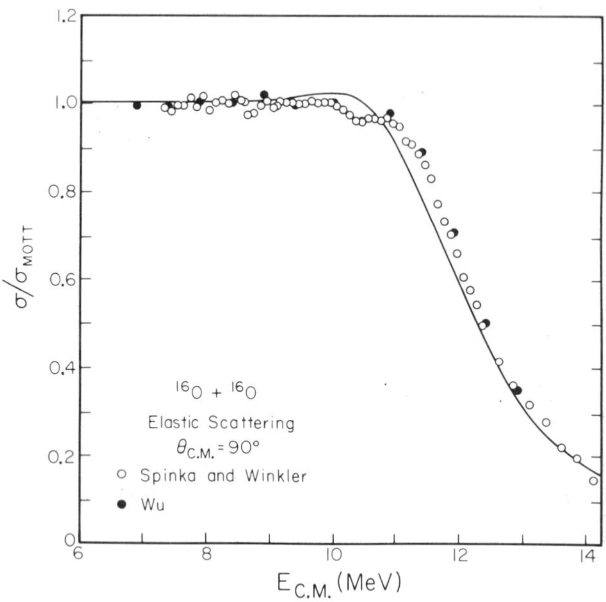

Figure 22. Optical model fit to the 90° $^{16}O + ^{16}O$ elastic scattering data of Spinka and Winkler (Sp 74) and Wu (Wu 78) with the Gobbi potential (Go 73).

(Go 73). This potential has the parameters, $U_0 = 17$ MeV, $W_0 = (0.8 + 0.2E)$ MeV, $r_r = 1.35$ fm, $a_r = 0.49$ fm, $r_i = 1.27$ fm, $a_i = 0.15$ fm. The fit to the elastic scattering is excellent; the fit to the fusion cross sections is good, but falls below the data at low energies. In that region, however, the fusion data sets disagree with one another by an amount similar to the disagreement between the data and the predicted cross sections. "Fine tuning" of the potential parameters could, no doubt, produce a better fit. The important point is that the optical model has so much flexibility, because of the large number of parameters that can be varied, that good fits can certainly be found to the data for the three reactions by "tailoring" the potential for each reaction.

5.4. The Optical Model Fit of Michaud

Michaud (Mi 72, Mi 73) noted that it was not possible to obtain acceptable optical model fits to the three reactions $^{12}C + {}^{12}C$, $^{12}C + {}^{16}O$, and $^{16}O + {}^{16}O$, with potential parameters varying *systematically* over the three reactions, for a potential of the form of equations (48), (49), and (50). At the time his fits were attempted, the only low-energy $^{12}C + {}^{12}C$ fusion data available at very low energies were those of Mazarakis and Stephens (Ma 73), the $^{12}C + {}^{16}O$ data did not extend below 4.54 MeV cm (Pa 71), and the only $^{16}O + {}^{16}O$ data available were those of (Sp 74). Michaud pointed out that, because $^{12}C + {}^{12}C$ and $^{16}O + {}^{16}O$ were identical spinless boson pairs, only even l-value partial waves would be allowed, while in the $^{12}C + {}^{16}O$ reactions both even and odd l values would be permitted. From the rapid rise in the $\tilde{S}(E)$ factor at very low energies exhibited by the $^{12}C + {}^{12}C$ data of (Ma 73), Michaud concluded that the imaginary potential must be quite shallow so that a shape resonance could be introduced to explain the rise. Similarly, the $^{12}C + {}^{16}O$ data also appeared at that time to exhibit a shape resonance at about 6 MeV, although, in a simple optical potential, even-l and odd-l shape resonances occur at different energies and thus tend to produce a smooth excitation function.

By adding a strongly repulsive central potential, Michaud was able to shift the odd-l shape resonances so that they occurred at about the same energies as the even-l resonances. The same kind of potential predicted a rapid rise in $\tilde{S}(E)$ [and also $S(E)$] at low energies in $^{12}C + {}^{12}C$, as then reported, and permitted a systematic variation of the potential parameters from $^{12}C + {}^{12}C$ to $^{16}O + {}^{16}O$, as shown in Table 5.

It was clear that the potential proposed by Michaud, to explain the steep upturn in $S(E)$ for $^{12}C + {}^{12}C$ at low energies, would also predict steep upturns for $^{12}C + {}^{16}O$ and $^{16}O + {}^{16}O$ at energies just below those for which data were then available. The measurements of Čujec and Barnes on $^{12}C + {}^{16}O$ (Cu 76) showed that the predicted rise did not occur in

Table 5. Best Fit Parameters for Michaud's Optical Potential[a]

Reaction	U_0 (MeV)	W_0 (MeV)	R_0 (fm)	a (fm)	c (fm^{-2})
$^{12}C + ^{12}C$	13.0	$0.22E$	6.20	0.55	0.100
$^{12}C + ^{16}O$	10.2	$0.14E$	6.55	0.50	0.156
$^{16}O + ^{16}O$	8.2	$0.123E$	7.50	0.35	0.190

[a] Michaud's repulsive-core optical potential has the form $V(r)(\mathrm{MeV}) = -(U_0 + iW_0)\{1 + \exp[(r - R_0)/a]\}^{-1} + 100\exp[-cr^2] + V_{\mathrm{Coul}}(r) + V_{\mathrm{Cent}}(r)$.

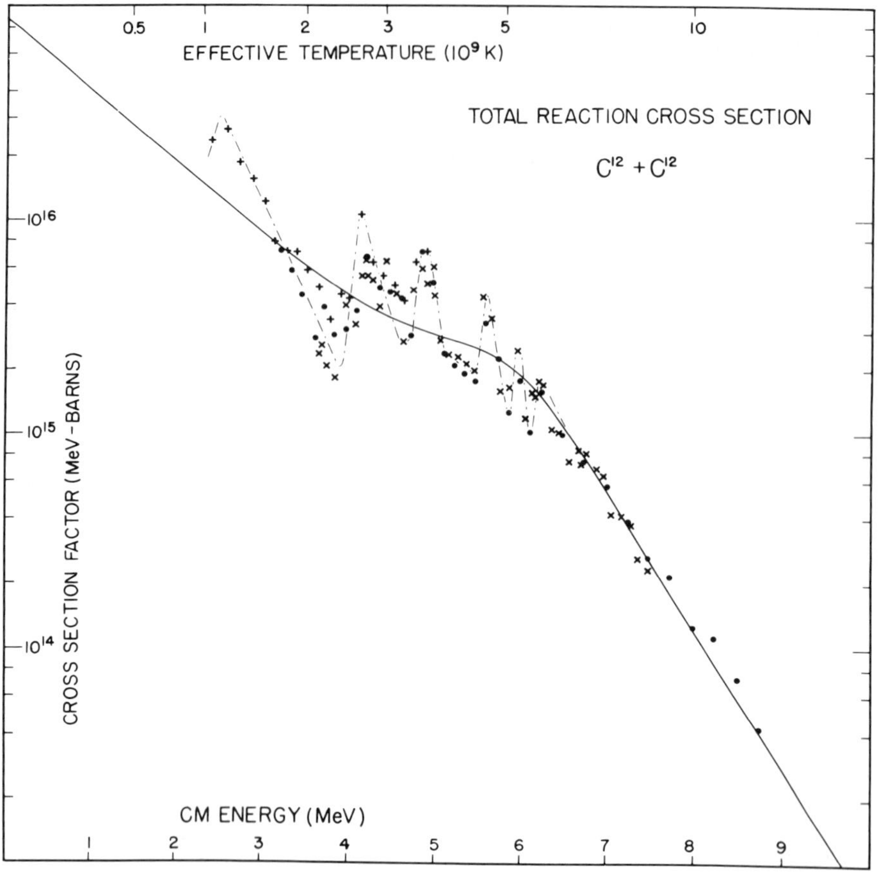

Figure 23. Equivalent square well model fits to the $^{12}C + ^{12}C$ total cross section with the potential parameters listed in Table 6. The solid circles are from Patterson *et al.* (Pa 69), the plus signs are from Mazarakis and Stephens (Ma 73), and the crosses are the NaI detector data of Spinka and Winkler (Sp 74).

Table 6. Best Fit Equivalent Square-Well Optical Parameters

Reaction	R (fm)	U (MeV)	W (MeV)
$^{12}C + {}^{12}C$	7.50	−5.8	0.75
$^{12}C + {}^{16}O$	7.25	−5.0	1.00
$^{16}O + {}^{16}O$	7.96	−7.5	0.5

$^{12}C + {}^{16}O$, and more recent work has shown that the $^{12}C + {}^{12}C$ reaction continues to show resonance structure rather than a steep rise in $S(E)$, down to the lowest energy for which measurements have been possible.

The predicted steep low-energy rise in $S(E)$ was named "absorption under the barrier" by Michaud. There is now experimental evidence in many light heavy-ion reactions that this effect does not exist, at least down to energies much lower than originally predicted. If there is a long tail on the imaginary potential, the optical model will predict such an effect; the failure to observe the effect experimentally must mean that fusion of two heavy ions does not occur for very distant collisions.

With the recent data, there is no longer any necessity to postulate a repulsive core of the type proposed by Michaud. However, absorption is so strong in the low-energy region for heavy-ion reactions that experiment does not really throw much light on the potential at small values of r. The attempt to find a systematically varying set of potential parameters as one progresses from $^{12}C + {}^{12}C$ to $^{16}O + {}^{16}O$, which was one of the motivations for Michaud's repulsive core potential, has so far been unsuccessful.

5.5. The Equivalent Square-Well Optical Potential of Michaud and Fowler

In 1970, Michaud and Fowler (Mi 70) developed a generalized "black" nucleus model, incorporating a complex, square-well potential. In this model, the radius, R, and the magnitude of the potential are adjusted to give results equivalent to a (presumably) more realistic Woods–Saxon potential. For a black nucleus, the wave function for $r < R$ is of the form of an ingoing wave, e^{-iKr}, where $K = [2m(E - V)]^{1/2}/\hbar$, m is the reduced mass, and $V = -[U + iW]$. U, W, and R are determined by fitting experimental data. The transmission functions are calculated according to the expression given by Vogt (Vo 68). This model has been mainly applied to calculating the cross sections of astrophysically important proton, neutron, and alpha-particle reactions on intermediate mass nuclei.

For the $^{12}C + {}^{12}C$, $^{12}C + {}^{16}O$, and $^{16}O + {}^{16}O$ reactions, this model yields a strongly repulsive real potential, which is quite unusual. The potential parameters determined by fitting the three reactions are shown in Table 6. The quality of the fits of this model to the data is excellent, as shown in Figures 23, 24, and 25. Here again, the potential parameters do not form a smooth sequence from $^{12}C + {}^{12}C$ to $^{16}O + {}^{16}O$.

Because the principal users of reaction rates as a function of temperature prefer analytic expressions where possible, Fowler, Caughlan, and Zimmerman have used an analytically integrable parametrization for the $S(E)$ factor which fits the curves calculated from their optical potentials to better than 2%:

$$S(E) = S(0)\exp(-\alpha E)[\exp(-\gamma E^m) + b\exp(+\beta E)]^{-1} \tag{52}$$

where $\alpha > 0$, $\beta > 0$, $\gamma > 0$, $m > 0$, and $0 < b < 1$. At very low energies, this expression is asymptotically proportional to $\exp(-\alpha E)$ and, at high energies,

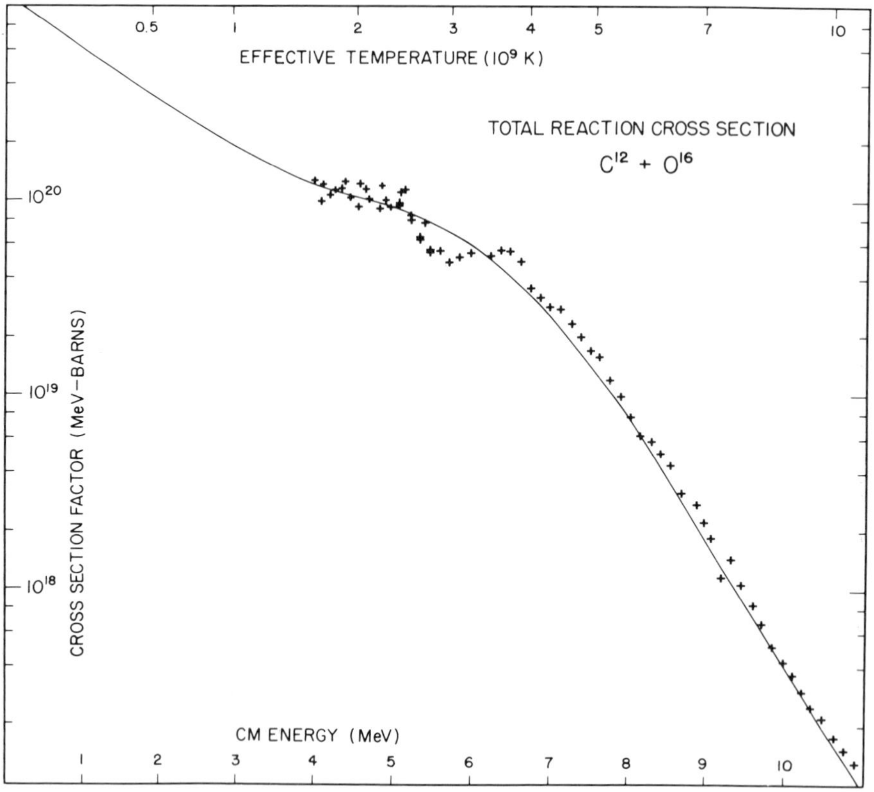

Figure 24. Equivalent square well model fit to the $^{12}C + {}^{16}O$ total reaction cross section with the potential parameters listed in Table 6. The data shown are from Čujec and Barnes (Cu 76).

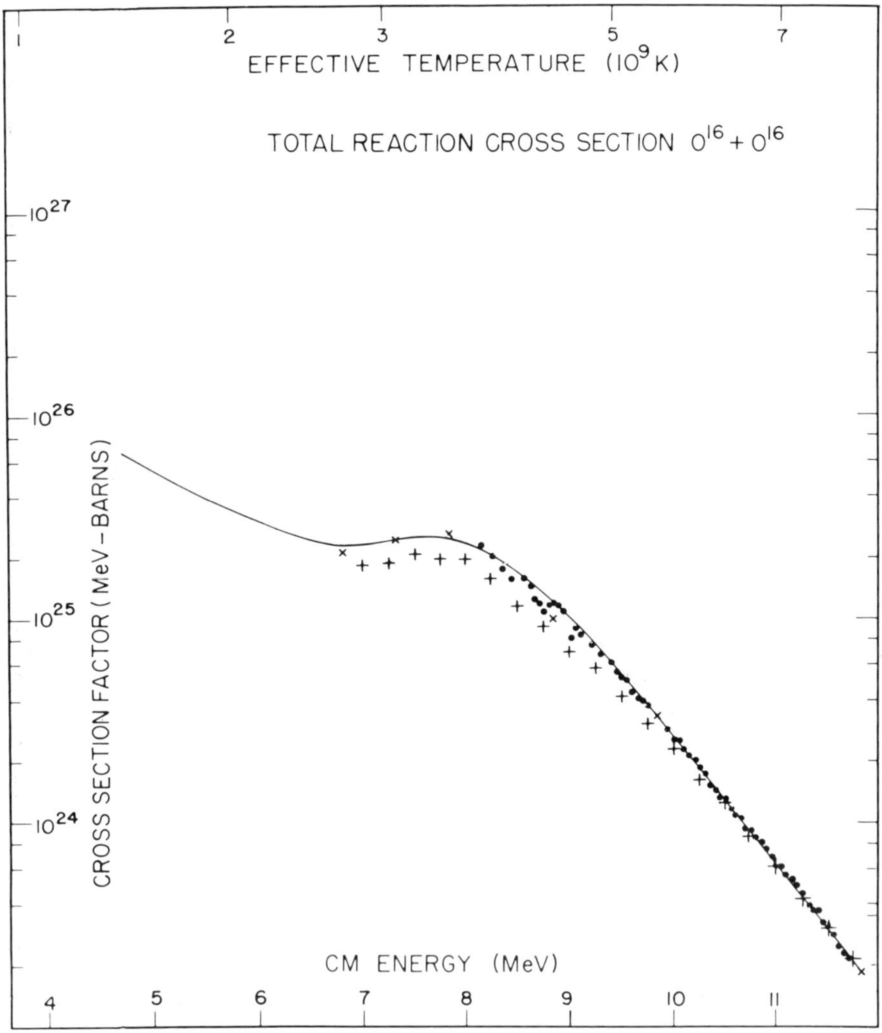

Figure 25. Equivalent square well model fit to the $^{16}O + {}^{16}O$ total reaction cross section with the potential parameters listed in Table 6. The solid circles and crosses are the data of Spinka and Winkler (Sp 74), and the plus signs are the data of Wu (Wu 78).

to $(1/b)\exp[-(\alpha+\beta)E]$. The parameters γ and m provide the flexibility to fit the transition region between the regions of the curve with the two different exponents. The values found for the various parameters by (Fo 75) on the basis of the data available at this time are given in Table 7. The new values for $^{16}O + {}^{16}O$ from Hulke *et al.* (Hu 80) have also been included in the table.

Table 7. Parameters for $S(E)$ Fits of Equation (52)

Reaction	$S(0)$ (MeV barns)	α (MeV^{-1})	b	β (MeV^{-1})	γ (MeV^{-m})	m
$^{12}C + ^{12}C^{a}$	8.83×10^{16}	0.772	5.56×10^{-3}	0.697	5.01×10^{-5}	6
$^{12}C + ^{16}O^{a}$	1.15×10^{21}	0.643	1.06×10^{-3}	0.837	6.26×10^{-3}	3
$^{16}O + ^{16}O^{a}$	2.31×10^{27}	0.780	3.89×10^{-4}	0.681	9.05×10^{-6}	6
$^{16}O + ^{16}O^{b}$	1.24×10^{28}	0.72	4.37×10^{-3}	0.68	9.8×10^{-6}	6

[a] From (Fo 75).
[b] From (Hu 80).

This model achieves fits to the measured data that are superior to those obtained from most other models. Although the physical justification of the model may not be as clear as it is for a more standard optical model, or for the IWBC model of the next section, the extrapolations from this model to the slightly lower energies required to describe hydrostatic carbon or oxygen burning are probably as reliable as those of the other models. An important point is that the low-energy behavior of the expression fitted to the curves seems to be consistent with our understanding of the physics of barrier penetration, and this is clearly essential in any expression to be used for extrapolation to low energy.

5.6. A Simple Coulomb Barrier Penetration Model

Because of the strong absorption evident in light heavy-ion reactions, and the apparent unimportance of transfer reactions for the three reactions of interest here, it is reasonable to assume that the reaction cross section will be directly given by the penetration of the Coulomb and angular momentum barriers. In the semiclassical Jeffreys–Wentzel–Kramers–Brillouin treatment of barrier penetration, which should be quite good for the massive heavy ions of interest here, the transmission coefficient for the Coulomb plus angular momentum plus nuclear potential barrier $V_l(r)$ is

$$T_l = (1 + I_l)^{-1} \tag{53}$$

where

$$I_l = \exp\left\{ + (2/\hbar) \int_{r_1}^{r_2} [2m(V_l - E)]^{1/2}\, dr \right\} \tag{54}$$

In this expression r_1 and r_2 are the classical turning points, i.e., $V_l(r_1) = V_l(r_2) = E$. For $I_l \gg 1$, the transmission function is well described by the more frequently encountered expression

$$T_l \to \exp\left\{ -(2/\hbar) \int_{r_1}^{r_2} [2m(V_l - E)]^{1/2}\, dr \right\} \tag{55}$$

The fusion cross section is then given, as usual, by

$$\sigma_f = \pi \bar{\lambda}^2 \sum_{l=0}^{\infty} (2l+1) T_l \tag{56}$$

Near the top of the barrier, $V_l(r)$ can be approximated reasonably closely by a parabola, for which function the transmission can be evaluated analytically (Hi 53):

$$I_l = \exp\{+2\pi[V_l(r_{\text{max}}) - E]/\hbar\omega_l\} \tag{57}$$

where $V_l(r_{\text{max}})$ is the potential at the vertex of the parabola and $\hbar\omega_l$ is the equivalent oscillator energy for the parabolic potential. With this expression the transmission function goes smoothly through the value 1/2 for E equal to the peak of the barrier, and approaches unity as the energy is raised well above the barrier.

We have found that I_l as given by Equation (57) agrees with that obtained from Equation (54) over a range of energies near $T_l \sim 0.1$. Clearly at energies far below the barrier the parabolic approach would be quite inadequate because the Coulomb portion of the potential falls off slowly with r. Thus, for low energies, we have used Equation (54) for I_l and switched over to Equation (57) in the neighborhood of $T_l \sim 0.1$. Dethier and Stancu (De 81) have approximated the shape of the barrier with a better but more complicated expression, for which an approximate analytic integration of Equation (54) can be carried out.

We have calculated σ_f for the three reactions of this paper by calculating the penetrabilities from Equations (54) and (57), and we find that the calculated cross sections are essentially indistinguishable from those derived from the IWBC model, for the same assumed nuclear potential. For this reason, separate figures for the simple barrier penetration model fits will not be presented.

5.7. The IWBC Model

The simple barrier penetration model does not provide, without further analysis, predictions for the elastic scattering as a function of energy and angle. A relatively simple model, which is closely related to the simple barrier penetration model, and which provides predictions for both the elastic scattering and the fusion cross sections, is the incoming wave boundary condition (IWBC) model [see, for example, (Ra 66), (Ei 74), or (Ch 77b)].

As for the barrier penetration model, it is assumed that direct reactions are negligible compared with fusion. One can therefore avoid using an imaginary potential altogether by requiring that, within a boundary radius,

R_b, only ingoing waves exist for each partial wave. R_b is normally chosen to be within the inner boundary of the interaction barrier, and the calculations are normally carried out with an optical model code, modified by including the IWBC. As is the case for standard optical model calculations, special care is required for very small T_l, i.e., at very low energies, to obtain adequate precision in the computations.

The attractive feature of the IWBC model is that it requires only a minimal number of parameters—the real well depth, the radius parameter, and the diffuseness. In addition, as the heavy-ion absorption is strong, the absorption depends only on the portion of the nuclear potential outside of R_b. Thus one can fix the depth of the potential at some reasonable value and define the tail of the nuclear potential in terms of only two adjustable parameters, e.g., the diffuseness, a, and the radius at which the s-wave barrier peaks, R_{barrier} (Ch 77b).

The excellent quality of the fits obtained by the IWBC method for $^{12}C + {}^{16}O$ fusion and elastic scattering cross sections are shown in Figures

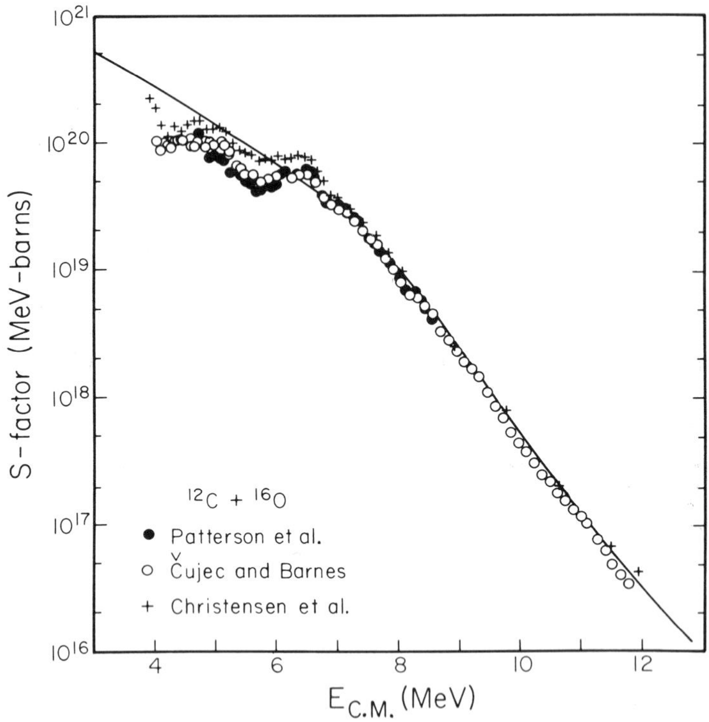

Figure 26. IWBC model fit for the $^{12}C + {}^{16}O$ fusion S-factor data of Čujec and Barnes (Cu 76) and Christensen and Switkowski (Ch 77a) with the potential parameters listed in Table 8.

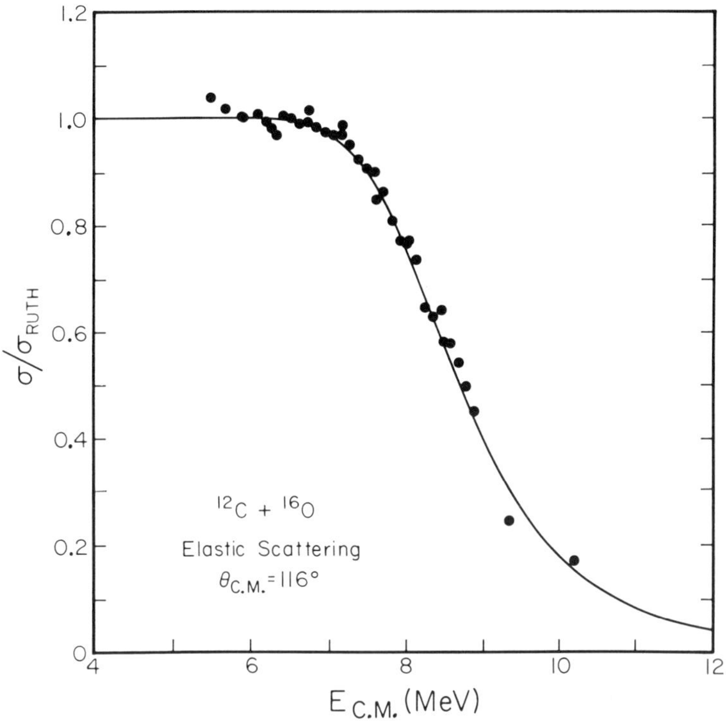

Figure 27. IWBC model fit for the 116° $^{12}C + ^{16}O$ elastic scattering data of Spinka and Winkler (Sp 74) with the potential parameters used in Figure 26.

26 and 27. For $^{12}C + ^{12}C$, judging the quality of the fits to the fusion and elastic scattering cross sections is somewhat complicated by the intermediate structure resonances. Nevertheless the fits obtained, of which examples are shown in Figures 28 and 29, are still reasonably good.

The quality of the simultaneous fits to the $^{16}O + ^{16}O$ fusion experiment of (Sp 74), (Wu 78), and (Hu 80), and the elastic scattering measurements of (Sp 74) and (Wu 78) is much poorer than for $^{12}C + ^{12}C$. Figures 30 and 31 show the fits with the parameters chosen to give a good fit to the elastic scattering data, with $R_{\text{barrier}} = 8.11$ fm. If R_{barrier} is increased substantially, to ~8.6 fm, a good fit to the fusion data is obtained, but at the price of a poor fit to the elastic scattering data. We have repeated the IWBC calculations and find almost the same parameters as found by Christensen and Switkowski (Ch 77b). We have also rechecked the $^{16}O + ^{16}O$ elastic scattering experimentally, and obtain good agreement with the detailed measurements of (Sp 74). The parameters for the IWBC fits shown in Figures 26 through 31 are listed in Table 8.

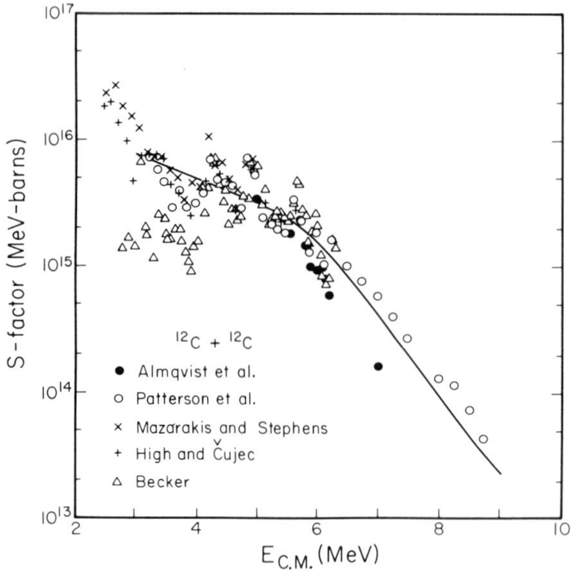

Figure 28. IWBC model fit for the $^{12}C + {}^{12}C$ fusion S-factor data of Almqvist *et al.* (Al 64), Patterson *et al.* (Pa 69), Mazarakis and Stephens (Ma 73), High and Čujec (Hi 77), and Becker (Be 78), with the potential parameters listed in Table 8.

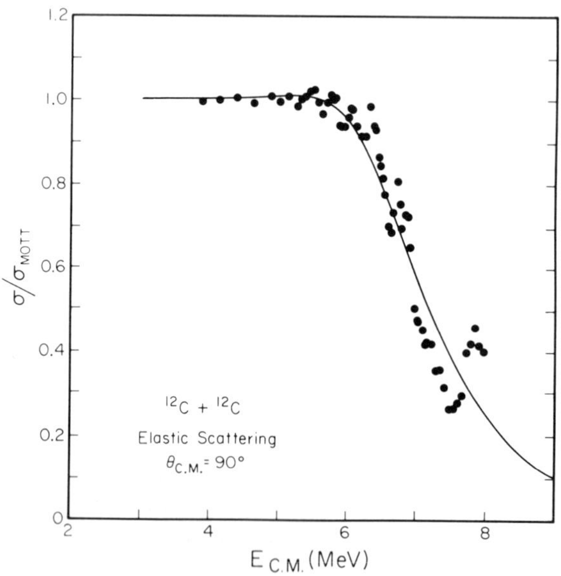

Figure 29. IWBC model fit for the 90° $^{12}C + {}^{12}C$ elastic scattering data of Spinka and Winkler (Sp 74) with the potential parameters used in Figure 28.

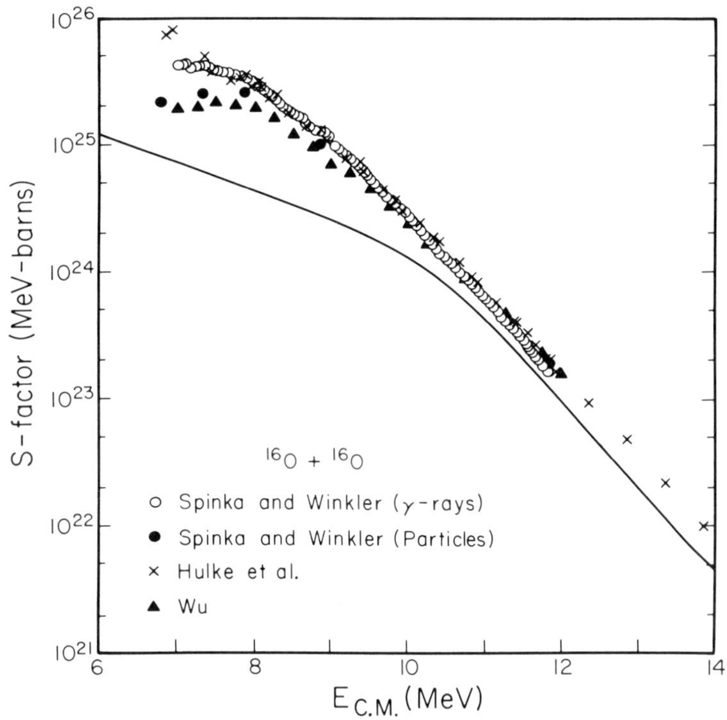

Figure 30. IWBC model fit to the $^{16}O + ^{16}O$ *S*-factor data of Spinka and Winkler (Sp 74), Wu (Wu 78), and Hulke *et al.* (Hu 80) with the potential parameters listed in Table 8. These parameters were chosen to give good fits for the 90° elastic scattering as shown in Figure 31. Better fits to the fusion data can be obtained at the price of poor fits to the elastic scattering, by changing the potential parameters.

Table 8. Parameters for IWBC Fits

Reaction	$R_{barrier}$ (fm)	a (fm)	R_0 (fm)[a]	Reference
$^{12}C + ^{12}C$	7.784	0.75	4.538	(Ch 77b)
$^{12}C + ^{16}O$	8.175	0.52	5.819	(Ch 77b)
$^{16}O + ^{16}O$	8.113	0.65	5.533	Present work

[a] The Woods–Saxon potential radius R_0 has been calculated from the fitted value of $R_{barrier}$ by assuming a real well depth of 50 MeV.

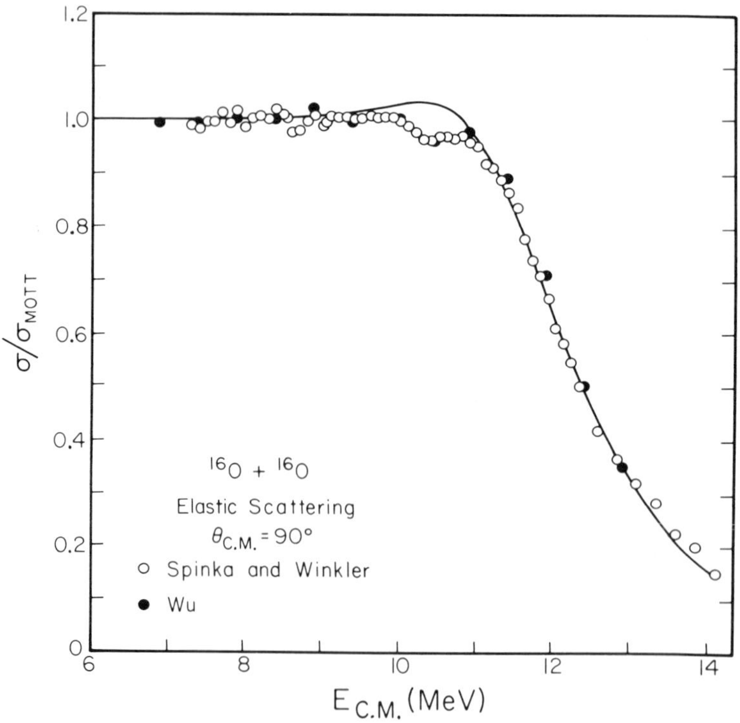

Figure 31. IWBC fit to the 90° $^{16}O + {}^{16}O$ elastic scattering data of Spinka and Winkler (Sp 74) and Wu (Wu 78) with the potential parameters listed in Table 8.

It is tempting to conjecture that the recurring problems in fitting the $^{16}O + {}^{16}O$ data, discussed in this and previous sections, could be eliminated by choosing $^{16}O + {}^{16}O$ fusion cross sections from a different source. However, in view of the agreement between the measurements of (Sp 74), (Wu 78), and (Hu 80), and the fact that the data of (Ch 79a) and (Ko 79a) have not been corrected for several important effects, there is no good reason for rejecting the concordant $^{16}O + {}^{16}O$ fusion data of (Sp 74), (Wu 78), and (Hu 80). As suggested earlier, perhaps the difficulties encountered with the $^{16}O + {}^{16}O$ fits are an indication that the optical potential between closed shell nuclei is really not the same shape as that for the other reactions. It is thoroughly unsatisfactory to have to leave the $^{16}O + {}^{16}O$ situation unresolved in this way. Further work, probably both experimental and theoretical, will be needed to resolve this problem.

5.8. Resonances in the Excitation Function

It has already been noted above that there are strong, relatively narrow resonances, ~50 keV wide, with spacings of a few hundred keV, in the $^{12}C + ^{12}C$ reactions, continuing down to the lowest energies for which measurements have been made. There also appears to be a suggestion of similar resonances at the lowest energies studied in the $^{12}C + ^{16}O$ reactions, with weaker structure persisting at higher energy. These resonances are much narrower and more closely spaced than the more frequently observed resonances, with widths of the order of 1 MeV and spacings of a few MeV, that occur in many heavy-ion reactions above the Coulomb barrier. The two kinds of resonance are probably closely related. Both kinds have larger elastic widths than would be expected on the basis of a statistical model, and are consequently often referred to as quasimolecular.

Although many explanations with increasing sophistication have been offered for the resonances, it is still not clear how they should be described. [See, for example, (Ci 78), (Im 69), (Sc 70), (Pa 77), (Ma 78), (Ab 78), and (Fe 76).] Some of the explanations involve excitation of particular collective excited states in the ions, such as the 2^+, 4.4 MeV, 0^+, 7.6 MeV, and 3^-, 9.6 MeV states in ^{12}C, or the 0^+, 6.0 MeV, 3^-, 6.1 MeV, and 2^+, 6.9 MeV states in ^{16}O. It is not clear that such explanations can provide an explanation for the enormous number of resonances in $^{12}C + ^{12}C$ (at least 27 resonances in a c.m. range of 9 MeV). Other explanations generally consider the resonances to be rotational states, based on excitations of the ions. The most recent suggestion is that of Iachello (Ia 81), and Erb and Bromley (Er 81) that these states form a complete vibration–rotation spectrum of the type familiar in diatomic molecules.

The occurrence of the relatively sharp, closely spaced resonances in $^{12}C + ^{12}C$ and, to a lesser extent, in $^{12}C + ^{16}O$, is almost certainly connected with the fact that the compound nuclear excitation energies, and hence, the compound nucleus level densities are lower in these cases than for most other heavy-ion reactions. It also seems likely that the microscopic structure of the ions, especially the presence or absence of "valence" nucleons outside of closed subshells, plays a role in determining whether narrow resonances appear.

It is not our purpose in this paper to enter the debate on the explanation of these resonances. They may be significant in the astrophysical context at low temperatures, however, because the Gamow peak becomes quite narrow at low temperatures, even though the effective energy may be 20–40 times kT. At the temperatures appropriate for explosive carbon burning, $(2\text{–}4) \times 10^9$ K, the most effective energy E_0 and the full width at $(1/e)$ of the maximum of the Gamow peak, ΔE_0, are 5 and 2.6 MeV, respectively, from Equations (9) and (10) in Section 2.2. The yield would be averaged over many resonances. However, hydrostatic carbon burning could perhaps occur at temperatures as low as 6×10^8 K, for which E_0 and ΔE_0 are 1.7 MeV and 0.7 MeV, respectively. There could be large differences in the reaction rate, depending on whether there were 0, 1, 2, or more resonances in the Gamow energy region ΔE_0. If there were no resonances in the Gamow peak region at 6×10^8 K, the temperature would have to rise until carbon could burn. A higher temperature would mean a higher neutrino luminosity and, hence, a shorter time scale. Higher temperatures would also increase the rates of other nuclear reactions which accompany carbon burning, although these should be much faster than carbon burning in any case. A major increase in temperature would increase the number of neutrons produced by the $n + {}^{23}\text{Mg}$ channel, with some effects on the nucleosynthesis of the neutron-rich nuclides.

6. Concluding Remarks

Although it will be clear by this point that the heavy-ion experimental data, and our unsophisticated attempts to explain them, leave much to be desired from a nuclear physics point of view, the fits are probably adequate for nuclear astrophysical purposes at the present incomplete state of knowledge of the locations, temperatures, and pressures at which the reactions occur. The difficulties in obtaining good simultaneous fits to the ${}^{16}\text{O} + {}^{16}\text{O}$ data, with parameters varying smoothly as one progresses from ${}^{12}\text{C} + {}^{12}\text{C}$, through ${}^{12}\text{C} + {}^{16}\text{O}$, to ${}^{16}\text{O} + {}^{16}\text{O}$, are especially disappointing, and this problem will have to be resolved by further theoretical and experimental work.

Although carbon burning, neon burning, and oxygen burning must surely succeed one another in the sequence of nuclear processes, because Coulomb barriers are the main determinants of nuclear cross sections at low energies, the actual cross section values play significant roles in several astrophysical contexts. The temperature dependence of the ${}^{12}\text{C} + {}^{12}\text{C}$ reaction rate may be decisive in deciding whether carbon detonation supernovae exist, as this question is a borderline situation. The question is whether the reaction will proceed far enough before electron degeneracy is lifted to explode the star or not. We have also shown that the ratio of the relative rates of ${}^{12}\text{C} + {}^{12}\text{C}$ and ${}^{12}\text{C} + {}^{16}\text{O}$ at explosive oxygen burning temperatures plays an important role in nucleosynthesis, as it determines the ratio of the heavier oxygen-burning products to silicon—essentially determining how far oxygen-burning nucleosynthesis proceeds. On a finer scale, the neutron-producing branches of the ${}^{12}\text{C} + {}^{12}\text{C}$, ${}^{12}\text{C} + {}^{16}\text{O}$, and ${}^{16}\text{O} + {}^{16}\text{O}$ reactions, especially the ${}^{12}\text{C} + {}^{12}\text{C} \rightarrow \text{n} + {}^{23}\text{Mg}$ reaction, partially determine the neutron excess during nucleosynthesis. The neutron excess determines the relative abundances of the neutron-rich isotopes produced in explosive burning, and is subject to experimental test by isotopic abundance measurements.

It is the nature of astrophysics that most of the processes and the objects that we are trying to understand are physically inaccessible to us. In this circumstance, it is imperative that those aspects of astrophysics that can be studied in the laboratory must be as well understood as possible. Only by developing a secure understanding of those areas of physics can we hope to advance in our understanding of the ever growing wealth of astrophysical observations. These now range over the entire electromagnetic spectrum, and also embrace galactic cosmic rays, and even neutrinos. Nuclear astrophysics is one area in which the laboratory data require hard work and perseverance to obtain, and careful theoretical work to understand, but those who are involved in this work are convinced that the quest is truly worth that effort.

Note Added in Proof

A recent study of the fusion of ^{16}O with ^{16}O and ^{18}O (J. Thomas, Y. T. Chen, S. Hinds, K. Langanke, D. Meredith, M. Olson, and C. A. Barnes, 1984, submitted for publication) essentially confirms the measurements of Wu and Barnes of the $^{16}O + {}^{16}O$ fusion cross section, (Wu 78) and (Wu 84).

The new measurements also show that the thicknesses of the $^{16}O + {}^{16}O$ and $^{16}O + {}^{18}O$ fusion barriers are essentially identical, when they are plotted as functions of $(B - E)$, the energy measured downwards from the top of the *s*-wave interaction barrier. This observation suggests that the two extra "valence" neutrons in the $^{16}O + {}^{18}O$ reaction do not significantly influence the fusion cross section for $^{16}O + {}^{18}O$, beyond their effect in increasing the interaction radius over that for $^{16}O + {}^{16}O$.

References

(Ab 78) Y. Abe, Y. Kondo, and T. Matsuse, *Prog. Theor. Phys.* **59**:1393 (1978).

(Al 64) E. Almqvist, J. A. Kuehner, D. McPherson, and E. W. Vogt, *Phys. Rev. B* **136**:84 (1964).

(Ba 66) J. N. Bahcall, *Astrophys. J.* **143**:259 (1966).

(Ba 71) C. A. Barnes, in *Advances in Nuclear Physics*, Vol. 4 (M. Baranger and E. Vogt, eds.), Plenum Press, New York (1971), p. 133.

(Ba 80) C. A. Barnes, in *Proceedings of the Erice School of Nuclear Physics*, March 25–April 6, 1980 (Sir Denys Wilkinson, F.R.S., ed.), Pergamon Press, Oxford (1980), p. 235.

(Ba 82) Chapters by R. W. Kavanagh, W. S. Rodney and C. Rolfs, C. A. Barnes, and T. A. Tombrello, S. E. Koonin, and B. A. Flanders, in *Essays in Nuclear*

Astrophysics (C. A. Barnes, D. D. Clayton, and D. N. Schramm, eds.), Cambridge University Press, New York (1982).

(Be 78) H.-W. Becker, Diplomarbeit, Institute für Kernphysik, Münster (1978).

(Be 80) R. H. Becker, S. S. Holt, B. W. Smith, N. E. White, E. A. Boldt, R. F. Mushotzky, and P. J. Serlemitsos, *Astrophys. J. Lett.* **235**:L5 (1980).

(Ch 77a) P. R. Christensen, Z. E. Switkowski, and R. A. Dayras, *Nucl. Phys.* **A280**:189 (1977).

(Ch 77b) P. R. Christensen and Z. E. Switkowski, *Nucl. Phys.* **A280**:205 (1977).

(Ch 78) R. A. Chevalier and R. P. Kirshner, *Astrophys. J.* **219**:931 (1978).

(Ch 79a) V. K. C. Cheng, A. Little, H. C. Yuen, S. M. Lazarus, and S. S. Hanna, *Nucl. Phys.* **A322**:168 (1979).

(Ch 79b) R. A. Chevalier and R. P. Kirshner, *Astrophys. J.* **233**:154 (1979).

(Ci 78) N. Cindro, *J. Phys. G* **4**:L23 (1978).

(Cl 68) D. D. Clayton, *Principles of Stellar Evolution and Nucleosynthesis*, McGraw-Hill, New York (1968).

(Cu 76) B. Čujec and C. A. Barnes, *Nucl. Phys.* **A266**:461 (1976).

(Cu 79) B. Čujec, S.-C. Wu., and C. A. Barnes, *Phys. Lett.* **89B**:151 (1979).

(Da 76) R. A. Dayras, R. G. Stokstad, Z. E. Switkowski, and R. M. Wieland, *Nucl. Phys.* **A261**:478 (1976).

(Da 77) R. Dayras, Z. E. Switkowski, and S. E. Woosley, *Nucl. Phys.* **A279**:70 (1977).

(Da 82) B. Dasmahapatra, B. Čujec, and F. Lahlou, *Nucl. Phys.* **A384**:257 (1982).

(De 81) J.-L. Dethier and F. Stancu, *Phys. Rev. C* **23**:1503 (1981).

(Di 65) R. H. Dicke, P. J. E. Peebles, P. G. Roll, and D. T. Wilkinson, *Astrophys. J.* **142**:414 (1965).

(Ei 74) Y. Eisen, E. Abramson, G. Engler, M. Samuel, U. Smilansky, and Z. Vager, *Nucl. Phys.* **A236**:327 (1974).

(Er 80) K. A. Erb, R. R. Betts, S. K. Korotky, M. M. Hindi, P. P. Tung, M. W. Sachs, S. J. Willett, and D. A. Bromley, *Phys. Rev. C* **22**:507 (1980).

(Ey 76) Y. Eyal, M. Beckerman, R. Chechik, Z. Fraekel, and H. Stocker, *Phys. Rev. C* **13**:1527 (1976).

(Fe 76) H. Feshbach, *J. Phys. Colloq.* **37C5**:177 (1976).

(Fo 59) K. W. Ford, D. L. Hill, M. Wakano, and J. A. Wheeler, *Ann. Phys.* **7**:239 (1959).

(Fo 67a) W. A. Fowler, in *High-Energy Physics and Nuclear Structure* (G. Alexander, ed.), North-Holland, Amsterdam (1967), p. 203.

(Fo 67b) W. A. Fowler, G. R. Caughlan, and B. A. Zimmerman, *Ann. Rev. Astron. Astrophys.* **5**:525 (1967).

(Fo 75) W. A. Fowler, G. R. Caughlan, and B. A. Zimmerman, *Ann. Rev. Astron. Astrophys.* **13**:69 (1975).

(Fr 80) H. B. French, *Astrophys. J.* **240**:41 (1980).

(Ga 77) W. Galster, W. Treu, P. Dück, H. Fröhlich, and H. Voit, *Phys. Rev. C* **15**:950 (1977).

(Go 73) A. Gobbi, R. Wieland, L. Chua, D. Shapira, and D. A. Bromley, *Phys. Rev. C* **7**:30 (1973).

(Hi 53) D. L. Hill and J. A. Wheeler, *Phys. Rev.* **89**:1102 (1953).

(Hi 77) M. D. High and B. Čujec, *Nucl. Phys.* **A282**:181 (1977).

(Ho 65) J. T. Holdeman and R. M. Thaler, *Phys. Rev. Lett.* **14**:81 (1965).

(Ho 79) S. S. Holt, N. E. White, R. H. Becker, E. A. Boldt, R. F. Mushotzky, P. J. Serlemitsos, and B. W. Smith. *Astrophys. J.* **234**:L65 (1979).

(Hu 80) G. Hulke, C. Rolfs, and H. P. Trautvetter, *Z. Phys.* **A297**:161 (1980).

(Im 68) B. Imanishi, *Phys. Lett.* **27B**:267 (1968).

(Im 69) B. Imanishi, *Nucl. Phys. A* **125**:33 (1969).
(Ke 77) K.-U. Kettner, H. Lorenz-Wirzba, C. Rolfs, and H. Winkler, *Phys. Rev. Lett.* **38**:337 (1977).
(Ke 80) K.-U. Kettner, H. Lorenz-Wirzba, and C. Rolfs, *Z. Phys.* **A298**:65 (1980).
(Ki 80) R. P. Kirshner and W. P. Blair, *Astrophys. J.* **236**:135 (1980).
(Ko 70) B.-Z. Kozlovsky, *Astrophys. Space Sci.* **8**:114 (1970).
(Ko 77) J. J. Kolata, R. C. Fuller, R. M. Freeman, F. Haas, B. Heusch, and A. Gallman, *Phys. Rev. C* **16**:891 (1977).
(Ko 79a) J. J. Kolata, R. M. Freeman, F. Haas, B. Heusch, and A. Gallman, *Phys. Rev. C* **19**:2237 (1979).
(Ko 79b) D. G. Kovar, D. F. Geesaman, T. H. Braid, Y. Eisen, W. Henning, T. R. Ophel, M. Paul, K. E. Rehm, S. J. Sander, P. Sperr, J. P. Schiffer, S. L. Tabor, S. Vigdor, B. Zeidman, and F. W. Prosser, Jr., *Phys. Rev. C* **20**:1305 (1979).
(Le 79) J. Lequeux, M. Peimbert, J. F. Rayo, A. Serrano, and S. Torres-Peimbert, *Astron. Astrophys.* **80**:155 (1979).
(Ma 73) M. G. Mazarakis and W. E. Stephens, *Phys. Rev. C* **7**:1280 (1973).
(Ma 74) H.-B. Mak, D. Ashery, and C. A. Barnes, *Nucl. Phys.* **A226**:493 (1974).
(Ma 78) T. Matsuse, Y. Kondo, and Y. Abe, *Prog. Theor. Phys.* **59**:1009 (1978).
(Mi 70) G. Michaud and W. A. Fowler, *Phys. Rev. C* **2**:2041 (1970).
(Mi 72) G. Michaud, *Astrophys. J.* **175**:751 (1972).
(Mi 73) G. Michaud, *Phys. Rev. C* **8**:525 (1973).
(Pa 69) J. R. Patterson, H. Winkler, and C. S. Zaidins, *Astrophys. J.* **157**:367 (1969).
(Pa 71) J. R. Patterson, B. N. Nagorcka, G. D. Symons, and W. M. Zuk, *Nucl. Phys.* **A165**:545 (1971).
(Pa 77) J. Y. Park, W. Greiner, and W. Scheid, *Phys. Rev. C* **16**:2276 (1977).
(Pe 65) A. A. Penzias and R. W. Wilson, *Astrophys. J.* **142**:419 (1965).
(Ra 66) G. H. Rawitscher, *Nucl. Phys.* **85**:337 (1966).
(Ro 78) C. Rolfs and H. P. Trautvetter, *Ann. Rev. Nucl. Part. Sci.* **28**:115 (1978).
(Sc 70) W. Scheid, W. Greiner, and R. Lemmer, *Phys. Rev. Lett.* **25**:176 (1970).
(Sp 74) H. Spinka and H. Winkler, *Nucl. Phys.* **A233**:456 (1974).
(Sp 80) W. M. Sparks and A. S. Endal, *Astrophys. J.* **237**:130 (1980).
(St 76) R. G. Stokstad, Z. E. Switkowski, R. A. Dayras, and R. M. Wieland, *Phys. Rev. Lett.* **37**:888 (1976).
(Sw 74) Z. E. Switkowski, R. M. Wieland, and A. Winther, *Phys. Rev. Lett.* **33**:840 (1974).
(Sw 76) Z. E. Switkowski, R. G. Stokstad, and R. M. Wieland, *Nucl. Phys.* **A274**:202 (1976).
(Tr 75) V. Trimble, *Rev. Mod. Phys.* **47**:877 (1975).
(Tr 80) W. Treu, H. Fröhlich, W. Galster, P. Dück, and H. Voit, *Phys. Rev. C* **22**:2462 (1980).
(Ts 78) I. Tserruya, Y. Eisen, D. Pelte, A. Gavron, H. Oeschler, D. Berndt, and H. L. Harney, *Phys. Rev. C* **18**:1688 (1978).
(Vo 68) E. Vogt, in *Advances in Nuclear Physics, Vol. 1* (M. Baranger and E. Vogt, eds.), Plenum Press, New York (1968), p. 261.
(Vo 77) H. Voit, W. Galster, W. Treu, H. Fröhlich, and P. Dück, *Phys. Lett.* **67B**:399 (1977).
(Wa 67) R. V. Wagoner, W. A. Fowler, and F. Hoyle, *Astrophys. J.* **148**:3 (1967).
(Wa 73) R. V. Wagoner, *Astrophys. J.* **179**:343 (1973).
(We 78) T. A. Weaver, G. B. Zimmerman, and S. E. Woosley, *Astrophys. J.* **225**:1021 (1978).
(We 80) T. A. Weaver and S. E. Woosley, *Ann. N.Y. Acad. Sci.* **336**:335 (1980).

(Wi 74) R. M. Wieland and C. A. Barnes, unpublished optical model calculations (1974).

(Wi 80) J. R. Wilson, *Ann. N.Y. Acad. Sci.* **336**:358 (1980).

(Wo 71) S. E. Woosley, W. D. Arnett, and D. D. Clayton, *Phys. Rev. Lett.* **27**:213 (1971).

(Wo 76) H. Wojciechowski, D. E. Gustafson, L. R. Medsker, and R. H. Davis, *Phys. Lett.* **63B**:413 (1976).

(Wo 82) S. E. Woosley and T. A. Weaver, in *Essays in Nuclear Astrophysics* (C. A. Barnes, D. D. Clayton, and D. N. Schramm, eds.), Cambridge University Press, New York (1982).

(Wu 78) S.-C. Wu, Ph. D. Thesis, California Institute of Technology (1978), see also (Wu 84).

(Wu 84) S.-C. Wu and C. A. Barnes, *Nucl. Phys.* **A422**:373 (1984).

2

FULVIO CACACE is a Professor of General Chemistry at the University of Rome, where he received his education and has been a member of the faculty since 1971.

Heavy Ions in Hot Atom Chemistry

FULVIO CACACE

1. Introduction

The classical approach to chemical kinetics is based on thermochemical methods, where the energy required to promote chemical change is provided by the thermal motion of the reacting molecules. Obvious experimental limitations have restricted such methods to the study of processes occurring near the classical Arrhenius activation energies. The past two decades or so have witnessed the development of a new field of kinetics, dealing with systems whose energy distribution is other than the classical Maxwell–Boltzmann, the reacting molecules possessing an excess of energy over that corresponding to thermal equilibrium. This field—often, and somewhat loosely, referred to as "hot atom chemistry"—initially relied upon nuclear techniques for the generation of translationally, electronically, and vibrationally excited reagents, frequently labeled with radioactive nuclides to allow identification of their reaction products (Ma 68; Ne 72; St 72; Ne 75; We 61).

Application of early hot atom techniques soon revealed the richness of the chemistry and the interest of reactions occurring at energies in the range 5–100 eV, well above the values typical of thermal equilibrium, and called for new and improved methods for the production of "hot" reagents. Since acceleration of charged species is much easier than that of neutral molecules, energetic ions have understandably played an important role in the development of hot atom chemistry.

Another major area of interest for heavy ions in hot atom chemistry has evolved as a result of extensive studies on the chemical consequences of

FULVIO CACACE • University of Rome, 00100 Rome, Italy.

spontaneous radioactive processes, such as β decay, isomeric transition, K capture, etc. It has been shown that such processes provide a unique route to radioactive hot ions, that have since been largely exploited, especially in the study of gaseous systems.

Finally, interaction of polyatomic ions, accelerated to energies of a few MeV, with solid or gaseous targets, and analysis of the scattering of the component atoms following the "Coulomb explosion" of the ions, have recently been developed into a powerful tool for probing the structural and stereochemical features of gaseous charged species.

In this chapter we will not attempt an exhaustive coverage of the applications of heavy ions in hot atom chemistry. Instead, a few significant examples of the results obtained in the three major areas of interest outlined above will be discussed in some detail, providing in addition leading references to the reader interested in a more comprehensive treatment of the wide, and somewhat ill-defined, title subject.

2. Chemical Reactions Promoted by Accelerated Ions

2.1. Early Studies

Early mechanistic studies on the chemical reactivity of electrostatically accelerated ions were largely focused on simple ^{14}C-labeled species, such as $^{14}C^+$, $^{14}CO^+$, $^{14}CO_2^+$, and must be viewed against a background represented by contemporary hot atom chemistry of nucleogenic carbon, obtained from the $^{14}N\ (n, p)\ ^{14}C$ reaction with a maximum kinetic energy of 45 keV.

The first investigation was reported by Giacomello and co-workers (Cr 54), who employed an isotope separator to produce a beam of $^{14}C^+$ ions to bombard their organic targets. Typical energies ranged from 30 to 40 KeV, with current densities from 10^{-10} to 10^{-7} A cm^{-2}. Analysis of the labeled products from relatively unvolatile targets, e.g., long-chain fatty acids, showed that a fraction of the $^{14}C^+$ ions promoted a typical hot process, i.e., a ^{14}C-for-^{12}C substitution, yielding ^{14}C-labeled substrate molecules. This pioneering study paved the way to subsequent mechanistic work, that confirmed the occurrence of the direct substitution process on much simpler substrates. Thus, bombardment of solid benzene with high-energy $^{14}C^+$ ions at -160°C, at current densities of ca. 10^{-9} A cm^{-2}, gave ^{14}C-labeled benzene with a yield of ca. 2%, via an unspecified mechanism (Le 56)

$$^{14}C^+ + C_6H_6 \rightarrow {}^{12}C + {}^{14}C_6H_6 \tag{1}$$

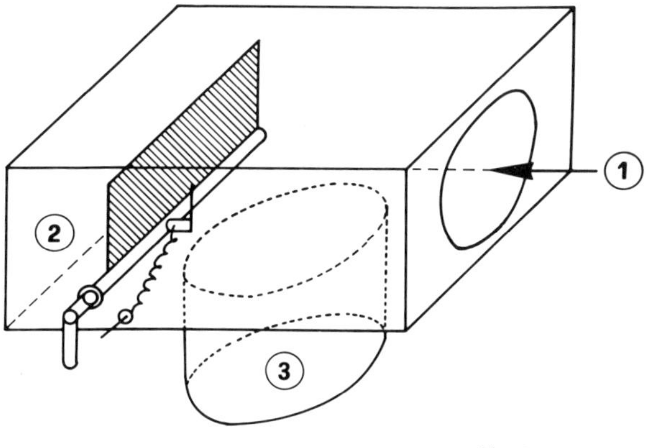

Figure 1. Apparatus for the irradiation of solid targets with $^{14}C^{+}$ ions (Al 65). 1, Ion beam from an isotope separator; 2, rotating target with a needle electrode; 3, 500 liter s^{-1} diffusion pump.

The identity of the labeled product was unequivocally established by the then-novel technique of radio glc. Extension of the study to other substrates, e.g., sodium benzoate (Al 56) and cholesterol (Al 56b), confirmed the occurrence of the ^{14}C-for-^{12}C substitution, although in general with low yields. Incidentally, the irradiation of solid organic targets with $^{14}C^{+}$ ions (polyatomic ions, such as $^{14}CO^{+}$ and $^{14}CO_2^{+}$, gave essentially the same results) represented a useful labeling technique, especially for structurally complicated molecules, hardly accessible in those days via conventional routes (Al 58).

A schematic illustration of the apparatus used for the labeling experiments is given in Figure 1. As a whole, these early studies demonstrated the advantages of the acceleration technique over nuclear methods, in particular the possibility of controlling the energy of the hot reagent and the absence of radiation damage from foreign radiation, such as the neutrons and the γ flux, typical of a nuclear reactor. Nevertheless, several serious drawbacks became soon apparent as well, associated in particular with the relatively high energy and the short range of the heavy ions, causing considerable radiation damage to the surface of the target. In addition, the possibility must be considered that volatile products could escape from the target into the vacuum system of the accelerator.

In this connection, very serious was the fact that the chemical fate of most (over 95%) of the incident $^{14}C^{+}$ ions remained obscure.

2.2. Improved Bombardment Techniques

In order to minimize the radiation damage, whose effects superimpose on the reactions of the heavy ions and complicate the interpretation of the results, improved techniques were introduced in several laboratories (As 68; Fi 69).

In particular, the beam was spread over a large target area, and fresh volatile substrate was continuously condensed onto the cooled bombarded surface, thus substantially reducing the radiation damage, and preventing the loss of volatile products from the surface of the substrate. In addition, specialized instruments ("chemical accelerators") were built, designed to provide intense beams of mass-resolved ions, e.g., up to 10^{-6} A of $^{14}C^+$ (Po 70). The instruments were working on the principle of the continuous condensation of fresh substrate onto a cold (−196°C) surface, and were equipped with electrostatic devices, designed to reduce the kinetic energy of the ions. In fact, the chemical reactions of interest occur at relatively low energies (<100 eV), and the high accelerating voltages used in early studies caused an unnecessary amount of radiation damage.

As an example, the decelerator described by Pohlit et al. (Po 70) makes it possible to continuously decrease the energy of the ions down to 5 ± 2 eV. The ratio of the number of incoming ions to that of the continuously deposited target molecules could be kept as low as 10^{-4}, and together with the low kinetic energy of the ions ensured a remarkably low level of radiation damage, and prevented appreciable losses of volatile reaction products. A schematic diagram of this chemical accelerator and its decelerating device is given in Figure 2.

2.3. Results

The techniques outlined in the previous paragraphs have been extensively used in the study of hot atom chemistry in a variety of systems, both inorganic and organic. Among the former ones, the reactivity of hot $^{51}Cr^+$ ions has been investigated in solid chromates (An 66), of $^{51}Cr^+$ and $^{51}Cr^{2+}$ ions in chromates, perchlorates and sulfates (An 65), of $^{75}Se^+$ ions in selenates (An 68; Wo 69; Co 72), of $^{60}Co^+$ ions in *trans*-$Co[C_2H_4N_2H_4]_2Cl_2 \cdot NO_3$ (Ra 60), of $^{56}Mn^+$ ions in $Cr(CO)_6$ (Je 72), of $^{32}P^+$ and $^{35}S^+$ ions in alkali chlorides (An 67; Fr 67; Ma 65; An 71; Ka 71; An 71b). Organic systems have also received much attention, the reactions studied including those of T^+ and T_2^+ ions with sodium benzoate, sodium *m*-iodobenzoate, and sodium phenylacetate (As 67), with benzene (Fi 68; Ku 69; Fi 73), with benzene–naphtalene solid solutions (Fi 69), with pyridine (Fi 69b), and with dibasic aromatic acids (As 68).

Other reactions investigated are those of hot I^+ and I^{2+} ions with benzene (Ac 73; Ca 73; Ca 75c), and of $^{14}C^+$ and $^{14}CO^+$ ions with water (Va 68).

The results of these studies have been of considerable value to hot atom chemistry, especially when compared with pertinent investigations

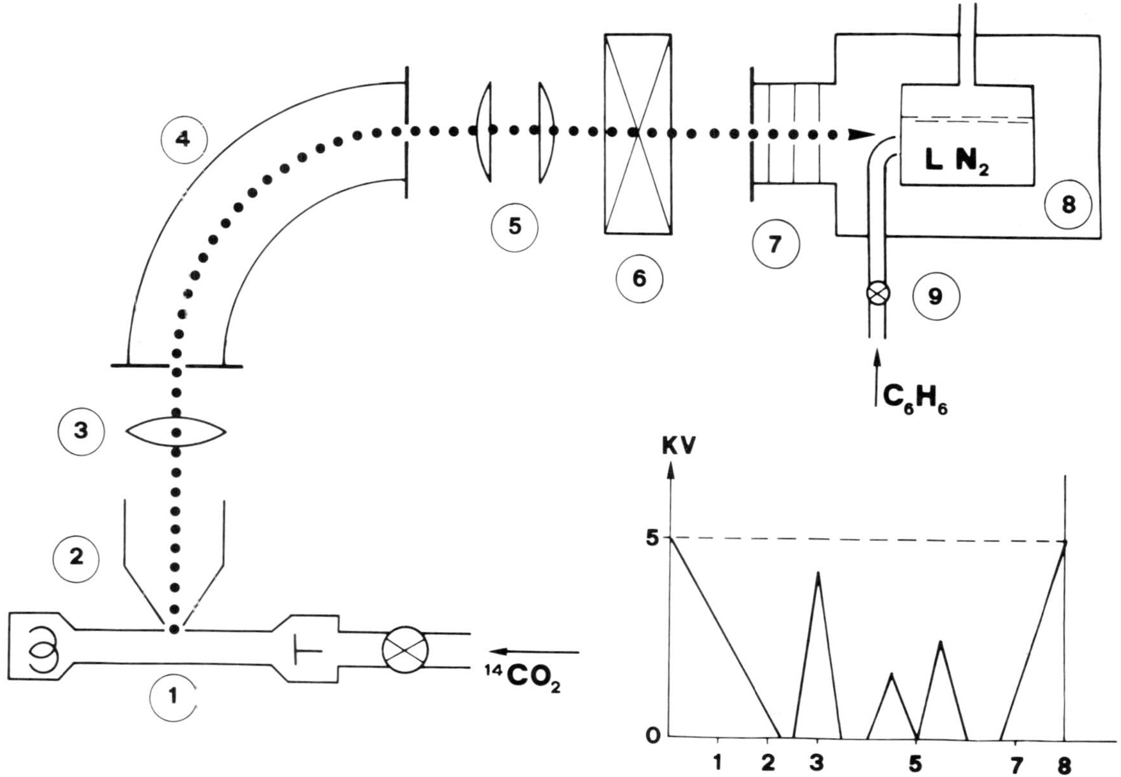

Figure 2. A "chemical" accelerator (Po 70). 1, Arc ion source; 2, extraction electrode; 3, electrostatic spherical lens; 4, magnetic analyzer; 5, cylindrical lens; 6, gate valve; 7, decelerating device; 8, cooled target; 9, substrate inlet tube. The potential distribution in the various sections of the apparatus is shown in the insert.

concerning the reactivity of the same substrates toward nucleogenic hot species.

As an illustrative example, we will discuss in some detail a sustained research effort by the Berkeley group leaded by R. Lemmon on the reactivity of $^{14}C^+$ ions accelerated onto solid benzene. The research has covered a span of over a decade, and its results have been reviewed (Le 73; Le 75). Early studies had demonstrated the formation of ^{14}C-labeled benzene and toluene from the bombardment of benzene with $^{14}C^+$, the latter product being labeled predominantly, but not exclusively (ca. 86%), in the methyl group (Le 61). Subsequent studies extended considerably the spectrum of products identified. Among them, seven hydrocarbons are formed in significant yields (i.e., exceeding 1%) as shown in Table 1, that illustrates as well the dependence of their yields on the energy of the impinging $^{14}C^+$ ions. Another major product is $^{14}C_2H_2$, formed with a 5% yield and accompanied by much smaller amounts of labeled styrene, allene, 1,1-butadiene, 1,3-butadiene, 1-butyne, and propyne. Many other radioactive products are formed in very small individual yields (below 0.1%) and only a painstaking analytical effort has made it possible to account for the fate of some 88% of the ions that strike the solid substrate (Li 72).

Such unusual variety of products aptly illustrates the complex reaction network and the rich hot atom chemistry of monoatomic carbon ions in a relatively simple organic matrix, such as benzene.

Table 1 shows that the yields of the products are rather insensitive to energy in the range from 5×10^3 down to 10^2 eV, while below 10^2 eV the yields of benzene and diphenylmethane are approximately constant, and those of toluene, cycloheptatriene, and phenyl acetylene decrease by factors of 5, 3, and 8, respectively.

Further insight has been obtained by Lemmon and co-workers measuring the intramolecular distribution of ^{14}C within the compounds formed, a feat that could be achieved via chemical degradation of the purified radioactive products (Po 71). The intramolecular ^{14}C distribution, and its dependence on the energy of the $^{14}C^+$ ions, are summarized in Table 2.

These results, and comparison with studies carried out with nuclear and photochemical techniques, made it possible to deduce reasonable mechanistic pathways for the formation of the major labeled products.

It should be noted in the first place that, consistent with the prediction of theory (Ma 52) and the large cross section for electron capture by $^{14}C^+$ ions at relatively low kinetic energies (Ha 52; Gi 56), it was concluded that the ions undergo complete neutralization following repeated collisions with the molecules of the substrate, before reaching the energy range where chemical combination is possible. This view is supported by further experiments, showing that the yields of the products do not change when the

Table 1. Yields of ^{14}C-Labeled Products from the Reaction of $^{14}C^+$ Ions with Solid Benzene at −196°C

Energy (eV)	Absolute yields[a] (%)						
	Benzene	Toluene	Cycloheptatriene	Phenylacetylene	Diphenyl	Diphenylmethane	Phenylcycloheptatriene
5000	3.4	1.0	2.5	1.6	1.3	1.5	5.3
1000	5.7	1.3	2.9	2.3	2.3	2.5	5.5
500	4.8	1.0	1.8	2.2	1.7	1.9	5.5
100	5.7	1.1	2.2	3.1	2.9	3.8	6.6
15	2.7	0.2	0.9	0.5	1.2	3.6	10.3
5	3.9	0.2	0.5	0.3	1.7	2.6	10.8
2	3.5	0.2	0.9	0.2	5.5	2.9	13.0[b]

[a] Fraction of the total $^{14}C^+$ activity in a given product.
[b] Data taken from (Li 72).

Table 2. Intramolecular ^{14}C Distribution in the Labeled Products[a]

Product	Energy					
	5 keV	31 eV	6 eV	5 eV	3 eV	2 eV
CH_3	85	83	86	63	92	94
C_6H_5	13	12	15	41	10	6
	55	70		66		
	45	30		34		
CH_2	91			76	87	
$(C_6H_5)_2$	9			18	11	
CH	30			33		30
C	66			62		63
C_6H_5	4			4		6
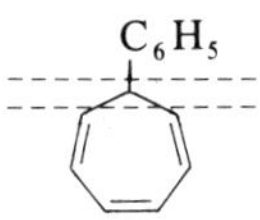	16			22		9
	52			57		85
	30			19		6
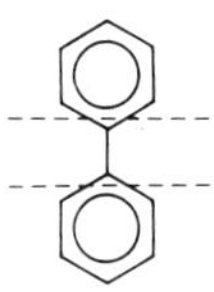	9		8		0	0
	81		84		100	100
	9		8		0	0

[a] From Le, 73.

charge of the $^{14}C^+$ ion is neutralized *before* the beam strikes the target surface (Po 71). Further "cooling" down allows the hot ^{14}C atoms to react with the substrate, either directly or following H abstraction, giving $:^{14}CH_2$ radicals, which still possess a few eV of kinetic and/or internal energy. Formation of toluene and cycloheptatriene is currently understood in terms of the reaction sequence illustrated in Chart I.

It is worth noting that the initial step of the reaction pathway is promoted by the insertion of a hot $:^{14}CH_2$ moiety either into a C–H bond,

Chart 1

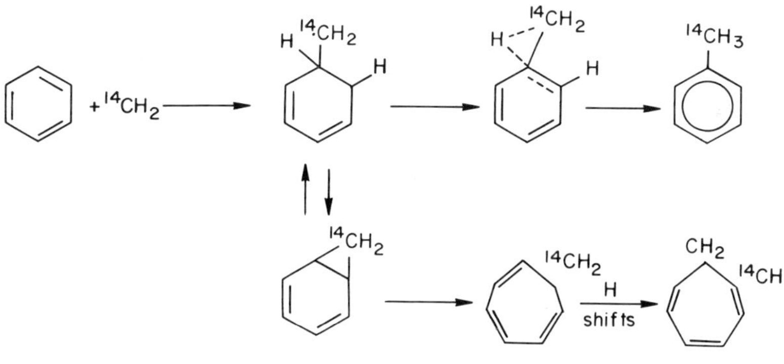

Formation of labeled toluene and cycloheptatriene from the reaction of hot methylene radicals with solid benzene. (Li, 72)

giving toluene, or into a C–C bond, yielding cycloheptatriene, in agreement with photochemical (Mü 63) and hot atom chemistry (Su 58) studies. The results suggest that at high $^{14}C^+$ kinetic energies (above ca. 15 eV), $:^{14}CH_2$ is the major hot reagent. At lower energies, however, there are less chances for the slowing down ^{14}C to pick two H atoms, and this can account for the drop of the yields of toluene and cycloheptatriene measured below 100 eV. In the lower energy range the bare ^{14}C atom becomes the major hot reagent, and its insertion into a C–C bond accounts for the activity found in the ring.

Chart 2

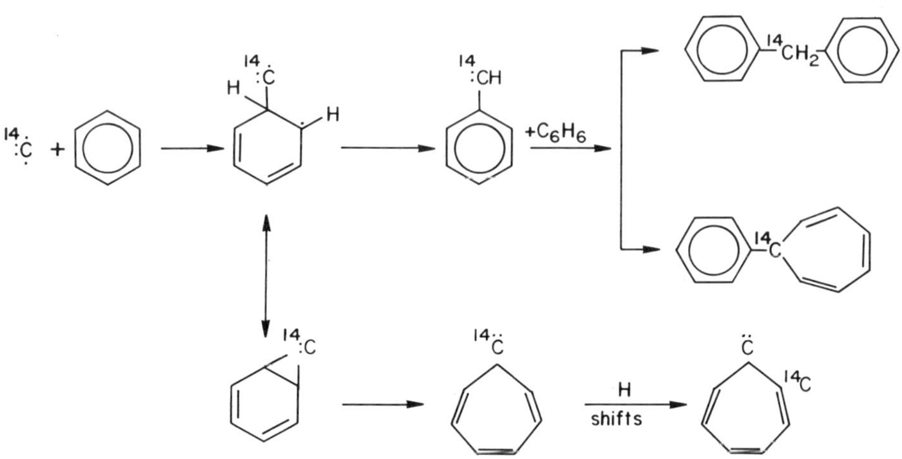

Formation of diphenylmethane and phenylcycloheptatriene from the reaction of hot ^{14}C atoms with solid benzene. (Li, 72)

The role of ^{14}C in this process becomes particularly significant near 5 eV, when it is still sufficiently energetic to overcome the activation energy for insertion, yet it does not deposit into the C_7H_6 intermediate formed enough energy to cause its dissociation. This accounts for the *increase* of the fraction of activity found in the ring when the $^{14}C^+$ energy is reduced to, but not much below, 5 eV. Independent support for the role of bare ^{14}C atoms comes from the observed formation of toluene and cycloheptatriene from the reaction of carbon vapor with benzene (Sp 65).

Chart II illustrates the reaction pathways responsible for the formation of the ^{14}C-labeled diphenylmethane and phenylcycloheptatriene.

So far, we have discussed exclusively the formation of C_7 and C_{13} hydrocarbons ("synthesis" products), without considering the formation of ^{14}C-labeled benzene, the product from ^{14}C-for-^{12}C substitution, whose isolation has been regarded from the very early experiments as a most intriguing result. Mechanistic analysis has so far only made it possible to discard several routes to ^{14}C-labeled benzene, e.g., those involving C_8 intermediates (Li 72), suggested by the isolation of acetylene and phenylacetylene:

$C_6H_6 + {}^{14}C_2H_x \longrightarrow$ 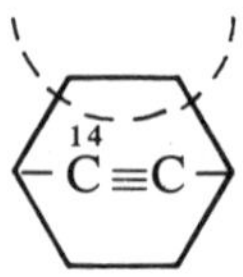 $\longrightarrow {}^{14}C_6H_6$ (2)

The available evidence, so far circumstantial and far from conclusive, points rather to a C_7 intermediate, according to the scheme

$C_6H_6 + {}^{14}C \longrightarrow$ 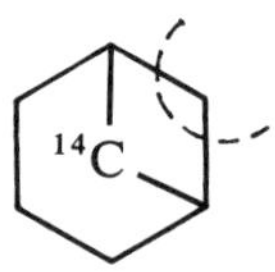 $\longrightarrow {}^{14}C_6H_6$ (3)

where hydrogen atom and electron distributions are not specified.

Undoubtedly, working out a detailed mechanism for the formation of ^{14}C-labeled benzene remains a major goal in the current development of this carefully investigated research area.

3. Heavy Ions from the Spontaneous Decay of Radioactive Precursors

The second major area of interest for heavy ions in hot atom chemistry concerns charged atomic and molecular species from the nuclear decay of radioactive precursors.

The subject has been the focus of theoretical and experimental interest, concerning both the *primary* effects of the nuclear transition, namely, the excitation, the ionization, and possibly the dissociation of *isolated* atoms and molecules, and the subsequent chemical interactions of the charged decay fragments with the surroundings. After a brief survey of the specific nuclear transformations of interest, examples of application to hot atom chemistry studies will be discussed in some detail.

3.1. β Decay of Isolated Tritiated Species

The excitation following the β decay of a T atom, and its molecular consequences when the atom was covalently bound, have been studied by theoretical (Su 40; Mi 41; Fe 41; Le 41; St 47; Sc 53; Ca 56; Wo 56; Gr 57; We 58; Ko 61; Ca 63; Ca 63b; Hs 65; Sh 73; Ik 77; Ik 77b; Ik 77c; Ik 79) and mass spectrometric (Sn 53; Sn 58; We 58; We 59; Ca 60; We 60; We 60b; We 61; Ca 62; We 62; Ca 63; Ni 73; Ik 79; Om 79) techniques.A typical apparatus for the experimental study of decay-induced ionization, the so-called "charge mass spectrometer," is illustrated in Figure 3. The ions, spontaneously formed from the decay of a radioactive gas in a large tank at pressure of ca. 10^{-5} Torr, are driven by an array of electrodes into the magnetic analyzer, where their m/e ratio is measured by conventional

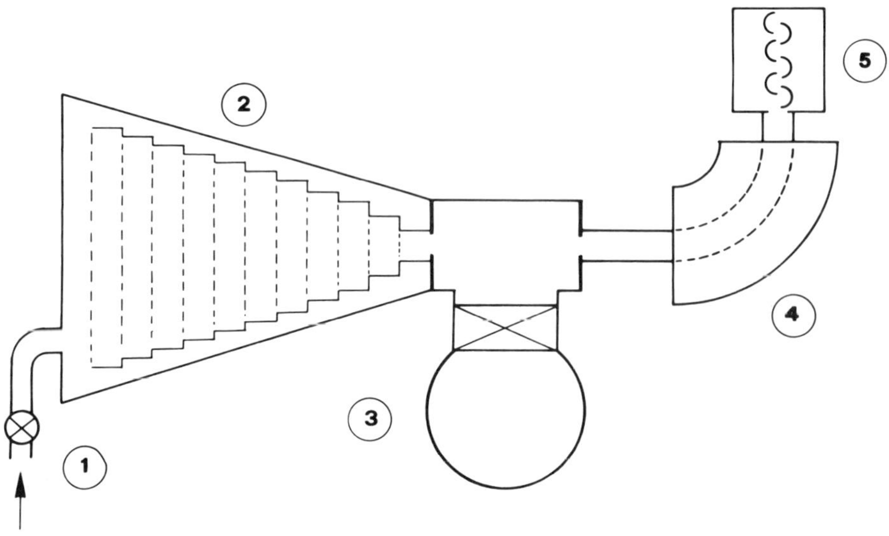

Figure 3. Scheme of a typical "charge" mass spectrometer. 1, Tritiated parent inlet; 2, large-volume ion source with extraction electrodes; 3, high-speed diffusion pump; 4, magnetic analyzer; 5, ion multiplier.

Table 3. Daughter Ions from the Decay of Gaseous Tritiated Molecules

Molecule	Daughter ion	% Abundance	
T_2	$^3HeT^+$	89.1 ± 0.8[a]	94.5 ± 0.6[b]
	T^+, $^3He^+$	10.9 ± 0.8	5.5 ± 0.6
HT	$^3HeH^+$	89.5 ± 1.1[b]	93.2 ± 1.9[c]
	$^3He^+$	8.2 ± 1.0	5.1 ± 0.3
	H^+	2.3 ± 0.4	1.5 ± 0.1
	$^3He_2^+$	—	0.14 ± 0.01
CH_3T	H^+	2.4 ± 0.1[d]	
	$^3He^+$, H_3^+	0.12 ± 0.01	
	C^+	4.9 ± 0.1	
	CH^+	4.0 ± 0.1	
	CH_2^+	4.9 ± 0.1	
	CH_3^+	82.0 ± 1.5	
	$CH_3{}^3He^+$	0.06 ± 0.01	
	C^{2+}	0.08 ± 0.01	

[a] (Sh 73).
[b] (We 60).
[c] (Ca 65).
[d] (Sn 58).

techniques. The theoretical and mass spectrometric results, confirmed by direct ir spectroscopic studies (Ra 74), indicate that the daughter $^3He^+$ ions from the decay of T receive, in a large fraction of the nuclear transitions, little or no kinetic recoil energy, and are formed in their electronic ground state.

When the T atom is covalently bound, its fate depends entirely on the partner. Thus, the ^{3}He–H bond is sufficiently stable to allow over 90% of the $^3HeH^+$ ions from the decay of HT to survive dissociation. On the other hand, the decay of tritiated hydrocarbons invariably leads to loss of neutral ^{3}He, giving an organic cation, as shown in Table 3, since the potential between C and He is repulsive at all distances (Ik 77).

The yields of the major daughter ions from the decay of representative tritiated molecules are given in Table 4.

3.2. Applications to Structural and Kinetic Studies

The results outlined in the previous paragraph represent the foundation of experimental techniques, introduced in 1964 (Ca 64) and since largely exploited in the study of the structure and reactivity of organic cations in gaseous, liquid, and solid systems. According to this technique, whose principles and applications have been reviewed (Ca 70; Ak 76; Ak 76b), a

Table 4. Yields of Major Daughter Ions from the Decay of Isolated Tritiated Molecules

Molecule	Daughter ion[a]	Abundance (%)
CH_3T	CH_3^+	82
C_2H_5T	$C_2H_5^+$	80
$CH_3CH_2CH_2T$	$C_3H_7^+$	41.0
CH_3CHTCH_3	$C_3H_7^+$	56.0
C_6H_5T	$C_6H_5^+$	72
o–$CH_3C_6H_4T$	$C_7H_7^+$	78
m–$CH_3C_6H_4T$	$C_7H_7^+$	79
p–$CH_3C_6H_4T$	$C_7H_7^+$	76
$C_6H_5CH_2T$	$C_7H_7^+$	79
c–C_4H_7T	$C_4H_7^+$	80
c–C_5H_9T	$C_5H_9^+$	75
CH_2IT	CH_2I^+ [b]	56

[a] Except when otherwise indicated, the values are taken from (Ca 75).
[b] (Ni 73).

tritiated precursor, containing more than one T atom in the same molecule, is allowed to decay in any systems of interest, from low-pressure gases to dense gases, liquids, and solids. The decay of one of the T atoms yields, independently of the environment, a daughter ion whose reactions can be conveniently followed, and the final products identified, owing to the presence of the other T atoms, e.g.,

$$CT_4 \rightarrow \beta^- + {}^3He + CT_3^+ \xrightarrow{+H_2O} CT_3OH_2^+ \xrightarrow{-H^+} CT_3OH \qquad (4)$$

In short, it is possible to introduce ions of known structure, composition, and initial location of the charge, in the system of interest. The method lends itself to application of the classical experimental techniques of physical organic chemistry, including the use of radical scavengers and ionic interceptors, competition kinetics, pressure dependence studies, and, of special interest, actual isolation of the final products, whose structure and stereochemical features provide otherwise inaccessible information on the structure and stereochemistry of their gaseous ionic precursors.

The major experimental difficulty arises from the need of multiply tritiated precursors, whose synthesis, purification, and storage require special cares, in view of their enormous specific activity, e.g., 1.2×10^5 Ci mol^{-1} for CT_4. Continuously improved radiochemical techniques have allowed the preparation of an increasing number of these precursors, including CT_4 (Jo 58; An 59; Ci 66; Le 74; Ca 75b; Si 80), ethane-1,2–T_2 (Al 67), propane–T_x (Ci 78), propane-1,2–T_2 (Ca 67), cyclobutane–T_x (Ca 79), cyclopentane–T_x (Ba 71), benzene–T_x (Ne 73), and benzene-1,4–T_2 (An 80).

The decay technique has been applied to the study of the reactions of CT_3^+ ions with gaseous methane (Ca 66), gaseous benzene, toluene, and the isomeric xylenes (Ne 70; Ne 73; Ne 76), with gaseous and liquid benzene and toluene (Ca 77; Ca 78), liquid alkylbenzenes (Ne 76b), *t*-butylbenzene (Gi 77), gaseous (Gi 79) and liquid halobenzenes (Gi 79b), with water and the lower alcohols (Ne 68), ethers (Ne 68b), gaseous water/ethylene mixtures (Ne 73b), aliphatic haloderivatives (Ne 68c), amines (Ne 73c), multidentate nucleophiles such as bromobenzene (Ak 76), nitrobenzene (Ne 71), and alkyl aryl ethers (Ne 74), carbon monoxide (Gi 77b), and gaseous CH_3F (Co 78).

Tritiated ethyl ions have been studied in ethane (Al 65; Al 67), propyl ions in propane (Ca 67), tritiated cyclopentyl ions in cyclopentane (Ba 71). Labeled phenylium ions from the decay of uniformly tritiated benzene have been allowed to react with gaseous benzene (Ne 73), several alkylbenzenes (Ne 76c), and halobenzenes (Ne 76d). The reactivity of OT^+, a labeled hydroxyl ion from the β decay of fully tritiated water, has been studied in gaseous benzene (Ne 76e).

The results of the extensive application of the decay technique have provided a wealth of data on the structure, the isomerization, the stability, and in general the reactivity of gaseous cations, often entirely inaccessible to other more conventional techniques, in particular to mass spectrometry. Of special value is the possibility of generating in widely different media, e.g., in a low-pressure gas and in a liquid system, *exactly the same* ionic species, in order to carry out a comparative study of their reactivity, and evaluate the influence of the environment on the structural and kinetic parameters of the charged species. While a detailed discussion of the results achieved is clearly beyond the scope of this chapter, representative examples will be discussed in some detail to enlighten the typical features and the potentiality of the technique.

The first example of application of the decay technique to a structural problem concerns the demonstration of the existence of cyclic $C_4H_7^+$ isomers, namely, the cyclobutyl and the cyclopropylcarbinyl cations, as stable species in the dilute gas state (Ca 79). Theoretical approaches led to somewhat conflicting conclusions, reliable STO-3G calculations showing that the cyclopropylcarbinyl ion in the bisected configuration (I)

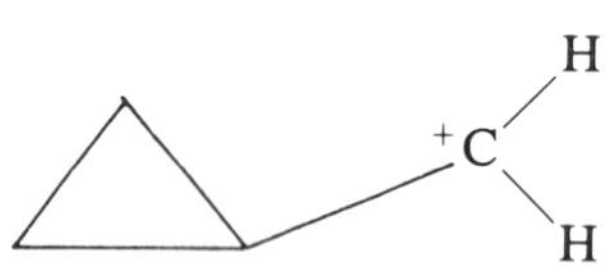

I

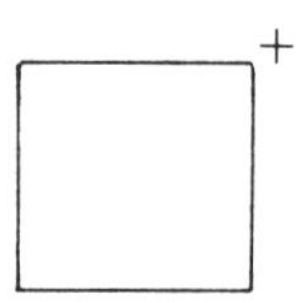

II

is the most stable species, to which other ones, including the cyclobutyl ion

(II), collapse immediately, without activation energy (He 74). Mass spectrometry did not provide significant evidence on the specific problem. Indeed, $C_4H_7^+$ ions had long been detected, but owing to the lack of structural discrimination, their structure (cyclic or open) could never be substantiated by mass spectrometry.

In order to apply the decay technique, multitritiated cyclobutane was allowed to decay for several months in gaseous systems whose major component was a suitable nucleophile, such as H_2O or NH_3, as illustrated in Table 5.

Preliminary ICR studies (Po 73) on the decay of monotritiated cyclobutane had shown that $C_4H_7^+$ ions (of unknown structure) are formed in over 85% of the decays, and survive undissociated for at least 10^{-5} s.

Consequently, isolation of *cyclic* tritiated products from the decay of multitritiated cyclobutane, $c\text{–}C_4X_8\,(X = H, T)$, in gaseous water and ammonia (Table 5) can be reasonably traced only to the condensation of *cyclic* cations with the nucleophile HY ($Y = OH$ or NH_2)

$$c\text{–}C_4X_8 \xrightarrow[\text{decay}]{\text{beta}} c\text{–}C_4X_7^+ + {}^3He + \beta^- \tag{5}$$

$$c\text{–}C_4X_7^+ + HY \rightarrow c\text{–}C_4X_7YH^+ \tag{6}$$

followed by proton transfer from the cyclic onium ion to a gaseous base, B, e.g.,

$c\text{–}C_4X_7YH^+ + B \longrightarrow$ 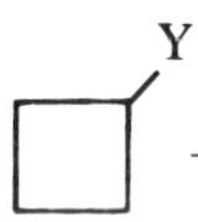 $+ BH^+$ (7)

From the above considerations, it follows that the observed formation of cyclobutanol (cyclobutylamine) and cyclopropylcarbinol (cyclopropylcarbinylamine) requires that cyclobutyl and cyclopropylcarbinyl cations, or at least some cyclic precursor of these ions, such as a bicyclobutonium ion, must exist in the dilute gas state at least for the time necessary to encounter a molecule of the nucleophile, with lifetimes in excess of 10^{-9} s. Accordingly, cyclic $C_4H_7^+$ ions must be regarded as fully legitimate ionic intermediates, characterized by significant minima on the $C_4H_7^+$ potential surface.

It should be noted that these conclusions have been subsequently confirmed by novel theoretical calculations (Le 79).

Another mechanistic application of the decay technique concerns the automerization of gaseous phenylium ions (Sp 80). Benzene-1,4–T_2 was

Table 5. Relative Yields of the Products from the Gas-Phase Reactions of $C_4X_7^+$ Decay Ions with Water and Ammonia[a]

System composition (Torr)		Relative yields of products (%)								
NH_3	H_2O	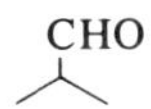	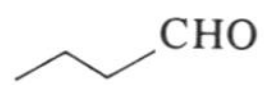	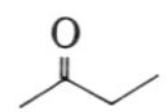	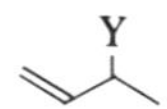	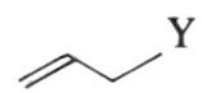		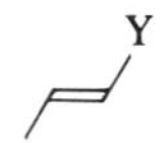	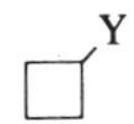	
400	—	n.d.[b]	n.d.	n.d.	—	—	28.8[c]	—	35.5[c]	35.7[c]
2	19	33.3	36.6	0.9	7.5[d]	4.1[d]	3.6[d]	4.3[d]	5.8[d]	4.0[d]
—	25	6.8	26.5	19.3	9.7[d]	11.0[d]	7.1[d]	4.9[d]	5.3[d]	9.3[d]
—	6	44.4	23.6	15.4	2.0[d]	1.8[d]	1.8[d]	2.4[d]	4.3[d]	3.4[d]

[a] Data from (Ca 79).
[b] Not detectable, corresponding to a relative yield of under 0.5%.
[c] Y = NH_2.
[d] Y = OH.

Table 6. Radioactive Product Distribution from Phenylium Ion Condensation with Methanol ($T = 25°C$)[a]

System composition[b]	Absolute yields of radioactive products (%)		
	Anisole	Others[c]	Total[d]
Liquid CH_3OH saturated with O_2	64.7	26.8	91.5
Gaseous CH_3OH (18 Torr) + O_2 (4 Torr)	44.9	23.9	68.8

[a] Data from (Sp 80).
[b] Standard deviation of data *ca.* 10%.
[c] Aromatic condensation products.
[d] Low-boiling fragmentation products account for residual activity.

allowed to decay dissolved in liquid methanol, or in gaseous methanol at 18 Torr, giving a high yield of tritiated anisole, as shown in Table 6.

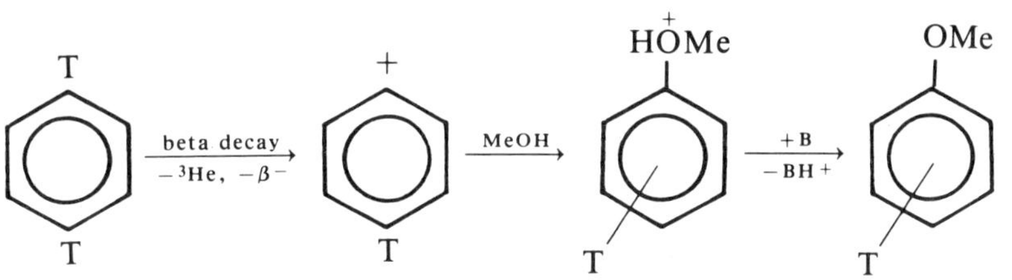

(8)

Measurement of the residual radioactivity of the product after gradual substitution of the ring H atoms with inactive groups, allowed determination of the T distribution within the labeled anisole:

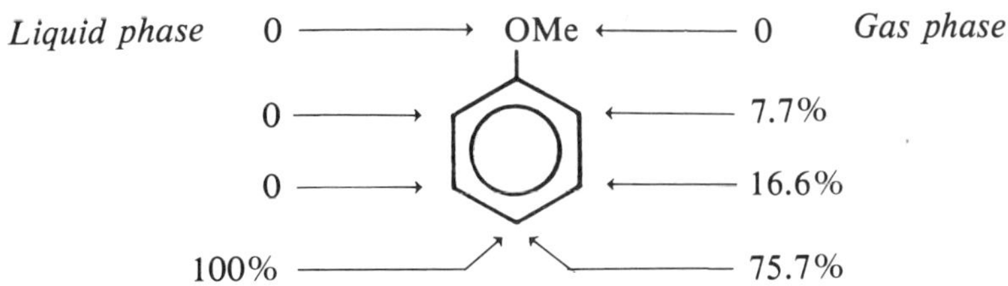

The label migration, observed only in the product from the gas-phase decay, provides the first experimental evidence for a degenerate rearrangement of the gaseous phenylium cation:

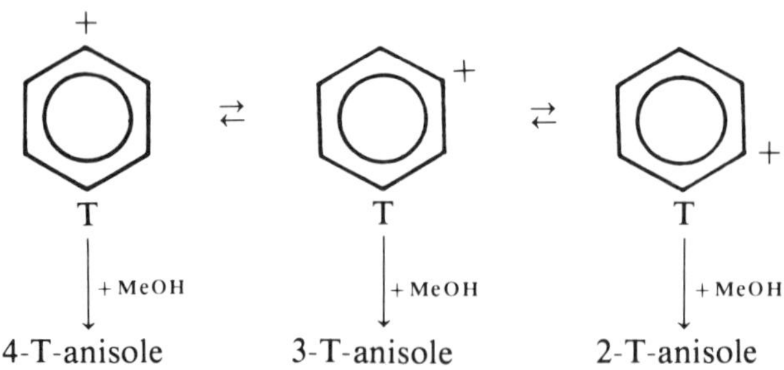

(9)

The complete retention of the original T position in the product from the liquid-phase reaction indicates that automerization is slow in comparison with the collision frequency in the condensed phase, in agreement with theoretical estimates (Di 76) of the activation energy for process (9), ranging from 44 to 77 kcal mol^{-1}. While no direct information is provided by the decay experiments as to the specific mechanism of automerization (H versus CH-group migration), the results nevertheless exclude any significant intervention of the nonclassical carbene structure

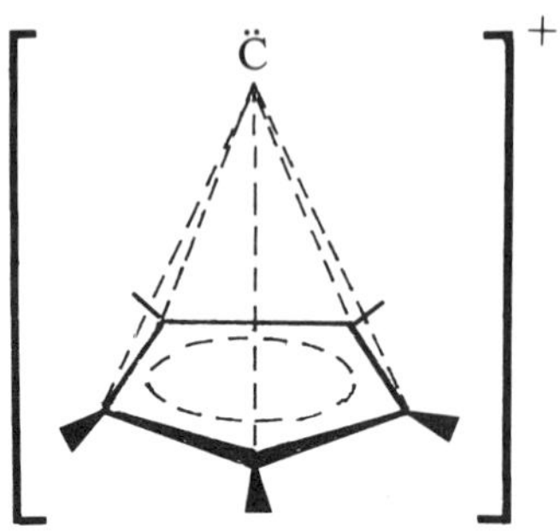

that would require a uniform distribution of T in all positions of the phenylium ion following its rearrangement.

3.3. Multicharged Ions from Nuclear Transitions Leading to Inner-Shell Ionization

Several nuclear transformations, including α and β decay, isomeric transition, electron capture, fission, etc. generate multiply charged daughter ions, often with a significant amount of kinetic energy. The same effects can sometimes be obtained by shining heavy atoms with X photons of sufficiently high energy to produce vacances in the inner shells.

The molecular consequences of these phenomena are a classical subject of hot atom chemistry. Even a cursory mention of the large body of theoretical, spectrometric, and radiochemical work related to deep ionization phenomena is outside the scope of the chapter, and the interested reader is directed to specialized reviews (We 65; Ma 68; Ne 72; Ne 75) for a comprehensive coverage. We shall briefly mention here the specific features of a particularly violent and disruptive molecular event, referred to as a *Coulomb explosion*, since it represent the foundation of the technique for the structural and stereochemical analysis of gaseous ions discussed in Section 4.

Among the various nuclear events leading to deep ionization, we shall concentrate on the $^{80m}Br \rightarrow {}^{80}Br$ 4.4*h* isomeric transition of chemically bounded ^{80m}Br atoms, since it not only represents a typical example, but is among the nuclear transitions whose chemical consequences have been most extensively investigated in hot atom chemistry.

The energies of the γ photons emitted, 49 and 36 keV, are too low to impart to the daughter nucleus sufficient recoil energy to rupture its chemical bond. However, since these photons are highly converted, they cause deep ionization (84% in the *K* shell and 16% in the *L* shell), which in turn triggers Auger cascades and leads eventually to multiple ionization, up to 13 electrons being expelled within 10^{-16}–10^{-15} s (We 60).

The fast "charging" of the ^{80}Br atom necessarily causes a correspondingly fast redistribution of the valence electrons through the molecule, and positive charges develop on several, or all, constituent atoms. In a time short compared to the period of its bond vibrations, the molecule literally explodes, as a consequence of multicenter electrostatic repulsion. Thus, the most abundant fragments from the isomeric transition of $CH_3{}^{80m}Br$ are CH_3^+ and Br^{7+}, formed with a total kinetic energy up to 53 eV (St 72).

The chemical consequences of the isomeric transition have been studied in many gaseous systems, in particular CH_4, CH_3Br, and HBr (Ni 67; Ni 68; Ok 69; Ta 69; Oa 70; Ta 70; Ta 70b; Ta 70c; Ta 70d; Sh 71; Sh 71b; Pe 72; Da 73; Nu 73; Sa 73; Ya 73; Da 74; Su 74; Ta 74; Ya 75; Su 76), as well as in liquid (Ka 71b) and solid systems (Jo 70). Recently, labeling techniques based on the isomeric transition of $CF_3{}^{80m}Br$ have been reported (Mu 79; De 80). As a typical example, we will briefly outline a technique using the Coulomb explosion of ^{80m}Br-labeled precursors as a tool for investigating aromatic substitution by free, unsolvated brominium ions in the dilute gas state. This approach can be outlined as follows. A tracer concentration of the precursor, e.g., $CH_3{}^{80m}Br$, is allowed to decay in a large excess of a rare gas whose ionization potential is intermediate between the first and the second ionization potential of Br. In such a system, the charge of the hot $^{80}Br^{n+}$ ions from the Coulomb explosion is rapidly reduced to +1 by fast charge-exchange processes, and their excess kinetic energy removed by moderating collisions before a reactive encounter with a molecule of the aromatic substrate does occur.

In such a way, the reactions of thermal, singly charged brominium ions can be investigated. The reactions can be followed, and their final products conveniently detected since the daughter nuclide from the isomeric transition is still radioactive, thus obviating the need of double-labeling methods.

The technique, introduced in 1972 (Ca 72), has been employed in a variety of aromatic substrates, making it possible to derive for the first time a comprehensive set of Hammet parameters for aromatic bromination by a charged, unsolvated reagent in the dilute gas state. The results obtained by Stöcklin and co-workers (Kn 74; Kn 77) in the study of the reaction

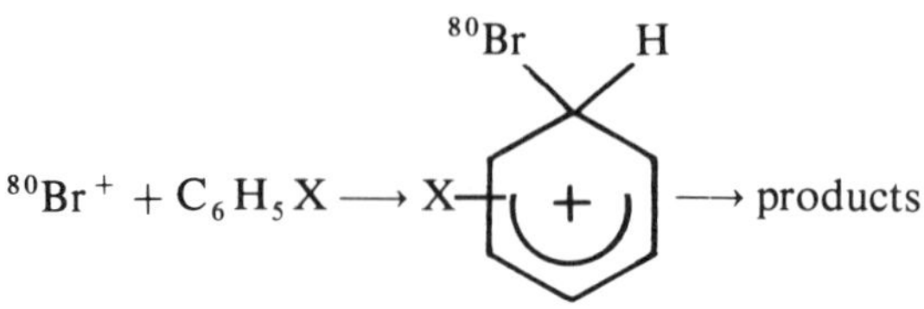

(10)

are summarized in Table 7. From these data, Knust has derived the excellent linear free energy correlation illustrated in Figure 4.

From these studies, it can be concluded that gaseous Br^+ ions obey the "classical" reactivity–selectivity relationship. This is mechanistically significant, in that it provides the first demonstration that linear free energy correlations retain their validity even for extremely reactive and unselective electrophiles, such as free, unsolvated cations, and that their application can be extended to the dilute gas state.

Table 7. Brown's Substituent Constants and Relative Reactivities in ^{80}Br-for-H Substitution[a]

X	σ^+	k/k_0	$\log(k/k_0)$
p–CH_3	–0.311	2.00	0.3010
m–CH_3	–0.066	1.02	0.0086
p–F	–0.073	1.46	0.1644
p–Cl	0.114	0.91	–0.0410
m–Cl	0.399	0.58	–0.2366
p–Br	0.150	0.93	–0.0315
m–Br	0.405	0.48	–0.3188
p–CF_3	0.612	0.40	–0.3979
m–CF_3	0.520	0.40	–0.3979
p–OCH_3	–0.778	7.24	0.8597
m–OCH_3	0.047	0.99	–0.0044
H	0	1.00	0.0000

[a] From (Kn 77).

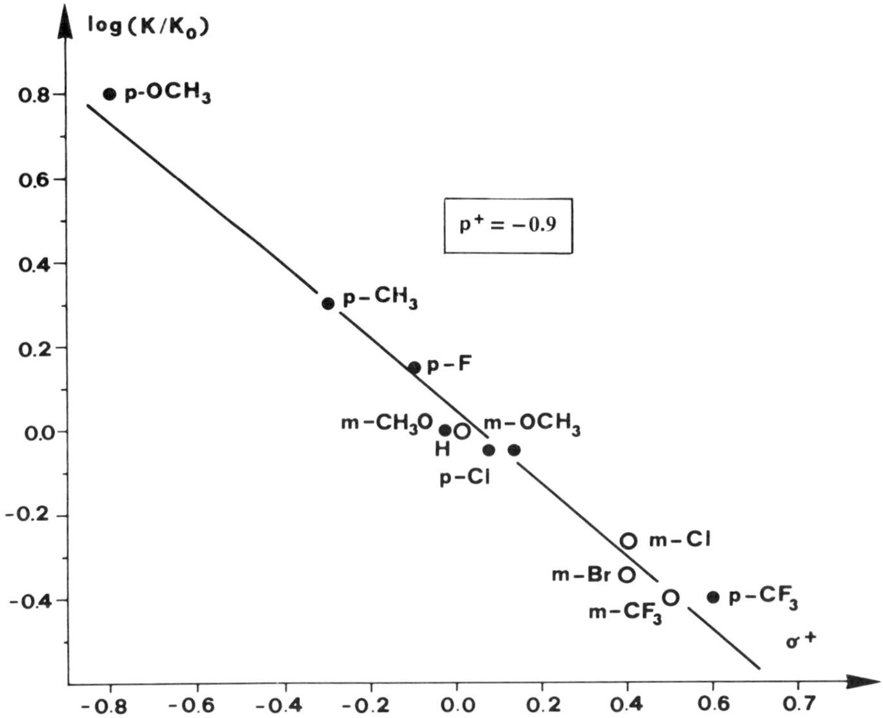

Figure 4. Free energy correlation for the aromatic substitution by gaseous $^{80}Br^{+}$ ions (Kn 77).

4. Collision-Induced Coulomb Explosion of Fast Molecular Ions as a Structural Probe

Recently, a technique has been developed which exploits a typical "hot" phenomenon, the Coulomb explosion of molecular ions, in order to determine their structure and their stereochemical features. The principles of this interesting approach, whose results have been recently reviewed (Ge 80), can be outlined as follows. Consider a molecular ion, AB^+, accelerated to several MeV, striking a stationary target, such as a thin graphite foil, the internuclear vector of AB^+ making an angle ϕ with the direction of the ion (Figure 5). At the high velocity of AB^+ many, if not all of its electrons are lost as a consequence of the collisions with the electrons of the target, in a time (ca. 10^{-17} s) which is short when compared to the dwell time of the ion within the foil, having a thickness of the order of 10^{-8} m.

As a consequence of the multiple positive charges that develop on the constituent atoms, the ion undergoes a Coulomb explosion similar to that

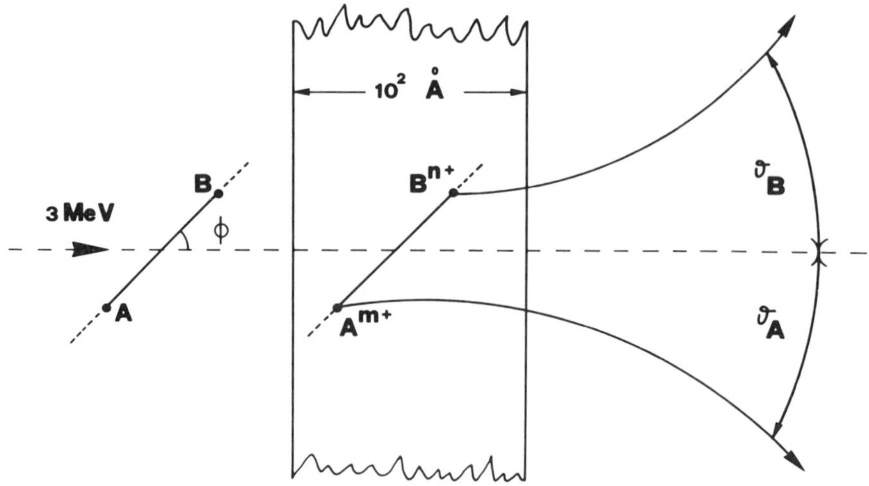

Figure 5. Coulomb explosion of a 3-MeV AB^+ molecular ion induced by its collision with a thin graphite foil (Ge 80).

described in a previous paragraph. The mutual repulsion between its A^{m+} B^{n+} fragments causes a measurable change in their direction and/or kinetic energy. The extent of such changes depends on the value of the angle ϕ. If $\phi = 0$, namely, the AB^+ ion strikes the target head on, no angular shift is observed, the effect of the mutual Coulomb repulsion between the two fragments being simply a change in their kinetic energy, the leading one being accelerated and the trailing one retarded. On the other hand, when $\phi = 90°$, no velocity changes occur in the beam direction, but the maximum value of the angular deflection is observed. Intermediate values of ϕ lead to both kinetic energy and direction changes and consequently, assuming an isotropic ϕ distribution, one obtains a typical "ring" pattern (Figure 6) when the intensity of the fragments is plotted as a function of the energetic and angular shift. Analogous considerations apply to experiments where the stationary target is a dilute gas rather than a scattering foil (Co 78b).

It is clear from the foregoing that joint measurements of the energetic and angular distribution of the fragments allow determination of the geometrical structure of the parent molecular ion, especially when the charged fragments from a given Coulomb explosion can be detected in coincidence.

From the experimental standpoint, the problem boils down to measuring kinetic energy changes of several keV, and angular deflections of a few mrad, of ionic beams characterized by an energy of several MeV. A typical experimental arrangement, currently operated at Argonne National Laboratory, is shown in Figure 7. The apparatus uses a magnetically

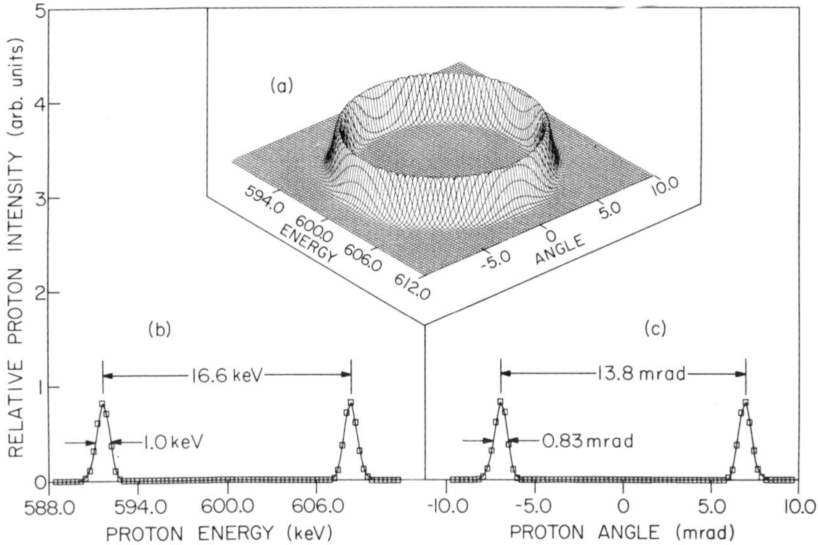

Figure 6. A typical ring pattern in the distribution (a) of H^+ from the Coulomb explosion of 3-MeV HeH^+ ions as a function of their energy shift (ΔE) and angular deflection (θ). Plot (b) shows ΔE for $\theta = 0$, plot (c) θ for $\Delta E = 0$. From (Ge 80), courtesy of D. S. Gemmell.

analyzed molecular ions beam and a 25° electrostatic analyzer with a relative energy resolution of ca. 2×10^{-4} and an angular resolution of 0.15 Mrad (Va 76; Ge 80b). Coincidence measurements of the fragments arising from a given Coulomb explosion are possible, and provide useful information. Alternative approaches are based on different detectors, including photographic plates (Ga 78), polycarbonate plastic foils etched after exposure to the fragments ions (Go 79), image intensifiers (Go 79), and charge-coupled semiconductor detectors (Al 80). The results must be corrected for several perturbing factors, especially multiple scattering, charge exchange, and polarization wake effects, arising from the perturbation of the target electrons by the passage of the fast ions.

The technique has been applied to the structural analysis of several simple cations, including CO_2^+, a linear ion, N_2O^+, an angular one (Ge 78), CH^+, CH_2^+, CH_3^+, and CH_4^+ (Ge 80c). Interestingly, the last species displays a highly asymmetric proton distribution around the central C atom, a result traced to the Jahn–Teller distortion of the ion (Di 71). Coincidence measurements have been carried out on CH_2^+, NH_2^+, and H_2O^+ molecular ions (Ge 80b). The results concerning ionized water give a bond angle of $110 \pm 2°$ and a bond distance of 1.0 ± 0.04 Å, in good agreement with the values (110.5° and 0.999 Å) deduced from an optical measurement (Le 76).

Perhaps the most significant application so far of the technique based

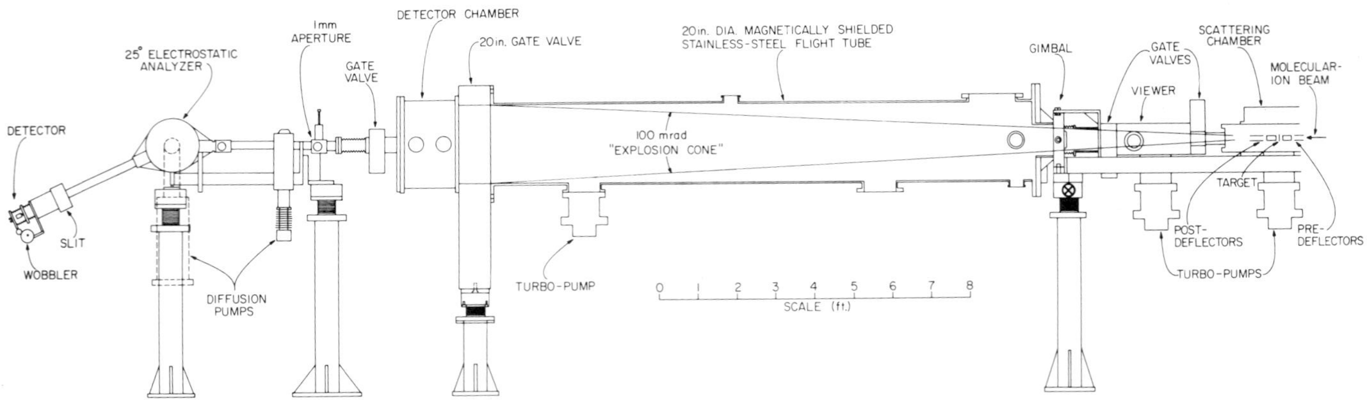

Figure 7. Experimental apparatus for the measurement of the energy shifts and angular deflection of heavy ions from Coulomb explosions. From (Ge 80), courtesy of D. S. Gemmell.

on the Coulomb explosion concerns the determination of the structure of H_3^+, jointly carried out at the Weizmann Institute, the Argonne National Laboratory, and the Lyon University (Ga 78). The results show that the molecule has indeed a triangular shape, as long predicted by theoretical approaches, and give the following values for the internuclear separation: 1.1 ± 0.2 Å (Weizmann), 0.95 ± 0.06 Å (Lyon), and 0.97 ± 0.03 Å (Argonne). These results are in excellent agreement with theoretical calculation, and taking into account that the measurements concern a free, gaseous ion, their significance and precision must be regarded as exceptional. Further refinement of the experimental techniques, and application to more complex species, will undoubtedly prove of great value to gas-phase ionic chemistry.

References

(Ac 73) H. Ackay, J. Cailleret, J. C. Abbe, J. M. Paulus, and Y. Llabador, *J. Chim. Phys.* **70**:970 (1973).

(Ak 76) G. P. Akulov, *Usp. Khim.* **45**:1970 (1976).

(Ak 76b) G. P. Akulov, *Russ. Chem. Rev.* **45**:1008 (1976).

(Al 56) B. Aliprandi, F. Cacace, and G. Giacomello, *Ric. Sci.* **26**:3029 (1956).

(Al 56b) B. Aliprandi and F. Cacace, *Ann. Chim. (Rome)* **46**:1204 (1956).

(Al 58) B. Aliprandi, F. Cacace, L. Cieri, G. Ciranni, R. Masironi, and M. Zifferero, *Radioisotopes in Scientific Research*, Pergamon Press, London (1958), Vol. 2, p. 146.

(Al 65) B. Aliprandi, F. Cacace, and A. Guarino, *Chemical Effects of Nuclear Transformations*, International Atomic Energy Agency, Vienna (1965), Vol. 2, p. 471.

(Al 67) B. Aliprandi, F. Cacace, and A. Guarino, *J. Chem. Soc. B*, 519 (1967).

(Al 80) M. Algranati, A. Faibis, R. Kaim, and Z. Vager, private communication in (Ge 80).

(An 59) Anonimous, British Patent 822, 617 (1959) granted to National Research Council of Canada.

(An 65) T. Andersen and G. Sörensen, *Nucl. Instrum. Methods*, **38**:204 (1965).

(An 66) T. Andersen and G. Sörensen, *Trans. Faraday Soc.* **62**:3427 (1966).

(An 67) T. Andersen and G. Sörensen, in *Interaction of Radiation with Solids* (A. Bifhay, ed.), Plenum Press, New York (1967).

(An 68) T. Andersen, T. Langvad, and G. Sörensen, *Nature (London)* **218**:1158 (1968).

(An 71) T. Andersen and A. Ebbessen, *Trans. Faraday Soc.* **67**:3540 (1971).

(An 71b) T. Andersen and J. L. Baptista, *Trans. Faraday Soc.* **67**:1213 (1971).

(An 80) G. Angelini, A. L. Segre, L. Altman, and M. Speranza, *J. Org. Chem.* **45**:3291 (1980).

(As 65) S. Ascoli and F. Cacace, *Nucl. Instrum. Methods* **38**:198 (1965).

(As 67) S. Ascoli, F. Cacace, G. Giacomello, and E. Possagno, *J. Phys. Chem.* **71**:427 (1967).

(As 68) S. Ascoli, R. Cipollini, and E. Possagno, *Ric. Sci.* **38**:1098 (1968).

(Ba 71) L. Babernics and F. Cacace, *J. Chem. Soc. B*, 2313 (1971).

(Ca 56) T. A. Carlson, *Phys. Rev.* **101**:1747 (1956).

(Ca 60) T. A. Carlson, *J. Chem. Phys.* **32**:1234 (1960).
(Ca 62) T. A. Carlson and R. M. White, *J. Chem. Phys.* **36**:2883.
(Ca 63) T. A. Carlson, *Phys. Rev.* **130**:2361 (1963).
(Ca 63b) T. A. Carlson, F. Pleasonton, and C. H. Johnson, *Phys. Rev.* **129**:2220 (1963).
(Ca 64) F. Cacace, in *Methods of Preparing and Storing Marked Molecules*, Euratom, Bruxelles (1964), p. 719.
(Ca 65) T. A. Carlson and R. White, *Chemical Effects of Nuclear Transformations*, International Atomic Energy Agency, Vienna (1965), Vol. 1, p. 1.
(Ca 66) F. Cacace, G. Ciranni, and A. Guarino, *J. Am. Chem. Soc.* **88**:2903 (1966).
(Ca 67) F. Cacace, M. Caroselli, and A. Guarino, *J. Am. Chem. Soc.* **89**:4584 (1967).
(Ca 70) F. Cacace, *Adv. Phys. Org. Chem.* **8**:79 (1970).
(Ca 72) F. Cacace and G. Stöcklin, *J. Am. Chem. Soc.* **94**:2518 (1972).
(Ca 73) J. Cailleret, J.-Ch. Abbe, and J. M. Paulus, *Radiochim. Acta* **20**:33 (1973).
(Ca 75) F. Cacace, in *Hot Atom Chemistry Status Report* (F. S. Rowland, ed.), International Atomic Energy Agency, Vienna (1975), p. 233.
(Ca 75b) F. Cacace and M. Schüller, *J. Labelled Compd.* **11**:313 (1975).
(Ca 75c) J. Caillaret, J. M. Paulus, and J. Ch. Abbe, *J. Chem. Soc. Faraday Trans. I*, 637 (1975).
(Ca 77) F. Cacace and P. Giacomello, *J. Am. Chem. Soc.* **99**:5477 (1977).
(Ca 78) F. Cacace and P. Giacomello, *J. Chem. Soc. Perkin Trans. II*, 652 (1978).
(Ca 79) F. Cacace and M. Speranza, *J. Am. Chem. Soc.* **101**:1587 (1979).
(Ci 66) G. Ciranni and A. Guarino, *J. Labelled Compd.* **2**:198 (1966).
(Ci 78) R. Cipollini and M. Schüller, *J. Labelled Compd.* **15**:703 (1978).
(Co 72) M. Cogneau, G. Duplatre, and J. L. Vargas, *J. Inorg. Nucl. Chem.* **34**:3021 (1972).
(Co 78) M. Colosimo and R. Bucci, *J. Phys. Chem.* **82**:2061 (1978).
(Co 78b) R. G. Cooks, *Collision Spectroscopy*, Plenum Press, New York (1978).
(Cr 54) U. Croatto and G. Giacomello, 45th Meeting of the Italian Society for the Progress of Science, Naples, October 16, 1954.
(Da 73) S. H. Daniels and H. J. Ache, *Radiochim. Acta* **19**:58 (1973).
(Da 74) S. H. Daniels, H. J. Ache, and G. Stöcklin, *J. Phys. Chem.* **78**:1043 (1974).
(De 80) O. De Jesus, J. F. Mustaklem, B. Djermouni, and H. J. Ache, *Radiochem. Radioanal. Lett.* **42**:297 (1980).
(Di 71) R. N. Dixon, *Mol. Phys.* **20**:113 (1971).
(Di 76) J. D. Dill, P. v. R. Schleyer, J. S. Binkley, R. Seeger, J. A. Pople, E. Haselbach, *J. Am. Chem. Soc.* **98**:5428 (1976).
(Fe 41) E. L. Feinberg, *J. Phys. (USSR)* **4**:123 (1941).
(Fi 68) L. P. Firsova and Y. S. Kuklin, *Radiokhimiya* **10**:607 (1968).
(Fi 69) L. P. Firsova, V. V. Murav'ev, and V. A. Kapitanluk, *Radiokhimiya* **11**:711 (1969).
(Fi 69b) L. P. Firsova and A. B. Kharono, *Radiokhimiya* **11**:581 (1969).
(Fi 69c) L. P. Firsova and D. K. Tiung, *Radiokhimiya* **11**:587 (1969).
(Fi 73) L. P. Firsova, *High Energy Chem.* **7**:96 (1973).
(Fr 67) J. H. Freeman, M. Kasrai, and A. G. Maddock, *J. Chem. Soc. Chem. Commun.* 979 (1967).
(Ga 78) M. J. Galliard, D. S. Gemmell, G. Goldring, I. Levine, W. J. Pietsch, J. C. Polzat, A. J. Ratkowski, J. Remillieux, Z. Vager, and B. J. Zabransky, *Phys. Rev. A* **17**:1797 (1978).
(Ge 78) D. S. Gemmell, E. P. Kanter, and W. J. Pietsch, *Chem. Phys. Lett.* **55**:331 (1978).
(Ge 80) D. S. Gemmell, *Chem. Rev.* **80**:301 (1980).

(Ge 80b) D. S. Gemmell, P. J. Cooney, and E. P. Kanter, *Nucl. Instrum. Methods* **170**:81 (1980), and references therein.

(Ge 80c) D. S. Gemmell, E. P. Kanter, and W. J. Pietsch, *J. Chem. Phys.* **72**:1402 (1980).

(Gi 56) H. B. Gilbody, *Proc. R. Soc. London Ser. A* **238**:334 (1956).

(Gi 77) P. Giacomello, *Radiochim. Acta* **24**:111 (1977).

(Gi 77b) P. Giacomello and M. Speranza, *J. Am. Chem. Soc.* **99**:7918 (1977).

(Gi 79) P. Giacomello, *Radiochim. Acta* **26**:185 (1979).

(Gi 79b) P. Giacomello, *J. Am. Chem. Soc.* **101**:4276 (1979).

(Go 79) G. Goldring, Y. Eisen, P. Thieberger, and H. Wegner, Physics Division Informal report No. ANL/PHY-79-3, p. 27.

(Gr 57) A. E. S. Green, *Phys. Rev.* **107**:1646 (1957).

(Ha 52) J. B. Hasted, *Proc. R. Soc. London Ser. A* **212**:235 (1952).

(He 74) W. J. Hehre and P. C. Hiberty, *J. Am. Chem. Soc.* **96**:302 (1974).

(Hs 65) C. Hsiung and A. A. Gordus, *Chemical Effects of Nuclear Transformations*, International Atomic Energy Agency, Vienna (1965), Vol. 2, p. 465.

(Ik 77) S. Ikuta, K. Okuno, K. Yoshihara, and T. Shiokawa, *J. Nucl. Sci. Technol. (Tokyo)* **14**:131 (1977).

(Ik 77b) S. Ikuta, K. Yoshihara, and T. Shiokawa, *J. Nucl. Sci. Technol. (Tokyo)* **14**:47 (1977).

(Ik 77c) S. Ikuta, K. Yoshihara, and T. Shiokawa, *Radiochem. Radionalyt. Lett.* **28**:435 (1977).

(Ik 77d) S. Ikuta, S. Iwata, and M. Imamura, *J. Chem. Phys.*, **66**:4671 (1977).

(Ik 79) S. Ikuta and S. Hashimoto, *Chem. Phys.* **42**:262–275 (1979); see also *Radiochem. Radioanalyt. Lett.* **39**:191 (1979).

(Je 72) G. M. Jenkins and D. R. Wiles, *J. Chem. Soc. Chem. Commun.* 1177 (1972).

(Jo 58) J. M. Jones, S. Goldblatt, *J. Mol. Spectrosc.* **2**:103 (1958).

(Jo 70) C. H. Jones, *Radiochim. Acta* **14**:1 (1970).

(Ka 71) M. Kasrai, A. G. Maddock, and J. H. Freeman, *Trans. Faraday Soc.* 2108 (1971).

(Ka 71b) N. A. Katsanos and I. L. Leontiadis, *J. Inorg. Nucl. Chem.* **33**:295 (1971).

(Kn 74) E. J. Knust, A. Halpern, and G. Stöcklin, *J. Am. Chem. Soc.* **96**:3733 (1974).

(Kn 77) E. J. Knust, *J. Am. Chem. Soc.* **99**:3037 (1977).

(Ko 61) W. Kolos, *Acta Phys. Polon.* **20**:175 (1961).

(Ku 69) Y. S. Kuklin and L. P. Firsova, *Radiokhimiya* **11**:729 (1969).

(Le 41) J. S. Levinger, *Phys. Rev. (USSR)* **4**:449 (1941).

(Le 56) R. M. Lemmon, F. Mazzetti, F. L. Reynolds, and M. Calvin, *J. Am. Chem. Soc.* **78**:6414 (1956).

(Le 61) R. M. Lemmon, R. T. Mullen, and F. L. Reynolds, *Chemical Effects of Nuclear Transformations*, International Atomic Energy Agency, Vienna (1961), Vol. 2, p. 27.

(Le 73) R. M. Lemmon, *Acc. Chem. Res.* **6**:65 (1973).

(Le 74) V. V. Leonov, E. E. Sinotova, and M. V. Korsakov, *Radiokhimiya* **16**:564 (1974).

(Le 75) R. M. Lemmon and W. R. Erwin, *Sci. Am.* 72 (1975).

(Le 76) H. Lew, *Can. J. Phys.* **54**:2028 (1976).

(Le 79) B. E. Levi, E. S. Blurock, and W. J. Hehre, *J. Am. Chem. Soc.* **101**:5537 (1979).

(Li 72) J. Lintermans, W. R. Erwin, and R. M. Lemmon, *J. Phys. Chem.* **76**:2521 (1972).

(Ma 52) H. S. W. Massey and H. H. S. Burhop, *Electronic and Ionic Impact Phenomena*, Oxford University Press, London (1952).

(Ma 65) A. G. Maddock and R. M. Mirsky, *Chemical Effects of Nuclear Transformations*, IAEA, Vienna (1965), Vol. 2, p. 41.

(Ma 68) A. G. Maddock and R. Wolfgang, in *Nuclear Chemistry* (L. Yaffè, ed.), Academic Press, New York (1968), Vol. 2, p. 185.

(Mi 41) A. Migdal, *J. Phys. (USSR)* **4**:449 (1941).

(Mu 79) J. F. Mustaklem and H. J. Ache, *Int. J. Appl. Radiat. Isot.* **30**:155 (1979).

(Mü 63) E. Müller and H. Fricke, *Justus Liebigs Ann. Chem.* **661**:38 (1963).

(Ne 68) V. D. Nefedov, E. N. Sinotova, G. P. Kulov, and V. P. Sass, *Radiokhimiya* **10**:600 (1968).

(Ne 68b) V. D. Nefedov, E. N. Sinotova, G. P. Kulov, and M. B. Korsakov, *Radiokhimiya* **10**:1586 (1968).

(Ne 68c) V. D. Nefedov, E. N. Sinotova, and G. P. Akulov, *Radiokhimiya* **10**:609 (1968).

(Ne 70) V. D. Nefedov, E. N. Sinotova, G. P. Akulov, and M. B. Korsakov, *Zh. Org. Khim.* **6**:1214 (1970).

(Ne 71) V. D. Nefedov, E. N. Sinotova, and G. P. Akulov, *The Isotope Tracer Method in Scientific Research and in Industry*, Atomizdat. Moskow (1971), p. 346.

(Ne 72) *Radiochemistry* (G. W. A. Newton, ed.), Vol. 1, The Chemical Society, London (1972).

(Ne 73) V. D. Nefedov, E. N. Sinotova, M. V. Korsakov, and E. G. Alekseev, *Radiokhimiya* **15**:635 (1973).

(Ne 73b) V. D. Nefedov, E. N. Sinotova, G. P. Akulov, and M. V. Korsakov, *Radiokhimiya* **15**:286 (1973).

(Ne 73c) V. D. Nefedov, E. N. Sinotova, V. V. Leonov, G. P. Akulov, and M. V. Korsakov, *Zh. Org. Khim.* **9**:1073 (1973).

(Ne 74) V. D. Nefedov, E. N. Sinotova, G. P. Akulov, and M. V. Korsakov, *Zh. Org. Khim.* **10**:1586 (1974).

(Ne 75) *Radiochemistry* (G. W. A. Newton, ed.), Vol. 2, The Chemical Society, London (1975).

(Ne 76) V. D. Nefedov, E. N. Sinotova, V. A. Shishkunov, *Zh. Org. Khim.* **12**:1491 (1976).

(Ne 76b) V. D. Nefedov, E. N. Sinotova, and V. A. Shishkunov, *Zh. Org. Khim.* **12**:1351 (1976).

(Ne 76c) V. D. Nefedov, E. N. Sinotova, and V. A. Shishkunov, *Zh. Org. Khim.* **12**:1495 (1976).

(Ne 76d) V. D. Nefedov, M. A. Toropova, V. V. Avrorin, N. E. Shchepina, and V. K. Vasil'ev, *Radiokhimiya* **18**:302 (1976).

(Ne 76e) V. D. Nefedov, G. P. Akulov, and S. V. Volkovich, *Zh. Obshchei Khim.* **46**:904 (1976).

(Ni 67) J. B. Nicholas, J. A. Merrigan, and F. P. Rack, *J. Chem. Phys.* **46**:1996 (1967).

(Ni 68) J. B. Nicholas and F. P. Rack, *J. Chem. Phys.* **48**:4085 (1968).

(Ni 73) K. Nishizawa, K. Narisada, H. Teramatsu, H. Iwami, and M. Shinagawa, *Mass Spectrosc.* **21**:199 (1973).

(Nu 73) K. Nukumura and E. Tachikawa, *Bull. Chem. Soc. Jpn.* **46**:346 (1973).

(Oa 70) D. W. Oates, R. L. Ayres, R. W. Helton, K. S. Schwartz, and E. P. Rack, *Radiochem. Radioanal. Lett.* **4**:123 (1970).

(Ok 69) J. Okamoto and E. Tachikawa, *Bull. Chem. Soc. Jpn.* **42**:1504 (1969).

(Om 79) T. Omori, T. Kikuchi, and T. Shiokawa, *Radiochem. Radioanal. Lett.* **37**:233 (1979).

(Pe 72) R. R. Pettijohn and E. P. Rack, *J. Phys. Chem.* **76**:3342 (1972).
(Po 70) H. M. Pohlit, W. R. Erwin, F. L. Reynolds, R. M. Lemmon, and M. Calvin, *Rev. Sci Instrum.* **41**:1012 (1970).
(Po 71) H. M. Pohlit, W. R. Erwin, Tz-Hong Lin, and R. M. Lemmon, *J. Phys. Chem.* **75**:2555 (1971).
(Po 73) L. G. Pobo, S. Wexler, and S. Caronna, *Radiochim. Acta* **19**:5 (1973).
(Ra 60) H. H. Rauscher, N. Sutin, and J. M. Miller, *J. Inorg. Nucl. Chem.* **12**:378 (1960).
(Ra 74) R. Raitz v. Frentz, K. Luchner, H. Micklitz, and W. Wittwer, *Phys. Lett.* **47A**:301 (1974).
(Ro 75) F. S. Rowland, *Hot Aatom Chemistry Status Report*, International Atomic Energy Agency, Vienna (1975).
(Sa 73) M. Saeki and E. Tachikawa, *Radiochim. Acta* **20**:27 (1973).
(Sc 53) H. M. Schwartz, *J. Chem. Phys.* **21**:45 (1953).
(Sh 71) T. Shiokawa, T. Kabayashi, K. Kondo, and M. Yagi, *Radiochem. Radioanal. Lett.* **7**:28 (1971).
(Sh 71b) T. Shiokawa, *Radiochem. Radioanal. Lett.* **7**:725 (1971).
(Sh 73) T. Shiokawa, G. Izawa, K. Nagashima, M. Hiraga, and K. Yoshihara, *Mass Spectrosc.* **21**:179 (1973).
(Sh 75) T. Shiokawa, *Hot Atom Chemistry Status Report*, International Atomic Energy Agency, Vienna (1975), p. 241.
(Si 80) E. N. Sinotova, M. V. Korsakov, and B. A. Shishkunov, *Radiokhimiya* **22**, 466 (1980).
(Sn 53) A. H. Snell and F. Pleasonton, *Phys. Rev.* **100**:1396 (1953).
(Sn 58) A. H. Snell and F. Pleasonton, *J. Phys. Chem.* **62**:1377 (1958).
(Sp 65) J. L. Spring, S. Winstein, and W. F. Libby, *J. Am. Chem. Soc.* **87**:1812 (1965).
(Sp 80) M. Speranza, *Tetrahedron Lett.* **21**:1983 (1980).
(St 47) H. Steinwedel and J. H. D. Jensen, *Z. Naturforsch.* **20**:125 (1947).
(St 72) G. Stöcklin, *Chimie des Atomes Chauds*, Masson et Ci, Paris (1972).
(Su 40) H. Suess, *Z. Phys. Chem. B* **45**:312 (1940).
(Su 58) B. Suryanarayana and A. P. Wolf, *J. Phys. Chem.* **62**:1369 (1958).
(Su 74) Y. Y. Su, Ch. Metcalfe, and H. J. Ache, *Radiochem. Radioanal. Lett.* **18**:349 (1974).
(Su 76) Y. Y. Su and H. J. Ache, *J. Phys. Chem.* **80**, 659 (1976).
(Ta 69) E. Tachikawa, *Bull. Chem. Soc. Jpn.* **42**:1504 and 2404 (1969).
(Ta 70) E. Tachikawa, *Bull. Chem. Soc. Jpn.* **43**:63 (1970).
(Ta 70b) E. Tachikawa and T. Kahara, *Bull. Chem. Soc. Jpn.* **43**:1293 (1970).
(Ta 70c) E. Tachikawa, *Radiochim. Acta* **13**:159 (1970).
(Ta 70d) E. Tachikawa, *Radiochem. Radioanal. Lett.* **5**:75 (1970).
(Ta 74) E. Tachikawa, K. Nukumura, *Bull. Chem. Soc. Jpn.* **47**:2749 (1974).
(Va 68) P. Vasudev, thesis, University of Cambridge (UK), 1968.
(Va 76) Z. Vager and D. S. Gemmell, *Phys. Rev. Lett.* **37**:1352 (1976).
(We 58) S. Wexler and D. C. Hess, *J. Phys. Chem.* **62**:1382 (1958).
(We 59) S. Wexler, *J. Inorg. Nucl. Chem.* **10**:8 (1959).
(We 60) S. Wexler and G. R. Anderson, *J. Chem. Phys.* **33**:850 (1960).
(We 60b) S. Wexler, G. R. Anderson, and L. A. Singèr, *J. Chem. Phys.* **32**:417 (1960).
(We 61) S. Wexler, *Chemical Effect of Nuclear Transformations*, International Atomic Energy Agency, Vienna (1965), Vol. 1, p. 115.
(We 62) S. Wexler and N. Jesse, *J. Am. Chem. Soc.* **84**:3425 (1962).
(We 65) S. Wexler, *Actions Chimiques et Biologiques des Radiations* (M. Haissinsky, ed.), Masson et Ci, Paris (1965), Vol 8, p. 105.
(Wo 56) M. Wolfsberg, *J. Chem. Phys.* **24**:24 (1956).

(Wo 69) G. K. Wolf and T. Fritsch, *Radiochim. Acta* **11**:194 (1969).

(Ya 73) M. Yagi, K. Kondo, and T. Kobayashi, *Radiochem. Radioanal. Lett.* **9**:123 (1973).

(Ya 75) M. Yagi and K. Kondo, *Radiochem. Radioanal. Lett.* **20**:299 (1975).

3

James F. Ziegler (top) is Manager of the Radiation Physics Group at the T. J. Watson Research Laboratories of IBM at Yorktown Heights, New York. He received his undergraduate and graduate education at Yale University and the Ph.D. degree in 1967 prior to joining IBM. He was awarded an IR-100 International Award for Industrial Research in 1975 and has served as Moshinsky Professor of Physics at the National University of Mexico in 1975, 1978, and 1980.

Jochen P. Biersack (bottom) is a member of the scientific staff of the Hahn-Meitner Institute for Nuclear Research in Berlin. He received both his undergraduate and graduate education at the Free University of Berlin, and the Ph.D. degree in 1960. During 1961–1963 he worked at the Oak Ridge National Laboratory as a NATO Fellow; in 1969 he joined the teaching faculty of the Free University of Berlin and during 1970–1971 served as an Associate Professor of Physics at New York University.

3

The Stopping and Range of Ions in Matter

James F. Ziegler and Jochen P. Biersack

Abstract

The purpose of this chapter is to review the calculation of the stopping and the final range distribution of ions in matter. During the last thirty years there have been published scores of tables and books evaluating the parameters of energetic ion penetration of matter. Rarely have the authors of these reference works included any evaluation of the accuracy of the tabulated numbers. We have chosen to show the development of ion penetration theory by tracing how, as the theory developed through the years, various parts have been incorporated into tables and increased their accuracy. This approach restricts our comments to those theoretical advances which have made significant contributions to the obtaining of practical ion stopping powers and range distributions. The Tables reviewed were chosen because of their extensive citation in the literature.

After the review of ion stopping and range tables, this chapter concludes with a detailed discussion of a modern calculation of ion stopping and scattering in a solid.

This review was written in 1981 and does not contain any of the significant advances made in the four years delay in the publishing of this volume. For a review of recent developments, see Reference 67.

J. F. Ziegler • Watson IBM Research Laboratories, Yorktown Heights, New York 10598 and Jochen P. Biersack • Hahn-Meitner Institute for Nuclear Research, Berlin, West Germany.

1. Introduction

For 75 years the stopping of energetic ions in matter has been a subject of great theoretical and experimental interest. The theoretical treatment of the stopping of ions in matter is largely due to the work of Bohr,[1-3] Bethe,[4-6] Bloch,[7,8] and Lindhard,[9-12] and it has been reviewed by Bohr,[3] Fano,[13] Jackson,[14] Sigmund,[15] Ahlen,[16] and Ziegler *et al.*[17-20]

Soon after the discovery of energetic particle emission from radioactive materials, there was interest in how these corpuscles were slowed down in traversing matter. In 1900, Marie Curie stated[21] the hypothesis that "les rayons alpha sont des projectiles materiels susceptibles de perdre de leur vitesse en traversant la matière." Early attempts to evaluate this were inconclusive for there was not yet an accurate proposed model of the atom. Enough experimental evidence was collected in the next decade to make stopping power theory one of the central concerns of those attempting to develop an atomic model. J. J. Thomson, director of the prestigious Cavendish Laboratory, and Niels Bohr, a fresh postdoctoral scientist at Rutherford's Manchester Laboratory, both published almost simultaneously[22,23] an analysis of the stopping of charged particles by matter, and each contained many of their divergent ideas on the model of an atom. Thomson ignored in his paper the Rutherford alpha-particle scattering experiment[24] of a year before. But the nuclear atom with a heavy positively charged core was the basis of Bohr's ideas.[25]

One of Bohr's conclusions was that the energy loss of ions passing through matter could be divided into two components: nuclear stopping (energy loss to the medium's atomic positive cores) and electronic stopping (energy loss to the medium's light electrons). He correctly deduced that the electronic stopping would be far greater than the nuclear stopping for energetic ions. He also realized that his accounting of the energy loss process was seriously limited by a lack of knowledge of the charge state of the ion inside the matter, i.e., its effective charge in its interaction with the target atoms.

These three topics will be the primary subjects of this chapter, and the reader should note the three basic terms defined above: *nuclear stopping*, *electronic stopping*, and the ion's *effective charge*.

In the three decades following Bohr's 1913 work, various theoretical advances were made in understanding the stopping of energetic light ions which could be assumed to be totally stripped of their electrons. But nothing could be done for most of the low-velocity stopping data from experiments except to tabulate it with limited attempts to search empirically for some common factor of a scaling law.

The use of scaling laws became quite active with various interests in the stopping of fission products in 1938–1941.[26-30] Attempts were made to find an empirical way to relate the various ions, energies, and targets. The most useful way appeared to be

$$\frac{S_{\mathrm{HI}}(v, Z_2)}{[Z_{\mathrm{HI}}^*(v)]^2} = \frac{S_p(v, Z_2)}{[Z_p^*(v)]^2} \tag{1}$$

where $S_{\mathrm{HI}}(v, Z_2)$ is the electronic stopping of a heavy ion of atomic number, Z_{HI}, at some velocity, v, in a target with atomic number, Z_{HI}, and with an effective charge at that velocity of Z_{HI}^*. This is related to similar quantities for protons, p, at the same velocity in the same material. This expression is not well justified theoretically. It assumes that the effective charge of an energetic ion is independent of the material it is in, and is a function only of ion velocity. It presumes that protons and heavy ions react identically to the same material. It asumes all dielectric effects between the heavy-ion electrons and the electrons in the solid do not significantly contribute to the stopping. Several authors[31,32] have listed many other objections. But this expression remains the basis of most scaling tables because it is the most accurate way which has been discovered to predict stopping for partially stripped energetic ions. In practice most authors use for values of Z_p^* the mean charge state of protons exiting from thin foils. The usual expression[33] for Z_p^* is of dubious accuracy, both because recent experiments[34,35] on the charge state of protons exiting from solid materials do not support this expression, and also because there is little basis on which to correlate the interior effective charge with the exterior mean charge state.

2. Stopping Power Tables

2.1. 1958: The Whaling Table

The first of the widely used stopping power tables was that of Whaling in 1958 in the *Handbuch der Physik*, Volume 34.[36] This table essentially presents experimental data for protons in various targets, with the curves extrapolated using the Bethe relation for high-energy stopping

$$S \sim (\ln E)/E \tag{2}$$

where E is the particle's energy. Whaling stated that so few data were available for ions heavier than protons that no comprehensive tables could be made. So at this point, tables were based only on data, with possible extrapolation to high energy using Equation (2), but no extrapolation could be made to lower energies or to heavy-ion projectiles.

2.2. 1970: The Northcliffe–Schilling Table

The next major table was the monumental effort by Northcliffe and Schilling published in *Nuclear Data Tables* in 1970.[37] Any detailed analysis of this work is difficult because it is a judicious mixture of experiment, interpolation, and a little theory. But in general, Northcliffe had spent a decade trying to organize in an empirical fashion the systematics of ion stopping in both solids and gases. This was done for solids by comparing the stopping of a wide variety of ions in A1, and, from this array of data he established the variation of stopping with ion type. Then the stopping of a few ions was determined in a wide variety of solid targets, especially C, Al, Ni, Ag, and Au. This established the target dependence of stopping; see Figure 1. These two sets of curves were then used to parameterize the scaling laws, Equation (1), so that any ion/target combination could be analyzed. The only basic theory introduced in this table was the use of LSS theory[38] for the low-energy stopping powers.

Northcliffe's work brought the scaling laws which were developed

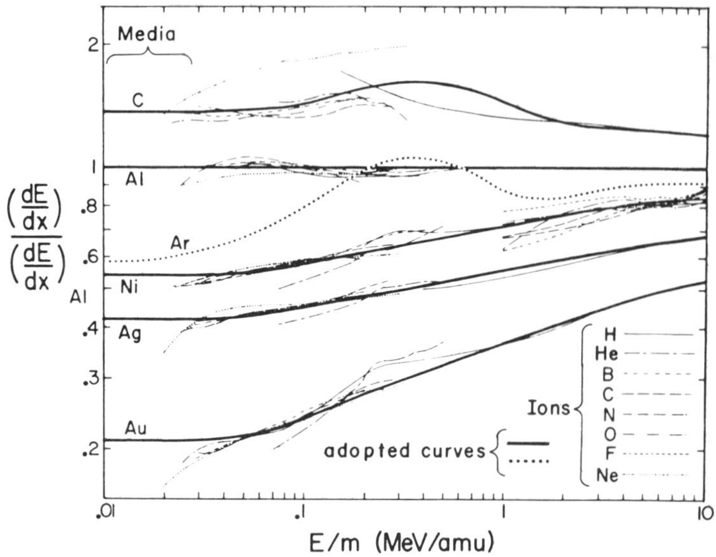

Figure 1. These curves show the primary stopping powers used for interpolation in the Northcliffe and Schilling Tables, 1970. The basic curve is for Al, which is a stable target which can be reliably made into thin pin-hole free foils which do not significantly oxidize or corrode with time. The other primary targets similarly are corrosion resistant. The thin lines are experimental values for the ions noted in the lower right of the figure. At the time it was not appreciated that the chemical nature of these stable targets (tightly bound atoms without loose electrons to form oxides) made their stopping powers anomalously low, and not the best targets for linear interpolation (see Figure 2). The one curve for Ar indicates that the stopping power of gas targets is quite different from solids.

between 1927 and 1941 to their most advanced state. The stopping powers predicted for high-energy ions, $E > 10$ MeV/amu, were quite good and in general were accurate to better than 20%. This is because Bethe–Bloch theory[4–8] is quite accurate if the ion is moving much faster than most of the electrons in the target. For this reason we quote energy in the units of MeV/amu. (A communications problem arises because nuclear physicists usually think in particle energies, while stopping theorists think in velocities, especially ion velocities relative to target electron velocities. The compromise usually made is to discuss nonrelativistic stopping in units of "energy/mass.")

As we consider ions of energy below 10 MeV/amu we find the Northcliffe–Schilling (NS) scaling laws begin to break down in special cases and this breakdown generates large errors of up to 400% at the lowest energies of the tables, 0.0125 MeV/amu. The reason for this breakdown is

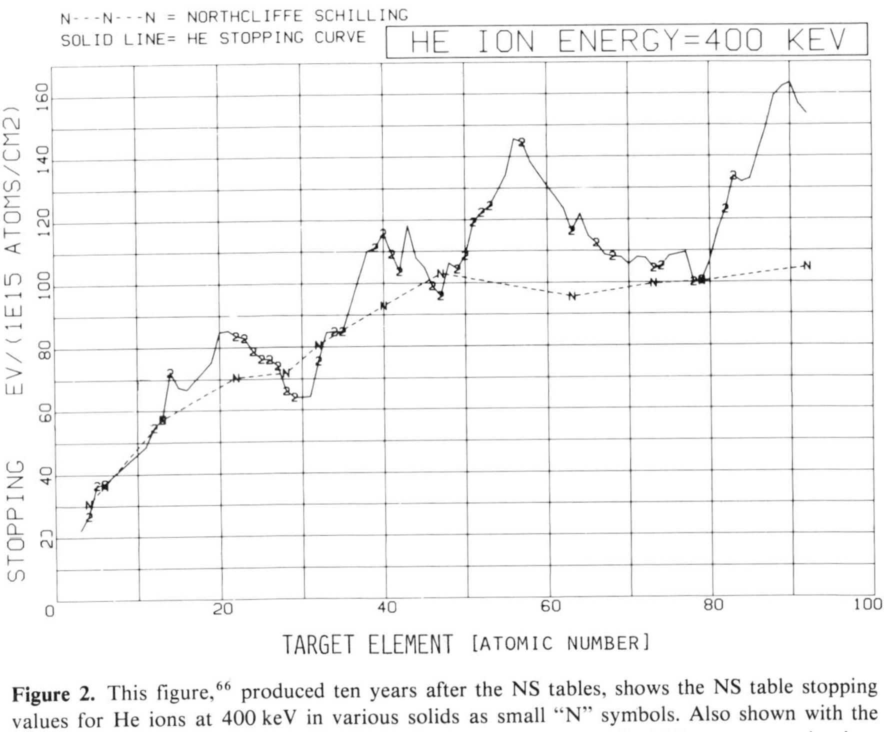

Figure 2. This figure,[66] produced ten years after the NS tables, shows the NS table stopping values for He ions at 400 keV in various solids as small "N" symbols. Also shown with the plot character "2" are experimental values for 38 solids, most of which are averaged values from three to six different papers. It can be seen that the stopping of ions in targets which do not oxidize are near the minima of these stopping values, and this can be directly related to the tightly bound nature of the outer electrons in these targets (see Figures 13 and 14). The use of nonoxidizing targets for the NS standards appears to be the main source of error to this magnificent work. The solid line is from a table made in 1978, Reference 18.

most easily explained by analyzing a plot such as in Figure 2. In this plot are shown the NS stopping values for He at 0.4 MeV in various targets. Their values change relatively smoothly from target to target as would be expected from the NS scaling laws. Also shown with small number 2's are experimental results for He stopping in 38 elemental targets (almost all of these experimental values were published after the NS tables). Each value shown is usually an average of three to six published experiments. These experimental values show wide oscillations with the target atomic number, and the NS values tend to occur at the minima of the oscillations. We shall show that this occurs because of an unfortunate choice of normalization targets for the NS scaling. Northcliffe concentrated his experimental efforts on targets which did not easily oxidize or corrode. Since the experiments extended over years, it was convenient to have targets which did not degrade with time. But the very reason these targets (Au, Ag, Ni, Al, and C) are materials which are chemically stable is the same reason they lie at the stopping minima! They do not have lightly bound outer-shell electrons. We shall see that any lightly bound electrons absorb a disproportionate amount of energy from a penetrating ion, and materials which oxidize readily have large stopping powers when compared to nearby tightly bound atoms.

Hence, since the NS tables are based on stable nonoxidizing targets, their stopping powers should be a lower limit of the stopping value for any ion/target combination, unless the target is one of those actually measured. As seen in Figure 2, for these targets, $Z_2 = 79$, 47, 28, 13, and 6, the values are quite good.

The solid line in Figure 2 represents the values of a 1978 table discussed below which is largely based on theoretical calculations.

2.3. 1972: The Bichsel Table

Two years after the NS tables were published, Bichsel published an abbreviated table in the *American Institute of Physics Handbook.*[38] This contribution is noteworthy because it includes for the first time details of atomic structure in the calculations. Bichsel had published many papers in the previous decade suggesting corrections to the Born approximation of the Bethe theory, which assumed the ion was moving much faster than the target electrons. Bichsel calculated the correction necessary to account for reduced absorption of energy by inner-shell electrons which often had classical velocities greater than the ion velocities (e.g., the innermost electrons of very heavy atoms such as Au have relativistic velocities, equivalent to ion velocities over 20 MeV/amu). He also introduced similar corrections to the mean ionization potential of atoms, which introduces into the stopping calculation the excited state level structure of the target atoms. His corrections are essential for the correct calculation of the stopping of high-energy particles, and his tables are quite accurate, better than 10%, for energies from 5 MeV/amu to ~2 GeV/amu. This upper energy limit of accuracy is caused by extreme relativistic effects which were not included in his formalism.

2.3.1. Recent Advances in Stopping and Ranges

The decade of the 1970s produced two new user groups which greatly expanded the interest and experiments in stopping theory. Ion beam analysis of materials proved to be a quantitative technique which gave detailed views of surfaces and thin films. Examples of this field are the nuclear backscattering analysis of integrated circuit technology, ion-induced x-ray analysis of environmental pollution, nuclear reaction analysis of the hydrogen embrittlement of metals, and a wide variety of age dating techniques. Most of these techniques require stopping powers of greater accuracy than was required by experiments in basic nuclear physics. Especially important was the stopping of He ions, and over 100 experimental papers for this one ion were published during the short span of 1972–1980.

A second major set of new users occurred in the field of ion implantation. In 1970 there were less than 10 accelerators in industry, while there were about 400 in academia. In 1980 the statistics were reversed, with about 1000 ion accelerators in industry, with those in academia shrinking to fewer than 200. Ion implantation provided the semiconductor industry with a new way to create materials. It provided an accurate and reproducible way to introduce impurities into a solid in a cold process. Each of these advantages—accurate, reproducible, and cold process—helped to make ion implantation into an indispensable manufacturing tool, and once accelerators were introduced into development laboratories they were used for many other materials besides semiconductors.

These new user groups substantially increased the amount of experimental data on ion stopping and range. Whereas in the 1960s there were only about 40 papers a year published in this field, by 1980 there were over 200 about ion implantation profiles in silicon alone, and over 600 per year on ion implantation in general.[39] Wow!

2.4. 1974: Ziegler and Chu Tables

The needs of this new community directly stimulated tables of He ion stopping powers published by Ziegler and Chu in the *Atomic and Nuclear Data Tables* in 1974.[40] He ions were used in much of ion beam analysis, and a major subject of such analysis in those years was ion implantation profiles. As with Northcliffe, Ziegler and Chu collected experimental data (from 11 papers) which established approximate stopping values for 47 elements over the narrow He ion energy range of 0.4–4 MeV (almost half of these 47 elements were represented by a single datum at one energy, so the experimental values were not really comprehensive).

The major new contribution of this work was to interpolate stopping powers using elaborate calculations based on Hartree–Fock atomic structure and using the detailed formalism of the Lindhard theory of electronic stopping.[41] The inclusion of these two concepts created a significant leap in accuracy over previous tables. Figure 3 shows a comparison of this new theoretical approach (originally demonstrated by Rousseau, Chu, and Powers[42]), and a wide variety of experimental data points for He ions at 1 MeV. There is remarkable agreement between pure theory and experimental values, and it is evident that only a slight normalization was necessary to create the semiempirical curves shown in Figure 3 which were the values used in the tables. This figure can be compared with the similar Figure 2 to see how dramatically stopping accuracy was improved by changing from an interpolation scheme based on scaling laws to detailed quantal calculations.

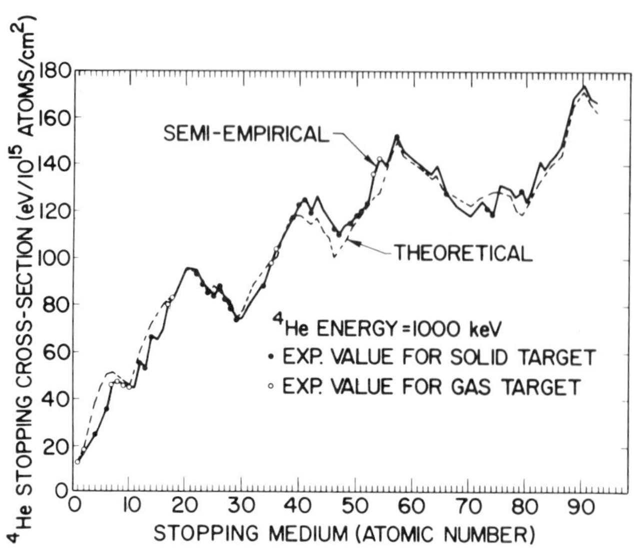

Figure 3. Extensive theoretical calculations were included in tables of the stopping of medium energy ions (~0.25 MeV/amu) for the first time in 1974. These calculations reproduce the large changes in stopping from one target to another as shown in Figure 2 (this effect was called the "Z_2 oscillation effect"). The dashed line is the theoretical calculations[42] based on the local density approximation and using Lindhard's theory[9] of particle stopping in a free electron gas. The dots show experimental measurements, and the solid line is the semiempirical tabulated values of Reference 40. The theoretical curve has been used to interpolate between the experimental values, giving accuracies of better than 10%.

2.4.1. Recent Advances in Stopping and Ranges

During the mid-1970s the outpouring of new experimental data overshadowed all previous data, both in quantity and more importantly in the increased understanding of ion penetration physics. It was shown that a large amount of the stopping and range data taken in the previous two decades was inaccurate as the experimental apparatus used did not measure average phenomena, but anomolous or unusual phenomena, or just plain garbage.

As an example, there is the phenomenon of ion channeling in crystals. This type of ion penetration takes place down major crystalline axes, and it has been shown that, as an extreme case, ions penetrating down loose salt structures can have stopping powers reduced by over 90%! Channeling was known since 1962, but another aspect of metal films (which was known by metallurgists) was not to penetrate to stopping power measurements until 1978. This was the "target texture" quality of metals in which polycrystalline metals do not have their microcrystallites randomly oriented, but in fact over 90% may be aligned within a degree or two about a common axis. This effect is particularly true of evaporated films, which are built up layer by layer, and atoms have time to move about the surface before they are buried. The stopping power experimenters had turned to evaporated targets in the 1960s because of their inability to make pin-hole-free rolled thin films. But the problem of target texture took a decade to catch up to the users. In Figure 4 is shown a typical result. This figure shows the collected stopping powers of H ions in Au.[17] As can be seen, there appear to be two independent curves which can be drawn through the data. One is much lower than the other at the stopping power peak. This is also the energy at which a large difference occurs between the energy loss of channeled and random-path ions (the maximum difference in random and channeled stopping powers occurs when the ion is at the same velocity as the electrons in the channel). An analysis of the experimental details of the 14 papers reported in Figure 4 is partially convincing. Three of the higher data sets were from experiments in which the ion beam was not perpendicularly incident to the target; three of the lower sets had perpendicularly incident beams; and 8 of the papers did not report the experimental conditions (which is typical of papers in *Physics Letters* or Soviet journals where the author has severe space limitations on his text).

In their paper on texture effects in ion penetration phenomena, Andersen, Tu, and Ziegler[43] pointed out how Au had significant texture problems for almost all techniques of preparation, and the major Au axis of the polycrystalline films were always within 1° of being perpendicular to the target plane.

A second problem in stopping power measurements prior to the 1970s

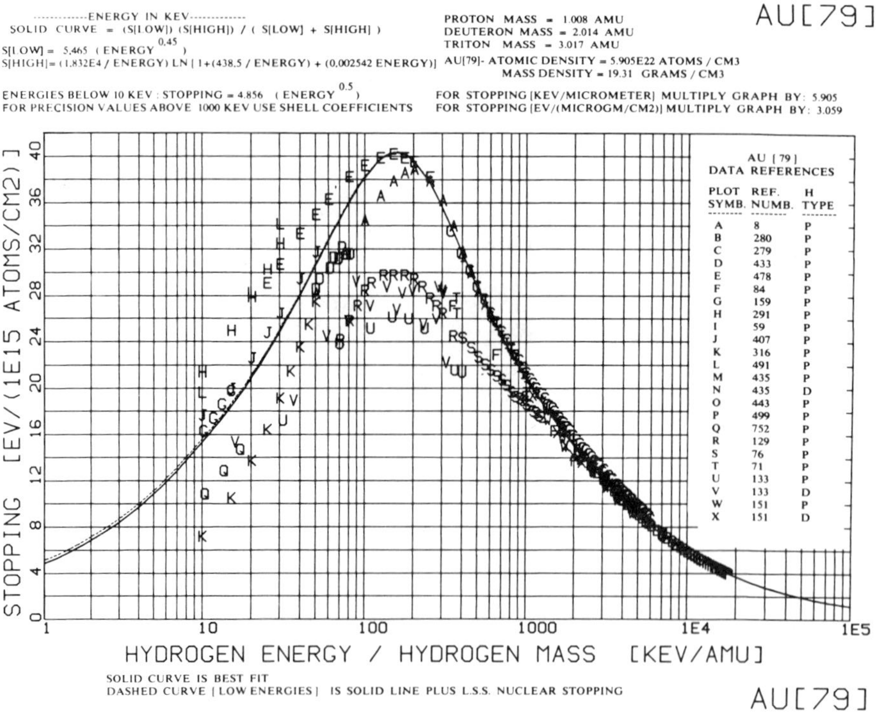

Figure 4. One of the unappreciated problems of early stopping power measurements was "target texture." This is the phenomenon especially found in evaporated films where the separate crystallites of the polycrystalline film have one axis in marked alignment, usually perpendicular to the substrate surface. For example, over 95% of the crystallites may have their $\langle 111 \rangle$ axis within 1° of each other, and this can lead to ion channeling and an apparent reduction of stopping powers. It has been suggested that the above H ion stopping plot[17] shows two distinct stopping curves, with the lower one being from experiments where the beam was inadvertantly aligned with the target texture. Later theoretical work[19] affirms that the upper measurements are probably correct. The subject of target texture is treated at length in Reference 43.

was the lack of appreciation of straggling effects, especially the angular scatter of the ion beam from its incident trajectory. During the 1950s and 1960s it was common to use a magnetic spectrometer to determine a particle's energy. Because of its high energy resolution, this was a natural way to measure energy loss. And so a technique was developed and widely used to simultaneously measure stopping powers and ion ranges. This was called the foil-stack method. An incident beam was made to directly enter the small entrance aperture of a magnetic spectrometer. Then thin foils were placed in the beam, and for each foil the energy loss was measured by using the magnet. This was continued until the energy spectrum of the emerging beams was so wide in energy distribution that its mean was hard to find. This kind of experiment produced a plot of ΔE vs. x, where ΔE is the energy loss in a film thickness x. A second curve was also made showing ΔI vs. x, where ΔI is the amount of current lost in the foil stack. The differential of this plot was the final ion distribution, for each incremental ΔI was presumed to have stopped in the last added foil.

This type of experiment became one of the standard techniques of the 1960s, then it disappeared as more advanced methods were developed. But the data stood in the literature, and they exasperated the theorists who were calculating good numbers which were sometimes in sharp disagreement with the "classic experiments." Finally, Biersack was persuaded to set up a Monte Carlo program to replicate one of the more famous experiments, and his results are shown in Figure 5.[55] He found in his simulation that the current detected was profoundly changed as the detector solid angle was increased (typical magnetic spectrometer entrance angles were less than 3°). When the data from various solid angles were differentiated, as shown in the lower figure, the profiles showed little resemblance to each other. Note the shaded area in the lower figure—this is the actual implantation profile from a full Monte Carlo calculation, and all the other solid lines indicate the relative errors from this ideal depending on the experimental setup.

A second problem with this experiment is very subtle and was overlooked by the scientists, but led to erroneous conclusions. The authors felt that the foil thickness which caused the transmitted fraction of the incident current to reach 50% was the mean depth of penetration of the ion beam. Half of the beam was stopped in the foil stack. But the mean range of an ion beam is defined as the mean depth in a target much thicker than the mean range. A little reflection may convince one that if a plane is inserted parallel to a solid's surface at the depth of the true mean range, almost an equal number of ions will be found scattering back towards the surface as penetrating deeper. This is especially true for light particles in heavy substrates. At this point of mean range the average vector momentum of the ions is zero. There is no way that this fact can be incorporated in a foil stack experiment, for here there can be no backreflection. So the experiment fails to be meaningful in determining implantation distribution. (Further problems with earlier measurements are found in Reference 20, pages 19–24.)

2.5. 1977: Andersen and Ziegler: H Tables

Because of all these problems, the next step in stopping tables was actually a necessary step back to the Whaling concept in 1957—just collect data and try to separate the good from the bad. This was aided by Andersen in his massive bibliography[44] (done before computer retrieval systems became widespread), and Andersen and Ziegler[44] published in 1977 their

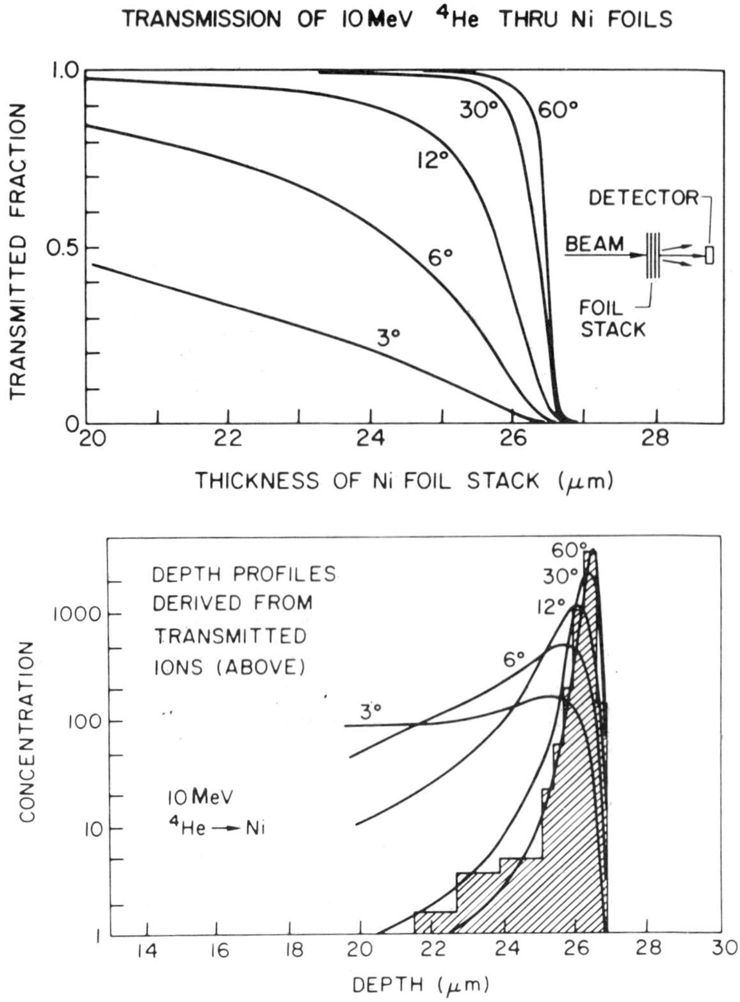

Figure 5. Two-dimensional Monte Carlo calculations (by J. P. Biersack) of ion beam ranges determined by transmission through foil stacks. This technique was the most popular method of measuring high-energy ion ranges. The upper plot shows the transmitted fraction of the beam for foil stacks of various thicknesses. The calculation shows how this fraction significantly depends on the solid angle subtended by the detector (the plot shows the detector half-angle in degrees). The lower plot shows the "range profile" obtained by differentiating the spectra of the upper plot. Also shown as a shaded area is the calculated final ion distribution of ions into a solid target. Large errors can occur unless the detector is arranged to collect all the transmitted ions. The common assumption that the "mean ion range" is where the transmitted beam falls to one half of the initial beam current is shown to be somewhat in error, and the estimated stragglings would be very erroneous. This plot is from Reference 20.

book on the stopping of H ions in all elements. By this time over 70 papers had been published on H stopping powers, and the authors made a data base of over 7000 experimental values. The special advance of this table was in the evaluation of high-energy stopping ($E < 1$ MeV/amu). Previously, we commented how Bichsel introduced theoretical mean ionization potentials (which brings in the atomic excitation structure) and also corrections for the fact that inner-shell electrons are not correctly treated in an impulse approximation like the Bethe theory (this correction is usually called a "shell correction"). Two other scientists, Janni[45] and Bonderup,[46] each also had produced significant papers on these two subjects. Andersen and Ziegler took these three works and normalized them to the data. Since the data base was comprehensive (36 elements were covered over wide energy spans) they were able to produce high-energy stopping curves which were believed to be accurate to better than 10% at 0.5 MeV/amu and better than 5% above 5 MeV/amu. Only a few experiments have proven to be reproducibly more accurate than this (although accuracies to 0.1% have been estimated by enthusiastic experimentalists).

For energies below ~0.5 MeV/amu these authors made only coarse estimates of interpolated stopping powers, basing their values mostly on the wiggles shown in the Ziegler–Chu tables (see Figure 3).

2.6. 1978: Ziegler: He Tables

At this point, about 1978, stopping power theory and experiment begin to come together. In the development of the next major table, "He Stopping Powers and Ranges in All Elements,"[18] a total of 121 experimental papers were included in the data base, and these were amplified by using scaling laws so that the H stopping powers of the previous book could be scaled to equivalent He values, as shown in Figure 6. A total of over 12,000 experimental data points thus became available to allow finally the weeding of the good data from the inaccurate. Using this data base, the shell corrections and mean ionization potentials for high velocity stopping were further refined. The accuracy of theoretical models of electronic stopping was further enhanced by the inclusion, for the first time, of actual solid-state charge distributions for solid targets (previously the charge distribution for isolated atoms was used). This type of calculation proved to be quite accurate for the energy span from where the Bethe theory broke down, ~1 MeV/amu, down to where a new problem arose, ~0.1 MeV/amu. Below this energy the He ion began to neutralize, and the target electrons had time to move and coalesce about the slow moving ion (called target polarization), and these effects were not includes in the theoretical approach.

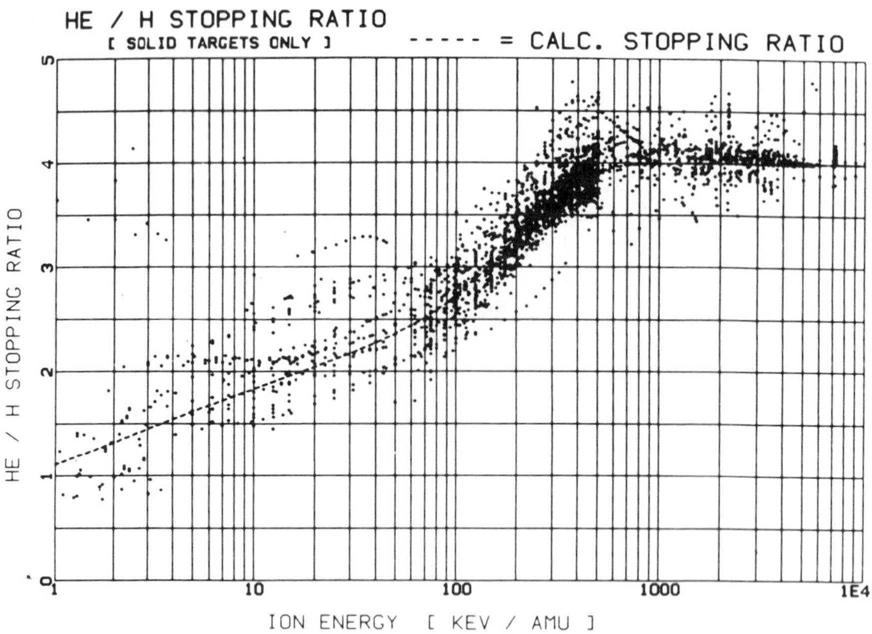

Figure 6. This plot[19] shows the ratio of the experimental stopping of He ions in elemental matter to the stopping of H ions at the same velocity and in the same material. Because of the extensive data available, a curve could be drawn (dashed line) which allowed the production of an accurate "Master Stopping" curve upon which scaling predictions for heavy ion stopping could be based (see Figure 7). This single curve presumes there is no target dependence of the He/H ratio and above about 80 keV/amu this appears to be accurate. Below this value this presumption breaks down.

2.7. 1980: Ziegler Energetic Ion Tables

In 1980 the next step in stopping tables was the publication of "*Handbook of Stopping Cross Sections for Energetic Ions*. This book used all the previous advances, and added one more improvement. First it was built on the H and He stopping books, and by the time this table was published there were over 15,000 experimental data points for these two ions alone. As noted before, considerable success was achieved in scaling He to H stopping powers, so together they formed what was called a Master stopping set. The next step was to scale this Master set to obtain the electronic stopping for heavy ions. Considerable theory had been done on this subject from 1941 to 1961, using statistical models of the atom for which scaling from one element to another was done by a simple multiplication factor. There was no shell structure in these atoms, but as long as it was concerned with only the stripping of outer shell electrons these approaches were adequate. In general, they led to equations such as

$$\frac{Z^*}{Z} \sim \exp -(v/v_0)^{2/3} \tag{3}$$

where Z^* is the "effective charge" of an atom of atomic number Z with a velocity v. This velocity is related to the Bohr velocity, v_0, about 2×10^8 cm/s. At very low velocities the atom is unstripped, but as its velocity increases beyond the Bohr velocity (a typical velocity of the ion's outer shell electrons) the ion's electrons begin to be stripped off. Another theory is to assume all the ion's electrons which have velocities below the ion's velocity will be stripped off (called the Bohr criterion).

The first extensive application of this approach for stopping power scaling was by Northcliffe in 1963, who used the form

$$Z^*_{\mathrm{HI}}/Z^*_{\mathrm{H}} \sim 1 - \exp -(v/v_0 Z^{2/3}_{\mathrm{HI}}) \tag{4}$$

which describes how the effective charge of a heavy ion, Z_{HI}, is related to the effective charge of a H ion (at the same velocity). This expression can be combined with the scaling law for stopping power, S,

$$S_{\mathrm{HI}} = S_{\mathrm{H}} \left(\frac{Z^*_{\mathrm{HI}}}{Z^*_{\mathrm{H}}} \right)^2 \tag{5}$$

From the experimental Master stopping set, Figure 6, which can give a S_{H} based on 12,000 data points, it is only necessary to establish the best form of Equation (4) to scale to heavy-ion stopping powers. Scores of papers have been written on this, but except for Northcliffe and Schilling[37] the previous authors used limited data bases for S_{H}. For the preparation of the Handbook, about 4000 heavy-ion data points were compiled, and a final parametrized variation of Equation (6) was established. The accuracy is shown in Figure 5, where the scaling law is tested by comparing actual experimental values to those predicted by using the variation of Equation (4) shown in the table. For plotting convenience, what is shown is reduced stopping, defined as $S_{\mathrm{HI}}/(Z^*_{\mathrm{HI}})^2$ for both the experimental and the predicted values. Thus, if the heavy ion is totally stripped, and stopping increases with the square of the ion's atomic number, then the reduced stopping value is one. For lower velocities where the heavy ion is only partially stripped, the reduced stopping must be less than one. The agreement in Figure 7 is about 5%.

A second advance of this 1980 table was the inclusion of a new nuclear stopping approach. Until this time, nuclear stopping was based on statistical atom interactions, especially the work of Lindhard *et al.*,[9–12] which considers

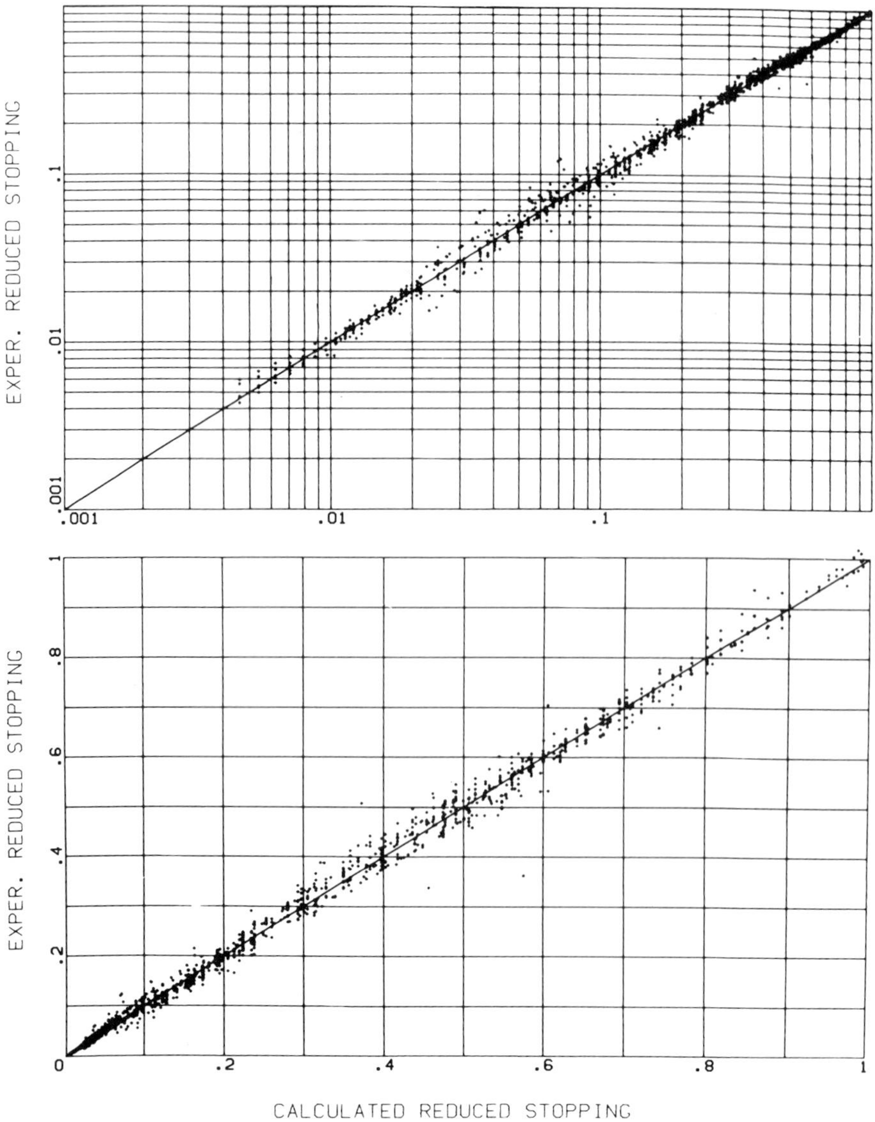

Figure 7. Shown are experimental versus calculated values of heavy-ion stopping to hydrogen stopping. The experimental values are from 127 ion–target combinations ranging from ions of C to U, in both gaseous and solid targets ranging from Be to Au, at ion energies from 0.22 to 22 MeV/amu. These experimental heavy-ion stopping values are reduced by dividing by equivalent hydrogen stopping values are reduced by dividing by equivalent hydrogen stopping as explained in the text. The experimental and calculated values agree to about 5%. The same plot is shown with both linear and log scales to display in detail various regions.

the energy transfer between two colliding nuclei shielded by a Thomas–Fermi distribution of electrons. The Thomas–Fermi atom has the difficulty that the electrons have unrealistically high densities far from the nucleus and hence allows energy transfers between widely spaced atoms. In general, these atoms generate too much energy transfer and also too much deflection of the ion during a distant collision. Consequently, stopping powers and straggling calculated with the Thomas–Fermi model were too high, and ion ranges were too small.

Then in 1976 nuclear stopping was calculated by computer numeric methods,[49] with results which agreed better with data. Whereas the Thomas–Fermi atom could predict nuclear stopping to within about a factor of 2, this new method appears accurate to 25%. This new formalism was adopted in these tables.

The Handbook thus introduced advanced heavy-ion scaling of electronic stopping powers based on a huge experimental data base. It also made a significant step in upgrading the calculations of nuclear stopping powers. Finally, it shows not only the final values, but plots all existing experimental stopping powers (see Figure 8), so users could review relevent data.

The book is limited by not giving stopping values for ion velocities below 0.2 MeV/amu. Below this velocity there is a direct interaction between the ion, Z_1, and the target, Z_2, and all scaling laws begin to break down.

2.8. The Current Accuracy of Stopping Tables

$E > 10$ MeV/amu: Well developed theoretical models with accuracy better than 5%.

$E > 1$ MeV/amu: Reasonably good combination of theory and scaling laws. Accuracy better than 10%.

$E > 0.2$ MeV/amu: Tables based on theory, scaling, and extensive data. Accuracy better than 20%.

$0 < E < 0.2$ MeV/amu: Pure theory[11] is accurate to a factor of 2. No published tables are any better except for H and He ions, where extensive experimental data are available.

3. Range Tables

There was less interest in tables of ion range distributions than in stopping powers until the industrial development of ion implantation. The first extensive table was in Northcliffe and Schilling[37] in 1970. After 1970 followed a flood of books all related to ion implantation. These books[50–55] were basically the same in both theoretical approach and accuracy. Range calculations can be broken into two parts: the basic physics of energy loss

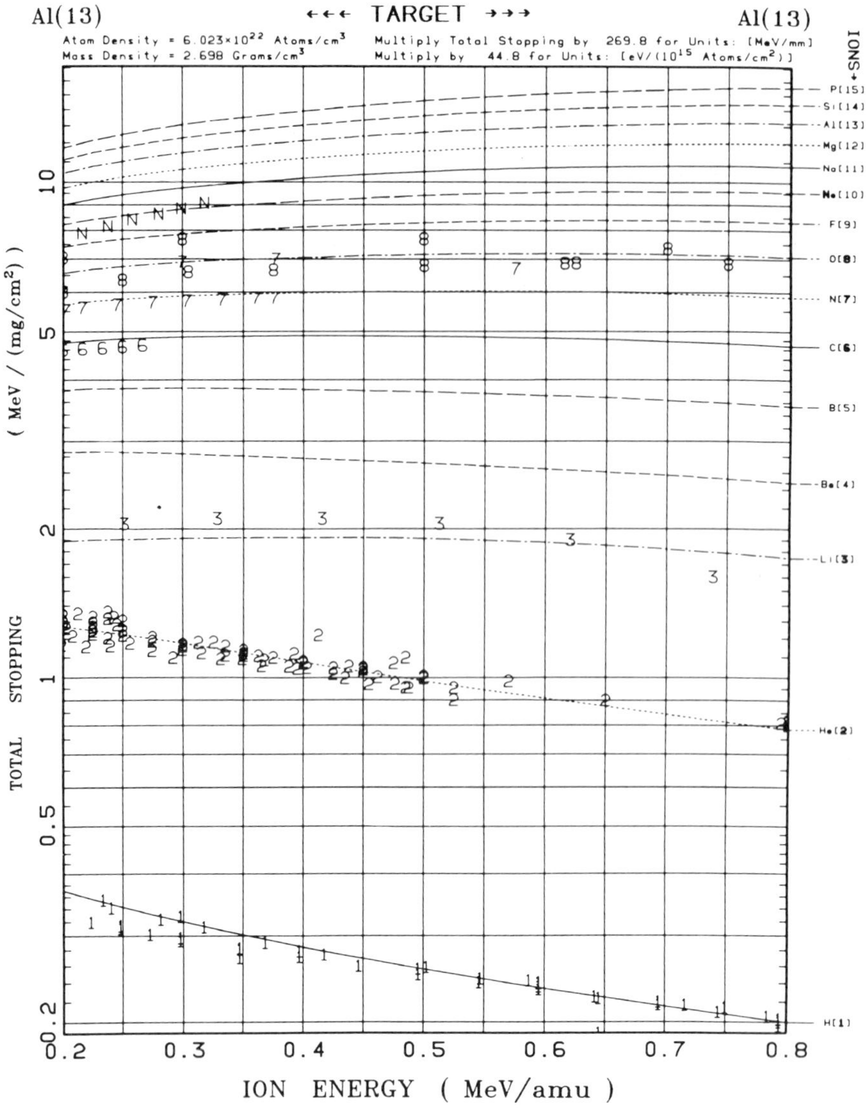

Figure 8. Shown is a page of Reference 19, which displays for a target of Al the calculated stopping of 15 ions. Also shown are the experimental results for each so that the accuracy of the table may be judged. The plot symbol for each ion, e.g., 7 = nitrogen ions, is indicated by printing that symbol in bold type in the right margin.

and scatter from atom–atom collisions, and the mathematics to follow these processes until the ion stops. Except for small points (noted later) all these books are based on the theories of Lindhard *et al.*, with particular importance being placed on what is called the LSS theory of ion ranges.[11] This theory develops analytic expressions for both the electronic and nuclear stopping of ions. It then takes these and sets up a Boltzmann transport equation to solve the statistical problem of the final ion distribution. For the numerical solution of the transport equations, expansions are made and thus what is obtained are the *moments* of the distribution, with the first moment being defined as the *mean range*, the second being the *straggling*, the third the *skewness*, etc. Given this theory, it is a direct computer step to go from the universal stopping concepts to the final moments.

What is the accuracy of LSS calculations? First, the range distribution will be as accurate as the collision physics, and we have discussed before that when one uses universal atoms such as the Thomas–Fermi atom used in LSS, the collisional accuracy is usually within a factor of 2. However, the first moment is usually better than this since the mean range is the integral of the stopping at all energies down from the initial energy, and, if the worst deviation is a factor of 2, it will possibly be compensated at other energies giving a more accurate range than stopping power. But the other moments are increasingly sensitive to errors in the interaction, and they are more inaccurate.

A second problem is equally important. The users of range tables do not wish to have *moments* of a range distribution, they wish to have the distribution itself, i.e., the *shape* of the concentration vs. depth. Given only two or four moments, can a shape be uniquely described? The answer is yes, sometimes, but lurking at innocent ion/target combinations can be hidden devils!

The usual method of obtaining an ion distribution shape from calculated moments is to fit an analytic expression which has a single peak in its shape. For example, a distribution used to give a peak beneath the surface, $x = 0$, with a distribution skewed toward the surface is

$$P(x) = ax^b \exp(-cx) \tag{6}$$

with a, b, and c being parameters which can be adjusted to give the correct first three moments. Usually, the distribution is a rising quadratic which is cut off by an exponential factor, forming a peak with an abrupt final edge. This type of shape is typical of light ions in heavy substrates where most of the ions go the full range with little deflection, but a few undergo a strong collision on the inward path and are distributed between the surface and the peak.

Winterbon considered the problem of converting moments to shapes and unearthed a significant worm. He took several typical ion/target combinations and solved for the moments. He then took all the popular shapes (analytic expressions) and found parameters for them which gave them the correct moments.

The problem arises when a three-parameter distribution like Equation (6) is used to fit a four-moment distribution. Winterbon took various distributions used by various authors and showed how much variation could

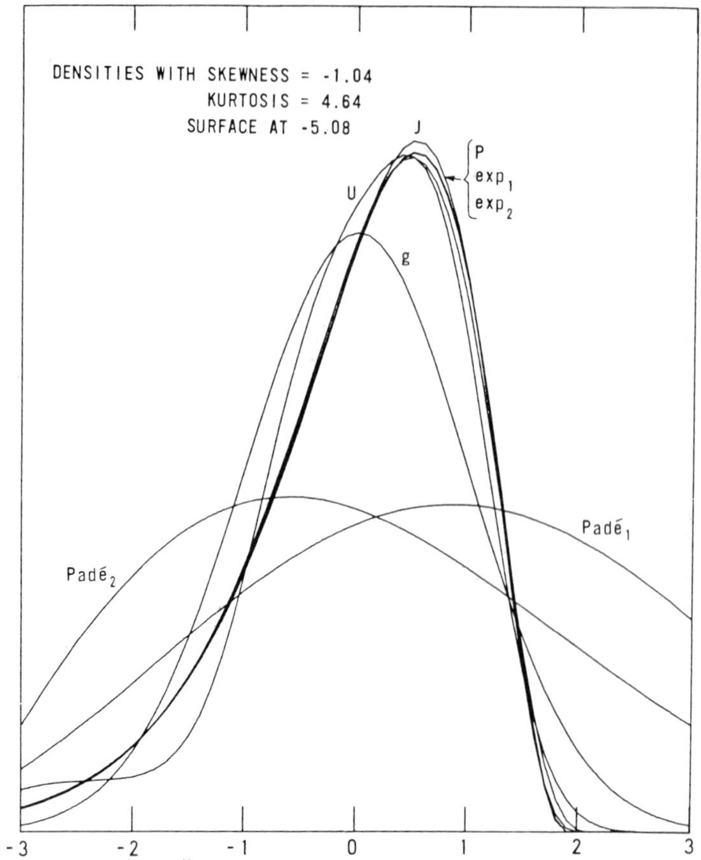

Figure 9. This plot shows the results of attempting to produce an ion range distribution from a series of distribution moments as calculated by a transport equation approach to range theory. The four moments are −5.08, 1, −1.04, and −5.08, with the x axis in units in which the standard deviation (the second moment) is one. Each of the curves comes from an analytic function with parameters which can make a nonnegative distribution with a single peak below the surface. Each function has been fitted to the four moments. This figure illustrates that there is not a unique range distribution, and without experimental data no clean conclusions can be reached about details of the distribution (from Winterbon, Reference 56).

be obtained; see Figure 9. It is clear that wild errors may occur when a shape is forced to fit a series of moments.

Some of the major range tables up to now are as follows.

3.1. 1970: Northcliffe and Schilling, "Range and Stopping Power Tables"[37]

Ranges calculated by the integration of stopping powers,

$$R(E) = \int_0^E (-dE/dx)^{-1}\, dE \qquad (7)$$

where R is the range of an ion with energy E, with a stopping power, dE/dx, which was calculated from the NS electronic stopping powers. A correction was applied for nuclear stopping and deflection using LSS theory.[11] Only mean ranges are tabulated. The results are good for energetic ions ($E > 0.1$ MeV/amu) and for heavy ions in light substrates $M_1/M_2 > 3$). The results are rather inaccurate for low energies and where the ion is not heavy ($M_1/M_2 < 3$), which is where the ion penetration is not a straight path and more detailed calculations are necessary. As an example, for 100-keV He ions in Al ($M_1/M_2 = .15$), the difference between the ion path length and its final mean range below the surface is about 43%.[57] The authors do not indicate what are the typical corrections which they apply to their ranges found using Equation (7), and the actual procedure they use is too complex to evaluate. For the case of He in Al, the NS results are about 35% higher than several measured ranges, and for 100-keV H ions into Al, their results are about 60% higher than experiments.

3.2. 1970: Johnson and Gibbons LSS Range Tables

The next major step in range tables was by Johnson and Gibbons[50] in 1970 where they produced the first *distribution* calculations. Their book was specifically oriented for ion implantation in semiconductors. They developed a computer code to calculate LSS theory,[11] and the authors were very generous in that they published the code in the book so it could be used by others for different ion/target combinations. Unfortunately, the code had a serious error, and only the calculation of the first distribution moment (the mean range) was reliable.[53]

By using LSS theory, the authors included a full statistical treatment of ion penetration of matter. The LSS theory incorporates an electronic stopping of the form

$$S_e = kE^{1/2} \qquad (8)$$

where k is a constant and the stopping is proportional to the ion velocity.

This stopping may approximate low-velocity stopping, but since electronic stopping goes through a peak at about 90 keV/amu, this limits the applicability of LSS theory to energies below this value. The nuclear stopping is based on the collision of Thomas–Fermi atoms and so is accurate to about a factor of 2 as discussed before. The result of this major step forward in range tables may be indicated by comparing the results to the previous NS tables. Recall that for 100-keV He in Al the NS tables predict a range of 0.8 μm. The Johnson and Gibbons (JG) tables show about the same, 0.89 μm, for the ion's total path length, but when the statistics of the penetration process are included the JG tables predict a mean range of 0.56 μm. The difference is in the multiple deflection of the He from its original trajectory. This result is reasonably close to the experimental data scattered from 0.5 to 0.6 μm.

He ions were used in the above example because the nuclear stopping component is small compared to the electronic component. For heavy ions, which have large nuclear stopping at low energies, the LSS theory is well established to be accurate to a factor of 2, and although many papers have found better agreement, these are coincidental in that some ranges will agree well, while others will be a factor of 2 away. Examples of this are shown in Figures 10 and 11. Figure 10[57] shows range and straggling measurements for 800-keV N^{14} ions in targets of atomic number 22 through 32. The solid lines are the LSS prediction and the dashed lines go through the experimental points. For nitrogen ions into a Cu target the agreement is excellent, while for a target of V ($Z_2 = 23$) the experimental values are about 1/2 the LSS

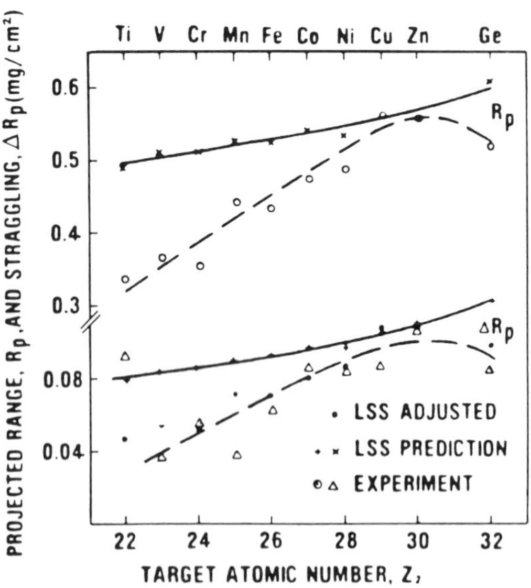

Figure 10. A comparison of LSS range theory with experimental results for N ions at 800 keV in various targets.[57] The solid lines are LSS predictions for the projected range, R_p, and the projected straggling, ΔR_p. The dashed lines are drawn through the experimental results shown as open symbols. For some targets such as Zn there is excellent agreement, while for the more loosely bound solids of Ti and V there is an error approaching a factor of 2. Similar results are shown in Figure 1 for many ions in a single target.

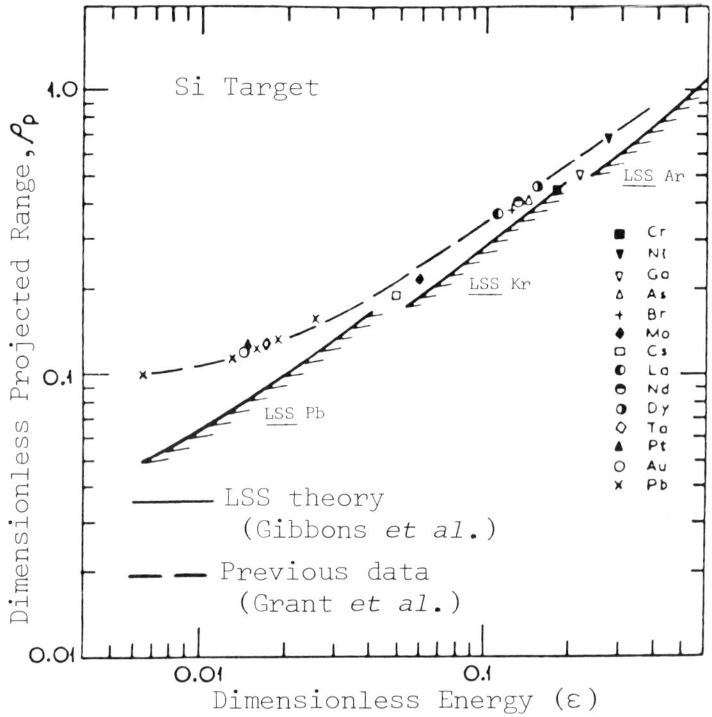

Figure 11. A comparison is shown between LSS range predictions[54] and the experimental results for many ions in silicon targets.[58] Various ion energies were used, but this factor can be mostly eliminated by plotting in LSS reduced coordinates, ε and ρ, as shown on the plot. The LSS values are shown as the solid lines, with a possible range of values depending on the ion mass being shown by the narrow cross-hatching. The dashed line shows previously published experimental values, while the dots show new measurements which essentially agree with the previous ones. The accuracy of LSS predictions for this single target ranges from 15% for high energy, ε, to about a factor of 2 for low values of ε.

prediction. Similar results are shown[58] for many ions into a single target, Si, in Figure 11. This plot shows the ranges of 14 ions with energies from 10 to 400 keV into silicon. The plot has coordinates which are called LSS reduced coordinates with the abscissa corresponding to energy and the ordinate being related to range. In these units the LSS range-energy predictions for all heavy ions falls about on a single line, shown with slight hatch-marks on the figure. Extensive prior experimental range data are shown as a dashed line and the results of Reference 58 are shown as various plot symbols. For the lowest energies there is a deviation between LSS theory and experiment by about a factor of 2, while at the higher energies the agreement is good, about 15%.

3.3. 1975: Gibbons, Johnson, and Mylroie Range Tables

The calculation errors in the above JG tables were corrected by the authors in the second Edition: Gibbons, Johnson, and Mylroie, *Projected Range Statistics*, Dowden, Hutchison and Ross, Inc., (1975).[54] This addition also included two major advances over their previous tables. First they realized that the LSS electronic stopping powers, S_e, could be improved by using experimentally determined parameters to create a similar power form of S_e

$$S_e = aE^b \tag{9}$$

where a and b were determined from experiments. This significantly increased the accuracy of the light-ion range tables. Secondly, they used a fit to the nuclear collisions of Thomas–Fermi atoms as suggested by Winterbon, Sigmund, and Sanders.[59–60] Since this correction remained in the spirit of Thomas–Fermi atoms, the basic nature of the nuclear stopping did not change, and the results do not significantly modify their previous results. However, GJM made the point that correcting S_e and S_n were the most important ways to improve the accuracy of LSS range calculations, and it is along these lines that the major improvements of later tables were accomplished.

3.4. 1975: Brice and Winterbon Range Tables

In the same year as the JGM tables were published, 1975, there were also published the books of Brice and Winterbon called *Ion Implantation Range and Energy Deposition Distributions*.[51,52]

The Brice book, Vol. 1, *High Ion Energies*, also indicated that S_e and S_n as developed by LSS in 1963 could be improved on, and Brice developed his own parametrized form of S_e so that the could calculate ranges for energies above the stopping power peak. His S_e equation is too complex to review, but it contained three parameters which allowed control over the initial (low energy) slope of S_e, the energy of the peak of S_e, and the down-slope of the high-energy S_e. The parameters were found by fitting the Northcliffe--Schilling[37] values of S_e for all ions except He. Since we have noted that the NS tables were empirical improvements over the LSS values, Brice used values which were probably better than the GJM values which were based on very limited experimental data. Brice used the Thomas–Fermi atom nuclear scattering of LSS for S_n.

But the major advance of this book was the inclusion of *damage distributions* of the ions. Brice had previously extended LSS theory to obtain the transport equations which govern the ion spatial distributions at all energies as the ions slow to a stop in the target. That is, if the incident ions

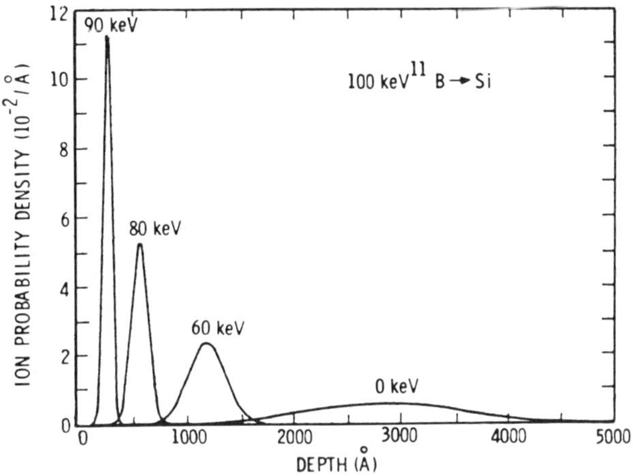

Figure 12. Illustration of the Brice technique to calculate the damage introduced in a target by an ion beam (see Reference 51 and citations within). Each curve shows the spatial distribution of ions at successively lower energies as 100 keV B ions penetrate into silicon. The original distribution is a delta function at zero depth, and the final ion distribution is shown for zero energy. The damage distribution is calculated by integrating the damage deposition rate at each energy as the projectile slows to a stop in the target. No account is made of displacement energy thresholds as discussed in the text.

are 100-keV boron atoms, then the distribution of 100-keV ions is a delta function at the target surface. As the ions penetrate the target the spatial distribution of ions at, say, 90 keV will be wider, as shown in Figure 12. Brice showed how to construct spatial distributions within the target for ions at any energy below the incident energy, with the distribution at $E = 0$ being the final implantation distribution. From these spatial distributions Brice developed a direct procedure to calculate the *energy deposition distributions* of ions in targets. The spatial redistribution of the energy deposited into recoiling target atoms, as well as its further partitioning into electronic and nuclear energy loss, is accounted for in these calculations in an approximate, but relatively accurate manner using the low energy results of WSS [16]. The nuclear stopping component of these energy deposition distributions was felt to describe the spatial distribution of target damage. Brice's innovative technique is still a fundamental way which is used to calculate damage profiles, and it has only one significant problem. That is, no account is made of the fact that solids have a displacement energy, *Ed*, which is defined as the minimum energy necessary to displace an atom from its lattice site, usually about 25 eV. If the collision transfers less than this energy, there will probably be no permanent damage (atomic displacement) and the energy may be dissipated as phonons. The neglect of the effect of these low-energy collisions will thus lead to overestimates of damage, although the effect is small for projectiles with energies greater than a few tens of keV.

Volume 2, by Winterbon, *Low Incident Ion Energies*, is different theoretically, and in the items calculated. Winterbon was the first to include calculations which were basic to other phenomena of ion implantation: (a) the fraction of the ion distribution which is outside the target surface, i.e., the fraction of the ions which are backscattered, (b) the energy which is in the backscattered ions and hence not injected into the target, (c) the damage distribution at the target surface which determines the sputtering yield, and (d) the number of sputtered target atoms per incoming projectile ion. This book is clearly the most sophisticated of the ones we have reviewed for it covers many of the phenomena of low-energy ion impact of solids. For the electronic and nuclear stopping Winterbon chose to use the traditional LSS formula, and the errors introduced by this choice have been previously discussed.

3.5. 1981: Littmark and Ziegler, Energetic Ion Range Tables

Finally, Littmark and Ziegler published in 1980 the *Handbook of Range Distributions for Energetic Ions in All Elements*.[55] This book contains the most advanced form to date of solving ion range distributions using transport equations. Recall that LSS theory in 1963 used a closed form of nuclear and electronic interactions between the ions and the target atoms. The authors in the range books reviewed above basically maintained the LSS formula approach. In the Littmark and Ziegler (LZ) book a new expansion was made to the transport equations which allowed relatively complete freedom in using whatever nuclear and electronic interactions were appropriate. Since great strides had been made in calculating stopping powers (as reviewed before), this new range solution allowed these to be used to obtain equally accurate range distributions (in general, the mean range will be more accurate than the accuracy of the stopping powers on which it is based).

The prime limitation in the LZ book was that only the first two moments of the final range distribution were given. They also only considered ranges for ion energies above 200 keV/amu, which was the lower limit of the accurate stopping powers of Reference 19.

4. Electronic Stopping of Ions

The total stopping cross section of ions in matter is usually taken as the sum of two contributions: the interaction of the ion with the target electrons (called electronic stopping) and with the target nuclei (called nuclear stopping). The nuclear stopping component can be separated because the heavy recoiling target nucleus can be considered to be unconnected to its lattice during the passage of the ion, and the interaction can be treated simply as the kinetic scattering of two screened particles (see section on Nuclear Stopping).

The electronic stopping of the ion in the target is treated within the local density approximation, wherein each infinitesimal volume element of the solid is considered to be an independent plasma: that is, the ion–target interaction can be treated as that of a particle with a density averaged free electron gas. The basic assumptions of this approximation are as follows:

- The electron density in the target varies slowly with position.
- Available electron energy levels and transition strengths are described by those in a free electron gas.
- There are no significant band-gap effects on electronic stopping.
- The charge of the ion can be reduced to a scalar quantity called the ion's effective charge.

The electronic stopping of an ion, using the local-density approximation, can be simply stated as

$$S_e = \int I(v, \rho)[Z_1^*(v)]^2 \rho \, dV \tag{10}$$

where S_e is the electronic stopping; I is the stopping interaction function of a particle of unit charge with velocity, v, with a free electron gas of density, ρ; Z_1^* is the effective charge of an ion with atomic number, Z_1; ρ is the electronic density of the target; and the integral is performed over each volume element, dV, of the target. The electronic density of an atom is normalized so that its atomic number $Z_2 = \int \rho \, dV$. Each of the three components of Equation (10) will be discussed below.

5. Interaction of a Particle with a Free Electron Gas

The interaction of a particle with a free electron gas has been studied by Lindhard[9] with the following assumptions:

- The free electron gas consists of electrons at zero temperature (single electrons are described by plane waves) on a fixed uniform positive background with overall charge neutrality.
- The initial electron gas is of uniform density.
- The interaction of the charged particle is a perturbation on the electron gas.
- All particles are nonrelativistic.

With these assumptions, Lindhard derived the interaction function, I, of Equation (10) as

$$I = \frac{4\pi e^4}{mv^2} \frac{i}{\pi\omega_0^2} \int_0^\infty \frac{dk}{k} \int_{-kv}^{kv} \omega\, d\omega \left[\frac{1}{\varepsilon^l(k, \omega)} - 1 \right] \tag{11}$$

where the longitudinal dielectric constant, ε^l, is derived to be

$$\varepsilon^l(k, \omega) = 1 + \frac{2m^2\omega_0^2}{\hbar^2 k^2} \sum_n \frac{f(E_n)}{N} \times \left[\frac{1}{k^2 + 2\mathbf{k}\cdot\mathbf{k}_n - (2m/\hbar)(\omega - i\delta)} + \frac{1}{k^2 - 2\mathbf{k}\cdot\mathbf{k}_n + (2m/\hbar)(\omega - i\delta)} \right] \tag{12}$$

where e and m are the charge and mass of an electron; ω_0 is the classical plasma frequency defined as $\omega_0^2 = 4\pi e^2 \rho/m$; E_n is the energy and $\mathbf{k}_n$ the wave vector of the electron in the nth state; $f(E_n)$ is the distribution function and is an even function of $\mathbf{k}_n$, and δ is a small damping factor.

This double integral equation has been used for the interaction term, I, in Equation (10) for the stopping of ions in matter using the local density approximation.

Typical results of these calculations are shown in Figure 13 for the complete interaction term, I, as defined by Equation (11) versus free electron gas density. Each curve has a flat section at low electron densities where the ion is going much faster than the mean electron velocity. Each curve bends down where the ion velocity becomes equal to the Fermi velocity, v_F, of the free electron gas, where

$$v_F = \left(\frac{\hbar}{m} \right) (3\pi^2 \rho)^{1/3} \tag{13}$$

At the higher electron densities, the electrons can respond adiabatically because of their higher velocities and stopping is reduced.

Insight into the physics of the ion stopping process is gained by considering the energy loss of a particle of unit charge in a solid. In Figure 14 we graphically evaluate the factors in the integrand of the basic stopping equation, Equation (10). First the dotted curve shows the charge distribution of solid Cu, plotted in units of $4\pi r^2 \rho$ versus r. The area under this curve equals the number of electrons of Cu, 29.The dashed curve is the interaction term, I, evaluated at each radius point by taking the electron density at that point and evaluating the Lindhard stopping for the interaction values (as shown in Figure 13). This curve shows that the interaction strength increases monotonically as the electron charge density decreases.

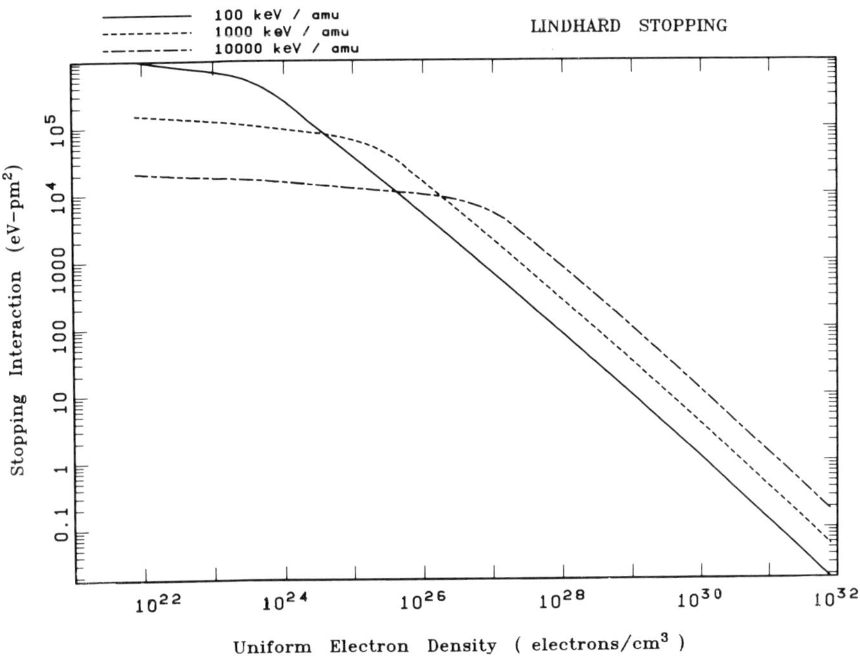

Figure 13. Plotted is the interaction function, I, of a unit charged particle with a uniform electron gas, as defined in Equation (12). For any electron density, the maximum interaction occurs with particles with velocities of about the Fermi velocity of the electron gas. For particle velocities below this value the electrons tend to undergo adiabatic collisions and the interaction is reduced. For velocities above this value the particle–electron interaction time becomes shorter, reducing the interaction magnitude. Since outer shell electrons of atoms which oxidize readily are loosely bound and are best represented by low electron densities, they absorb much more energy from an ion than the tightly bound and high-density outer electrons of noble metals such as Cu, Ag, and Au. See Figures 2 and 3.

The product of these two curves is the stopping of an energetic unit charge, and is the integrand of Equation (10). This integrand is shown as the solid line. The area under this curve, which is the energy absorbed by the electrons of Cu, is the electronic stopping.

Note that for the high-velocity ion in Figure 14 there is almost no energy loss to the inner k-shell electrons, but there is significant loss to the next two shells. For low-velocity ions the energy loss is quite different with little loss to the inner three shells, and the interaction is dominated by the outer-shell and conduction electrons. An accurate description of these electrons is critical in evaluating low-velocity stopping.

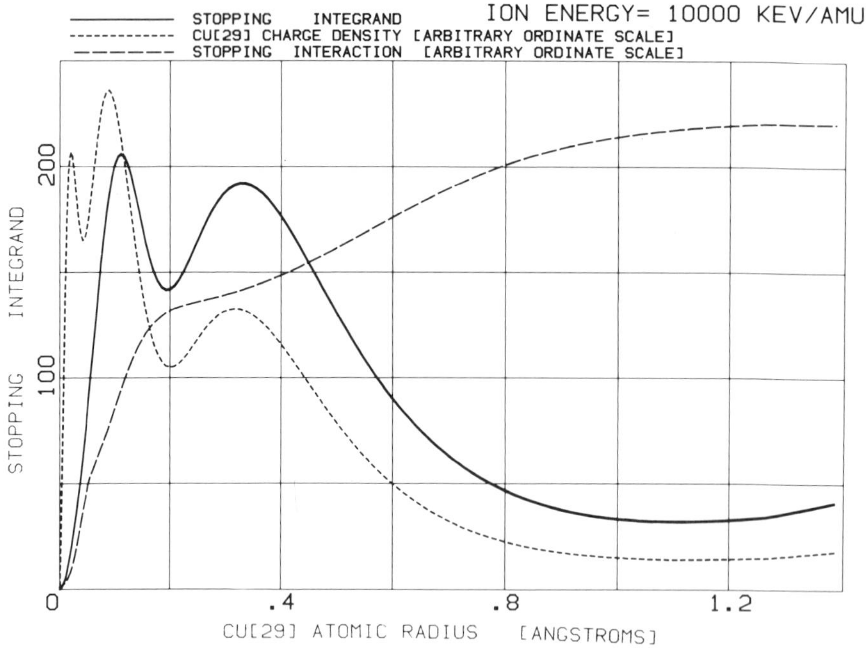

Figure 14. The interaction of particles with solid Cu is shown in a manner which isolates the various factors. The dotted line is the Cu solid-state distribution of electrons. The dashed curve shows for each electron charge density the value of the stopping interaction of Figure 13. When these two factors are multiplied together, the solid line is determined which is the integrand of the electronic stopping formula, Equation (11), for a unit charge. The area under this solid curve is the stopping. Note the *distribution of stopping* among the various electron shells of Cu, with the *L*-shell electrons interacting strongly with the particle.

6. Nuclear Stopping of Ions

The total stopping cross section of ions in matter can be divided into two parts: the interaction of the ion with the target electrons (electronic stopping, discussed above) and with the target nuclei (nuclear stopping).

The nuclear stopping component is very small at ion energies above 200 keV/amu, with a typical value being below 1% of the electronic stopping and the maximum value rarely exceeding 24% (such as for U ions in Au). The basic theory of nuclear stopping was due to Bohr,[1-3] which led to a simple formula[11] called LSS nuclear stopping which has been widely used for the last 15 years. Several significant improvements have been made in the calculation of nuclear stopping both from theoretical considerations[62] and from an empirical approach using experimental studies of ion stopping and ranges.[63,64] A detailed evaluation[65] of these approaches concluded that they were more accurate for energetic ions than the LSS formula. One reasonably accurate form (called the Kr-C in Reference 62) has the simple reduced stopping form s_n:

$$s_n = 0.5 \ln(1+\varepsilon)/(\varepsilon + 0.10718\varepsilon^{0.37544}) \tag{14}$$

which may be converted to units of $e\text{V}/(10^{15}\ \text{atom/cm}^2)$ by

$$S_n = s_n(8.462 Z_1 Z_2 M_1)/[(M_1 + M_2)(Z_1^{2/3} + Z_2^{2/3})^{1/2}] \tag{15}$$

The reduced energy of an ion, ε, is defined as

$$\varepsilon = 32.53 M_2 E/[Z_1 Z_2 (M_1 + M_2)(Z_1^{2/3} + Z_2^{2/3})^{1/2}] \tag{16}$$

where the ion energy, E, is in keV, Z_1 and Z_2 are the ion and target atomic numbers, and M_1 and M_2 are the ion and target masses in amu.

7. Range Theory

In general the range distribution for a particle injected into the surface of a target is defined as the probability density for finding the particle at rest at a given position inside the target at a later time. Statistical notation enters the description because the single ion history is generally not representative for the experimental situation, where a *beam* of ions is injected, and only the *mean history* of many ions is observable. Furthermore, for amorphous and random targets even the calculation of the trajectory of a single ion involves statistics, because the scattering centers which cause the stopping of the ion and quantum mechanical probabilities for electronic excitation and ionization are statistically distributed.

In range theory, the range distribution is regarded as the end effect of a transport problem describing the motion of the ions during their slowing down to zero energy. The final range distribution has been calculated using analytic transport theory, Monte Carlo calculations, or the direct simulation of ion trajectories. The common problem in these treatments has been to include reliable descriptions for the interactions between the moving ion and the surrounding target particles. The interactions are usually separated in binary elastic collisions with the screened nuclei of the target atoms and in inelastic collisions with the entire electron system of the target.

The elastic interaction between the ion and target nuclei is described by means of a detailed interaction potential in direct simulations and by means of scattering cross sections in transport and Monte Carlo calculations. The elastic collisions cause both angular deflections and stopping (energy loss) of the ions. The inelastic interactions are conveniently collected into mean values such as the electronic stopping cross section and electronic energy loss straggling because of the large variety of possible discrete collisions between the ion and the target electrons. Ion slowing down is included this way, but ion scattering is neglected because of the electron mass.

According to the definition of ranges, both elastic as well as inelastic collisions are statistical events. Therefore, the "range" must be considered as a *distribution* and not just a specific number. Even neglecting the angular deflections, the elastic collisions alone would result in a range distribution because there is a spectrum of elastic energy losses. A similar result holds true for the inelastic collisions, but has less quantitative importance. Insufficient knowledge of the higher-order moments of the electronic energy loss spectrum has made it necessary to assume that the interaction of the ion with the target electrons is a simple energy loss, which depends only on the ion's energy. With this assumption, the electronic interaction alone leads to one specific range (a δ-function distribution).

Transport theory is concerned with the following problem: A beam of ions with atomic number Z_1, mass M_1 (amu), and energy E is directed towards the plane surface of a monoatomic amorphous target along a direction $\bar{e}$, which makes an angle θ to the surface normal $\bar{n}$; see Figure 15. We wish to determine what is the probability $F\,dx$, for the ion to be slowed down to zero energy in a target-slab of width dx, at a depth x beneath the surface (measured along $\bar{n}$).

The assumption of a random target limits the applicability of the calculations to targets without pronounced crystalline structures. The assumption of an infinite target allows ion–target interactions outside the real target surface (negative depths), so that, for example, for light ions at low ion energies where the range distribution may overlap with the target

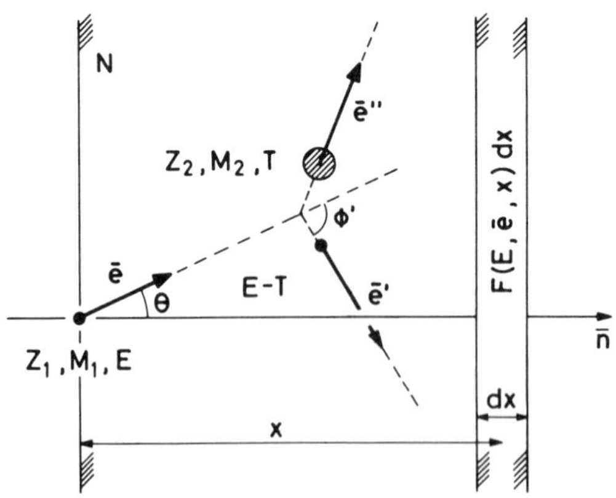

Figure 15. Schematic picture of the creation of a recoil atom in an elastic collison, the basic process leading to the ion deposition of magnitude $F(x)\,dx$ in a slice of material of thickness dx at depth x. See the text for the notations. This figure represents the heart of calculating range distributions using transport theory.

surface this may lead to erroneous results. For high energies such effects are negligible. A further assumption is that the ranges determined are low fluence ranges, i.e., so few ions are injected that their presence does not change the target significantly and saturation effects can be neglected. The target may now be characterized by the atomic number, mass, and density, Z_2, M_2, and N, respectively.

As discussed above, the interaction between incident ions and the target complex is considered to be made up of two effects which are uncorrelated: (1) binary elastic collisions with target atoms are described by the scattering cross section $d\sigma_n(E, T_n, \bar{e}, \bar{e}')$ for changing the ion state from $(E, \bar{e})$ to $(E - T_n, \bar{e}')$, where T_n is the energy transfer to the target recoil atom and $\bar{e}'$ is the direction of motion of the scattered ion (see Figure 15); and (2) the continuous slowing down by electronic interactions which will be characterized by a corresponding cross section $d\sigma_e(E, T_e)$. This interaction does not affect the ion's direction. The ion is traced through a differentially small pathlength $dx/\bar{e} \cdot \bar{n}$ and, by use of the transformation invariance of the target and the complete separability of $d\sigma_n$ and $d\sigma_e$, the following linearized Boltzmann equation ("backward" type) is obtained:

$$-\cos\theta \frac{\partial F(E, \bar{e}, x)}{\partial x} = N \int d\sigma_n [F(E, \bar{e}, x) - F(E - T_n, \bar{e}', x)] + N \int d\sigma_e [F(E, \bar{e}, x) - F(E - T_e, \bar{e}, x)] \tag{17}$$

where $\cos\theta = \bar{e} \cdot \bar{n}$.

Equation (17) has been solved under various assumptions concerning $d\sigma_n$ and $d\sigma_e$. The simplest realistic results are obtained by neglecting the inelastic interactions and describing the nuclear collisions by a power cross section.

As long as the electronic energy transfer T_e is small compared to E we may expand the lower term of Equation (17) in terms of T_e/E and obtain

$$\sum_{k=1} \frac{\partial^k F}{\partial E^k} \frac{(-1)^k}{k!} \int N \, d\sigma_e \, T_e^k \tag{18}$$

The integral terms are the electronic energy loss moments. $k = 1$ gives the mean electronic energy loss per unit path length and $k = 2$ is the mean electronic energy straggling per unit path length.

Usually the terms $k = 2$ and higher are neglected in range calculations. Since the electronic stopping power is

$$S_e = \int N \, d\sigma_e \, T_e \tag{19}$$

(where we have neglected electronic scattering and have used a sharp energy loss function) the transport equation (17) then reads

$$-\cos\theta\frac{\partial F}{\partial x}-NS_e(E)\frac{\partial F}{\partial E}=N\int_0^E d\sigma[F(E,\bar{e},x)-F(E-T,\bar{e}',x)] \quad (20)$$

where we have now omitted the lower index n on the nuclear cross section and energy transfer, and the integration limits have been inserted. See References 61 and 55 for further details.

References

1. N. Bohr, *Phil. Mag.* **25**(6) 10 (1913).
2. N. Bohr, *Phil. Mag.* **30**(6) 581 (1915).
3. N. Bohr, *Mat. Fys. Medd. Dan. Vid. Selsk.* **18**, No. (8) (1948).
4. H. A. Bethe, *Ann. Phys. (Leipzig)* **5**:325 (1930).
5. H. A. Bethe, *Z. Phys.* **76**:293 (1932).
6. H. A. Bethe and W. Heitler, *Proc. R. Soc. London* **A146**:83 (1934).
7. F. Bloch, *Ann. Phys. (Leipzig)* **16**:287 (1933).
8. F. Bloch, *Z. Phys.* **81**:363 (1933).
9. J. Lindhard, *Mat. Fys. Medd. Dan. Vid. Selsk.* **28**, No. (8) (1954).
10. J. Lindhard and M. Scharff, *Mat. Fys. Medd. Dan. Vid. Selsk.* **27**, No. (15) (1952).
11. J. Lindhard, M. Scharff, and H. E. Schiott, *Mat. Fys. Medd. Dan. Vid. Selsk.* **33**, No. (14) (1963).
12. J. Lindhard and Aa. Winther, *Mat. Fys. Medd. Dan. Vid. Selsk.* **34**, No. (4) (1964).
13. U. Fano, *Ann. Rev. Nucl. Sci.* **13**:1 (1963).
14. J. D. Jackson, *Classical Electrodynamics*, Wiley, New York (1962, 1975), Chap. 13.
15. P. Sigmund, *Radiation Damage Processes in Materials* (C. H. S. Du Puy, eds.), Noordhoff, Leyden (1975), Chap. 1.
16. P. Sigmund, *Rev. Roum. Phys.* **17**:823 (1972); **17**:969 (1972); **17**:1079 (1972).
17. H. H. Andersen and J. F. Ziegler, *Hydrogen Stopping Powers and Ranges in All Elements*, Pergamon Press, New York (1977).
18. J. F. Ziegler, *Helium Stopping Powers and Ranges in All Elements*, Pergamon Press, New York (1978).
19. J. F. Ziegler, *Handbook of Stopping Cross Sections for Energetic Ions in All Elements*, Pergamon Press, New York (1980).
20. U. Littmark and J. F. Ziegler, *Handbook of Range Distributions for Energetic Ions in All Elements*, Pergamon Press, New York (1980).
21. Mme. Pierre Curie, *C. R. Acad. Sci. (Paris)* **130**:76 (1900).
22. J. J. Thompson, *Phil. Mag.* **6-23**:449 (1912).
23. N. Bohr, *Phil. Mag.* **25**:10 (1913).
24. E. Rutherford, *Phil. Mag.* **21**:212 (1911); **21**:699 (1911).
25. N. Bohr, *Phil. Mag.* **30**:581 (1915).
26. L. A. Turner, *Rev. Mod. Phys.* **12**:1 (1940).
27. W. E. Lamb, *Phys. Rev.* **58**:696 (1940).
28. N. Bohr, *Phys. Rev.* **59**:270 (1941).
29. J. Knipp and E. Teller, *Phys. Rev.* **59**:659 (1941).
30. J. H. M. Brunnings, J. Knipp, and E. Teller, *Phys. Rev.* **60**:657 (1941).
31. L. C. Northcliffe, *Ann. Rev. Nucl. Sci.* **13**:67 (1963).
32. H. D. Betz, *Rev. Mod. Phys.* **44**:465 (1972).

33. W. Booth and I. S. Grant, *Nucl. Phys.* **63**:481 (1965).
34. J. C. Ashley, R. H. Ritchie, and W. Brandt, *Phys. Rev. A* **8**:2402 (1973).
35. R. Behrisch, W. Eckstein, P. Meischner, B. M. U. Scherzer, and H. Verbeck, *Atomic Collisions in Solids*, Plenum Press, New York (1975), Vol. 1, p. 315.
36. W. Whaling, *Handbuch der Physik*, Vol. XXXIV, 13, Springer-Verlag, Berlin (1958).
37. L. C. Northcliffe and R. F. Schilling, *Nucl. Data Tables* **7**:233 (1970).
38. H. Bichsel, *American Institute of Physics Handbook*, 8-142, McGraw-Hill, New York (1972).
39. A. H. Agajanian, *Ion Implantation: A Bibliography*, Vol. 6 (Covering 1977–1980), IBM Technical Comm., Report No. TR22.2374, East Fishkill, New York (1981).
40. J. F. Ziegler and W. K. Chu, *At. Data Nucl. Data Tables* **13**:463 (1974).
41. J. Lindhard, *Mat. Fys. Medd. Dan. Vid. Selsk.* **28**, No. (8) (1954).
42. C. C. Ronsseau, W. K. Chu, and D. Powers, *Phys. Rev. A* **4**:1066 (1970).
43. H. H. Andersen, K. N. Tu, and J. F. Ziegler, *Nucl. Instrum. Methods* **149**:247 (1978).
44. H. H. Andersen, *Bibliography and Index of Experimental Range and Stopping Power Data*, Pergamon Press, New York (1977).
45. J. F. Janni, AFWL-TR 65-150 (1966) (now out of print).
46. E. Bonderup, *Mat. Fys. Medd. Dan. Vid. Selsk.* **35** No. (17) (1967).
47. F. W. Martin and L. C. Northcliffe, *Phys. Rev.* **128**:1166 (1962).
49. W. D. Wilson, L. G. Haggmark, and J. P. Biersack, *Phys. Rev. B* **15**:2458 (1977).
50. W. S. Johnson and J. F. Gibbons, *Projected Range Statistics in Semiconductors*, Stanford University Bookstore, Stanford, California (1970) (now out of print).
51. D. K. Brice, *Ion Implantation Range and Energy Deposition Distributions*, Vol. 1, *High Energies*, Plenum Press, New York (1975).
52. K. B. Winterbon, *Ion Implantation Range and Energy Deposition Distributions*, Vol. 2, *Low Energies*, Plenum Press, New York (1975).
53. B. Smith, "*Ion Implantation Range Data for Silicon and Germanium Device Technologies*, Research Studies Press, Forest Grove, Oregon (1977).
54. J. F. Gibbons, W. S. Johnson, and S. W. Mylroie, *Projected Range Statistics: Semiconductors and Related Materials*, 2nd Edition, Halsted Press, Stroudsbury, Pennsylvania (1975).
55. U. Littmark and J. F. Ziegler, *Handbook of Range Distributions for Energetic Ions in All Elements*, Pergamon Press, New York (1980), Vol. 6.
56. K. B. Winterbon, Chalk River Laboratory Report No. CR-1817 (unpublished).
57. D. G. Simons, D. J. Land, J. G. Brennan, and M. D. Brown, *Phys. Rev. A* **12**:2383 (1975).
58. W. A. Grant, D. Dodds, J. S. Williams, G. E. Christodoulides, R. A. Baragiola, and D. Chivers, *Proc. 5th Int. Conf. on Ion Implantation in Semicond. and other Matl.*, Plenum Press, New York (1977).
59. K. B. Winterbon, P. Sigmund, and J. B. Sanders, *Mat. Fys. Medd. Dan. Vid. Selsk.* **37**, No. (14) (1970).
60. K. B. Winterbon, *Rad. Effects* **13**:215 (1972).
61. U. Littmark and J. F. Ziegler, *Phys. Rev. A* **23**:64 (1981).
62. W. D. Wilson, L. G. Haggmark, and J. P. Biersack, *Phys. Rev. A* **15**:2458 (1977).
63. W. K. Chu, *Proceedings of the 1978 International Conference on Ion Beam Modification of Materials*, Hungarian Academy of Sciences, Budapest, Hungary (1979), p. 179.
64. S. Kalbitzer, *Proceedings of the 1978 International Conference on Ion Beam Modification of Materials*, Hungarian Academy of Sciences, Budapest, Hungary (1979), p. 3.
65. J. F. Ziegler, *Appl. Phys. Lett.* **31**:544 (1977).
66. J. F. Ziegler, *Nucl. Instrum. Methods* **149**:129 (1978).
67. J. F. Ziegler, J. P. Biersack, and U. Littmark, *The Stopping and Range of Ions in Solids*, Pergamon Press, New York (1985).

4

JOHN M. POATE is a Department Head at AT&T Bell Laboratories, Murray Hill, New Jersey. He received his B.Sc. and M.Sc. degrees from Melbourne University and his Ph.D. in 1967 from the Australian National University. Prior to joining the Bell Laboratories in 1971, he was a Harwell Fellow at AERE Harwell. He has served as President of the Materials Research Society and is a Fellow of the American Physical Society.

Ion Implantation

JOHN M. POATE

1. Introduction

Ion implantation is one of the most important techniques for modifying the surface properties of solids. The development of the subject represents a fascinating interplay of solid-state, atomic, and nuclear physics. There is little doubt that the driving force behind such development has come from the scientific and technological imperatives of solid-state and materials science. Nevertheless there is also little doubt that the rapid advances in the applications and understanding of implantation have come predominantly from the fields of nuclear and atomic physics. Implantation offers the possibility of introducing atoms into solid-state environments without the usual constraints of equilibrium thermodynamics. Novel states of matter can thus be produced. In practice, however, ion implantation has found greatest acceptance in silicon integrated circuit technology where the technique is primarily used because of its precise spatial and numerical control.

The first studies of implantation in semiconductors appear to be due to Ohl (Oh 52) at Bell Laboratories. In 1952 he described the improvement in the electrical characteristics of Si point contact diodes produced by bombardment with hydrogen, helium, nitrogen, and argon. The improvements were, in fact, caused by surface damage and not specific chemical dopings. The concept of chemical doping by implantation is spelled out in the original patent by Shockley (Sh 54) in 1954. This patent, entitled "Forming Semiconductive Devices by Ionic Bombardment," has proved remarkably prophetic, and it is interesting to quote some of the sections of this patent:

JOHN M. POATE • AT&T Bell Laboratories, Murray Hill, New Jersey 07974.

> In accordance with this process, a semiconductive body of one conductivity type is bombarded with a monoenergetic beam of ions of a significant impurity element characteristic of a conductivity type opposite to that of the body in a manner to convert the conductivity type of a thin layer in the interior of the body....
> In the process of the invention, the energy level of the bombarding beam is adjusted so that the projected beam will penetrate into the interior of the semiconductive body and be localized there for converting that region to opposite conductivity type. By utilization of a monoenergetic beam, the ions are all made to penetrate to a fairly uniform depth whereby the thickness of the region of converted conductivity type is kept small. The thickness of this region can be controlled by variations in the energies of the bombarding ions. Thereafter, the semiconductive body is treated to repair the radiation damage done to the surface region penetrated. This is done advantageously by annealing at a temperature sufficiently low that there results little migration from the region of deposition of the significant impurity ions introduced by bombardment.

The review by Gibbons (Gi 68) details some of the early developments in implantation.

Many books (Ma 70, De 73, Hi 80) and conference proceedings (Ei 71, Ca 76, Ch 77, Gy 80, and Pr 80) deal exclusively with ion implantation modification of materials. In this review we will concentrate on the salient physical processes. We will firstly consider the dynamics of the implantation process. The mechanisms by which the impinging ion loses energy to the solid are now quite well understood. The atoms of the solid return to their equilibrium or metastable configurations approximately 10^{-11} s after the initial violent disturbance. The state, or lattice site location, that the implanted atom occupies after this period is perhaps the most interesting and fundamental aspect of implantation. We have much data on lattice site location and some understanding of the parameters that determine the final site. The process is so complicated, however, that many questions remain.

The final state of the solid after bombardment clearly depends upon its initial state. For example, the response of semiconductors to implantation can be very different from the response of metals. This behavior is mainly due to the differences in the chemical bonding. We will discuss implantation, therefore, in terms of the materials and will concentrate on the metals and semiconductors. Finally we will discuss an emerging area of ion implantation—ion beam mixing. The concept underlying ion beam mixing is to deposit thin films of the materials of interest and then mix or react them at the interface by the collision cascade generated, for example, by an inert gas atom. We will present examples of novel metallic and intermetallic phases produced by ion beam mixing.

2. Dynamics

2.1. Energy Loss, Range, and Damage

The Coulomb interactions by which heavy ions impart energy to the electrons and atoms of a solid are discussed in detail in Chapter 3, by Ziegler and Biersack. Two aspects of energy loss are of primary interest in implantation studies. Firstly, there is the range of the incident ion, which is clearly determined by the rate of energy loss. Figure 1 shows the projected ranges of As in Si and Au as a function of incident energy. The hatched areas represent the standard deviations in the range. The figure illustrates several features of ion penetration. The ranges are quite shallow over presently accessible energies; in fact the ranges coincide nicely with the thicknesses of many device and thin film structures of interest. The standard deviations in the ranges are large because, for most heavy-ion–target combinations, nuclear stopping predominates and the path of the ion through

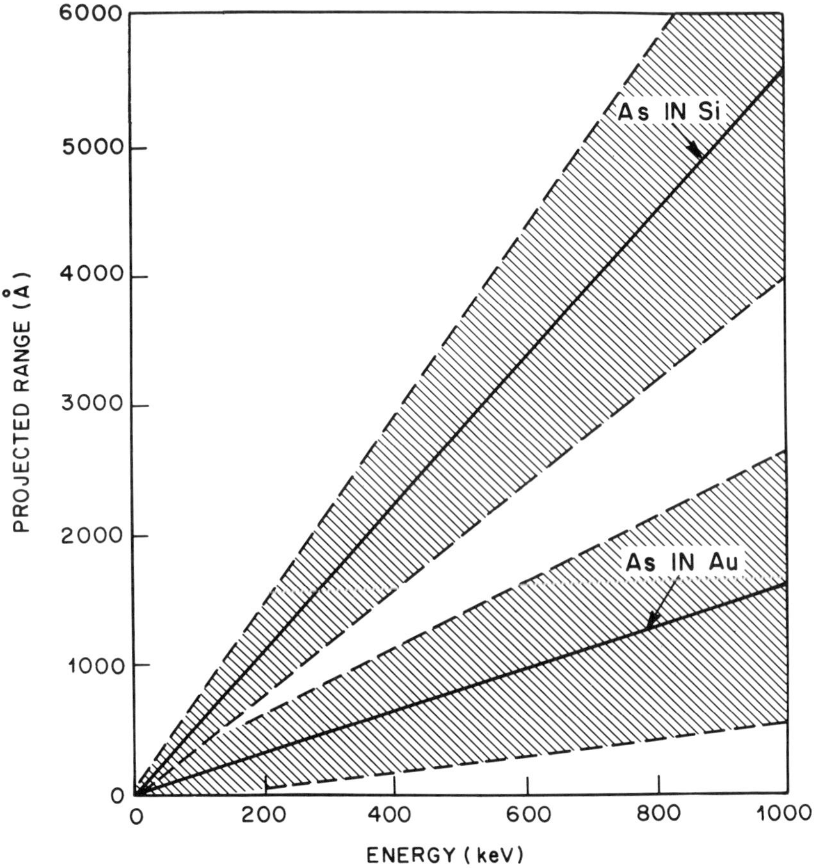

Figure 1. Range and straggling of As in Au and Si. The solid lines are the projected ranges and the hatched areas are the straggling.

the solid can resemble a random walk process. This behavior is quite contrary to the ideas expressed by Shockley, where he assumed that the ions would be deposited in extremely thin layers.

The second aspect of energy loss of importance to implantation studies is the amount of deposited energy that ultimately ends up in damage or disorder in the lattice. For all the solids we consider here, energy loss to the electrons of the solid will not result in any structural damage. However, the atomic displacements caused by the nuclear (or atomic) stopping can result in many forms of damage. We will discuss later a particularly violent and collective form of atomic displacement: the collision cascade.

The first-order approach to calculating the number of displaced atoms is given by the Kinchin–Pease formalism (Th 69). In this formalism, the total number of displacements ν is given, to first approximation, by $\nu = E_N/2E_d$, where E_N is the total energy deposited in elastic nuclear processes and E_d is the displacement energy. Of course the number of defects observed after implantation or bombardment is not necessarily related in any straightforward fashion to the number of atoms displaced. In metals, for example, each host atom can be displaced many times and the lattice can still retain its basic structural integrity after the implantation. This resistance to damage is due to the lack of directionality of the metallic bond and the fact that damage anneals out easily. The situation is different for covalent semiconductors, such as Si, which can be made amorphous relatively easily by ion bombardment. The tetrahedral bonding is still essentially maintained, but the atoms are assembled in a metastable amorphous network. The whole subject of damage induced by heavy-ion bombardment is discussed in detail in Chapter 8, by English and Jenkins.

2.2. Replacement Collisions

A heavy ion traversing a solid with an initial kinetic energy ~100 keV will come to rest rapidly in 10^{-13} s (Si 74). It will be retarded by electronic and elastic atomic energy-loss processes at the start of its trajectory. Toward the end of range, at low energies, most of the stopping will be due to the elastic atomic scattering effects. At the end of the ion track there is a certain probability that the incident ion may displace a host atom but have insufficient energy to escape the newly produced vacancy. This collision process is termed a replacement collision and represents the simplest concept of the formation of a substitutional solution by implantation.

We will present two figures due to Brice (Br 76) which nicely demonstrate the features of a replacement collision for an incident ion of energy E and mass M_1. A struck target atom, of mass M_2, is considered to be displaced if it receives a kinetic energy, T, greater than E_d, the displacement threshold energy. The scattered atom is then assumed to be captured by the resultant vacancy if its remaining kinetic energy, $E - T$, is less than E_c, the capture threshold energy. The maximum value which T can assume is $T_m = \gamma E$, where $\gamma = 4M_1M_2/(M_1 + M_2)^2$. A replacement collision cannot occur if $E < E_d/\gamma$ as the target atom cannot be displaced. Conversely if $E > E_c(1-\gamma)$ the scattered ion cannot be captured. The hatched energy region of Figure 2 shows where replacement collisions are possible: this figure shows a portion of the $\gamma - E/E_d$ plane where $E_c = E_d$. Consider Au atoms impinging on Al or Cu targets and $E_c = E_d = 25$ eV. The cut for Au–Al ($\gamma = 0.42$) shows that replacement collisions are impossible whereas for Au–Cu ($\gamma = 0.73$) replacement collisions are possible for Au energies between 30 and 130 eV.

What is the probability of a replacement collision occurring? Brice has calculated the replacement probability P for various ions on Cu and this is shown as the smooth curve in Figure 3, where $E_d = 25$ eV. For incident atoms of $Z_1 \sim 40$ there is an $\sim 80\%$ chance of a replacement collision occurring and the atoms residing on a Cu substitutional site. Plotted on this curve are the substitutional fractions of implanted atoms in Cu. Clearly the collisional arguments (i.e., replacement collision probabilities) presented above are not sufficient by themselves in explaining substitutionality.

2.3. Collision Cascades

In the previous section we presented the concept of the replacement collision as a way of lodging foreign atoms on lattice sites. However, in most cases of implantation this concept will be too simplistic. If the incident ion is reasonably energetic it can displace many atoms, which in turn may displace other atoms; this process is termed a collision cascade. Collision cascades have been actively studied by workers in the radiation damage field for over 20 years (Di 57, Th 69).

There are various ways of viewing the collision cascade. We will firstly consider the concept of thermal spikes as it appears particularly relevant to implantation and metastable alloy formation. Although thermal spike phenomena were postulated some 30 years ago, they received relatively little attention until recent years, when there has been a resurgence of interest. Attention was rather focused upon descriptions of collision events in terms of weak or linear collision cascade theory (Li 63). Theoretical interest in thermal spikes was apparently rekindled by Sigmund (Si 74), who considered collective atomic interactions within the cascade. This renewed interest is also due to sputtering and implantation damage experiments involving dense collision cascades which present evidence for collective interactions.

We will follow the early treatment of Seitz and Koehler (Se 56). Figure 4 shows the calculation of Dienes and Vineyard (Di 57) for the temperature dependence of a thermal spike in Cu. In this calculation the

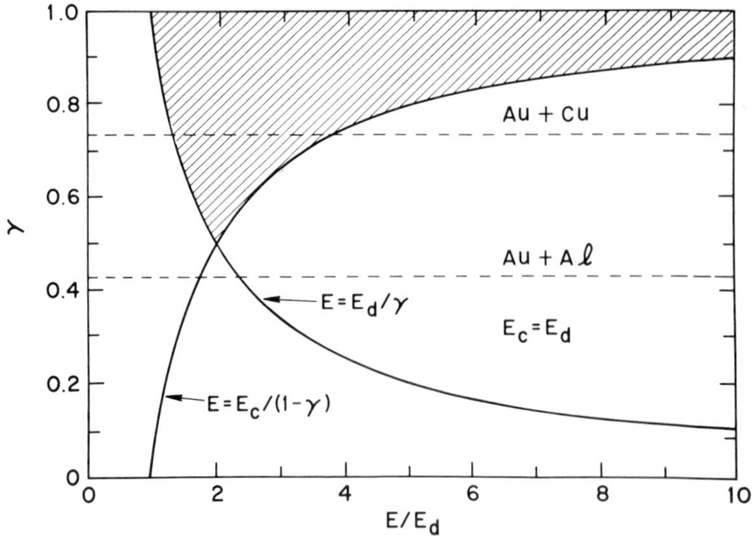

Figure 2. Portion of γ–E plane showing regions (hatched area) where replacement collisions are possible. From Brice (Br 76).

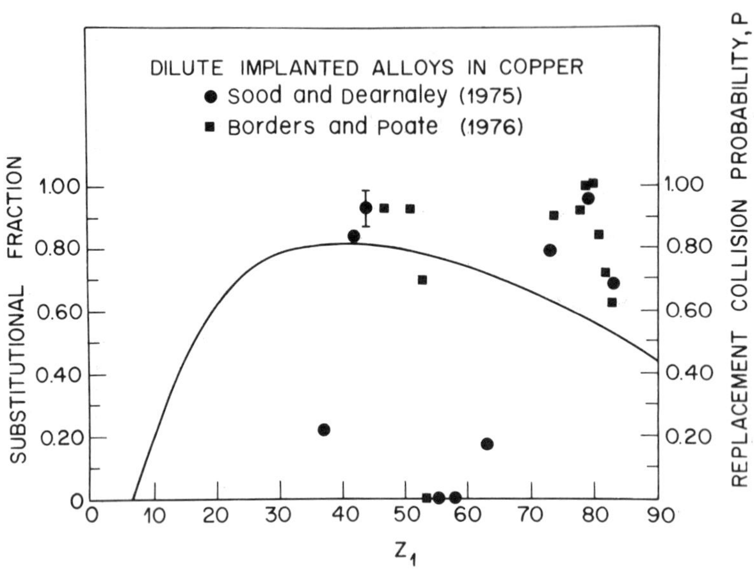

Figure 3. Replacement collision probability (smooth curve) in Cu and measured substitutional fraction of various ions in Cu. From Brice (Br 6) and Sood and Dearnaley (So 78).

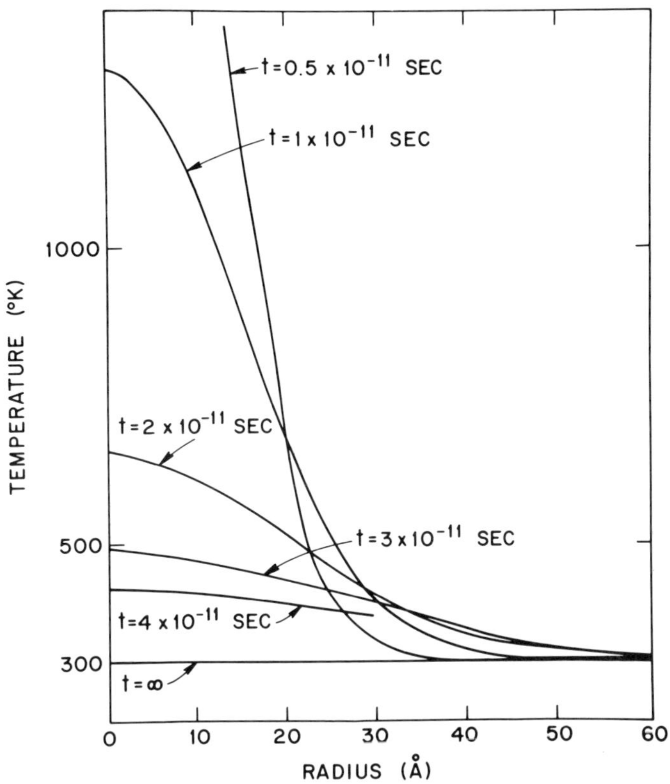

Figure 4. Temperature distributions in a thermal spike at various times. From Dienes and Vineyard (Di 57).

bombarding particle transfers an energy of 300 eV to a Cu lattice atom. It is then assumed that this energy is liberated instantaneously as heat energy. The thermal diffusivity ($0.001\ cm^2/s$) is obtained from the lattice thermal conductivity alone as there will be little coupling between the lattice atoms and electrons during the spike's existence. It can be seen that the spike heats a sphere 20 Å in diameter to temperatures greater than the melting point of Cu, 1086°C, in a time of 5×10^{-12} s. The temperature then decays to less then 500°C in 3×10^{-11} s. There will be some 1000 atoms in this rapidly quenched region.

If many atoms are displaced in the initial excitation the process is sometimes called a displacement spike. Figure 5 shows Brinkman's (Br 54) picture of the initial stage of the displacement spike where atoms along the axis of the track would be driven violently outward (marked "interstitial" in the figure). The displacement cascade must then immediately collapse,

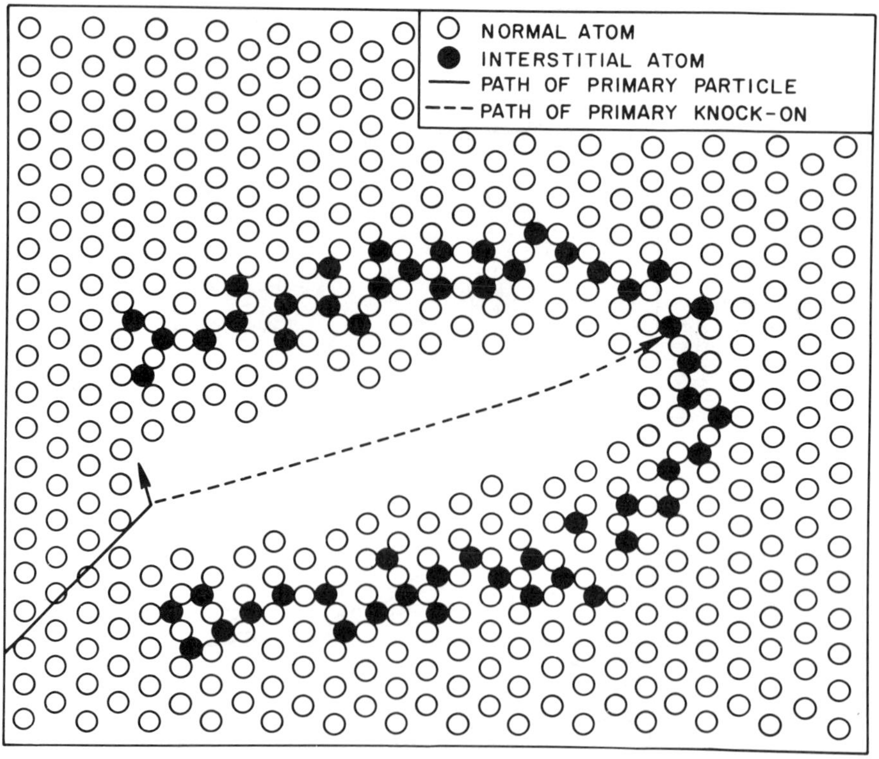

Figure 5. Displacement spike. From Brinkman (Br 54).

leaving such defects as vacancy loops from the collapse of the vacancy-rich regions. Dienes and Vineyard (Di 57) suggest that as this highly confused or molten zone freezes from the outside, most of the solidified material will be in register with the host lattice.

One of the more interesting questions of implantation metallurgy at the present time is whether we can usefully consider the cascade process as a rapid quench. The thermal spike calculations show that the quench rates are of the order 10^{14}°K/s. It is probably too simplistic to consider the collision cascade in terms of a temperature as the excited region is never in equilibrium. Nevertheless, in the following sections on metals and semiconductors we will present results on the formation of metastable phases by implantation that bear close relationship to similar phases produced by more conventional rapid quenching techniques.

2.4. Sputtering

A noticeable effect of energy deposition from ion bombardment is sputtering, the process by which an incident beam of energetic particles removes surface atoms. Surface morphologies can be changed markedly by sputtering. In fact the eroded surface can have the appearance of a surface which has been chemically etched. Measurements of the sputtering yields of monoelemental targets under carefully controlled conditions are in good agreement with theoretical predictions (An 74). The agreement between theory and experiment is shown in Figure 6, which shows the sputtering yield of Si ($S =$ number of atoms removed per incident ion) as a function of Ar ion energy. The calculations are based upon the collision cascade theory of sputtering (Si 69). The incident ion makes multiple collisions with target atoms, which in turn collide with other target atoms. Momentum can thus be returned to the surface atoms and, as a result, those atoms which obtain sufficient energy to overcome the surface binding energy will be sputtered. Consequently, the sputtering yield is proportional to the amount of energy deposited in atomic motion and is inversely proportional to the surface binding energy.

The results of Figure 6 show that sputtering must be considered when medium or high dose implantation experiments are carried out. For example Ar implanted in Si at a dose of $10^{16}/\text{cm}^2$ and energy of 20 keV will sputter away 1.8×10^{16} Si/cm^2 or approximately 20 monolayers of Si. In many cases the amount of sputtered host material can equal or exceed the range of the incident ion. Range distributions can therefore be grossly changed by sputtering.

Another important consequence of sputtering is that the sputtering of

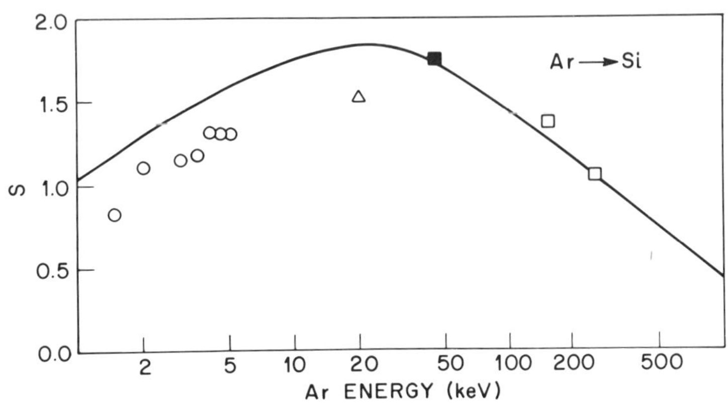

Figure 6. Sputtering of Si by Ar. The curve is from Sigmund (Si 69) and the data from Andersen (An 74) and Poate *et al.* (Po 76).

alloys or compounds can produce surface layers whose compositions are different from the starting material (Li 81). We have (Po 76, Li 78) observed in studies of metal alloys, silicides, compound semiconductors, and oxides that the surface layer becomes enriched in the heavier element. The compositional changes can be large and extend over hundreds of angstroms when energetic ion beams are used for the sputtering. Thin film silicides are useful model systems for studying alloy sputtering phenomena. They can be produced with well-defined stoichiometry and a high degree of purity. In our measurements we measured the surface compositional changes using Rutherford backscattering. Figure 7 shows the depth profiles (Li 78) of a PtSi film after sputtering with an 80-keV Ar beam. The composition has changed to a depth of nearly 800 Å. The surface composition can be obtained by extrapolation to the surface. Other measurements were made with 10-, 20-, and 40-keV Ar and the depth of the Pt enrichment was found to be related to the Ar energy but the steady-state enrichment was independent of Ar energy.

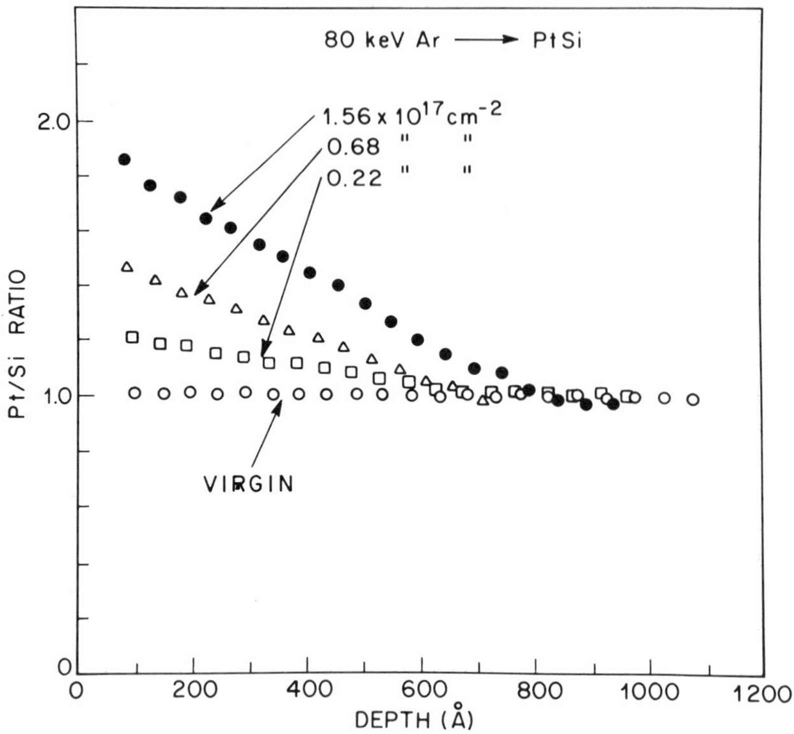

Figure 7. Depth profiles of Pt/Si concentration ratios for 80-keV Ar sputtering of PtSi samples. From Liau *et al.* (Li 78).

From these measurements and other considerations one can build a plausible *ad hoc* model of the alloy sputtering process (Li 78). At steady state the actual number of Si atoms sputtered off must equal the number of sputtered Pt. However, the initial sputtering yield of Si is higher than that of Pt. As sputtering continues, the sputtering yields approach equality as the surface becomes enriched in Pt. Measurements of the number of Pt and Si atoms sputtered off the surface do, in fact, show this behavior. Figure 8 shows our measurements of the sputtering yields of Pt and Si as a function of dose. Steady state has not been reached for the 80-keV bombardment but for the 40-keV bombardment, at steady state, the sputtering yields are equal. These are apparently the first direct measurements of this expected phenomenon. We have presented these results in detail as a cautionary tale because, in many cases, the effects of implantation *per se* can be completely masked by sputtering induced compositional charges. Moreover, sputtering ultimately imposes the limit on the maximum concentration of material that

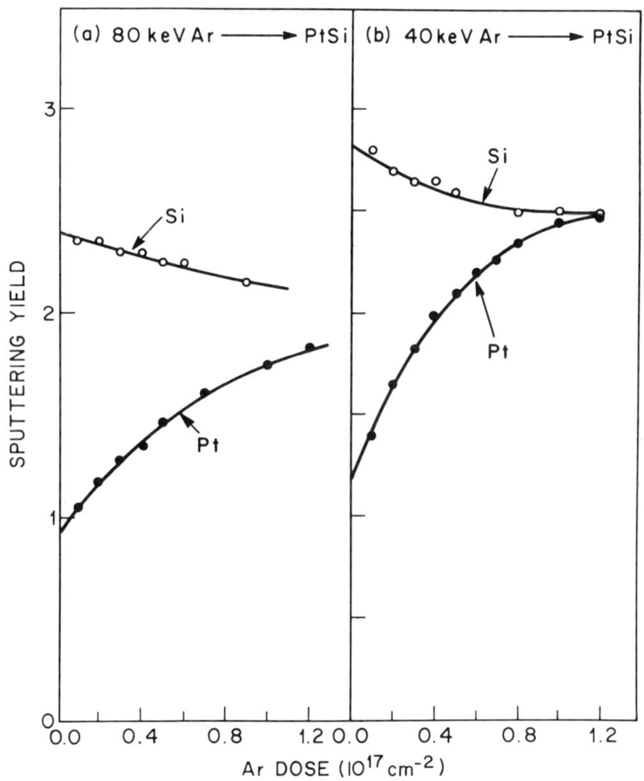

Figure 8. Sputtering yields of Si (○) and Pt (●) atoms sputtered from PtSi at energies of (a) 80 and (b) 40 keV. From Liau *et al.* (Li 78).

can be implanted. At steady state the maximum surface concentration of an implanted atom will be approximately equal to the inverse of the sputtering yield. For example if the sputtering yield of A on B is 5, then the maximum retained concentration of A in B will be 20 at. %.

3. Metals

3.1. Dilute Alloys

Although ion implantation is a well-developed technique for modifying surface layers, it is only recently that the concepts of metastable phase formation during the implantation process have been recognized. This is perhaps surprising as ion implantation is a grossly nonequilibrium process. It can be understood, however, in light of the development of the subject. Implantation into semiconductors is by far the most widely studied and utilized branch of the subject. Typical implants into Si will result in the production of an amorphous surface layer, itself a metastable phase. As the implanted dopant concentrations are usually much less than 1 at. %, the amorphous implanted layer cannot be usefully regarded as a metastable alloy but rather as doped amorphous Si. Crystallinity can be restored by annealing and the dopant atoms may then reside on regular lattice sites depending on equilibrium solid solubilities. Indeed equilibrium solid solubilities can be exceeded in this case, but the systems are such exceedingly dilute solutions that they have not attracted much attention as metastable solid solutions.

In the past few years, however, implantation into metals at low and high concentrations has produced alloys of unique metallurgical interest (Po 80). Such alloys are produced by implantation at room temperature without any heat treatment of the host. The motivation for the early experiments on implantation into metals appears to have come predominantly from the fields of impurity hyperfine interactions and channeling (deW 74). With hindsight it is obvious that the interaction between these communities, while stimulating channeling and lattice site location measurements, led to some interesting bottlenecks in the development of implantation metallurgy. For example de Waard and Feldman in their 1974 review emphasize the point that, up to then, there were no examples of implantations giving 100% substitutionality (i.e., complete solubility). This lack of solubility was not all that surprising as the systems studied were rather bizarre (e.g., Br implanted in Fe, Al 74) in terms of conventional alloys. When more conventional binary allows were studied (e.g., Au in Cu) complete substitutionality was observed (Po 74b). The analysis of dilute implanted alloys has been greatly helped by Rutherford backscattering and channeling. We will present some channeling results which help show the development of the subject. Figure 9

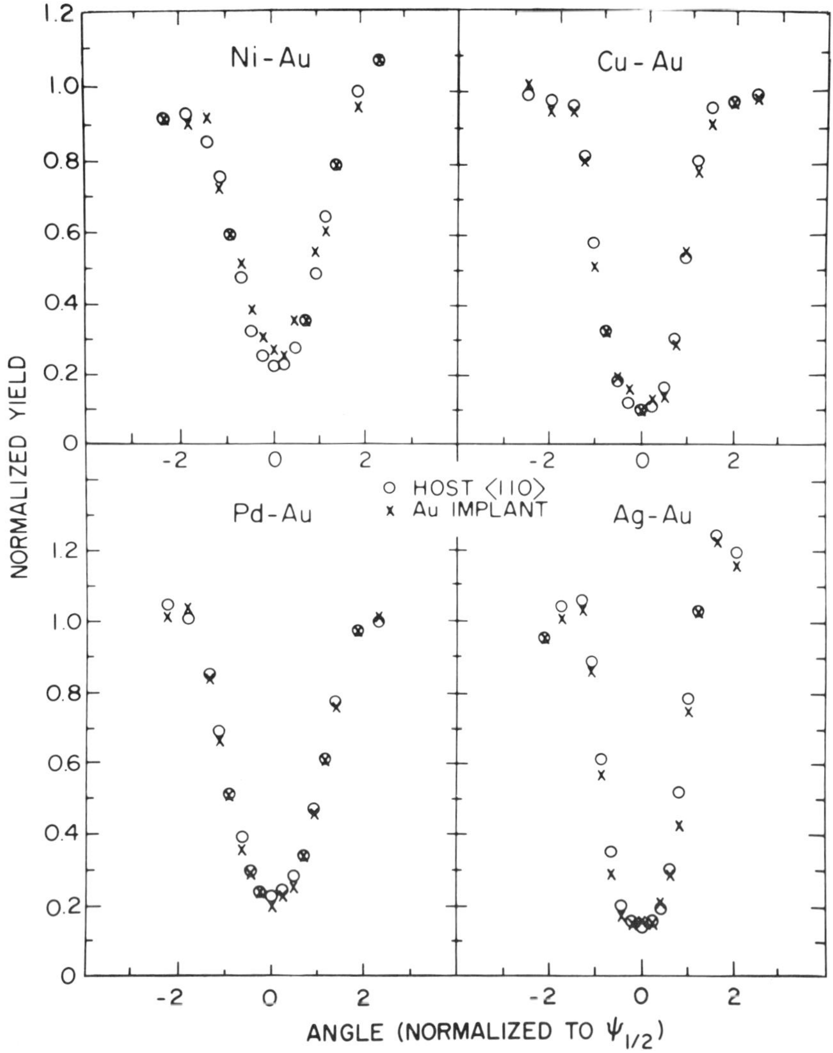

Figure 9. Normalized channeling ⟨110⟩ angular scans for Ni, Cu, Pd, and Ag single crystals implanted to peak Au concentrations of ~1 at. %. From Borders and Boate (Bo 76).

shows channeling angular scans of Au implanted into single crystal Ni, Cu, Pd, and Ag single crystals at concentrations less than 1 at. % (Po 74b, Bo 76). Gold is completely soluble in these f.c.c. metals at these concentrations at room temperature. Indeed the complete overlap of the Au and host normalized angular scans shows that the implanted Au is also substitutional. These results were the first measurements of complete solubility formed by implantation but were to be expected if equilibrium phase arguments are applicable to the implantation process.

However, when we implanted Ag, Sb, W, and Pt into single-crystal Cu we obtained results which were not interpretable in terms of equilibrium concepts. Figure 10 shows the impurity and host angular scans which overlap completely thus indicating complete substitutionality. The Cu–Pt

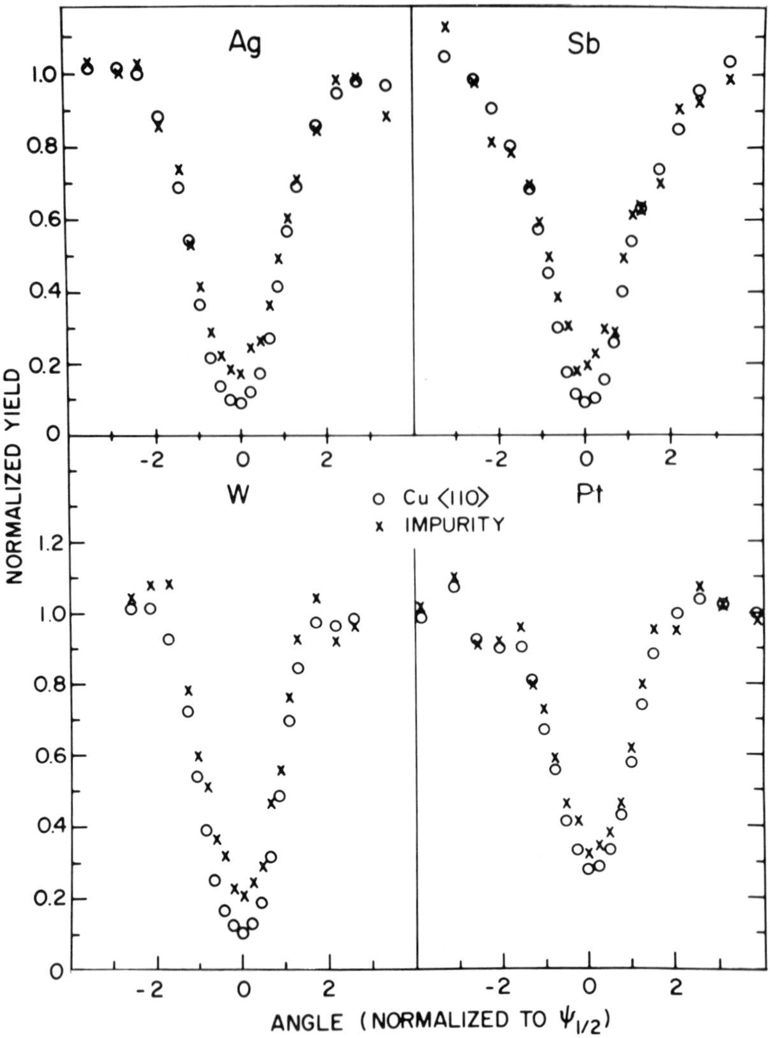

Figure 10. Normalized channeling ⟨110⟩ angular scans for Cu single crystals implanted with Ag, Sb, W, and Pt, respectively, to peak concentrations ~1 at. %. From Borders and Poate (Bo 76).

system forms a complete series of solid solutions but Ag, Sb, and W have very limited solubilities in Cu at room temperature. In fact W and Cu are immiscible in the melt. These are examples, therefore, of metastable solid solutions formed by ion implantation. How do we systematize these alloys formed by implantation? The most widely used rules of metallic alloying are those of Hume–Rothery (Hu 69) which state, in their most simplified form, that if the elements have the same crystal structure, and small differences in atomic size and electronegativity, they should form complete series of solid solutions. One way of displaying these conditions is by a Darken–Gurry plot. Figure 11, taken from Sood and Dearnaley (So 78), of all the room-temperature low dose (<1 at. %) implantations in Cu shows such a plot. If, for example, the Hume–Rothery rules are obeyed, then those elements which form substitutional allows with the host should lie within an ellipse or circle whose major axis or radius is ≈15% of the host radius (and electronegativity difference of ±0.4). Clearly, for the case of Cu, many elements lie outside this circle but are substitutional—all the elements within the dashed box. Similar behavior is also observed for implantation in Ni and Fe.

One explanation for these zones of substitutional metastability is that the final site is mainly determined by the probability of a replacement collision (see Figure 3). For most of the atom–target combinations the probability will be quite high. Indeed, channeling measurements on implantations into Al, where such wide ranges of replacement collisions are not

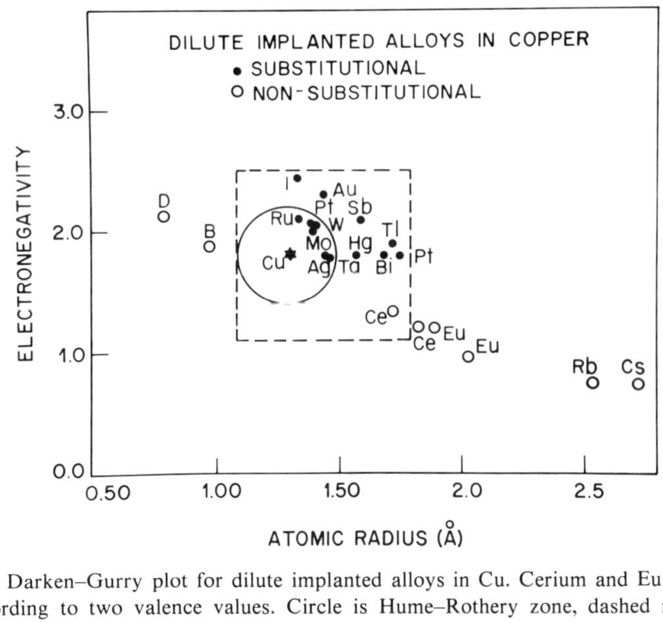

Figure 11. Darken–Gurry plot for dilute implanted alloys in Cu. Cerium and Eu are plotted twice according to two valence values. Circle is Hume–Rothery zone, dashed rectangle is modified Hume–Rothery zone. From Sood and Dearnaley (So 78).

possible, do not show extended zones of metastability. However, we should not pursue the replacement collision arguments too far for these examples, as a dense collision cascade can probably erase most memories of the replacement collisions and subsequent defect diffusion will surely influence the final lattice site location.

There is one metallic host, Be, which proves to be a nearly ideal model system. Its light mass guarantees that replacement collisions are impossible for most projectiles. The light mass is also helpful in guaranteeing a minimum defect density in the collision cascade. In particular, heavy projectiles will create no vacancies at the end of range. Moreover the defects are not thought to be very mobile at room temperature. The final impurity lattice site should therefore represent the metallurgical or chemical propensities of the species involved.

Kaufmann and his colleagues, in a now classic series of experiments (Ka 77, Vi 80), implanted 25 metallic elements into Be at concentrations of ~0.1 at. %. The implanted atoms were found to occupy one of three sites in the host lattice: the regular substitutional or one of the regular octahedral or tetrahedral interstitial sites. These sites are metastable by any usual alloying definition. Figure 12 shows the channeling angular scans for Gd in Be. The impurity appears "substitutional" along the $(11\bar{2}0)$ planar channel but "interstitial" along the $\langle 11\bar{2}0\rangle$ axial direction as witnessed by the two flux peaks. The flux peaking phenomenon arises from the fact that the channeled particle flux is concentrated in the central regions of channels and backscattered yields from atoms at these central positions can be greater than random scattering probabilities. Indeed, the channeling data indicate unambiguously that Gd is occupying an interstitial site.

The first attempt to systematize the Be data involved a Darken–Gurry plot. A rough separation was obtained between the substitutional and interstitial cases, with the larger atoms being forced to remain in the interstices of the lattice. However, no separation between the octahedral and tetrahedral sites was discerned. There has been a recent advance in the theory of alloying due to Miedema and co-workers (Mi 73, Mi 76). They have developed thermochemical parameters for the elements with which signs of the heats of formation of over 500 alloy systems have been successfully predicted. Miedema's scheme is based on two parameters, ϕ^* and n_{WS} related, respectively, to the electronic work functions of the binary alloy and the electron density in the metal at the boundary of the Wigner–Seitz cell. Both these parameters can be related to the heats of formation for binary alloys. A difference in ϕ^* between the constituents of the binary alloy, in the cellular model, results in a charge transfer across boundaries between dissimilar cells, thus providing a negative contribution to the heat of formation. A difference in n_{WS} implies the need to provide energy to smooth a discontinuity in electron density at that boundary, thus giving rise to a

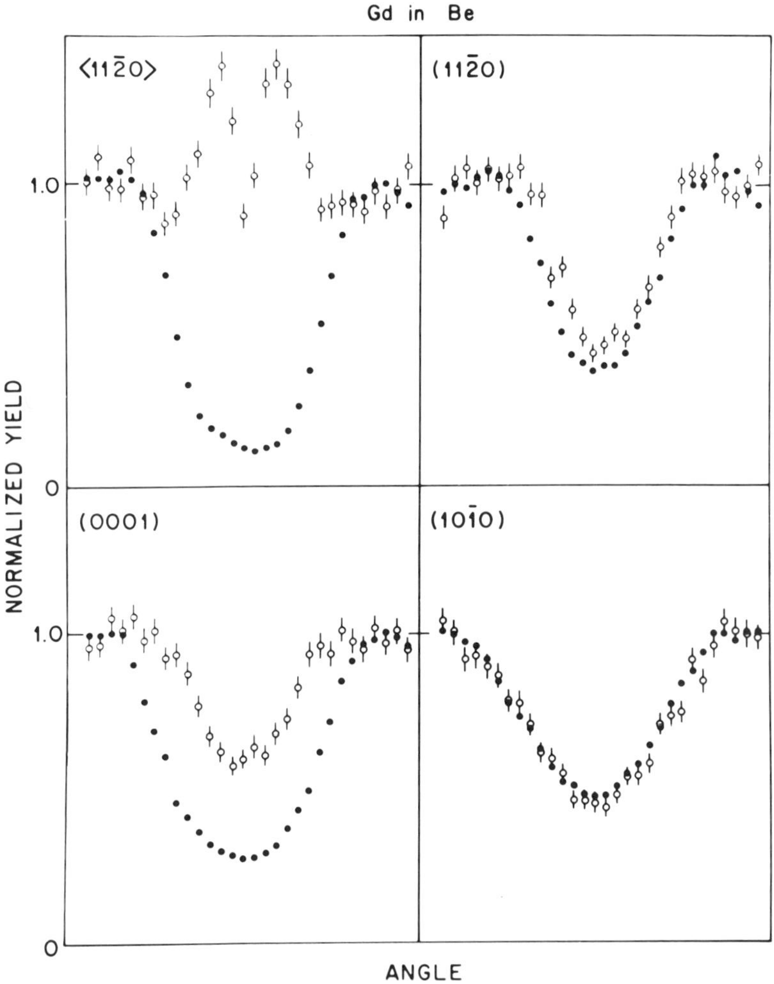

Figure 12. Measured channeling angular scans of Gd implanted in Be. From Kaufmann *et al.* (Ka 77).

positive contribution to the heat of formation. The parameter n_{WS} can be related, in a more macroscopic sense, to the surface tension of the elemental liquid metal. Miedema has determined the variables ϕ^* and n_{WS} for more than 50 metallic elements by fitting the signs of ΔH_f for more than 500 binary metallic alloys with a formula of the form

$$\Delta H_f = -P(\Delta\phi^*)^2 + Q(\Delta n_{WS}^{1/3})^2$$

with the constant P, $Q > 0$.

Figure 13 shows a plot of the Miedema variables relative to the Be host. It is immediately obvious that the substitutional, octahedral, and tetrahedral sites lie in separate domains. This is quite convincing evidence that the Miedema approach is applicable to these metastable implanted configurations. Kaufmann *et al.* have used the Miedema coordinates in a modified Landau–Ginsberg expansion of the site energy differences for each element to determine the boundaries of the site regions. Thus the substitutional region is an ellipse and the octahedral and tetrahedral regions

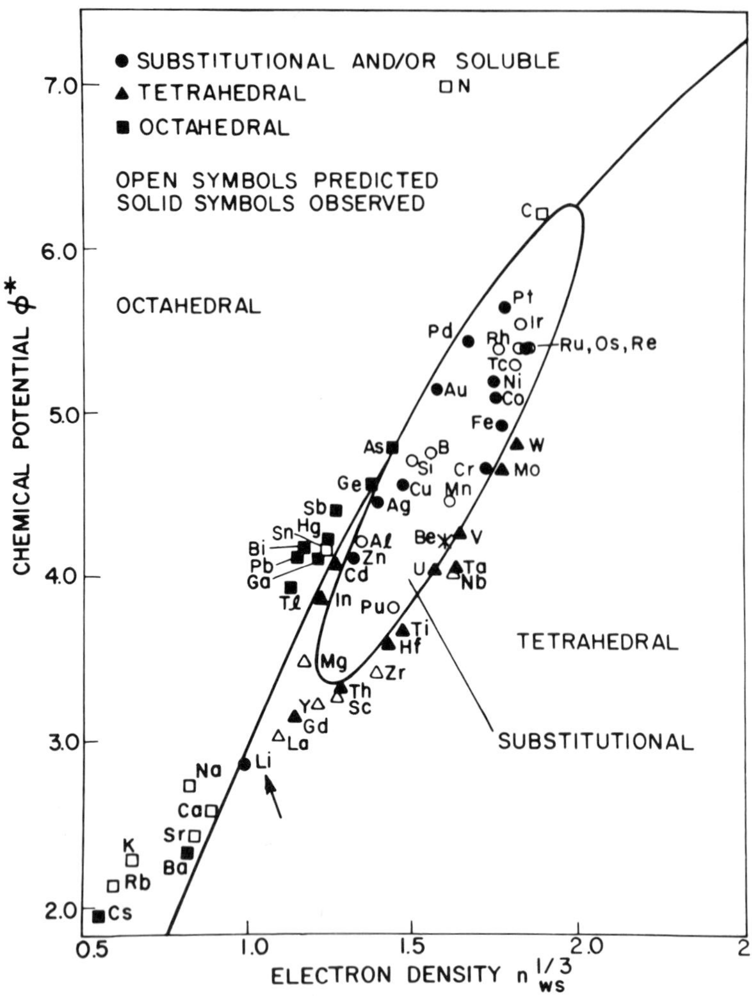

Figure 13. Miedema plot. The boundaries for the site regions were obtained from a Landau–Ginsberg expansion of the site energy differences. From Kaufmann *et al.* (Ka 77).

are separated by a hyperbolic contour which passes through the substitutional region.

The one exception to this scheme is Li (Ka 79), which resides on a substitutional site and not the predicted tetrahedral site. This behavior is not surprising as Li implanted in Be is a case where ballistic considerations can play a significant role; the replacement collision probability is high and vacancies can be produced at the end of the range.

These experiments and theoretical developments of Kaufmann, Vianden, Chelikowsky, and Phillips represent significant developments in the field of implantation and alloying. It would appear that they have chosen an ideal model system where the implanted atoms finally reside in sites determined by metallurgical constraints and not by the dynamics of the implantation process. These sites are metastable by any usual alloying definition. It is remarkable, therefore, how well the Miedema model, based on heats of formation obtained from equilibrium alloys, fits the data.

3.2. Concentrated Alloys

There is much current interest concerning high dose implantation into metals. The interest is due to several reasons. The surface composition and microstructure of metals can control such important properties as corrosion, hardness, and oxidation and there is a strong technological driving force to investigate the obvious applications of implantation into metals. This subject is reviewed in the book by Hirvonen (Hi 80). Another reason for the interest is that the concentrated, metastable alloys formed by implantation can be compared with the alloys prepared by the more conventional rapid quenching techniques (Po 80).

In 1959, Duwez, Willens, and Klement (Du 60) developed the first rapid quenching techniques (splat cooling) and were able to produce a complete series of Ag–Cu solid solutions. The technique consisted of shooting molten alloys onto cooled Cu strips, and, in this manner, quenching rates of 10^{6}°K/s were achieved. The equilibrium alloys of Ag–Cu represent something of an anomaly as the maximum concentration of Ag in Cu is 4.9 at. % although these elements are prime candidates for obeying the Hume–Rothery rules. Duwez *et al.* reasoned that if the Ag–Cu melt could be quenched rapidly enough, solid solubility could be extended across the composition range of the phase diagram.

In a series of experiments (Po 77, Cu 78) we attempted to determine whether we could produce similar metastable solid solutions by implantation. Figure 14 shows the channeling angular distributions for Ag implanted into Cu as a function of concentration. The maximum concentration of Ag that could be implanted was 16.6 at. %. At this concentration the Ag is completely substitutional even though the Cu lattice is considerably

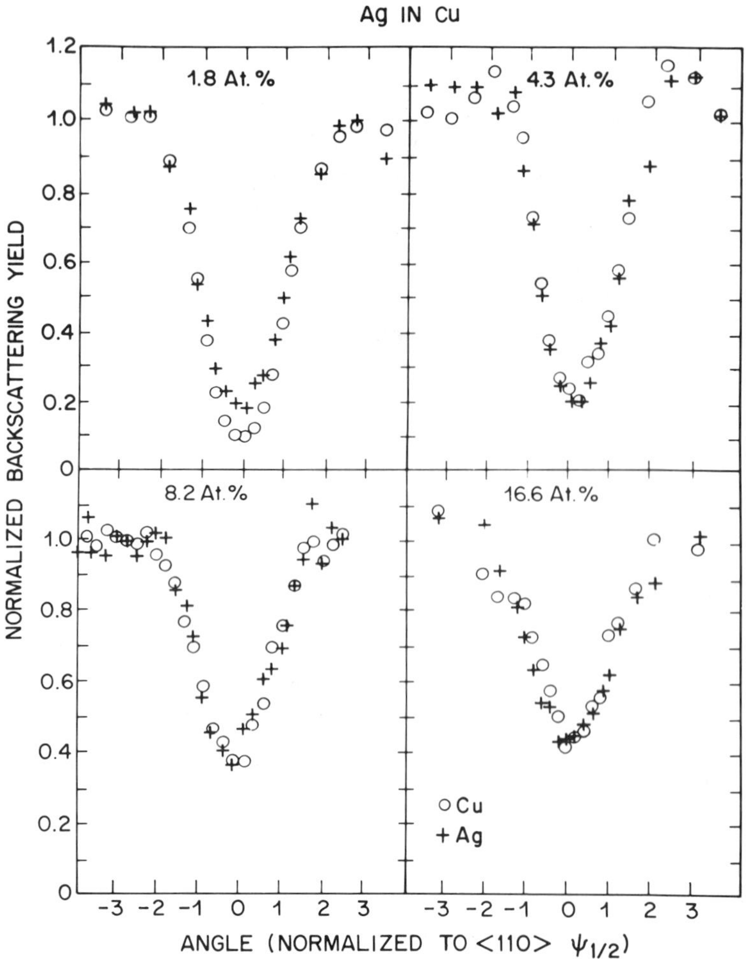

Figure 14. Normalized channeling ⟨110⟩ angular scans for Cu single crystals implanted to peak Ag concentrations of 1.4, 4.3, 8.2, and 16.6 at. %, respectively. From Poate *et al.* (Po 77).

damaged. The implanted Ag is clearly in a metastable configuration. This metastability is demonstrated by the TEM diffraction patterns (Cu 78) of Figure 15 of the 8 at. % implanted Ag in Cu sample. The as-implanted Cu specimen exhibited high-quality single-crystal diffraction poles. The diffraction pattern also exhibited weak subsidiary spots (A) due to epitaxial Cu_2O and (B) very small amounts of oriented metallic Ag present within the Cu matrix, although the lattice parameter of this Ag was reduced by about 1% from its equilibrium value. When the sample was annealed at 400°C for

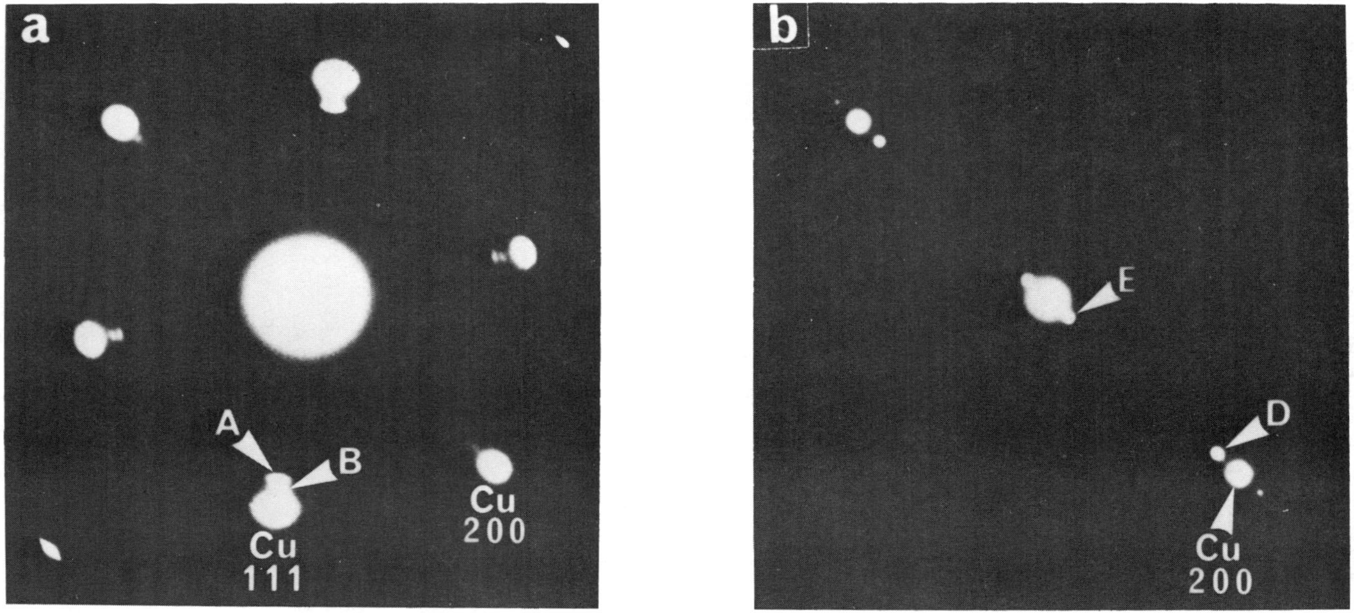

Figure 15. TEM diffraction patterns; (a) of as-implanted Ag in Cu at concentrations of 8 at. %. Note well-defined Cu pole pattern with weak subsidiary spots from small amounts of crystalline Cu_2O (A) and Ag (B); (b) sample following annealing at 400°C for 15 min. Note strong metallic Ag spots (D) and double diffraction (E). From Cullis *et al.* (Cu 78).

15 min the Cu_2O diffraction spots disappeared. Moreover the diffraction patterns (Figure 15b) exhibited extremely strong Ag spots (D) and when the corresponding Cu matrix spots were excited, pronounced double diffraction effects (E) were also evident. The Ag precipitates had dimensions of the order 200 Å with their lattices parallel to that of the Cu matrix. These TEM results convincingly demonstrate that the as-implanted Ag is in a metastable solid solution which returns to the equilibrium two-phase structure of Ag precipitates in Cu upon heating at 400°C. Many other concentrated, equilibrium, and metastable alloys have been produced by implantation, and these are reviewed elsewhere (Po 80).

It might reasonably be argued that a rapid quench from the liquid should freeze in the liquidlike structure and produce an amorphous solid. The first metallic system to be made amorphous by splat-cooling from the melt was Au–Si (Kl 60). Many other metallic systems have been made amorphous by rapid quenching but they usually involve one nonmetallic element (Du 76). These structures are termed glassy metals as their electrical conductivity is metallic. One of the first metallic systems to be made amorphous by implantation appears to be W in Cu (Cu 76). Tungsten and Cu are immiscible in the melt; however, at low implantation concentrations metastable solid solutions are formed. At concentrations $\geqslant$10 at. % amorphous surface W–Cu alloys are formed. There now exists quite a large amount of data on the formation of amorphous surface alloys by the implantation of metalloids (B or P) into Fe, Co, and Ni polycrystalline foils (Gr 78, Gr 81). There are clear similarities between the metallic glasses formed by implantation and those formed by splat cooling.

We can picture the formation of quenched metastable alloys using the concept of rapidly collapsing spikes or cascades. The collapsing spikes need not revert to a crystalline structure but can retain a high degree of disorder if assisted by the presence of a large concentration of implanted "impurity" atoms. Moreover, spikes will spatially overlap at high implant concentrations. At low implant concentrations the spike can collapse to a crystalline structure and the implanted ion will find itself in a stable or metastable configurations. The phase change from metastable solid solutions to amorphous alloys can therefore be explained within the context of collapsing spikes. The role of sputtering can also be seen in the high concentration regime. The surface is continually receding, permitting the spikes to collapse in regions of high impurity concentration.

4. Semiconductors (Si)

4.1. Amorphous Silicon and Epitaxy

The effects of implantation into covalently bonded semiconductors such as Si is completely different than implantation into metals. For example, at energies of 100 kV, an implantation of 10^{16} Si/cm^2 into Si will produce an amorphous layer some 2000 Å thick, whereas it is unlikely that a fcc metal such as Cu can ever be made amorphous by self ion bombardment. This difference is a manifestation of the differences in chemical bonding. It is somewhat meaningless, therefore, to try to systematize even the most dilute as-implanted alloys in Si. At the lowest doses, the lattice site locations and alloying propensities will be completely dominated by the damaged environment. There is, however, increasing interest in the positions the implanted atoms occupy after the lattice has undergone a rapid annealing process. We will discuss this subject in the next section.

The very fact that Si can be amorphized so readily has been used to advantage to study the amorphous–crystalline interface and epitaxial regrowth processes. There are two reasons why ion implantation is such an apposite technique for producing amorphous layers for the study of epitaxial regrowth processes; firstly the layers are free of impurities and voids, and secondly, the interface between the amorphous layer and underlying single substrate is very clean. Figure 16 shows ⟨100⟩ channeling spectra (Cs 75,

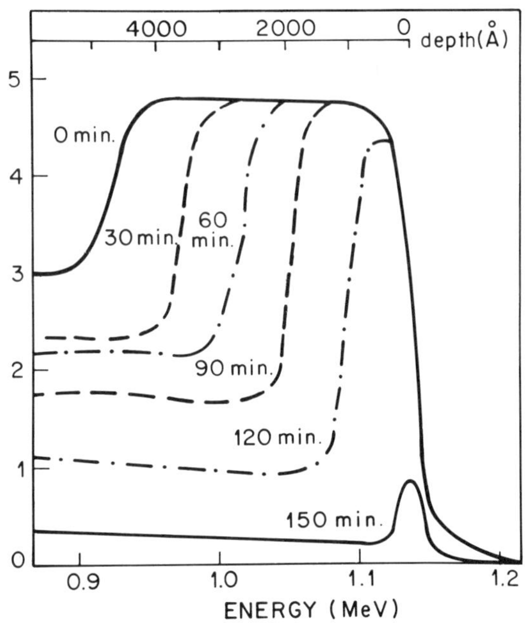

Figure 16. Aligned ⟨100⟩ spectra for 2-MeV He ions on (100) amorphized Si. Samples annealed at 550°C. From Csepregi *et al.* (Cs 75, Cs 76).

Cs 76) of the epitaxial regrowth of a 4500-Å thick amorphous Si film on (100) Si after heating at 550°C. The amorphous layer, of course, shows no channeling behavior, but the beam is channeled in the underlying single crystal material. In this fashion the crystalline–amorphous interface can be delineated with accuracy. After the 150-min anneal the amorphous layer has been consumed by the epitaxial regrowth process and the channeling spectra show essentially single crystal material.

By performing similar measurements at different temperatures, the Caltech group were able to extract regrowth rates and activation energies. Figure 17 shows these regrowth rates plotted as a function of reciprocal temperature for different orientations in Si and Ge. The activation energies are 2.4 eV for Si and 2.0 eV for Ge. The intriguing feature of these data is the large difference in the rates for ⟨100⟩ and ⟨111⟩ regrowth. This intrinsic difference in the regrowth rates has been interpreted (La 80) in terms of the interface model proposed by Spaepen (Sp 78). The growth kinetics of ⟨100⟩ Si are particularly simple with the amorphous–crystal interface moving with a laterally uniform front toward the surface. The regrowth requires no long-

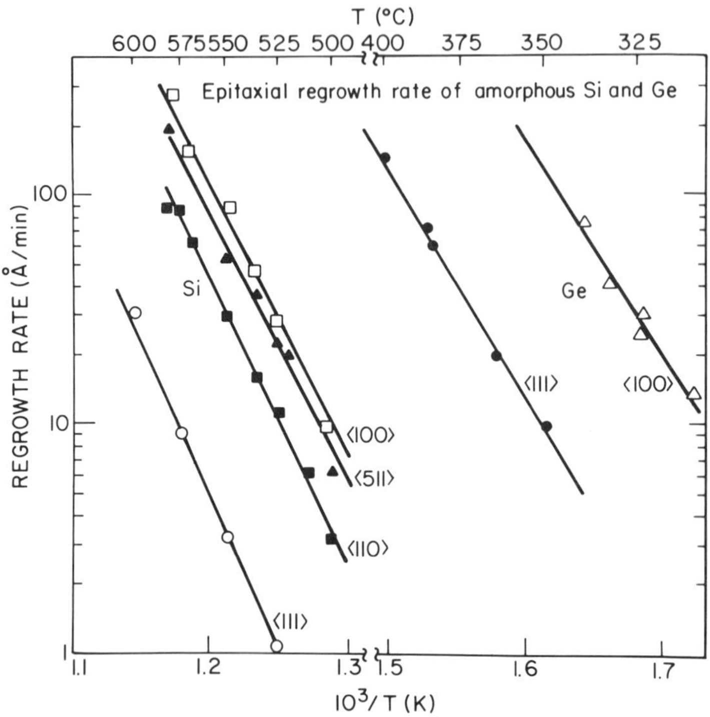

Figure 17. Regrowth rates of various Si and Ge orientations vs reciprocal temperature. From Csepregi *et al.* (Cs 75, Cs 76).

range transport of Si but only a rearrangement of the bonds at the amorphous–crystal interface. The process can be visualized as one broken bond propagating along the interface and causing reconstruction in its wake. Growth is not so easy in the ⟨111⟩ direction because atoms in (111) planes have one bond connecting them to an atom in the adjacent (111) plane. This connecting bond provides no rotational orientation information for the next set of covalent bonds. Consequently growth occurs along ledges rather than normal to the (111) plane (Sp 78, Sp 79).

4.2. Supersaturation

In the section on metals we presented examples of supersaturation produced by direct implantation. This phenomenon does not occur in Si because the implanted ion causes so much local damage. Crystallinity is restored by heating the Si typically at temperatures of 1000°C for 30 min. At such temperatures and time the implanted atom can diffuse quite large distances and its local concentration will probably be that predicted by equilibrium solid solubility. However, there has been much recent interest (Fe 79, Wh 80a) in the use of transient sources of heat to rapidly heat the surfaces of semiconductors. Pulsed lasers are commonly used.

An interesting facet of laser annealing is that implanted atoms can be trapped on lattice sites at concentrations far above those predicted by equilibrium solid solubilities. Table I shows the data of While *et al.* (Wh 80b), where various implants in Si were irradiated with a Q-switched ruby laser (wavelength of 0.694 μm) at an energy density of 1.5 J/cm^2 and pulse duration of 15 ns. It can be seen, for example, that the solid solubility of Bi is exceeded by a factor of 500.

There is quite good physical understanding of this trapping process. The surface is melted to a depth of 3000 Å or so and then epitaxially recrystallizes with a velocity of a few meters per second [this velocity should be compared with the (100) solid phase velocity of 1.5×10^{-10} m/s at a temperature of 550°C]. The dopants diffusing in the liquid Si are confronted

Table 1. Solubility Limits[a]

Dopant	Equilibrium (cm^{-3})	Laser annealing (cm^{-3})
As	1.5×10^{21}	6×10^{21}
Sb	7×10^{19}	1.3×10^{21}
Ga	4.5×10^{19}	4.5×10^{20}
In	8×10^{17}	1×10^{20}
Bi	8×10^{17}	4×10^{20}

[a] After (Wh 80b).

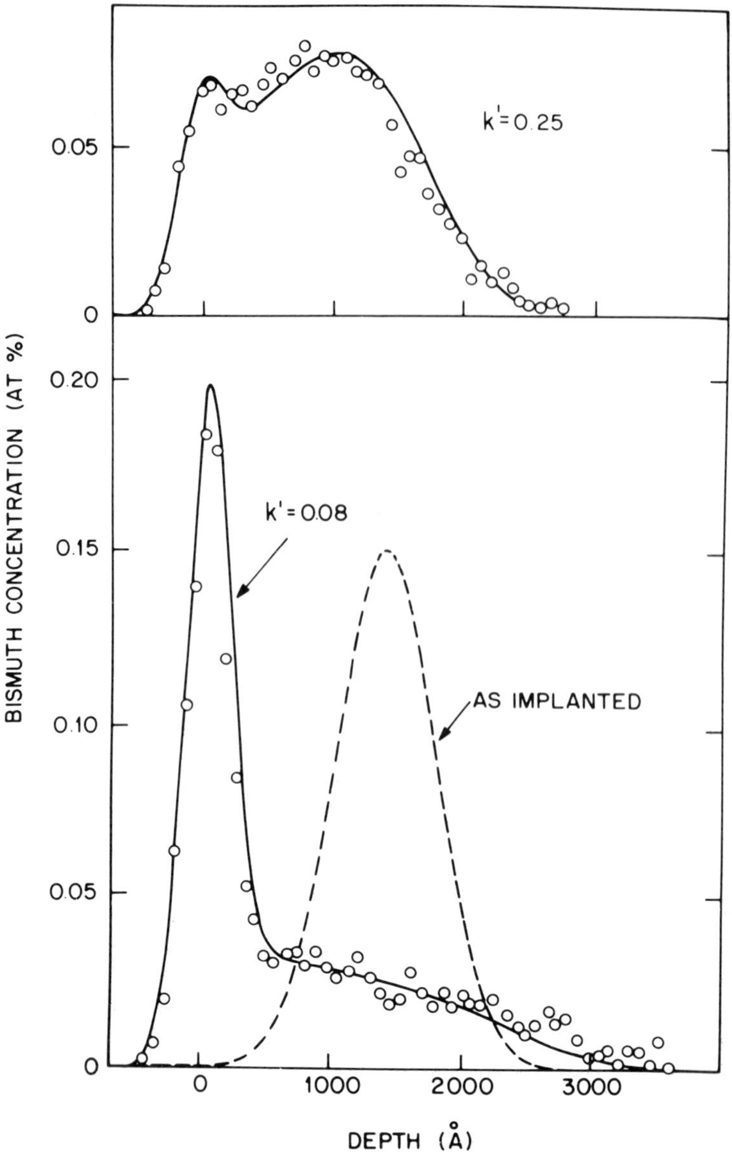

Figure 18. Bi depth profiles for irradiation of (111) Si. Velocity of regrowth in upper part is 2.3m/s and 0.8m/s in lower. Smooth curves are k' fits. From Baeri *et al.* (Ba 81).

by the rapidly moving interface, and if their diffusive velocity, in the liquid, is less than the interface velocity they will be trapped. We (Ba 81) have demonstrated the physical basis of this argument by changing the velocity of the liquid–solid interface. The lower part of Figure 18 shows Rutherford backscattering spectra of as-implanted Bi in (111) Si and after liquid-phase regrowth at a velocity of 0.8 m/s. Some Bi is trapped in the lattice but a good fraction is zone-refined to the surface. When the velocity is increased to 2.3 m/s, upper part of the figure, most of the Bi is trapped within the lattice. The smooth lines represent fits to the data assuming unique interface segregation coefficients, k'.

Figure 19 shows the segregation coefficients as a function of velocity and orientation for Bi in (100) and (111) Si. Various models are now being developed (Gi 81) to explain these data. It is hoped that those studies of impurity segregation and trapping will lead to a better understanding of liquid-phase epitaxy as the amorphous regrowth studies have led to a better understanding of Si solid-phase epitaxy. One interesting aspect of the laser work is that it has stimulated interest in supersaturation in the solid phase

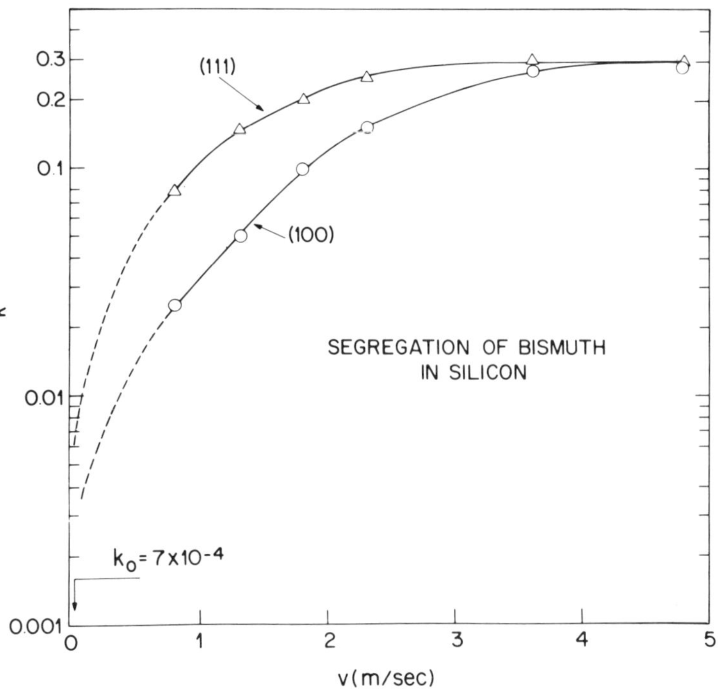

Figure 19. Bi interface segregation coefficients in Si as a function of velocity and (100) and (111) orientation. From Baeri *et al.* (Ba 81).

(Wi 81). This supersaturation can be accomplished by heating in the solid phase at elevated temperatures but for very short times so that diffusion is inhibited.

5. Ion Beam Mixing

An exciting development in ion-beam and thin-film science in recent years has been the reaction and interdiffusion of thin film structures by energetic ion beams. The emergence of the ion-beam mixing techniques is largely due to J. W. Mayer and colleagues at Caltech (Ts 80a, b, Ma 81). The concept behind this technique is to deposit the materials of interest in the form of thin films and then to react the films using the collision cascades generated by a heavy ion. It turns out in many cases that this process is remarkably efficient and the number of mixed atoms can greatly exceed the number of bombarding atoms. In this technique concentrations are not limited by sputtering and one of the fundamental limitations of implantation

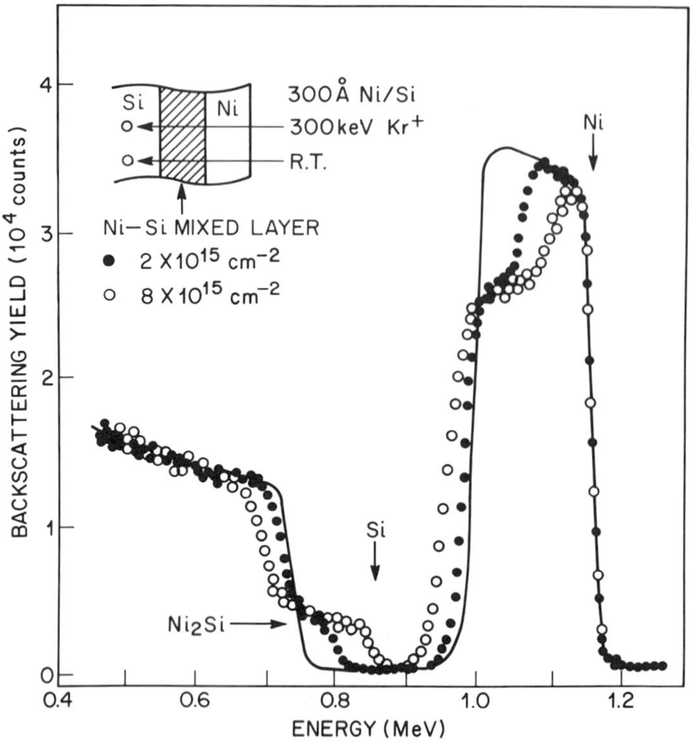

Figure 20. Formation of Ni_2Si by ion beam mixing. From Tsaur (Ts 80b).

can, therefore, be circumvented. One of the first demonstrations of the ion beam mixing phenomenon was the formation of Pd and Pt silicides (vdW 74, Po 74a). More recent measurements by Mayer and colleagues on the silicide reactions have helped elucidate some of the mechanisms of ion beam mixing. Figure 20 shows the backscattering spectra of the reacted zone progressing through a Ni film evaporated on Si. The reacted zone consists, in fact, of the Ni_2Si phase which is the first phase to form under usual thermal annealing. The thickness of the silicide phase increases with the square root of the ion dose.

The potentialities of the technique are demonstrated by the reaction of Ag–Cu films (Ts 80b). Figure 21 shows the backscattering spectra of a series

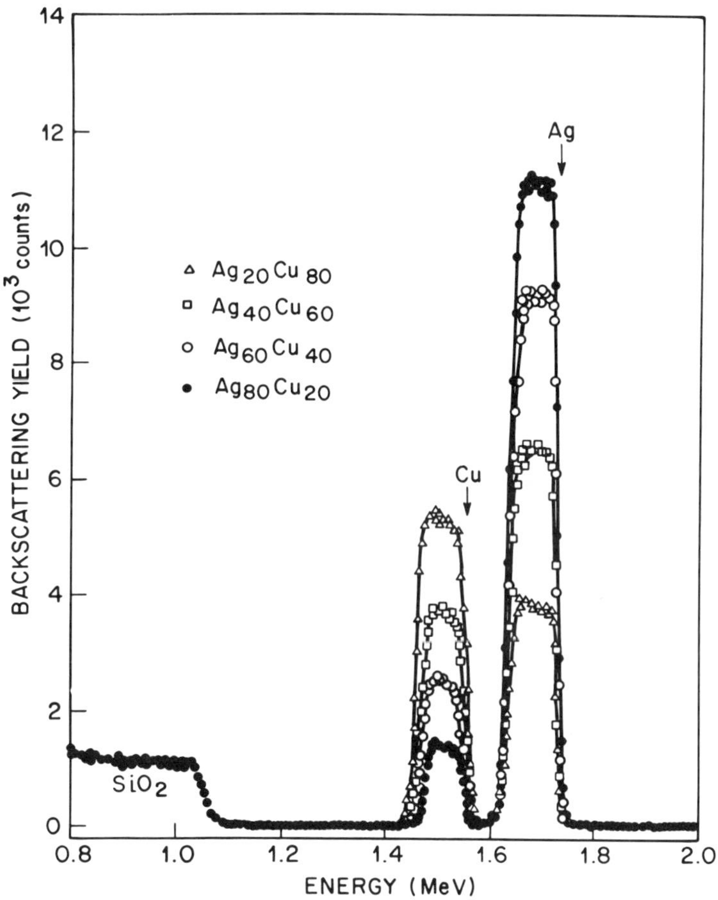

Figure 21. Rutherford backscattering spectra of metastable solid solutions of Ag–Cu (approximately 1000 Å in thickness) formed by ion beam mixing with 300-keV Xe beam at a dose of 2×10^{15} cm^{-2}. From Tsaur *et al.* (Ts 80b).

of Ag–Cu alloys produced by ion beam mixing. Multiple layers of Ag and Cu (of approximate thickness 200 Å) were first deposited on SiO_2 and the thicknesses adjusted so that the average film composition varied between $Ag_{20}Cu_{80}$ and $Ag_{80}Cu_{20}$. The films were then mixed with a 300-keV Xe at a dose of $2 \times 10^{15}/cm^2$. During bombardment the sample was held at room or liquid nitrogen temperatures. X-ray examination of the films following bombardment showed a single phase fcc structure without any elemental Ag or Cu reflections. The lattice parameter of this metastable solid solution is shown in Figure 22. The maximum deviation from linearity (Vegard's law for ideal solid solutions) is about 1%. These measurements of the lattice parameters of the Ag–Cu solutions produced by mixing agree with the splat cooling results of Duwez *et al.* (Du 60). It is remarkable that this single-phase metastable solid solution was produced by bombardment at liquid nitrogen temperatures. Moreover, a concentrated 1000-Å thick structure was produced by a bombardment dose of only 2×10^{15} Xe/cm^2. This result should be compared with the implantation of Ag into Cu, discussed previously, where the maximum attainable Ag concentration was 16.6 at.% over a depth of some 500 Å. To achieve this concentration, Ag doses in excess of $10^{17}/cm^2$ were used.

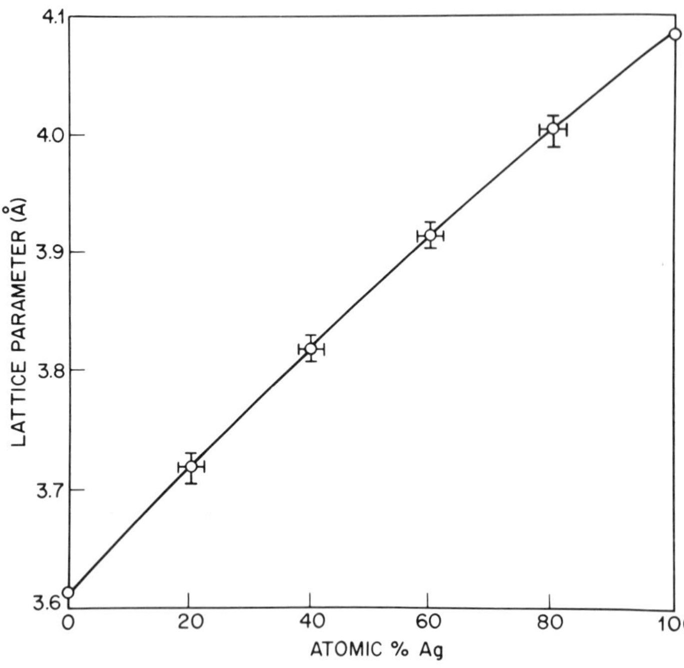

Figure 22. Lattice parameters as function of composition for Ag–Cu solid solution formed by ion beam mixing. From Tsaur *et al.* (Ts 80b).

The efficiency of the mixing process on the density of the collision cascade (i.e., nuclear stopping) is illustrated in Figure 23 which shows the number (Q_{Si}) of intermixed Si atoms cm^{-2} versus the square root of Ar, Kr, or Xe ion dose for 450-Å Pt films on Si for 300-keV bombardment of LN_2 temperatures. The square root dependence suggests a diffusionlike process. Mayer *et al.* (Ma 81) argue that as dose is proportional to time, an effective diffusion coefficient ($D_{EFF} \propto Q^2/t$) can be defined which is proportional to the square of the slope for each straight line shown in Figure 23. The ratio of the D_{EFF}'s for Xe, Kr, and Ar respectively is 7.3 : 4.4 : 1, which is very close to the ratio of the average nuclear energy loss.

The amount of mixing or reaction as a function of temperature for Cr films on Si is shown in Figure 24. The mixing is fairly insensitive to temperature below room temperature. But above room temperature the reaction increases very rapidly with temperature. This division into athermal and temperature activated regimes can be explained in terms of the different driving forces. In the athermal regime the mixing is dominated by the dynamic collision processes such as recoil mixing. The dynamic processes depend mainly on the efficiency of the nuclear stopping (i.e., on ion energy and mass) and should be fairly insensitive to temperature. In the high-temperature regime, however, chemical driving forces and enhanced diffusion

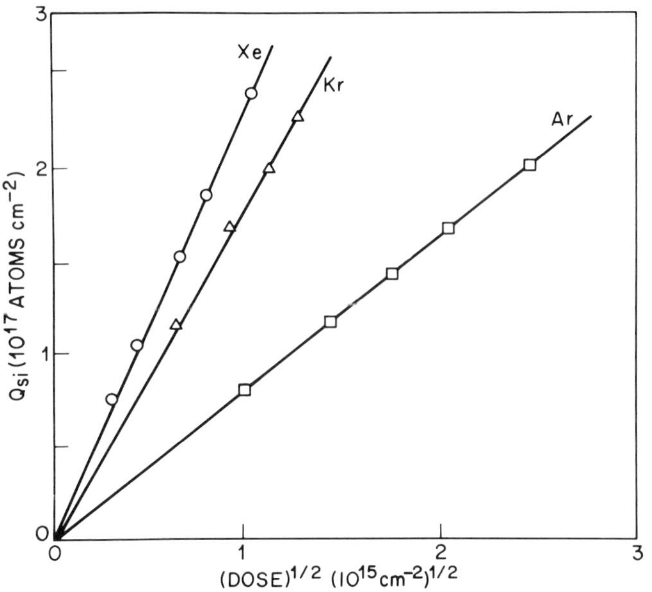

Figure 23. Number of intermixed Si atoms, Q, as a function of the square root of Ar, Kr, or Xe ion dose. From Tsaur (Ts 80b).

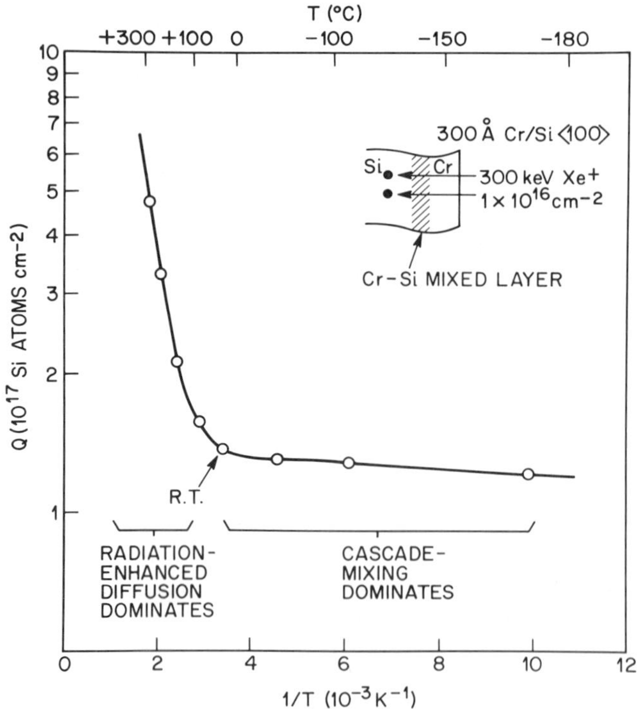

Figure 24. The number of Si atoms, Q, contained in the Cr–Si mixed layers versus reciprocal implantation temperature. From Mayer *et al.* (Ma 81).

become dominant. Discrete phases can thus be formed. The interaction between the cascade processes and temperature of the local solid-state environment are clearly synergetic.

Novel metastable phases have been observed in the Au–Si, Au–Ge, and Pt–Si systems. At present it does not appear possible to predict from the equilibrium phase diagrams which systems will show metastable phase formation. The large concentration gradients, however, impose quite different thermodynamic driving forces than those present in implantation.

Ion beam mixing is a powerful new tool for the formation of metastable and equilibrium thin film structures. Compositions can be controlled accurately by depositing films of different thickness. The mechanisms of the mixing process are not fully understood, but the fact that mixing essentially occurs in the solid phase means that the process is very controllable. In some ways the technique is more versatile than either direct high dose implantation or the melt–quench techniques.

References

(Al 74) R. B. Alexander, P. T. Callaghan, and J. M. Poate, *Phys. Rev. B* **9**:3022 (1974).

(An 74) H. H. Andersen, in *Physics of Ionized Gases 1974* (V. Vujnovic, ed.). Institute of Physics, Univ. Zagreb, Yugoslavia (1974).

(Ba 81) P. Baeri, G. Foti, J. M. Poate, S. U. Campisano, and A. G. Cullis, *Appl. Phys. Lett.* **38**:800 (1981).

(Bo 76) J. A. Borders and J. M. Poate, *Phys. Rev. B* **13**:969 (1976).

(Br 54) J. A. Brinkman, *J. Appl. Phys.* **25**:961 (1954).

(Br 76) D. K. Brice, in (Ca 76), p. 334 (1976).

(Ca 76) G. Carter, J. J. Colligon, and W. A. Grant (eds.), *Applications of Ion Beams to Materials, 1975*, Conf. Series No. 28, Institute of Physics, London (1976).

(Ch 77) F. Chernow, J. A. Borders, and D. K. Brice (eds.), *Ion Implantation in Semiconductors, 1976*, Plenum Press, New York (1977).

(Cs 75) L. Csepregi, J. W. Mayer, and T. W. Sigmon, *Phys. Lett. A* **54**:157 (1975).

(Cs 76) L. Csepregi, J. W. Mayer, and T. W. Sigmon, *Appl. Phys. Lett.* **29**:92 (1976).

(Cu 76) A. G. Cullis, J. M. Poate, and J. A. Borders, *Appl. Phys. Lett.* **28**:316 (1976).

(Cu 77) A. G. Cullis, J. A. Borders, J. K. Hirvonen, and J. M. Poate, *Inst. Phys. Conf. Ser.* **36**:181 (1977).

(Cu 78) A. G. Cullis, J. A. Borders, J. K. Hirvonen, and J. M. Poate, *Phil. Mag.* **37**:615 (1978).

(De 73) G. Dearnaley, J. H. Freeman, R. S. Nelson, and J. Stephen, *Ion Implantation*, North-Holland, Amsterdam (1973).

(deW 74) H. de Waard and L. C. Feldman, in *Applications of Ion Beams to Metals* (S. T. Picraux, E. P. EerNisse, and F. L. Vook, eds.), Plenum Press, New York (1974), p. 317.

(Di 57) C. J. Dicncs and G. H. Vineyard, *Radiation Effects in Solids*, Interscience Publishers, London (1957).

(Du 60) P. Duwez, R. H. Willens, and W. Klement, Jr., *J. Appl. Phys.* **32**:1136 (1960).

(Du 76) P. Duwez, *Ann. Rev. Mater. Sci.* **6**, 83 (1976).

(Ei 71) F. Eisen and L. T. Chadderton (eds.), *Ion Implantation*, Gordon and Breach, New York (1971).

(Fe 79) S. D. Ferris, H. J. Leamy, and J. M. Poate, eds., *Laser–Solid Interactions and Laser Processing—1978*, AIP Conf. Proc. No. 50 (1979).

(Gi 68) J. F. Gibbons, *Proc. IEEE* **56**:295 (1968).

(Gi 81) G. H. Gilmer, private communication (1981).

(Gr 78) P. J. Grundy, A. Ali, and W. A. Grant, in *Thin Film Phenomena* (J. E. E. Baglin and J. M. Poate, eds.), The Electrochemical Society Proceedings, Vol. 78-2, p. 244 (1978).

(Gr 81) W. A. Grant, *Nucl. Instrum. Methods* **182/183**:809 (1981).

(Gy 80) J. Gyulai, T. Lohner, and E. Pasztor, eds., *Proceedings International Conf. on Ion Beam Modification of Materials*, Vol. III (1980).

(Hi 80) J. K. Hirvonen (ed.), *Ion Implantation*, Vol. 18 in *Treatise on Materials Science and Technology*, Academic Press, New York (1980)

(Hu 69) W. Hume-Rothery, R. E. Smallman, and C. W. Haworth, *The Structure of Metals and Alloys*, Institute of Metals, London (1969).

(Ka 77) E. N. Kaufmann, R. Vianden, J. R. Chelikowsky, and J. C. Phillips, *Phys. Rev. Lett.* **39**:1671 (1977).

(Ka 79) E. N. Kaufmann, R. Vianden, T. E. Jackman, J. R. MacDonald, and L. C. Haggmark, *J. Phys. F* **9**:L23 (1979).

(Kl 60) W. Klement, R. H. Willens, and P. Duwez, *Nature* **187**:869 (1960).

(La 80) S. S. Lau, J. W. Mayer, and W. Tseng, in *Handbook on Semiconductors* (T. S. Moss, ed.), North-Holland, Amsterdam (1980), Chap. 3.

(Li 63) J. Lindhard, M. E. Scharff, and H. E. Schiott, *Kl. Dan. Vid. Selsk. Mat. Fys. Medd.* **13**, No. 14 (1963).

(Li 78) Z. L. Liau, J. W. Mayer, W. L. Brown, and J. M. Poate, *J. Appl. Phys.* **49**:5295 (1978).

(Li 81) Z. L. Liau and J. W. Mayer, Chap. 3 in (Hi 80) (1980).

(Ma 70) J. W. Mayer, L. Ericksson, and J. A. Davies, *Ion Implantation in Semiconductors, Silicon and Germanium*, Academic Press, New York (1970).

(Ma 81) J. W. Mayer, B. Y. Tsaur, S. S. Lau, and L. S. Hung, *Nucl. Instrum. Methods* **182/183**:1 (1981).

(Mi 73) A. R. Miedema, F. R. deBoer, and P. F. deChatel, *J. Phys. F: Metal Phys.* **3**:1558 (1973).

(Mi 76) A. R. Miedema, *Philips Tech. Rev.* **36**:217 (1976).

(Oh 52) R. S. Ohl, *Bell. Syst. Tech. J.* **31**:104 (1952).

(Po 74a) J. M. Poate and T. C. Tisone, *Appl. Phys. Lett.* **24**:391 (1974).

(Po 74b) J. M. Poate, W. J. DeBonte, W. M. Augustyniak, and J. A. Borders, *Appl. Phys. Lett.* **25**:698 (1974).

(Po 76) J. M. Poate, W. L. Brown, R. Homer, W. M. Augustyniak, J. W. Mayer, K. N. Tu, and W. F. van der Weg, *Nucl. Instrum. Methods* **132**:345 (1976).

(Po 77) J. M. Poate, J. A. Borders, A. G. Cullis, and J. K. Hirvonen, *Appl. Phys. Lett.* **30**:365 (1977).

(Po 80) J. M. Poate and A. G. Cullis, Chap. 4 in (Hi 80) (1980).

(Pr 80) C. M. Preece and J. K. Hirvonen, *Ion Implantation Metallurgy*, A.I.M.E. (1980).

(Se 56) F. Seitz and J. S. Koehler, *Solid State Phys.* **2**:305 (1956).

(Sh 54) W. Shockley, U. S. Patent 2787564.

(Si 69) P. Sigmund, *Phys. Rev.* **184**:383 (1969).

(Si 74) P. Sigmund, *Appl. Phys. Lett.* **25**:169 (1974).

(So 78) D. K. Sood and G. Dearnaley, *Radiat. Eff.* **39**:157 (1978).

(Sp 78) F. Spaepen, *Acta Mat.* **26**:1167 (1978).

(Sp 79) F. Spaepen and D. Turnbull in (Fe 79), p. 73 (1979).

(Th 69) M. W. Thompson, *Defects and Radiation Damage in Metals*, Cambridge University Press, London (1969).

(Ts 80a) B. Y. Tsaur, in *Thin Film Interfaces and Interactions* J. E. E. Baglin and J. M. Poate, eds.), E.C.S. Proceedings Vol. 80-2, p. 205 (1980).

(Ts 80b) B. Y. Tsaur, Ph.D. thesis California Institute of Technology (1980).

(vdW 74) W. F. van der Weg, D. G. Sigurd, and J. W. Mayer, in *Applications of Ion Beams to Metals* (S. T. Picraux, E. P. EerNisse, and F. L. Vook, eds.), Plenum Press, New York (1974), p. 209.

(Vi 80) R. Vianden, E. N. Kaufmann, and J. W. Rodgers, *Phys. Rev. B* **22**:63 (1980).

(Wh 80a) C. W. White and P. S. Peercy, eds., *Laser and Electron Beam Processing of Materials*, Academic Press, New York (1980).

(Wh 80b) C. W. White, S. R. Wilson, B. R. Appleton, and F. W. Young, Jr., *J. Appl. Phys.* **51**:738 (1980).

(Wi 81) J. S. Williams and R. G. Elliman, *Nucl. Instrum. Methods* **182/183**:389 (1981).

5

SHELDON DATZ (top) is currently a Distinguished Research Scientist and Section Chief for Atomic Physics at the Oak Ridge National Laboratory. He received his B.S. and M.S. degrees from Columbia University and his Ph.D. from the University of Tennessee in 1960. He has been a member of the Oak Ridge staff since 1951 and during this time has held guest appointments at the FOM Institute for Atomic and Molecular Physics in Amsterdam, the Physics Institute of Aarhus University, the Max Planck Institute for Plasma Physics at Garching, and the Physics Department and James Franck Institute of the University of Chicago.

CHARLES D. MOAK (bottom) is a Senior Scientist at the Oak Ridge National Laboratory. Both his undergraduate and graduate education were received at the University of Tennessee, from which he received the Ph.D. degree in 1954. He has been a member of the Oak Ridge staff since 1944 and during that time has held guest positions at AERE, Harwell, the Physics Institute of Aarhus University, and the Daresbury Laboratory. He holds an honorary doctorate from the University of the Witwatersrand.

Heavy-Ion Channeling

SHELDON DATZ AND CHARLES D. MOAK

1. Introduction

Energetic heavy ions penetrating random solids undergo a rapid series of collisions which degrades their energy and causes excitation and ionization of both the ion and the atoms in the medium. Because of the chaotic nature of the collisions and the short time between relatively violent events it is difficult to study isolated collisions in dense media or to determine the states of ions as they penetrate solids.

Since 1962 studies of directional effects ("channeling") on, e.g., ion trajectories, ranges, and energy losses have done much to elucidate the detailed nature of the processes experienced by the heavy ion. That directional effects should exist can easily be visualized from the lattice model pictured in Figure 1. If a crystal is viewed along a low-index direction, the lattice appears to be highly transparent. For the face-centered cubic lattice of Figure 1, the most transparent direction is $\langle 101 \rangle$, the face diagonal on the front of Figure 1b. The second best direction is $\langle 010 \rangle$. Figure 1a, a perspective view of the crystal along this axis, shows the $\langle 010 \rangle$ atomic rows. Four neighboring rows form an axial $\langle 010 \rangle$ channel. Rotating the crystal of Figure 1a about the $\langle 100 \rangle$ axis leads to configurations of the type shown in Figure 1c. The transparency does not disappear, because open planes are maintained between the densely packed sheets of atoms.

The degree of "transparency" depends on the radii of the spheres representing the lattice atoms, which in turn depend on the physical effect under investigation. The interaction between an energetic particle and each

SHELDON DATZ AND CHARLES D. MOAK • Physics Division, Oak Ridge National Laboratory, P.O. Box X, Oak Ridge, Tennessee 37831. Research sponsored by the U. S. Department of Energy, Division of Basic Energy Sciences under contract No. W-7405-eng-26 with Union Carbide Corporation.

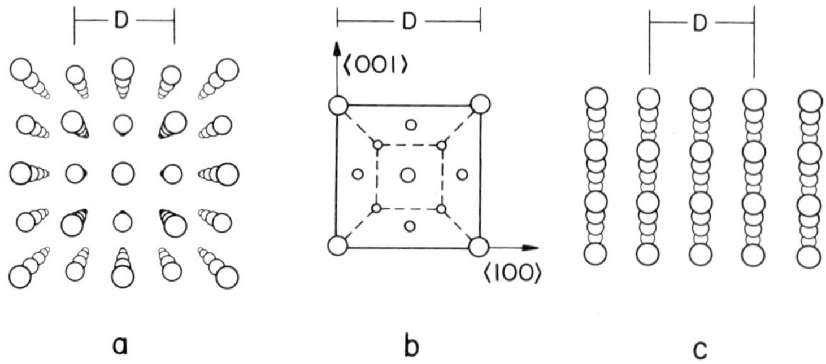

Figure 1. (a) Perspective view of a *fcc* lattice along ⟨010⟩. (b) Elementary cube of a *fcc* lattice. (c) Rotating the crystal about the ⟨100⟩ direction shows (100) planar channels.

lattice atom can be represented by a repulsive potential, and the hard-sphere radius may be defined by the distance of closest approach for a head-on collision. Since this distance decreases with increasing energy, an incoming particle of higher energy sees an emptier lattice.

For crystals of finite thickness, the angular aperture for transparency is vanishingly small, and one would not expect that any appreciable effects would be observed in penetration experiments involving beams with measurable angular divergence. However, when energetic particles enter the lattice with small angles to either the rows or planes, they undergo a set of correlated small-angle collisions which tend to "channel" their directions so as to avoid close collisions with lattice atoms.

This is illustrated in Figure 2. Consider a projectile atom aimed for a hard collision with an atom contained in a closely packed atomic row in a crystal. The projectile will be slightly deflected by the repulsive potential of the first atom, again by the second, *etc.* The final collision will thereby be considerably softened, and, depending upon the angle with respect to the atomic *row*, it may not collide at all with the original target atom. This is the basis of channeling; i.e., within an angle with respect to an atomic row or plane, determined by the interatomic potential and spacing of atoms, the particle will be deflected by a continuum potential made by properly summing the atomic potentials. These angles can be on the order of several

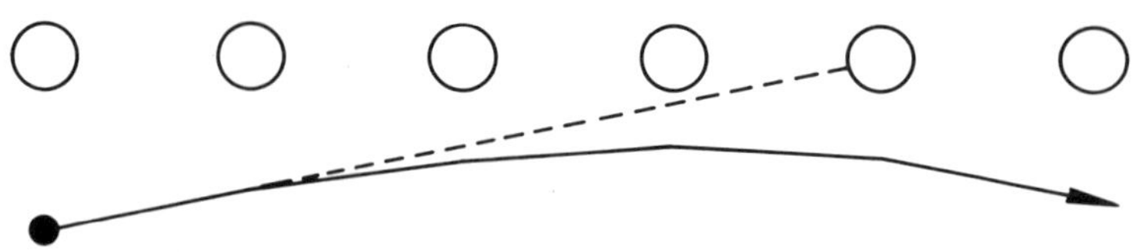

Figure 2. Effect of correlated collisions with a row of atoms in a crystal on penetrating ion trajectories ("channeling").

degrees and, as can be seen in Figure 1, a considerable part of all directions of incidence in the crystal can in fact be effective in channeling.

These effects were actually anticipated by Stark and Wendt in 1912 (St 12) and rediscovered in the computer calculations of Robinson and Oen (Ro 63) in 1963. They investigated the slowing down of 1–10-keV heavy ions in a lattice model which included realistic repulsive interaction, and obtained very large penetrations for particles with initial velocities nearly parallel to close-packed directions. Experimentally, this effect was first seen in studies on the penetration of keV ions into crystals by Piercy *et al.* (Pi 63) and by Lutz and Sizmann (Lu 63). They measured the distribution of penetration distances and found that a significant fraction of the ions incident along low-index directions had anomalously long range. At about the same time, Nelson and Thompson (Ne 61) observed anomalously high transmission of 50–75-keV protons and helium ions in low-index directions of thin gold single crystals.

In 1965, Lindhard (Li 65) demonstrated that for heavy particles ($M \gg M_e$) classical analysis is applicable to the collision series involved in channeling. Using a continuum string approximation for the potential of a row of atoms he showed that for particles entering close to an axial direction (e.g., Figure 1a) at high velocity, hard collisions with lattice atoms are avoided for an angle of incidence ψ if

$$\psi \lesssim (2Z_1 Z_2 e^2/E \cdot d)^{1/2} \tag{1}$$

for a Coulomb potential where Z_1 and Z_2 are the incident and target atomic numbers, E is the ion's energy, and d is the lattice spacing between atoms (see Figure 2). This high-energy condition is met when

$$E > 2Z_1 Z_2 e^2 d/a_{\mathrm{TF}}^2 \tag{2}$$

where a_{TF} is the Thomas–Fermi screening length. If the incident ion is fully ionized

$$a_{\mathrm{TF}} = 0.4685 Z_2^{-1/3}\ (\text{Å}) \tag{3}$$

and for a nonfully ionized ion

$$a_{\mathrm{TF}} = 0.4685(Z_1^{1/2} + Z_2^{1/2})^{-2/3}\ (\text{Å}) \tag{4}$$

For planar channeling, i.e., confinement of the ion's motion between sheets of planar density n_p (atoms cm^{-2}) (see Figure 1), the critical angle ψ_p is

$$\psi_p = (2\pi n_p Z_1 Z_2 e^2 a_{\mathrm{TF}}/E)^{1/2} \tag{5}$$

The primary result of channeling is to prevent small impact parameter collisions from occurring. Hence, such effects as the elimination of nuclear reactions, Rutherford scattering, and the reduction of inner-shell ionization are observed. The effect also leads to the elimination of nuclear stopping and restricts close electronic interactions to only valence and conduction electrons.

The results have been applied to studies of lattice disorder, surfaces and epitaxial layers, location of dopant and impurity atoms, and measurement of nuclear lifetimes.

The work done by many groups up to 1973 has been reviewed by Gemmell (Ge 74) in an article which contains more than 700 references. Work done since that time may often be found in a series *Atomic Collisions in Solids* (Da 75, Sa 76, Bu 81, Ja 80). However, most of the work reported has employed proton and helium ion beams and is concentrated on studies of the properties of the crystal being penetrated. In this chapter we confine our discussion to heavy ions ($E > 1$ MeV) and accent the study of the motion, the collisions, and the states of such ions as they move through crystal channels.

Before entering into a detailed discussion of specific characteristics of channeling it would be well first to consider briefly the set of processes experienced by a heavy ion penetrating a dilute gas, a random solid, and a crystal channel.

An ion entering a dilute gas experiences collisions at all impact parameters. If it is neutral or in a low charge state it will rapidly lose outer-shell electrons with orbital velocities v_e lower than the ion velocity v_i. Excitation of electrons with velocity $v_e \simeq v_i$ will occur with high probability, but in the absence of further collisions radiative relaxation will occur. As the ion charge increases, the electron loss cross section, σ_l, decreases and the electron capture cross section, σ_c, increases. Capture may occur to ground, excited, or continuum states. For swift highly stripped heavy ions, most capture occurs to excited states, which in a dilute gas may decay to ground level.

Considering only a single electron capture and loss, the fractional population of a given ion charge state ϕ_q after penetration of a given target thickness x will be given by

$$d\phi_q/dx = \phi_{q-1}\sigma_l(q-1) + \sigma_{q+1}\sigma_c(q+1) - \phi_q[\sigma_l(q) + \sigma_c(q)] \qquad (6)$$

Ultimately a steady state will be reached $(d\phi_q/dx) = 0$, and if σ_c increases with q at about the same rate σ_l decreases, the most probable charge state $\bar{q}$ is that for which $\sigma_c \sim \sigma_l$. Inner-shell electrons of the moving ion (i.e., those lying below the shell in which rapid charge and excitation exchange is occurring) must, by definition, have velocities $v_e > v_i$ and will in general have low cross sections for ionization by Coulomb excitation. However, if promotion through quasi-molecular orbitals (MO) is possible, relatively large cross sections for inner-shell vacancy formation may occur, and may lead to an increase in the ion charge.

The electronic stopping power of the ion will be proportional to the square of the screened nuclear charge ("effective" ion charge) and the angular straggling of the ion beam will be determined by multiple scattering at all impact parameters.

In a random solid many of the above considerations are thought to remain unchanged but additional effects enter because of the density of the medium. Binary impact parameters must always be less than half a lattice spacing. The time between collisions is short compared to the relaxation time of excited states, so that we would expect that the presence of unrelaxed excited states arising either from capture or from a collision which excited but did not ionize a bound electron would lead, in the subsequent collision, to an effectively lower capture cross section and effectively larger loss cross section. The result of these processes would be an increase in the mean equilibrium charge in the solid, and indeed charge state distributions in heavy-ion beams emerging from solids are significantly higher than those emerging from dilute gas targets. The electronic stopping power of the ion is found to be proportional to the square of an effective ionic charge but, to the dilemma of many, the effective charge adduced from stopping powers is approximately equal to that obtained from equilibrium in gases. Inner-shell ionization by quasi-molecular-orbital promotion may be enhanced because of exit channel openings due to greater ionization of outer shells, and the fluorescence yield for inner-shell relaxation may be affected by differing outer-shell configurations.

A true description of the state of the moving ion in the solid must also include its effect on its immediate environment. Conduction electrons scattered by the moving ion affect the electron density. For velocities $v_i \lesssim v_0$ the density is enhanced in the region of the nucleus of the ion and effects of this are seen, e.g., in transient field effects on hyperfine interaction studies; for velocities $v_i > v_0$, the density enhancement trails behind the ion, and troughs of lowered electron density as well follow in the wake. The details of the wake will depend on the ion charge and velocity and the plasmon spectrum of the material.

A state of equilibrium is said to be reached when all the processes of outer- and inner-shell electron capture and loss processes and ion wake formation have reached a steady state.

Upon emergence from the solid, the steady-state excitation may be relaxed either radiatively or nonradiatively by autoionization in the outer shell or by Auger processes in the inner shells. The nonradiative processes will increase the ion charge but the charge state may be decreased by capture of electrons from surface states or from electrons trapped in the ion's wake.

Capture of electrons to continuum states ["convoy electrons" (Br 82)] also occurs but appears to have considerably different characteristics from those observed with gas targets.

Channeling can considerably reduce the complexity of the multiple collision processes occurring in solids. Planar channeled ions are trapped in a one-dimensional potential (Section 2.1) and follow highly ordered trajectories. The determination of these trajectories gives direct information on the planar interatomic potentials. Trajectories of hyperchanneled ions (Section 2.2) give information on axial potentials. Crystal channeling leads to a limited range in the distance of closest approach $a/2 \geqslant r > a_{\mathrm{TF}}$, where a is the lattice constant. In this range of relatively large impact parameters some ions with $v_i \gg v_e$ can maintain a fixed charge state over significant distances (Section 4.1) and permit the determination of screening effects (Section 3.1) and higher-order Z_1 effects (Section 3.2) on stopping powers. The fact that some ions maintain a fixed charge may also be used to aid studies of radiative electron capture (Section 4.2) and the origin of convoy electrons (Section 4.3). Channeled ions also experience a periodic coherent perturbation by the lattice atoms which can lead to a resonant excitation of the penetrating particle (Section 5). The frequency spectrum leading to such excitation may then be used to adduce the detailed state of energetic ions as they penetrate solids.

2. Trajectories and Interaction Potentials

Positive ions moving in crystal channels will be subjected to an accumulation of small-angle scattering events which act to turn the path of the ion toward the center of the channel. In planar channeling the particle will execute an oscillatory motion transverse to the direction along the planes. Axial channeling is more complex because, in this case, the particle may execute transverse planar oscillations in two directions perpendicular to the direction of motion; in this case, the particles will be gently deflected away from the atomic strings which define the axial channel. If the particle has insufficient transverse energy to pass the electrostatic saddle point between two strings, it will be confined to stay within the boundary formed by a single set of strings, and this special case of axial channeling is called "hyperchanneling."

Particle trajectories are related to the electrostatic potentials through which the particle travels. The potentials are related to the spatial distribution of charges in the crystal in the vicinity of the particle path. The energy loss of the particle is related to this same distribution of charges in the crystal in the vicinity of the particle trajectory. The interrelation of trajectory, potential distribution, and particle stopping power combine to provide a method which will test various models for the potentials in the crystal.

2.1. Planar Channeling

Consider a planar channel consisting of two parallel closely packed lattice planes. In this channel a charged particle of energy E experiences a repulsive force from the atoms in each wall. We assume a smooth "planar potential," $V(x)$, where x is the distance from the midplane. The particle motion reduces to a two-dimensional problem in which the direction of incidence into the channel lies in the xz plane. The walls are assumed to be rigid. Since for channeled particles the maximum angle between particle trajectory and the plane is 1° or less, the z component of the velocity is assumed to be constant. The transverse motion is that of an undamped anharmonic oscillator. The consequences can be seen in Figure 3, which shows trajectories for three different impact parameters with respect to the channel midplane. Ions entering close to the midplane are deflected only weakly (path a); the wavelength λ of their oscillatory motion is comparatively long and their amplitude A is comparatively small. Those entering closer to the edge of the channel (paths b and c) are repelled more strongly and have paths of larger amplitude and shorter wavelength. For any realistic potential the anharmonicity decreases with decreasing amplitude so that in the region of minimum amplitude, i.e., for parallel entrance close to the center of the channel, the wavelength should be independent of amplitude. We designate this the A_0 region.

The energy loss suffered by particles in this energy range is primarily caused by inelastic collisions with electrons and, since the electron density close to the atomic planes is higher than at the channel center, the stopping power $s(x)$ increases with x. The energy loss of particle b (Figure 3) should be larger than that of a because it has penetrated the atomic planes more deeply and more often. We can therefore associate (for a given crystal thickness) an energy loss with a trajectory and a given entrance point and hence with a given wavelength and amplitude.

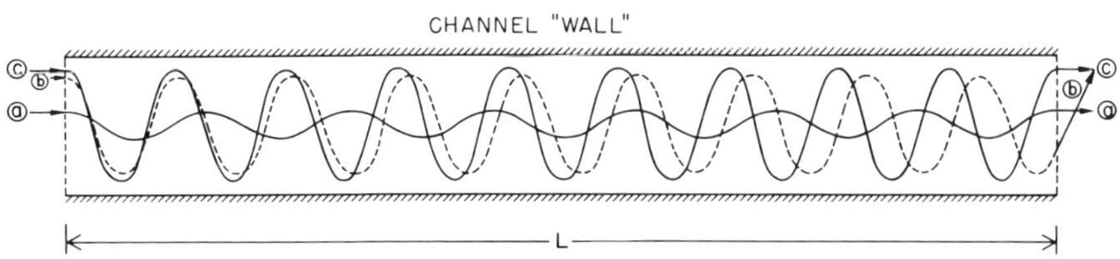

Figure 3. Trajectories of three entrance points. Particles a and c will reach a detector in line with the beam.

As the entrance point is moved from the midplane toward the channel wall, the energy loss increases, and if the energy spectrum of all the emerging particles were measured it would be structureless. However, if the detector subtends a small angular aperture, only those trajectories which leave the crystal within that angle would be detected. This introduces a boundary condition for the wavelength of the detected particles

$$\lambda_a = (L \pm z_0 \mp z_e)/n \quad (7a)$$

$$\lambda_b = (L \pm z_0 \pm z_e)/n + 1/2) \quad (7b)$$

where n is an integer, L is the length of the crystal, and z_0 and z_e are the lengths associated with the phase shifts of incidence angle ψ_0 and exit angle ψ_e. [For sinusoidal motion $z = \lambda \cos^{-1}(\lambda\psi/A)$.] An early example of a real spectrum is shown in Figure 4. Let us first consider the information obtainable from an ideal experiment in which we employ a flawless crystal of perfectly uniform thickness. If the experiment were arranged with the detector and beam in line and the planar channel aligned with the detector beam axis, i.e.,$\psi_0 = \psi_e = 0$, then $\lambda_a = L/n$, $\lambda_b = L/(n + 1/2)$. If the path length is changed, the wavelengths (and associated energy loss) of each detected group shift until at $\Delta L = L_2 - L_1 = \lambda$: $(dE/dz)_{n,L} = (dE/dz)_{(n+1),L}$.

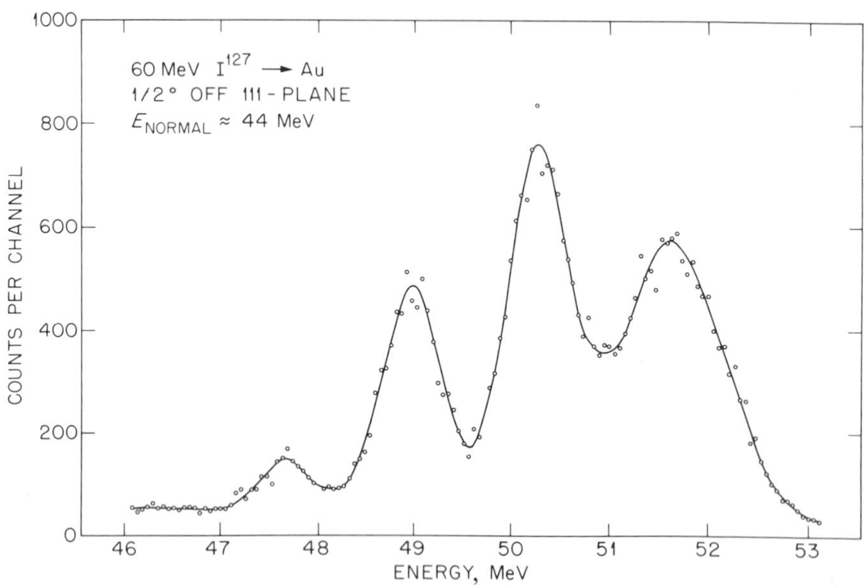

Figure 4. Energy spectrum observed in 60-MeV ^{127}I ions transmitted through {111} planar channels in a 0.7-μm-thick Au crystal. Beam incident 0.5° from the plane. Detector colinear with beam.

A measurement of the ΔL required to return to a given wavelength group is a direct measure of the wavelength, and the dependence of λ upon stopping power is established. Tilting the crystal plane with respect to the beam axis (i.e., $\psi_0 = \psi_e \neq 0$) splits the λ_b solutions, but the λ_a solutions are unchanged as long as the detector is kept on the beam axis.

The population of a given group depends on the length of the region δx_0 which leads to exit angles included within the detector aperture $\delta\psi_e$. With $\psi_0 = \psi_e = 0$ and a fixed small aperture detector, two effects combine to suppress the intensity of high-amplitude groups. First, the variation in exit angle with entrance point $\delta\psi_e/\delta x_0$ increases with both decreasing wavelength and increasing amplitude; and, second, a steeper potential gradient at larger amplitude causes a more rapid decrease in wavelength within a given x_0 zone.

The amplitude of an oscillation is a function of the transverse energy,

$$V(A) = V(x_0) + E\sin^2\psi \tag{8}$$

so that for each amplitude there is a critical angle $\psi_c(A)$ corresponding to $x_0 = 0$ beyond which the intensity at A vanishes. When the crystal is tilted, the population of a given sidegroup increases until it drops abruptly to zero at $\psi_c(A)$. Here again two effects are responsible for the increase. As ψ is increased from 0 to $\psi_c(A)$ the phase shift passes from $\frac{1}{4}\lambda$ to 0 and $d\psi/dz$ goes from maximum to minimum. The second effect again is related to the steepness of the potential. As ψ is increased the entrance point x_0 for a given A is moved to smaller values and since $\partial V(x_0)/\partial x_0$ decreases with x, a larger range of x_0 values will be included in $\delta V(A)$. The width of the energy-loss group is related to the gradient of the stopping power function within the δA region contributing to the peak.

If the functional form of the trajectory is known, a measurement of the curoff angle $\psi_c(A)$ for a given wavelength group would give the amplitude, and the exact path is determined. For example, if the wave were sinusoidal, we should have

$$A = \lambda\psi_c(A) \tag{9}$$

Even if the functional form is not known, it could be measured by an experiment in which the detection angle (ψ_e) is changed to compensate for a thickness change which deflects the emergence angle for a given dE/dz group; e.g., start with $L = L_1$ and $\psi_0 = \psi_e = 0$ and focus attention on one dE/dz group. Now if the pathlength is changed slightly, the group emergence angle also changes. By rotating the detector to find the emergence angle, it could be followed up to $\psi_c(A)$. This procedure determines the detailed trajectory shape over a quarter wavelength.

In practice, the spectra are strongly affected by even small amounts of mosaic spread in the specimens used. The principal effect is to eliminate structure attributable to the half-integral groups [Equation (7b)]. For example, when the detector is in line with the beam direction, $\psi_0 = \psi_e$ and $z_0 = z_e$. The phase factors for the integral solutions cancel and the structure is unaffected by a spread in ψ; but for the half-integral solutions the phase factors add, and slight variations in ψ cause drastic changes in the transmitted wavelength. Mosaic spread (variation in ψ) is equivalent to variation in entrance and exit points. For the integral solutions, a shift in entrance angle is exactly compensated by a shift in exit angle. On the other hand, with the half-integral solutions, a decrease in entrance angle causes an increase in exit angle, and since $\psi_c(A) < 0.5°$, a distribution in ψ of greater than $0.1°$ can completely eliminate the observation of structure attributable to half-integral solutions.

As we have seen, the path length through a given planar channel can be varied within a single specimen by simply varying the tilt angle of the crystal with respect to the beam axis. This method (Lu 66, Da 69, Da 70) is exemplified in Figure 5, where we have plotted $\Delta E/\Delta z$ spectra for ^{127}I ions channeled in a $\{111\}$ plane for several different thicknesses. (These spectra were taken at various rotation angles with respect to the plane and no significance should be attached to the relative group populations.) The position marked A_0 corresponds to the minimum-energy-loss group. The particles in this group have been confined to trajectories near the channel midplane where the potential is almost harmonic. The definition of the A_0 group depends on the detector aperture $\delta\psi_e$; e.g., for $\psi_0 = \psi_e = 0$ all amplitudes having $\psi_c(A) \leqslant \delta\psi_e$ will be detected without regard to integral wavelength conditions. Thus, in the 1-μm wavelength region for $\psi_e \cong 0.001$ rad, all particles with $A < 0.15$ Å will be detected [see Equation (9)]. The peak position of the A_0 group should not vary, and the upper right-hand leading edge of this group represents the energy loss for particle paths very close to the midplane. The peaks of the intermediate groups move with changing pathlength as predicted by Equation (7a).

To the degree that the parallel and transverse energies are separable, there are two effects of energy loss on the path. First, since the particle is oscillating in a transverse potential with a fixed frequency, a decrease in longitudinal velocity will decrease the wavelength. On the other hand, a loss in transverse energy will decrease the amplitude and increase the wavelength. A procedure which avoids the effect of wavelength shrinkage involves the use of the frequency of transverse oscillation. The frequency does not vary as the particle velocity changes (Ro 69).

A schematic drawing of the experimental arrangement for recording data such as those shown in Figure 5 is shown in Figure 6. The experimental data consist of observations of the energies, E, of ions of incident energy E_0, transmitted through a path length z, in the thin crystal target. From these, it

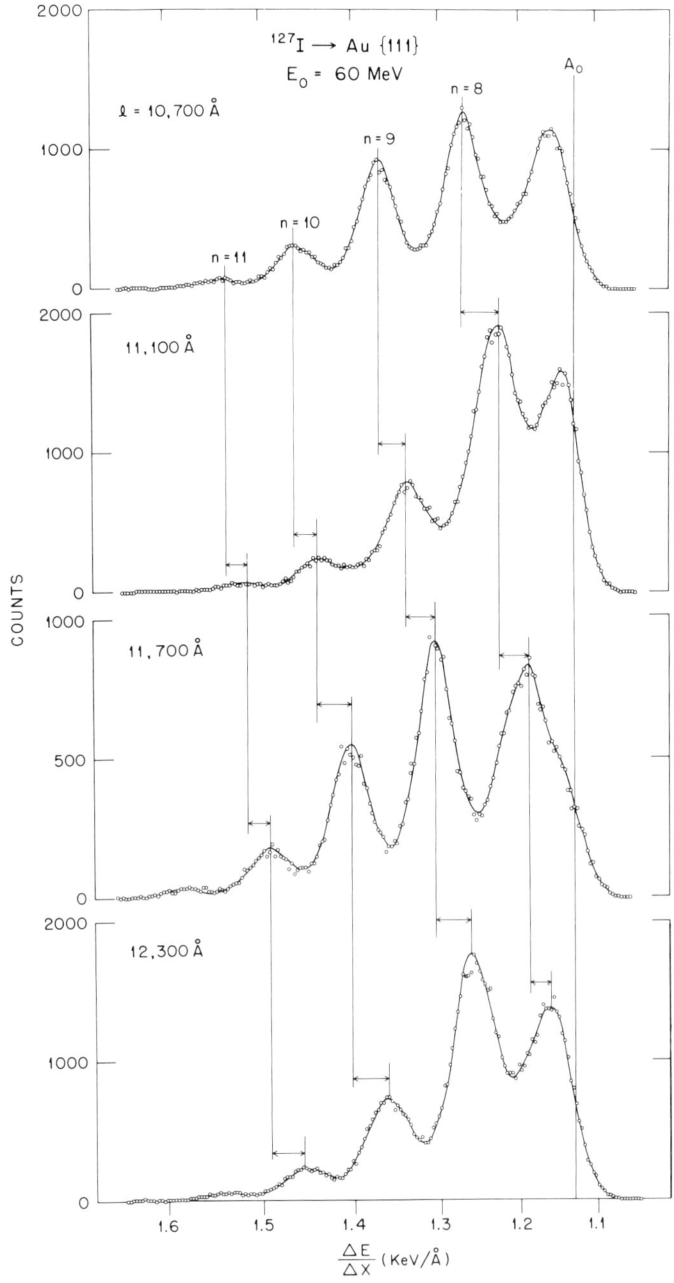

Figure 5. Energy spectra observed in ^{127}I ions transmitted through the {111} channels of a Au crystal as a function of pathlength.

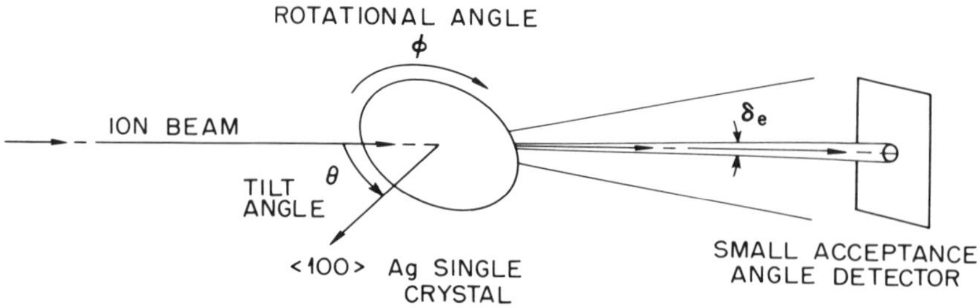

Figure 6. Schematic drawing of experimental arrangement.

is desired to deduce the frequency with which the ions oscillate in the channel and the stopping power, averaged over the oscillatory motion, but corrected to the incident ion energy. The latter quantity will be termed the mean initial stopping power. Assuming it to vary with energy as E^p, the initial stopping power is

$$\left(-\frac{dE}{dz}\right)_{E=E_0} = \begin{cases} \dfrac{E_0^p}{(1-p)z}(E_0^{1-p} - E^{1-p}), & p \neq 1 \\ \dfrac{E_0}{z}\ln\dfrac{E_0}{E}, & p = 1 \end{cases} \tag{10}$$

and the channel oscillation "frequency" is

$$\omega = \begin{cases} \dfrac{n(1-2p)(E_0^{1-p} - E^{1-p})}{(1-p)(E_0^{1/2-p} - E^{1/2-p})}, & p \neq 1/2, 1 \\ \dfrac{4n}{z}\dfrac{(E_0^{1/2} - E^{1/2})}{\ln(E_0/E)}, & p = 1/2 \\ \dfrac{n\ln(E_0/E)}{z(E^{-1/2} - E_0^{-1/2})}, & p = 1 \end{cases} \tag{11}$$

where n is the (integral) number of oscillations which the ion makes in passing through the channel. Note that ω is not a true frequency, since a factor $(2m)^{1/2}$, where m is the mass of the ion, has been omitted from its definition. The values of n may be estimated from the data by noting the difference in path length between successive appearances of particular values of the stopping power (Da 69, Ro 69). The wavelengths of the oscillations, though not constant, are approximately $2E_0^{1/2}/\omega$; the values observed in the experiments range from 0.5 to 0.18 μm, depending on the ion, its energy, and its stopping power.

In Figure 7, the initial stopping power is plotted as a function of

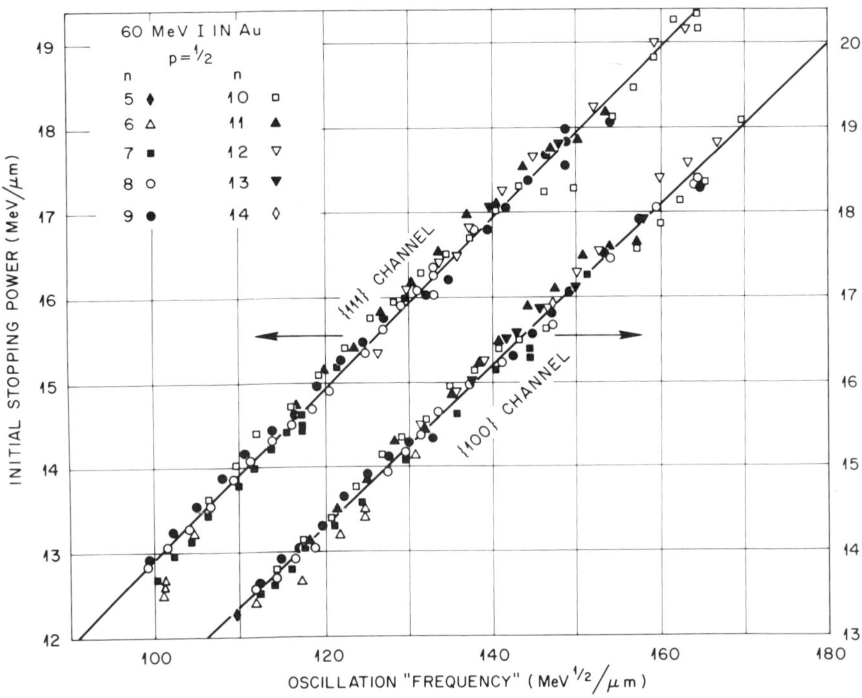

Figure 7. The dependence of the stopping power of 60-MeV ^{127}I ions on their transverse oscillation frequencies in two planar channels of Au.

channel oscillation frequency for 60-MeV ^{127}I ions in two planar channels. In Figure 8 15- and 60-MeV ^{127}I ions are compared for the Au {111} channel. Figures 9 and 10 display the results obtained for 21.6-MeV ^{127}I and 10-MeV ^{16}O ions, respectively, transmitted through Au {111} and {100} channels. In all cases, the initial stopping power is accurately proportional to the channel oscillation frequency. This proportionality is quite general and applies to planar channeling in gold, irrespective of the ion or of its energy. In every case, then, the initial stopping power is well represented by the straight line

$$(-dE/dz)_{E=E_0} = \alpha + \beta\omega \tag{12}$$

where the parameters α and β and their uncertainties can be evaluated from the data by the method of least squares. The values obtained for these parameters depend rather strongly on the value chosen for the stopping power energy exponent p. However, as long as Equation (12) is correct, the curvature of the planar channel potential is given by

$$y = 2\pi^2(s_0 - \alpha)^2/\beta^2 l \tag{13}$$

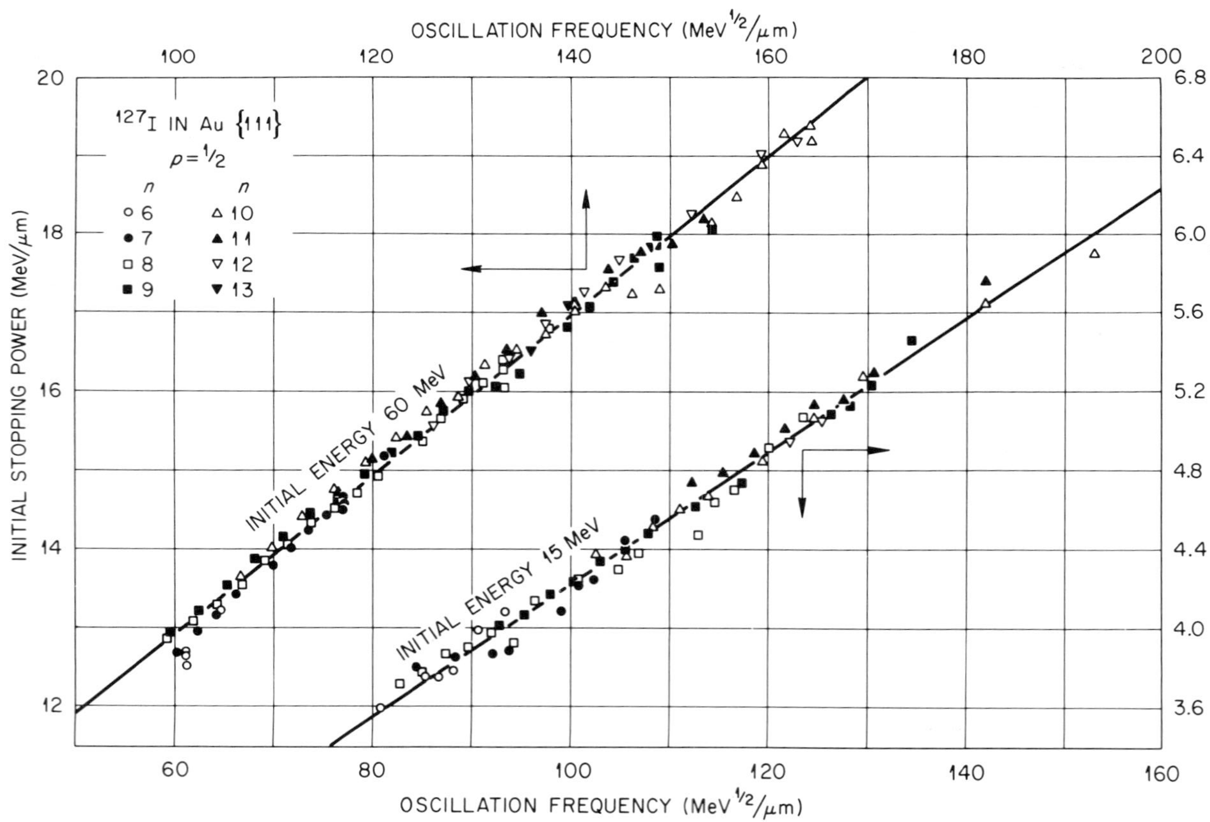

Figure 8. Comparison of stopping powers versus frequency in the {111} channels of Au for 60- and 15-MeV ^{127}I ions.

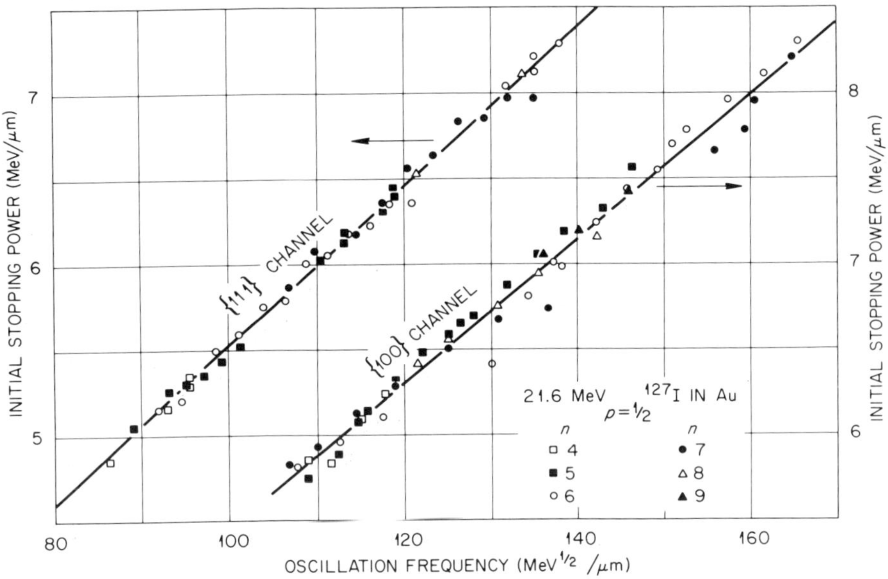

Figure 9. 21.6-MeV ^{127}I ions in the {100} and {111} channels of Au.

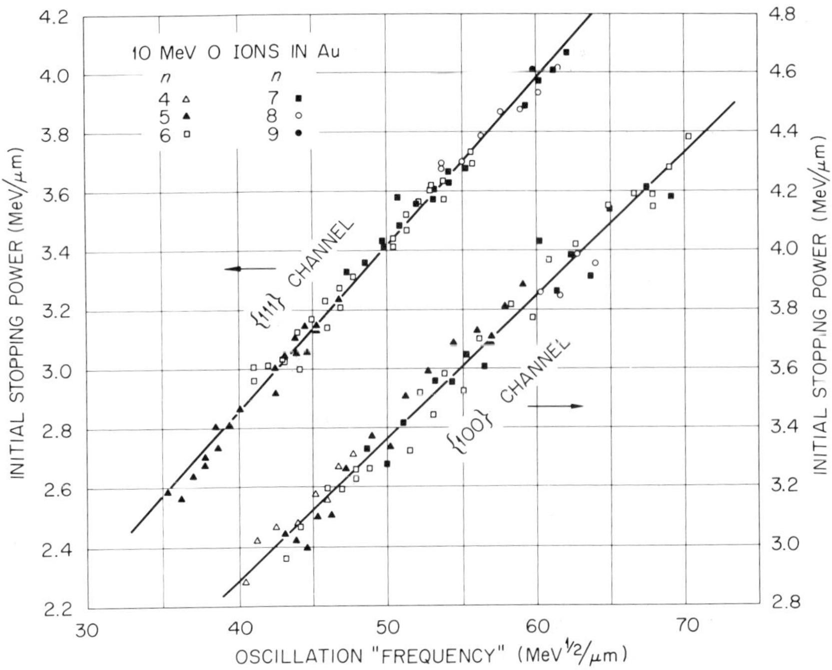

Figure 10. 10-MeV O ions in the {100} and {111} channels of Au.

where l is the half-width of the channel and s_0 is the stopping power at the center of the channel, determined experimentally from the high-energy edge of the energy-loss spectrum. According to the theory, the value of y is

$$y = V_2''(0)/l \tag{14}$$

where $V_2(x)$ is the planar channel potential, the primes represent differentiation with respect to x, and x is the displacement of the ion from the channel center. The experimental curvature parameters are listed in Table 1 for each combination of ion, energy, and channel, using all data reported before (Ro 69). The values of y are strongly dependent upon the ion and its energy, showing that the planar channel potential must also depend on these variables. However, the ratio of the values for the two channels appears to be essentially independent of the ion and its energy. If the interaction potential between the channeled ion and the individual lattice atoms is of the screened Coulomb type, then the planar channel potential may be written as

$$V_2(x) = 4\pi\rho l Z_1 Z_2 e^2 \left[\int_{l-x}^{\infty} \phi(br)\, dr + \int_{l+x}^{\infty} \phi(br)\, dr \right] \tag{15}$$

where ρ is the atomic density of the target, $Z_1 e$ and $Z_2 e$ are the (effective) nuclear charges of the two particles, $\phi(br)$ is the screening function, and b is a constant. Combining Equations (14) and (15),

$$y = -8\pi\rho Z_1 Z_2 e^2 b\phi'(bl) \tag{16}$$

and the ratio of the values for two channels determines the value of the screening constant, b. The constancy of the ratio y_{111}/y_{100} in Table 1 implies a similar constancy of b, which is thus shown to be a property of the gold crystal. Physically, the implication is that the nuclear charge of the ion is

Table 1. Empirical Charge Products and Ionic Charges for Ions Moving in the Planar Channels of Gold Crystals

Ion	Energy (MeV)	Charge product, $Z_1 Z_2$ {111}	Charge product, $Z_1 Z_2$ {100}	Ionic charge $q_1 = Z_1 Z_2/79$	Equilibrium ionic charge[a]
^{4}He	3	118 ± 19	120 ± 17	1.5 ± 0.2	2
^{16}O	10	285 ± 47	281 ± 48	3.6 ± 0.6	5.5
^{127}I	15	1540 ± 260	—	19 ± 3	13
	21.6	1690 ± 280	1770 ± 260	22 ± 3	15
	60	2370 ± 380	2300 ± 310	29 ± 4	22

[a] From (Mo 68) and (Da 72).

always screened by the same number of electrons, whatever its position in the channel. As it approaches a target atom, it never gets so close that this screening is reduced significantly. Thus, it acts always as a simple charged particle which does not much perturb the medium. That is, all three ions may be considered as test charges.

In the last column of Table 1, screening constants are listed on the assumption that the screening function is

$$\phi(br) = 0.35 \exp(-br) \tag{17}$$

that is, the long-range term of the Molière potential (Mo 47). The use of this screening function, which yields $V_2(x) \sim \cosh bx$, has been justified previously (Ro 69). Accepting the mean screening constant of Table 1, the empirical values of the charge product Z_1Z_2 may be deduced. These are listed in Table 2. If the ions are indeed acting as test charges, then the ionic charge would be

$$q_1 = Z_1Z_2/79 \tag{18}$$

values of which are also shown in Table 2. These may be compared with the equilibrium charge states observed for ^{127}I ions (Mo 68) and ^{16}O ion (Da 72); for ^{4}He, the α-particle charge is expected. For the two light ions, q_1 is about 30% less than expected, while for I, it is about 40% greater at each energy. Any simple alteration in the screening function, say by adjusting the numerical factor in Equation (17), that improves the agreement in the former case will increase the discrepancy in the latter, and vice versa. Nevertheless, the values of q_1 obtained in this way are close enough to expectation to be quite encouraging for the viewpoint on which the analysis was based.

Table 2. Curvature Parameters Observed for Ions Moving in the Planar Channels of Gold Crystals

Ion	Energy (MeV)	Curvature parameter (eV/Å^3) y_{111}	y_{100}	Ratio y_{111}/y_{100}	Screening constant,[a] b (Å^{-1})
^{4}He	3	64.0 ± 2.8	108.1 ± 5.8	0.592 ± 0.041	3.32 ± 0.44
^{16}O	10	154.4 ± 8.9	253 ± 29	0.610 + 0.078	3.12 ± 0.82
^{127}I	15	837 ± 48	—	—	—
	21.6	915 ± 53	1600 ± 114	0.572 ± 0.053	3.55 ± 0.58
	60	1288 ± 50	2077 ± 92	0.620 ± 0.037	3.03 ± 0.37
			Weighted mean:	0.640 ± 0.017	3.22 ± 0.18

[a] For a hyperbolic cosine planar channel potential $b = [\ln(y_{111}/y_{100})]/(l_{100} - l_{111})$, $l_{100} = 1.0197$ Å and $l_{111} = 1.1774$ Å.

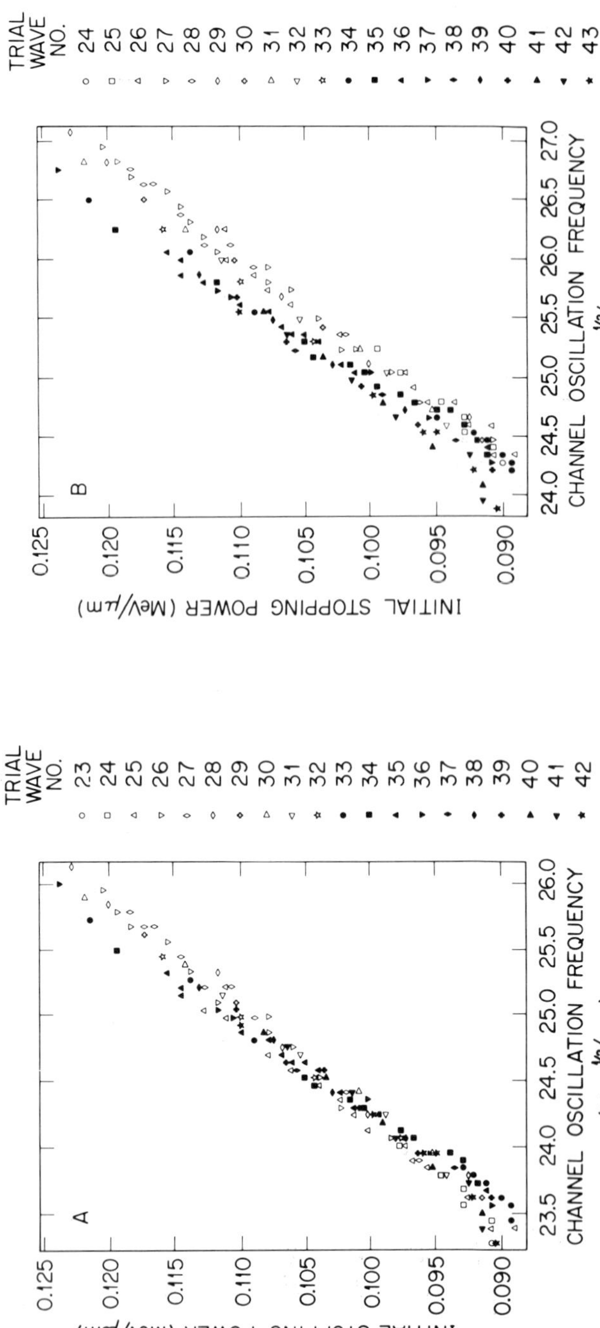
A
B
INITIAL STOPPING POWER (MeV/μm)
0.125
0.120
0.115
0.110
0.105
0.100
0.095
0.090
CHANNEL OSCILLATION FREQUENCY (MeV$^{1/2}$/μm)
23.5
24.0
24.5
25.0
25.5
26.0
26.5
27.0
TRIAL WAVE NO.
23
24
25
26
27
28
29
30
31
32
33
34
35
36
37
38
39
40
41
42
43

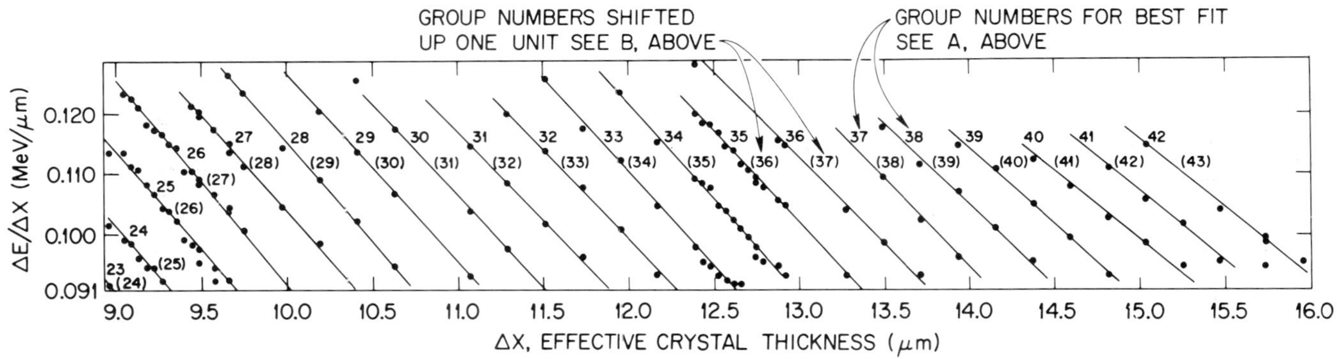

Figure 11. Stopping power versus planar channel thickness. 21-MeV ^{6}Li in {100} channel of Si. (A) Oscillation frequency for trial first wave number 23. (B) Oscillation frequency versus stopping power for trial first wave number 24.

Most recently it was decided to perform a set of experiments where the velocities of the ions would be high enough to assure that the ions would be completely bare and to assure that the value of p [in Equation (11)] would be -1, corresponding to the Bethe–Bloch region of stopping powers (Mo 81). In this way, two parameters of the analysis would be nonadjustable and thus a very sensitive test could be applied to the exact form of the interplanar potential.

The ions chosen were ^3He, ^4He, ^6Li, and ^7Li, all at 3.5 MeV/amu, to provide two values of Z_1 and a variation of the charge-to-mass ratio. The crystal chosen for this case was Si with a thickness of 8.87 ± 0.13 μm, and measurements were made over a very large change in the number of cycles of transverse oscillation. Data were taken for the $\{111\}$ and $\{110\}$ planar channels. Energy-loss data extracted from the peaks of the energy spectra were plotted against the effective thickness (for that tilt angle) Δx. The values of $\Delta E/\Delta x$ were converted to $(dE/dx)_{E_0}$ using Equation (10) and values of the oscillation frequency were derived with the use of Equation (11). When large numbers of cycles of oscillation occur in the crystal, an ambiguity concerning the starting wave number must be resolved. A least-squares fitting procedure was used to determine the value of n which produced the best fit to all the data. This is illustrated in Figure 11A and B, which display the values of the initial stopping power versus ω for the best fit $n_0 = 23$ in A and for a value of n_0 increased by one unit in B. A similar deterioration of fit occurs if n_0 is decreased by one unit. Once again the linear relationship between $(-dE/dx)_{E_0}$ and ω is demonstrated. In these cases α and ω were derived by the same least-squares computer fit to the data. Table 3 shows the

Table 3. Slopes, Intercepts, and Curvatures Derived from the Data

Ion	$\alpha \pm \Delta\alpha$ (eV/Å)	$\beta \pm \Delta\beta$ (eV$^{1/2}$)	$s_0 \pm \Delta s_0$ (eV/Å)	$y \pm \Delta y$ (eV/Å^3)
		Experimental values for (110)		
^3He	$-8.49 \pm .32$	$6.34 \pm .16$	$3.65 \pm .03$	75.4 ± 5.6
^4He	$-9.69 \pm .28$	$7.12 \pm .14$	$3.59 \pm .03$	71.6 ± 4.2
^6Li	$-21.58 \pm .34$	$13.01 \pm .14$	$8.35 \pm .08$	108.8 ± 3.4
^7Li	$-22.74 \pm .38$	$13.50 \pm .15$	$8.00 \pm .08$	106.8 ± 3.6
		Experimental values for (111)		
^3He	$-7.68 \pm .22$	$6.94 \pm .14$	$2.68 \pm .03$	37.5 ± 2.2
^4He	$-7.14 \pm .23$	$6.86 \pm .15$	$2.74 \pm .03$	34.9 ± 2.1
^6Li	$-16.94 \pm .22$	$12.60 \pm .11$	$6.36 \pm .07$	55.6 ± 2.2
^7Li	$-16.60 \pm .21$	$12.40 \pm .10$	$5.96 \pm .06$	55.5 ± 2.1

results for the α and β coefficients and the respective uncertainties. Also in Table 3 the stopping powers of the best channeled particles (leading edge values) s_0 are given. An important parameter in the analysis is the curvature (y) of the planar potential [Equation (13)]. The values of l, the half-width of the channel, are 0.96 and 1.1757 Å for the {110} and {111} planes, respectively. The extracted values of y from the experimental data are given in the last column of Table 3.

The units of ω are such that a factor $M^{1/2}$ is suppressed. If two real oscillation frequencies (units t^{-1}) differ by the factor $(M_a/M_b)^{1/2}$, then in the absence of any isotope effect the values of ω would be the same. We expect that the values of α, β, s_0, and y should be the same for two isotopes. Errors in the value of the intercept α have a very small influence upon the value of the slope β derived from the data. The values of α for ^3He and ^4He lie outside their error values, especially for the {110} channel data; the discrepancy does not significantly alter the agreements between the values of s_0 and y. Except for this disagreement the data seem to prove that there is no isotope effect when comparing the results for ^3He and ^4He for the ^6Li and ^7Li.

The previous analyses, as we have seen, had the basic objective of extracting the planar potential from the experimental data and comparing this with the theory for a potential function of the Molière or Hartree–Fock type. The extraction of the interaction potential could be made only by comparing the ratios of the experimental y values for two different channels. This procedure cancels out the factor Z_1Z_2 appearing in $V(r)$, since Z_1Z_2 (the product of the effective charge states) is usually not known in channeling experiments with low-velocity ions. In this work for 3.5 MeV/amu He and Li ions, the energy is high enough to ensure that the ion charge is the (totally stripped) nuclear charge. Using the Molière and Hartree–Fock potentials and calculating the ratio of the y values for the {111} and {110} planes, the extracted inverse screening lengths (b) are 2.14 ± 0.26 Å^{-1} and 1.95 ± 0.13 Å^{-1} for He and Li, respectively, for the Molière potential; for the Hartree–Fock potential, the values are 2.17 ± 0.26 Å^{-1} for He and 2.00 ± 0.13 Å^{-1} for Li. If we are to reproduce the absolute values of y, it would be necessary to arbitrarily normalize the Molière potential by a factor of 1.4 and the Hartree–Fock by 1.02. For the earlier data, similar normalization factors were used for the Molière potential; however, for that case of channeling of low-velocity ions, these factors were incorporated into the uncertainties in Z_1, the effective charge, and therefore an independent check on the potential could not be made. For the 3.5 MeV/amu He and Li ions, Z_1 is the bare nuclear charge and therefore the normalization factors must be related to the potential alone, and we conclude that the Molière potential has to be normalized by a large factor. This factor, if used in a low-velocity regime, would predict unrealistically high effective charge states for

He and Li. It should be noted that the inverse screening lengths b show an apparent decrease from He to Li (9.5% for the Molière and 8.5% for the Hartree–Fock potentials), but in both cases the differences are within our experimental uncertainty. A better way to test the adequacy of the analysis is to use a potential with no adjustable parameters, and this is achieved by using the $V(r)$ for the Hartree–Fock potential (Lu 71).

For the case of channeling in the Si {111} planes, Equation (9) had to be modified since this channel is bordered by pairs of closely spaced planes. The modification is (Ro 71)

$$V_2(x) = V_1(l+x) + V_1(l-x) + V_1(\tfrac{5}{3}l+x) + V_1(\tfrac{5}{3}l-x) \tag{19}$$

where l is the half-width of the larger planar spacing (1.1757 Å).

Table 4 shows comparisons between the experimental y values for He and Li (averaged over mass) and the predictions of Equations (15) using the Hartree–Fock potential. The comparisons are shown for both planar channels, {110} and {111}, and the agreement is excellent for the {110} planes and consistent with the experimental uncertainties for the {111} planes.

Table 4 shows the comparisons of the σ' factors. The column labeled $\sigma'(\text{th})^b$ corresponds to the calculated values without thermal corrections, and as can be seen they exceed the experimental values by about 2% for He {110}, by about 4% for Li {110}, and by 18% for He and Li in {111}. These discrepancies can be interpreted as due to effects of thermal vibrations of the atoms in the lattice. An approximate method to account for such effects is given as follows: Equation (12) can be rewritten (Ei 71):

$$V_1(\bar{x}) = 4\pi\rho\kappa l \int_0^\infty r\, V(r)\, dr \int_{\bar{x}-r}^{\bar{x}+r} P(u)\, du \tag{20}$$

In Equation (20) u is the thermal displacement perpendicular to the plane and $P(u)$ is a Gaussian distribution given by

$$P(u) = (2\pi\langle u^2\rangle)^{-1/2} \exp[-u^2(2\langle u^2\rangle)^{-1}] \tag{21}$$

where the root mean square amplitude $\langle u^2\rangle^{1/2}$ is 0.076 Å for Si. This value

Table 4. Comparisons Between the Experimental and Theoretical Curvature Parameters y

Ion	y_{110}(Exp)	y_{110}(th)	y_{111}(Exp)	y_{111}(th)
He	73.5 ± 4.9	70.0	36.2 ± 2.1	33.2
Li	107.8 ± 3.5	105.0	55.5 ± 2.2	49.5

Table 5. Comparisons Between the Experimental and Theoretical σ′ Factors

Ion	$\sigma'(\text{Exp})^a$	$\sigma'(\text{th})^b$	$\sigma'(\text{th})^c$
He (110)	76.4 ± 7	77.8	69.38
He (111)	66.8 ± 6	78.3	62.94
Li (110)	91.7 ± 6	95.2	85.0
Li (111)	83.0 ± 5	95.9	77.10

[a] Experimental values extracted from the data of Table 1 using the relation $\sigma'(\text{Exp}) = (\hat{s} - \alpha)\pi/\beta$, when the random stopping $\hat{s}$ is 7.25 ± 0.07 eV/Å for He ions and 16.32 ± 0.16 eV/Å for Li ions.
[b] Theoretical values without thermal corrections.
[c] Theoretical values with thermal effects included.

has been deduced from x-ray data (Ba 62) and is calculated at a temperature of 301.7 K. Using Equations (21) and (20) in (11) and the relation $\sigma' = V_2[(l) - V_2(0)]^{1/2}/l$, the resulting values of σ' with thermal effects included are given in the last column of Table 5, and although they are smaller than the experimental values, they agree within the experimental uncertainties.

The results demonstrate that planar channeling energy loss groups provide a very sensitive test of the form and size of the interplanar potentials, provided that the particles are totally stripped of electrons. Almost all of the data indicate that, as expected, different isotopes behave identically in planar channeling oscillations [i.e., $\nu_A(m_A)^{1/2} = \nu_B(m_B)^{1/2}$, where ν is the oscillation frequency and m the ion mass]. The data prove that the Molière potential, as it is commonly used, does not fit the correct form or size of the real interplanar potentials in silicon. In contrast, the Hartree–Fock potential provides an accurate fit to the data. The absence of any differences between He and Li indicates that, within the experimental errors, both nuclei behave as good test charges for the crystalline potentials in silicon.

2.2 Hyperchanneling

Ordinary axial channeling occurs when the particles are deflected away from strings of atoms in the crystal and, in general, the particles may wander from one set of strings to the next, etc. If the particle has insufficient transverse energy to surmount the potential saddle point between a pair of strings, it will be confined to a single set of string; this subclass of axial channeling is called "hyperchanneling" (Ap 72a). The phenomenon was recognized in early computer calculations by Robinson and Oen (Ro 63) and was observed by Eisen (Ap 72b) as a low-energy loss "tail" in axial energy loss distributions. A set of "paths" generated by computer is shown in Figure 12. In this figure, particles are traveling perpendicularly to the figure and only transverse displacements are shown. The upper left particle is hyperchanneled.

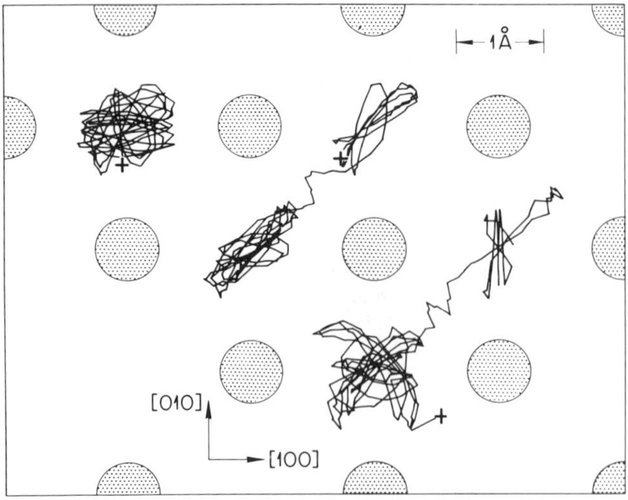

Figure 12. Computer simulation of transverse motion of particles moving in axial channels. The particle at upper left is confined to a single set of four strings.

Hyperchanneled ions are expected to have at least two distinctive characteristics, a smaller acceptance angle than ordinary axial channeling and low energy losses because they sample larger average impact parameters than ions which wander from channel to channel. This effect was studied (Ap 72a, Ap 72b, Ba 75) using a beam of 21.6-MeV I ions collimated to $\leqslant 0.01°$ full width onto a 0.60-μm-thick $\langle 001 \rangle$ Ag crystal which had a mosaic spread of 0.05° FWHM. The energy spectra of those ions transmitted within $\pm 0.012°$ of the incident beam direction were recorded with an energy resolution of 185 keV FWHM and normalized by a beam monitor system. The arrangement is shown in Figure 13. The data shown in Figure 14 illustrate the fact that hyperchanneled ions show much less energy loss than particles channeled in the $\{111\}$ planes. Energy spectra measured with the beam incident at various angles from $\langle 011 \rangle$ in the $\{111\}$ direction are shown in Figure 15. The prominent group on the right, lying within a narrow angular range and centered on a dE/dx value of 3.15 MeV/μm, is due to hyperchanneling. It should be emphasized that the angular extent of this series of spectra ($\pm 0.45°$) was well within the calculated critical angles for ordinary axial channeling [i.e., $\psi_{1/2} \cong 1.2°$ from Barrett (Ba 71) and $\Psi_1 \cong 1.9°$ from Lindhard (Li 65)]. The major peaks in the energy-loss distributions in Figure 15 taken at $\pm 0.45°$ relative to $\langle 011 \rangle$ are nearly identical to chose obtained at angles 5 to 10° from $\langle 011 \rangle$ in $\{111\}$. This indicates that when the beam is incident parallel to a low index plane many of the transmitted ions have distributions characteristic of planar channeling even

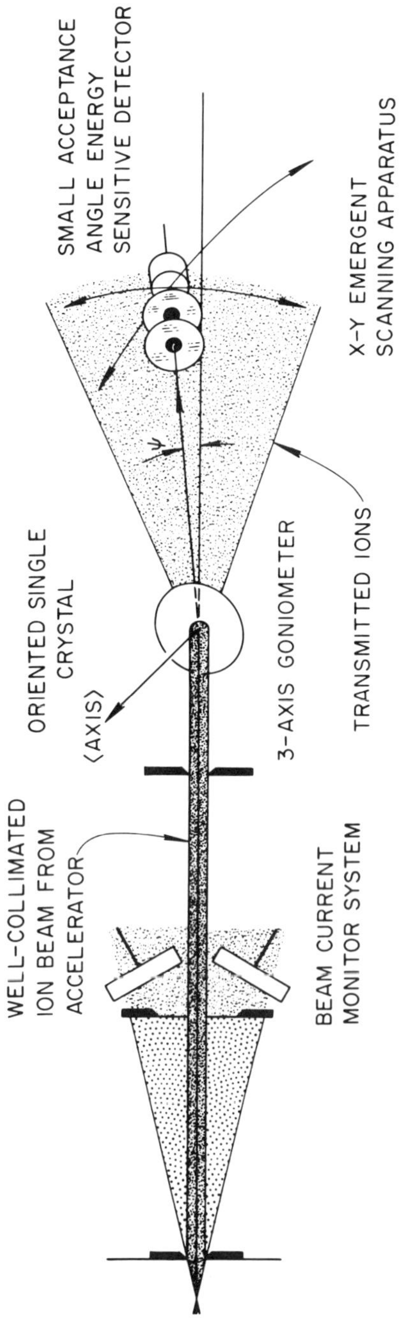

Figure 13. Experimental apparatus.

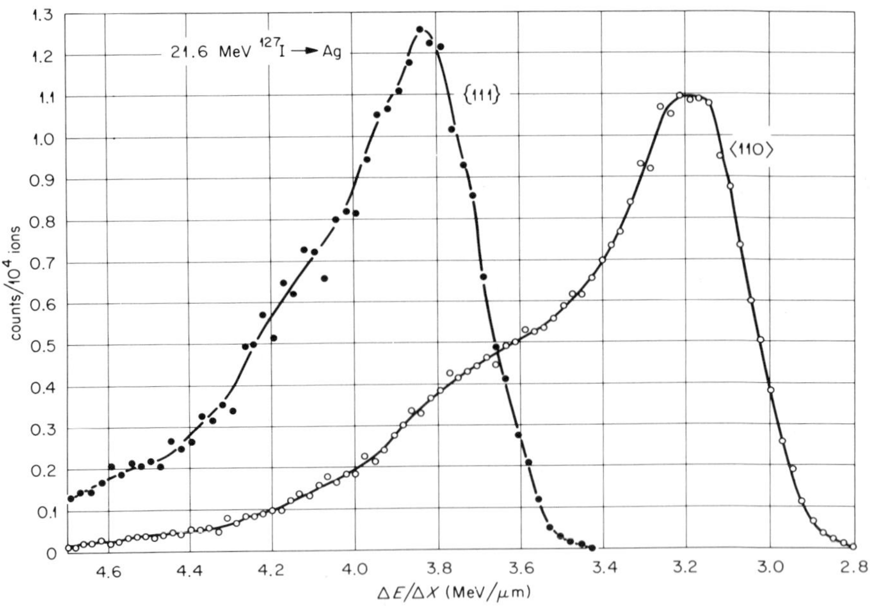

Figure 14. Comparison of energy losses for hyperchanneled ions and ions moving in planar channels.

at angles of approximately 0.25 ψ_1. As the incident angle with respect to $\langle 011 \rangle$ is decreased, the energy-loss distributions in Figure 15 lose the characteristic $\{111\}$ shape at about 0.30°, and a new group of ions with significantly lower energy losses is dominant within approximately 0.12° of $\langle 011 \rangle$. This low-loss group has the two characteristics expected of hyperchanneled ions. Related emergent patterns recorded photographically agreed well with these spectra. Where planar energy losses were evident, a characteristic elongated planar spot centered on the incident beam was observed superimposed on a ring of image intensity centered on $\langle 011 \rangle$. Within the range of hyperchanneling the ring had collapsed into an intense circular spot centered on $\langle 011 \rangle$.

The data were analyzed to determine a characteristic angle and absolute fraction for hyperchanneling. It was estimated from the shapes of the population curves at various energy losses, such as those for 3.15 and 3.95 MeV/μm in Figure 15, that only ions with loss rates less than or equal to 3.7 MeV/μm were hyperchanneled. The fraction of all ions recorded with energy losses below a specified value was extracted as a function of incidence angle and the results are the data points in Figure 16. The solid curves were calculated using the following model. As an ion of energy E entered the crystal at an angle ψ_i to the axial direction, it possessed a

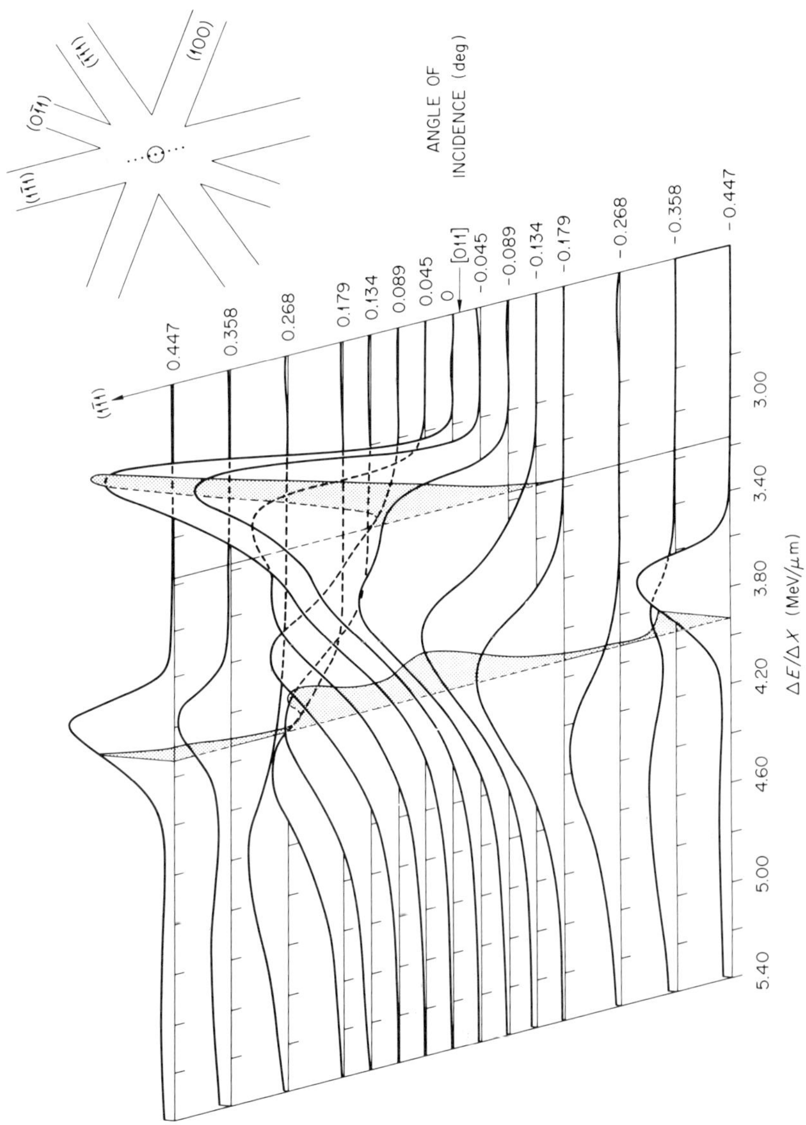

Figure 15. Normalized energy-loss distributions at various incidence angles with respect to the ⟨011⟩ axis for 21.6-MeV I ions through 0.85 μm of Ag. The insert shows the small angular extent of the measurements (dotted line) and the hyperchanneling region (shaded central spot).

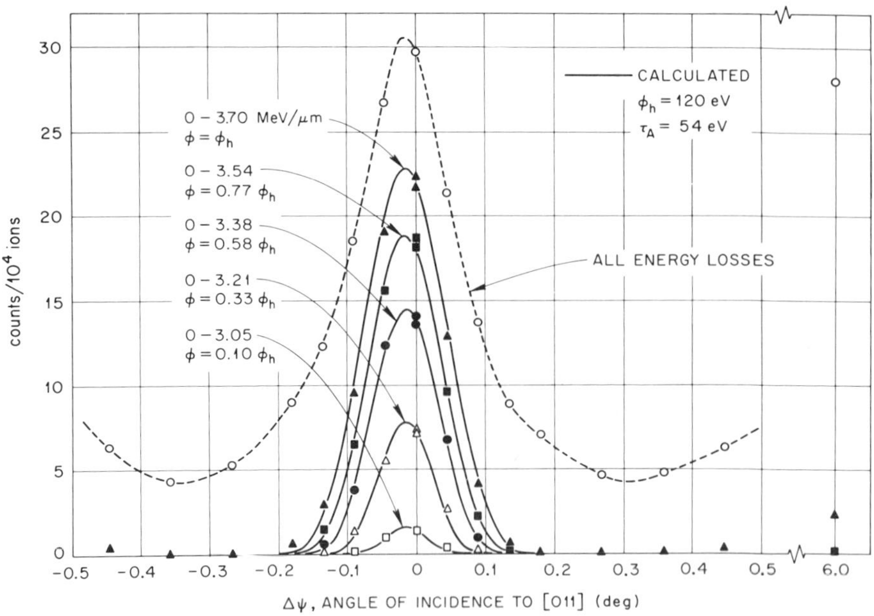

Figure 16. Comparison of measured (data points) and calculated (solid lines) fractions of the beam reaching the detector as a function of angle of incidence to $\langle 011 \rangle$. The dashed curve is not calculated but can be understood qualitatively in terms of emergent intensity distribution and detector geometry.

transverse kinetic energy $\tau_i = E\psi_i^2$. In addition, its entry point gives it a transverse potential energy ϕ_i chosen in the contours of Figure 17. These were calculated using a potential from planar channeling experiments. As the ion penetrates the crystal, it tends to traverse the entire space within the contour corresponding to its total transverse energy. It was assumed that the crystal used was of sufficient thickness for the ion to achieve statistical equilibrium over this contour, which implied that the probability of finding it was constant inside the contour and zero outside (Li 65). In moving through the crystal, the ion acquires additional transverse energy ε_a from multiple scattering by electrons, lattice vibrations, and defects. The values of ε_a for the trajectories bounded by a particular contour were assumed to have an exponential distribution with mean value ε_A.

Those ions which had total transverse energies less than the critical contour, $\phi_h \cong 60$ eV, in Figure 17 were treated as hyperchanneled. The fraction of the area enclosed within ϕ_h is $\mathscr{F}_h = 0.30$. Calculations using these values gave curves which agreed qualitatively with the experimental data in Figure 16 but had peaks about three times too high and were about 25% too narrow at the base. Varying ϕ_h influenced only base width and ε_A only the

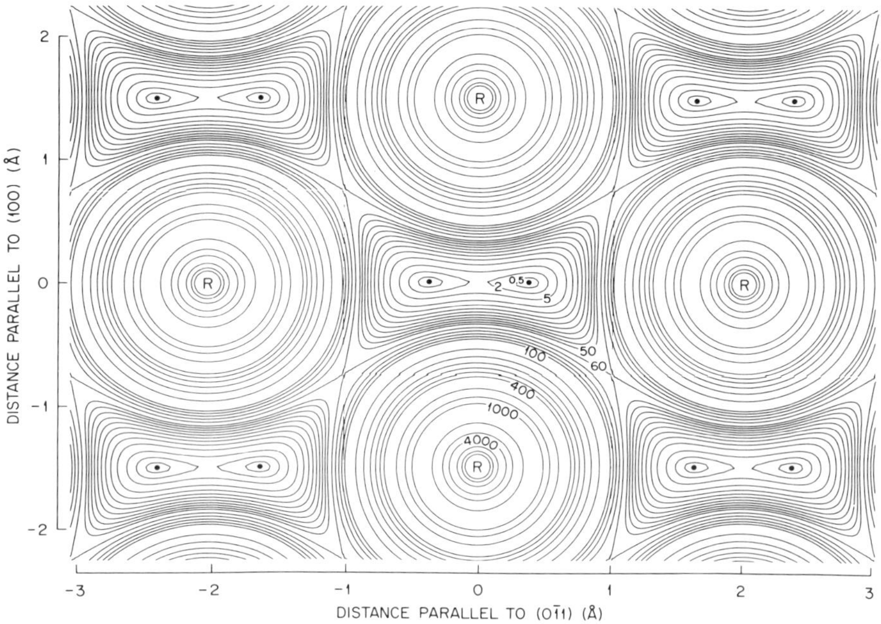

Figure 17. Potential energy contours of I ions in the continuum potentials of ⟨011⟩ rows (R) of Ag atoms. All values are in eV and are relative to the minimum (•), which has an absolute value of 72 eV.

peak height, while ψ_M, the mosaic spread, influenced both. Values of 120 eV for ϕ_h and 54 eV for ε_A provided the best fit to the 0–3.70-MeV/μm data and are shown as the solid lines in Figure 16. Since all combinations of parameters which gave a good fit of theory to experiment had values of ε_A several times the estimated value, it was concluded that there was considerably more multiple scattering than could be estimated for the mechanisms believed to be important. Such extra multiple scattering should also affect other experiments involving dechanneling rates. The results also indicated that hyperchanneling investigations provide information on potentials in solids at impact parameters not accessible to other channeling experiments. The value of ϕ_h used in calculating the solid curves in Figure 16 was twice as large as the value obtained from the contours of Figure 17. This may be an indication of the effects of lattice structure on the ion–atom potential. A simple way of taking this into account would be the application of Wigner–Seitz rather than free atom boundary conditions to the electron wave functions. Robinson (Ro 71a) has examined the resulting potentials and finds that they differ only near the cell boundary where the Wigner–Seitz potential drops more rapidly. It appears that the Wigner–Seitz potential will give better agreement with the hyperchanneling experiments than the

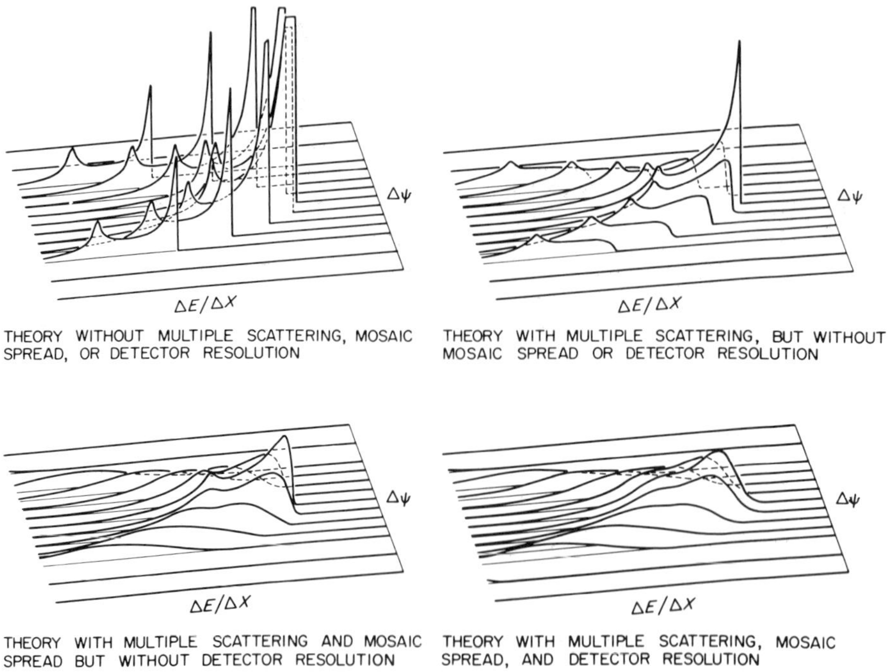

Figure 18. Computer simulations of the energy spectra predicted for hyperchanneled particles.

potential used in Figure 17 and at the same time preserve agreement with earlier planar channeling experiments.

As illustrated in Figure 18, computer simulations of the hyperchanneling phenomenon seem to reproduce all of the features found in the energy spectra for ⟨011⟩. Similar data were aquired for [001] hyperchanneling by Barrett *et al.*, who give an extended analysis (Ba 75). Some features observed in this case are not reproduced in the computer analysis and are most likely explained by wave effects of the kind observed for planar channeling.

3. Energy Loss in Channels

It is not surprising that particles moving in crystal channels will lose energy more slowly than would be the case for particles moving in some random direction in a crystal. An unchanneled particle will lose energy by exciting and ionizing electrons which are tightly bound to the target atoms; a channeled particle does not approach a target atom closely enough to produce such effects. In addition, a slow unchanneled ion will impart some of its momentum to entire target atoms so that target atom displacements will occur; this low-velocity effect is called nuclear stopping although "nuclear" in this case does not imply any nuclear reactions. For channeled particles nuclear stopping is almost completely eliminated even at very low ion velocities. An illustration of this effect is found in the behavior of a Si solid state detector when heavy ions enter along a channel direction of the detector. Normally, if the ions are not channeled, they will fall to a certain velocity and then begin to displace Si atoms in the detector. Radiation damage is produced near the end of the track as the ion stops. The damaged part of the track fails to produce collected electron–hole pairs because of the traps formed in this part of the range. Solid-state detectors are known to have a "pulse height defect" due to this localized damage. Figures 19 and 20 illustrate the effect which occurs when some of the particles are channeled. In Figure 19 the detector response is shown for a sharp series of iodine ion energies wherein the particles are unchanneled. For Figure 20 the detector is oriented with the $\langle 110 \rangle$ axis parallel to the beam. Some of the particles are channeled and these exhibit larger pulse heights and much narrower peaks because nuclear stopping and its attendant localized damage is almost completely eliminated (Mo 66). At high velocities the target atoms do not have time to move appreciably and thus only electronic stopping occurs.

A complete discussion of the subject of stopping power can be obtained elsewhere (Zi 77). Generally, stopping powers increase with energy to a maximum and then slowly decrease with energy. The low-energy part of the curve is particularly interesting since most heavy ions being used as of the time of this writing are at velocities at or below the stopping power maximum. At low velocity, heavy-ion stopping powers have been expected to be proportional to ion velocity. Theoretical discussions of velocity-proportional stopping have been given by various authors (Bo 40, 41. La 40, Kn 41, Br 41, Fi 59, Li 63). The theories of Lindhard, Scharff, and Schiøtt (Li 63) and of Firsov (Fi 59) do not agree as to the magnitudes of the predicted stopping powers but they do agree that, if the nuclear stopping component is subtracted out, the remaining electronic stopping power should have the form $dE/dx = k\sqrt{E}$. The relationship appeared to work well for the lighter heavy ions but, as shown in Figure 21, the data indicate a form $dE/dx = k\sqrt{E} + b$, where the constant b becomes more important for heavier ions (Br 72). Some doubt existed as to whether the nuclear stopping corrections were responsible for the discrepancy. It was decided that channeling would be used to eliminate the nuclear component of the stopping power. A special channel electron multiplier position sensitive detector was developed so that the emerging channeled iodine ions could be observed downward in energy to 0.6 MeV. The results shown in Figure 22 (Mo 76) show clearly that the relation $dE/dx = k\sqrt{E}$ is not correct at very low energy. No theoretical picture of stopping correctly predicts this behavior at

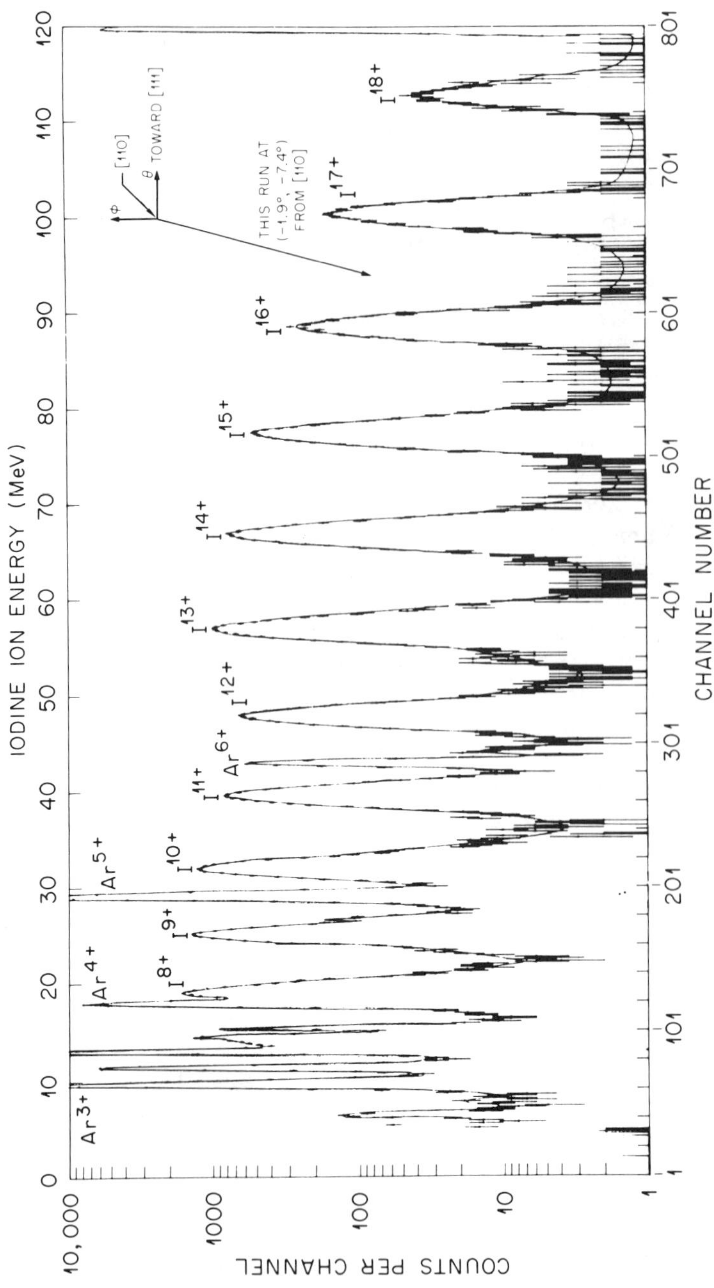

Figure 19. Si detector response to iodine and argon ions entering the detector near its ⟨110⟩ axis.

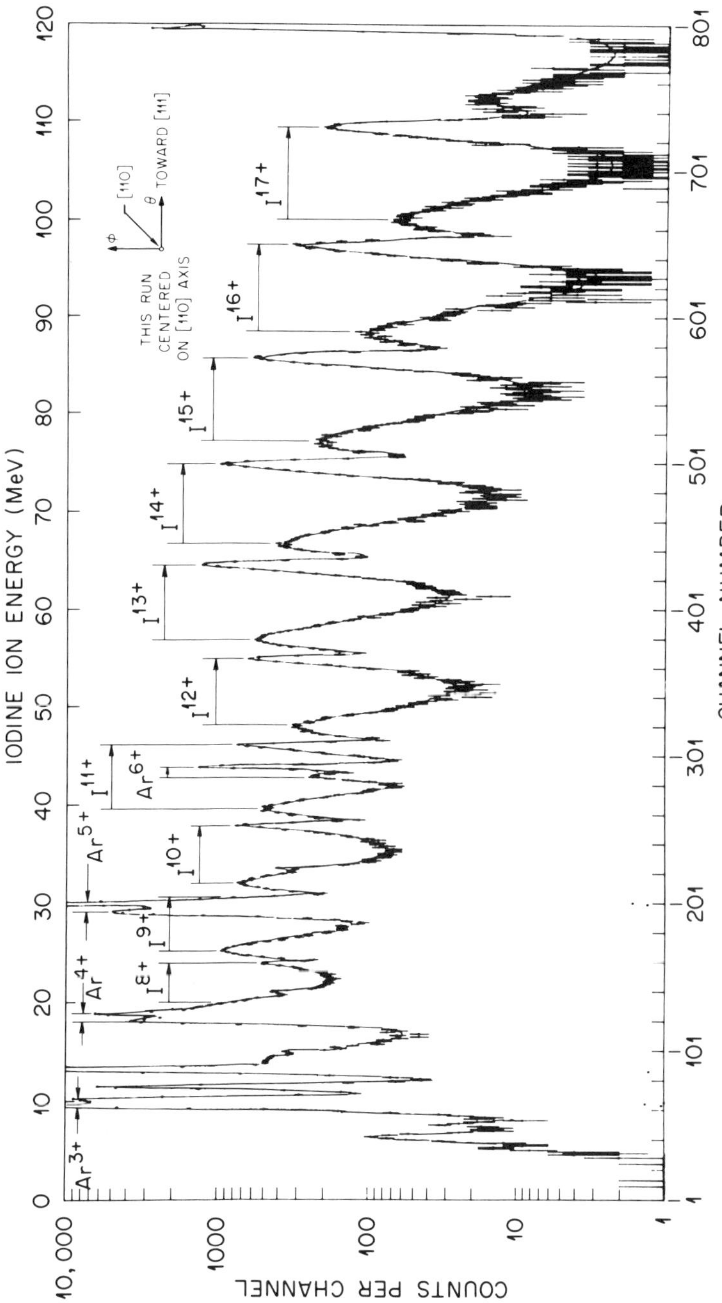

Figure 20. Si detector response to iodine and argon ions entering the detector along its ⟨110⟩ axis.

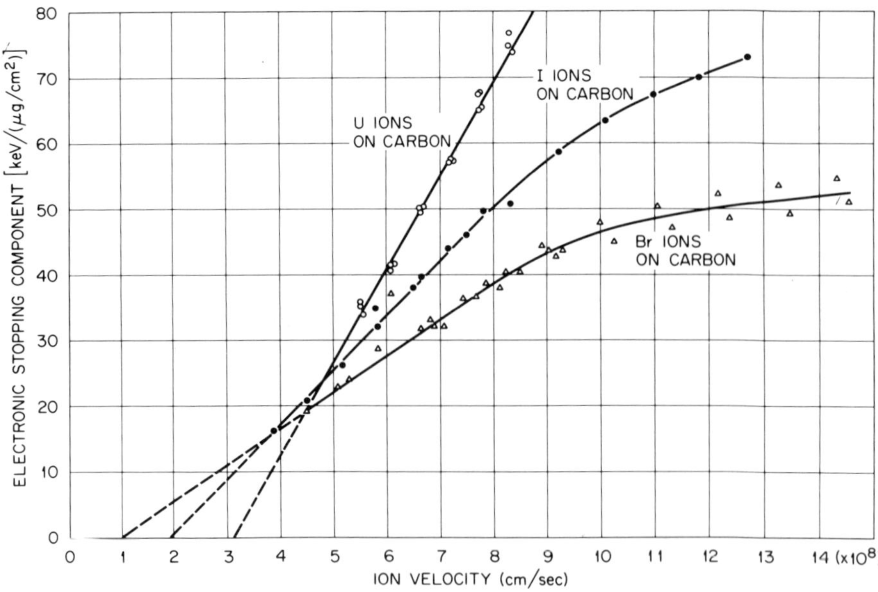

Figure 21. Stopping powers of U, I, and Br ions in carbon.

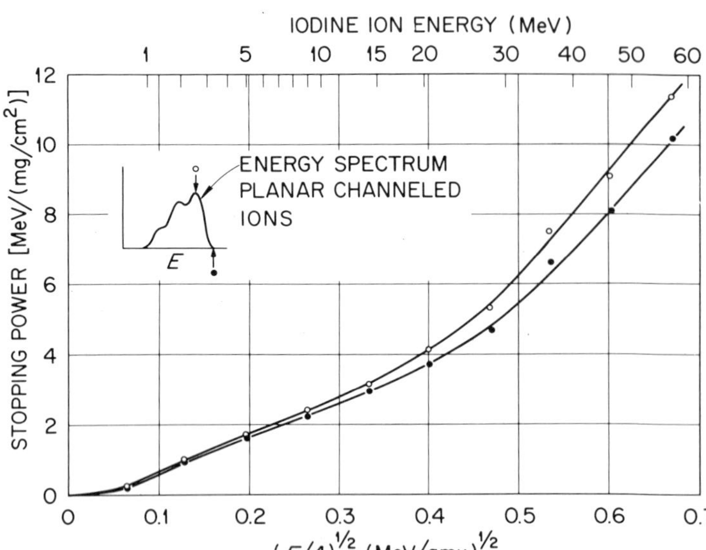

Figure 22. Stopping power of iodine ions in the (111) planar channels of Ag. 0.6 to 60 MeV.

the time of this writing. Stopping powers can be predicted by means of empirical formulas which make use of a parameter called the effective charge for stopping (Br 72).

3.1. Screening Effects on Energy Loss

The relation of ionic charge to electronic stopping powers for non-fully-stripped ions in solids is, as indicated above, not entirely clear and many of the uncertainties have been encompassed in the empirical concept of "effective charge," $Z_{\text{eff}} < Z_1$ which increases with increasing velocity. The electronic stopping power in any velocity region will then ostensibly be proportional to Z_{eff}^2 and Z_{eff} can be determined from the measured stopping power by comparing it to the stopping power of the proton at the same ion velocity.

An additional complication arises if one considers the effect of screening by electrons in the medium. Because electrons are attractively scattered by the moving ion the electron density is enhanced in the region of the ion and can act to screen the ionic charge and thereby reduce the stopping power (i.e., for $v_i \leqslant v_0$, the center of the enhanced charge is near the ion nucleus).

Usually it is difficult to tell whether an ion moving in a solid is highly stripped but highly screened or less highly stripped and screened. In the case of channeling, however, experiments have been devised and carried out which enable the measurement of these effects. The experiments are based on the observations described in Section 4.1 of this chapter that under channeling conditions electron capture and loss are sufficiently suppressed that ions may pass through a crystal without ever undergoing charge change.

In the first experiments done to measure these effects it was shown that O^{8+}, O^{7+}, and O^{6+} ions at 27.8 and 40 MeV passing through 8000 Å of Ag in the [011] axial direction and emerging in the same charge state in which they entered suffered energy losses which were closely proportional to the square of the ion charges [i.e., 8^2, 7^2, and 6^2, respectively (Da 75b, Mo 74)].

In further work using higher energy resolution and an array of ions, the stopping powers in Au {111} planar channels were measured for totally stripped nuclei of $Z_1 = 5$–9, nuclei having a single bound electron $Z_1 = 6$–9, and ions having two bound electrons for $Z_1 = 7$–9, all for velocities corresponding to 2 MeV/amu (Da 77).

In all cases reported here the energy-loss spectra were recorded for the emergent charge state equal to the input charge state. Such spectra for F^{7+}, F^{8+}, and F^{9+} are shown in Figure 23. Ions entering parallel to a planar channel follow oscillatory trajectories whose amplitude depends upon the entrance point with respect to the midpoint between planes (see Section 2.1). Ions entering close to a plane (high amplitude) penetrate regions of higher

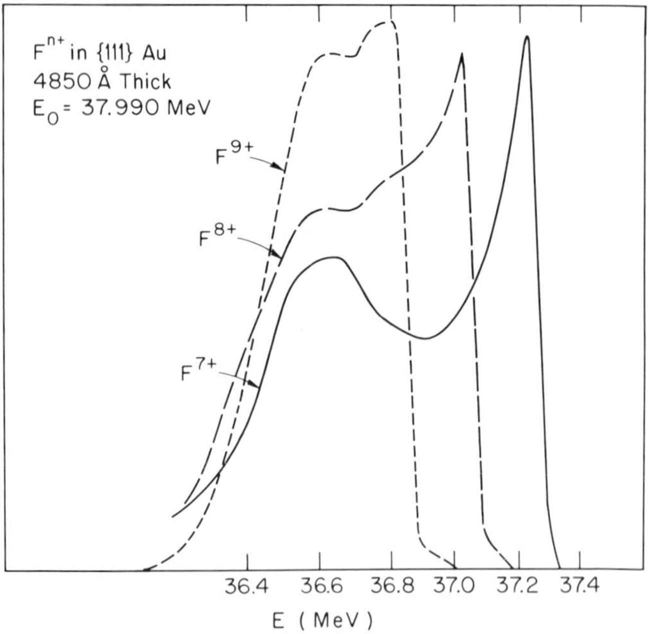

Figure 23. Energy-loss spectra for F^{q+} ions ($E_0 = 37.99$ MeV) emerging from Au {111} channels. The spectra recorded are for $\Delta q = 0$ (e.g., 9+ in, 9+ out).

electron density. These particles lose more energy and also equilibrate in charge state (Da 72). Hence the high energy-loss region is similar for all charge states. The width and the position of the centroid of the peak energy loss can be affected by wavelength restrictions of the emergent particles. Particles whose trajectories lie closest to the channel midplanar axis experience the minimum energy loss which contributes to the right-hand leading edge of the spectrum. These are most likely to pass through with no charge exchange. It is these particles which we treat. (To be sure, the energy losses of particles which have even perfectly defined trajectories will not be identical because of statistical straggling, δE. But, insofar as $\delta E \propto \Delta E$, relative comparison of energy losses for center-channeled particles, obtainable from the extrapolated leading edge, is a valid procedure.) For the energy losses reported here for $Z_1 \geqslant 5$, this assumption should be reasonable. For the lower $Z_1 = 1$–3 losses, also reported in (Da 77) but not reported here, Vavilov straggling complicates matters.

The upper leading edges of all of the spectra in Figure 23 are extremely sharp down to the ~10% level. The tails observable in the F^{9+} and F^{8+} spectra arise from particles which have captured, then lost an electron along their path. During the time of electron attachment their energy losses are less. The lack of such a tail in the F^{7+} spectrum reflects the improbability of capture into the L shell to form F^{6+}.

The energy loss at 1/10 height on the leading edge of 0, 1, and 2 electron ions ($Z_1 = 5$–9) divided by q^2 is shown in Figure 24 for ions at 2 MeV/amu as a function of Z_1. In this region the stopping powers for the bare nuclei are found to scale exactly as Z_1^2 (see Section 3.2 of this chapter). If the bound electrons screened the nuclear charge perfectly, then, e.g., S/q^2 for B^{5+}, C^{5+}, and N^{5+} should be identical. The higher values of S/q^2 for the clad ions reflects imperfect screening and an "effective" charge $q' = (SZ_1^2/S_z)^{1/2}$ can be obtained by comparison with the stopping power, S_z, of the bare nucleus with the same charge. The values of q' are shown in the figure. In all cases studied, the screening per electron is ~0.9. Other methods of treating the data, e.g., using the 1/2 height or extrapolating the leading edge to zero by shape deconvolution, will give different values for S/q^2 (see below), but the values of q' obtained are closely the same as those shown in Figure 24.

Center-channeled ions interact directly only with conduction (6s and 5d) electrons of the Au ($E_b < 10$ eV) and at longer range with 5p, 5s, and 4f electrons. The reduction in screening arises from contributions to stopping from conduction electrons which penetrate within the radius of the bound projectile K electron. It might be expected that the effective screening per

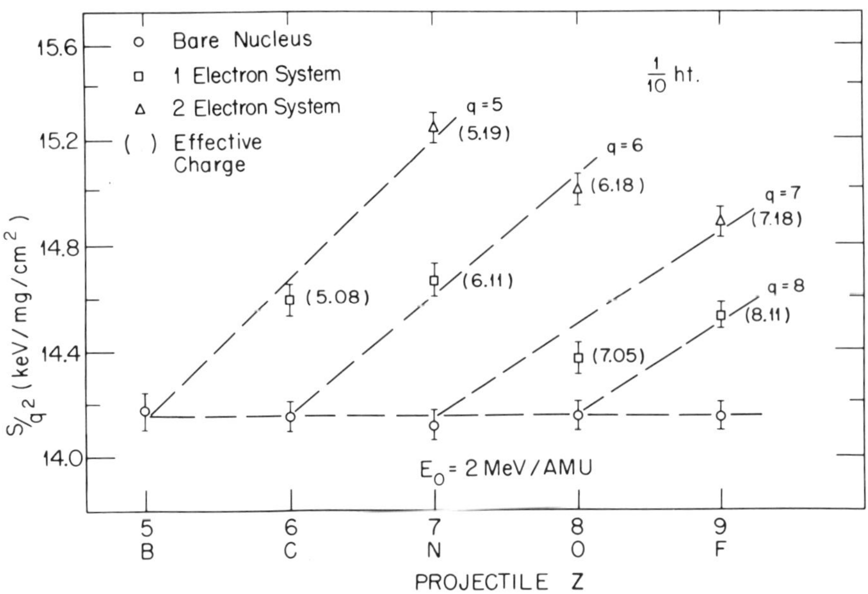

Figure 24. S/q^2 at 1/10 height for 2.0 MeV/amu channeled ions with 0, 1, and 2 bound electrons. The dashed lines indicate the connection between ions of the same q.

electron would increase with increasing Z_1 (decreasing orbital radius); however, no such trend can be adduced within the precision of the experiment.

Using a simple perturbation calculation for the difference in energy loss rate for a bare ion and one containing one or two bound electrons, Golovchenko *et al.* (Go 81) obtained reasonable agreement with the results of Figure 24 and with additional data they obtained for F, Mg, and Si at 3 MeV/amu.

3.2. Higher-Order Corrections for Electronic Stopping Powers of Heavy Ions

At high velocity, first-order quantal perturbation calculations yield expressions which show that heavy-ion stopping powers are proportional to Z_1^2, for both channeled (Es 78a) and unchanneled (Fa 63) particles, where Z_1 is the nuclear charge of the projectile. As the velocity is decreased, terms with higher-order Z_1 dependence are expected to become important. At the same time, as the ion velocity decreases towards $Z_1^{2/3}e^2/\hbar$, the ion no longer remains bare because electron capture and loss begin to be important. This reduction of ion charge affects the stopping power and prevents a direct measurement of underlying higher-order Z_1 dependences. In most cases the charge state distributions of heavy ions moving inside solid targets are unknown. The only exception to this is the case where heavy ions move through crystal channels. Well-channeled heavy ions have been shown to pass through thin crystals with no change of charge. As shown in Section 4.1 of this chapter, the electron capture and loss probabilities are significantly reduced for ions traveling down the low electron density, central region of crystal channels, and as demonstrated in the previous Section 3.1, such trajectories have a signature feature in channeled energy-loss spectra in that they are associated with the least energy loss for a given charge state. This special class of particles emerging from the crystal has been used to investigate higher-order Z_1 effects upon stopping power for heavy ions.

In addition to the Z_1^2 dependence for stopping power, terms in higher orders of Z_1 have been found experimentally by Barkas (Ba 63) and theoretical descriptions have been presented by Jackson and McCarthy (Ja 72), Ashley *et al.* (As 72), and by Lindhard (Li 76). Lindhard pointed out that terms must exist in even powers of Z_1 as well as the odd power terms proposed earlier, and that the Z_1^3 term and the Z_1^4 term can be comparable in magnitude and opposite in sign. Measurements of stopping powers of protons, α particles, and Li ions in random solids have been reported by Andersen *et al.* (An 77). These measurements indicate the presence of both Z_1^3 and Z_1^4 terms in the stopping power formulas. To measure these higher-order Z effects requires that the ion be totally stripped during its entire passage, hence the limitation to $Z_1 \leqslant 3$ in random media.

By utilizing the charge state integrity of heavy ions moving near the centers of crystal channels, stopping power measurements on totally stripped ions have been extended to higher Z_1 in two experiments. In the first by Datz *et al.* (Da 77) the range was $Z_1 = 5$–9 ($E/A = 2$ MeV/amu) channeled in $\{111\}$ planes of Au, and in the second by Golovchenko *et al.* (Go 81) the range was $Z_1 = 9$–17 ($E/A \approx 3$ MeV/amu) channeled in the $\langle 110 \rangle$ axis of Si (1.39 μm thick).

The results of Datz *et al.* (Da 77) for $Z = 5$–9 are shown as part of Figure 24 in Section 3.1. The values of S/Z_1^2 for the bare ions appear to be totally independent of Z_1 even though Z_1^3 terms might be expected to cause an ~10% increase.*

The data for leading edge stopping powers obtained by Golovchenko *et al.* (Go 81) for $\langle 110 \rangle$ axial channeling in Si are shown in Figure 25. Within very high accuracy the value of S/Z_1^2 is independent of Z_1 ($Z_1 = 9$–17). The result is $S/Z_1^2 = 21.2 \times (1 \pm 0.05)$ keV/mg/cm^2 and it is only within these very narrow limits that higher-order terms might exist.

An extended theoretical treatment and possible explanation has been developed (Go 81) as follows. In the absence of channeling effects and assuming that heavy ions, of atomic number Z_1, pass through the target totally stripped, we expect

$$\left(\frac{dE}{dx}\right)_r = \frac{4\pi Z_1^2 e^4}{mv^2} Z_2 N(L_0 + Z_1 L_1 + Z_1^2 L_2 + \cdots) \tag{22}$$

with

$$L_0 = \ln \frac{2mv^2}{\hbar\bar{\omega}} \tag{23}$$

$$L_1 = \frac{3\pi e^2 \bar{\bar{\omega}}}{2mv^3} \ln \frac{mv^2}{\hbar\bar{\bar{\omega}}} \tag{24}$$

$$L_2 = -\left(\frac{e^2}{\hbar v}\right)^2 \sum_{n=1}^{\infty} \frac{1}{n[n^2 + (e^2 Z_1/\hbar v)^2]} \tag{25}$$

The subscript (r) on the left-hand side of Equation (22) indicates we are interested in stopping in a random direction. The ion velocity is v, electronic charge and mass are e and m, and the target atomic number and density Z_2 and N, respectively.

* Data for H, He, and Li were also reported in this work, but because of the low energy losses, stripping procedures to obtain minimum energy losses which assumed Gaussian straggling distributions leave some doubt about the accuracy. For $Z \leqslant 5$, however, the losses are large enough to justify the Gaussian assumption.

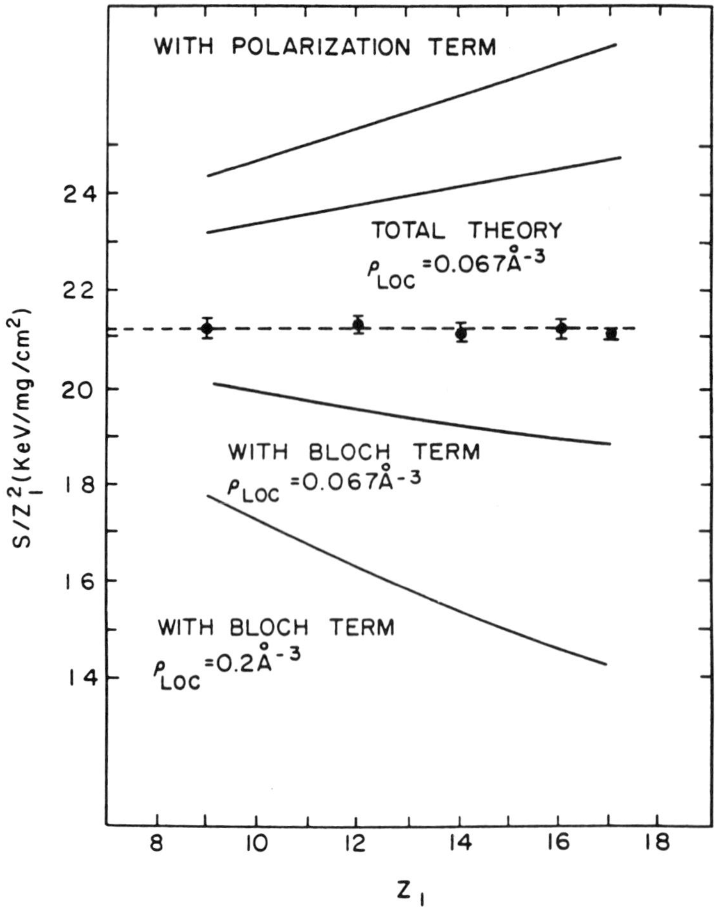

Figure 25. S/Z_1^2 versus Z_1 for bare F, Mg, Si, and Cl ions compared with theoretical estimates including Bloch correction and/or polarization term.

The contribution to Equation (22) from the term containing L_0 was deduced by Bethe (Be 30) via a quantal perturbation calculation to first order. Here $\hbar\bar{\omega}$ represents a particular weighted average of atomic excitation energies of the target. It is generally recognized that L_0 receives similar contributions from distant and close collisions with target electrons. [Actually included with L_0 in Equation (23) should be a term representing shell effects. This term has no Z_1 dependence and for our case has a value 2%–3% that of L_0 (Bo 67).]

The additional contribution to Equation (22) from the term containing

L_1 has been discussed at some length in the literature recently (Ja 72, As 72, As 73, Hi 74) and is generally known by the name of "Barkas effect" (Ba 73) after its discoverer. The form we have adopted for L_1 in Equation (24) has been suggested by Lindhard (Li 76), who has called attention to the roughly equal role played by close and distant collisions in the Barkas effect. In all of the above, the quantity $\kappa = Z_1 e^2/\hbar v$ is assumed to be much smaller than unity and as a result the term in Equation (22) containing L_1 yields the only Z_1^3 dependence in the stopping. This part of the stopping may be looked upon as resulting from the inclusion of certain nonlinear features in the response of the atomic system to the penetrating ion. For κ increasing toward unity, it is reasonable that this contribution becomes more significant. Note that the "average" energy $\hbar\bar{\bar{\omega}}$ in L_1 is different from $\hbar\bar{\omega}$ in L_0.

Finally as κ increases towards unity and beyond there are certain features of the Coulomb scattering of bound charges that may only be handled with great difficulty within the framework of a completely perturbation quantal calculation. Bloch (Bl 33) discussed this problem at some length in providing a connection formula between the regions $\kappa \ll 1$, where Bethe's formulas applies, and $\kappa \gg 1$, where a classical calculation due to Bohr (Bo 13) is expected to apply. Let it be sufficient here to note that the contribution to Equation (22) from Equation (25), the so-called "Bloch correction," arises from the inability of the quantal perturbation approach to adequately handle the small distance singular part of the Coulomb interaction. Collisions with atomic electrons specified by small angular momenta dominate this contribution, which is therefore only associated with the close collision part of the stopping power. For small κ, L_2 is independent of Z_1, which accounts for the particular form of expansion in Equation (22), with the Bloch correction yielding a Z_1^4 power correction to the stopping (Li 76).

In order to interpret the data we must modify the above discussion to incorporate the channeling effect and the large values of $\kappa (\kappa > \sim 1)$ associated with the experiments. Only where this is accomplished can it be shown that the processes associated with Barkas and Bloch corrections can play an important role in the experiments in spite of the nearly pure Z_1^2 dependence of the measured channeled stopping power.

For channeled particles, several theories for the analogous contribution represented by L_0 in Equation (22) have been discussed in the literature (De 75, Ap 67, Bo 70, Ko 74, Es 78a). The results generally may be applied when $\kappa \ll 1$, i.e., they are based on first-order quantal perturbation theory. These theories all predict therefore a Z_1^2 stopping dependence with a reduction factor of order 2–3 for well-channeled particles as compared to random direction penetration. A large part of this reduction factor, α, may be heuristically explained as due to a reduction in close collision contributions to L_0 associated with the drop in electron density at the center of the channel. One can set the best channeled energy-loss rate

$$(dE/dx)'_{\text{ch}} = \alpha \frac{4\pi e^4}{mv^2} Z_1^2 Z_2 N L_0 \tag{26}$$

and emphasize that this is the value expected from a "Bethe-like" first-order quantal calculation.

The channeling effects on the Z_1^3 contribution to the stopping may be estimated as follows. Since for random penetration roughly similar contributions from close and distant collisions are expected for both the Bethe and Barkas terms and since even the incremental contributions from a range of impact parameters is similar for both of these terms we suspect that at least roughly the Z_1^3 contribution to the stopping from the Barkas effect will be reduced for channeled particles again by the factor α.

Finally, the Bloch correction must be considered. Firstly, since this is a local effect we expect that for channeled particles it will depend on the average electron density on the path in the channel rather than the average total electron density NZ_2. Secondly, for the range of κ values in the experiment L_2 is no longer independent of Z_1. In fact, $Z_1^2 L_2$ is plotted in Figure 26 and it is virtually linear in Z_1 for the region applicable to our

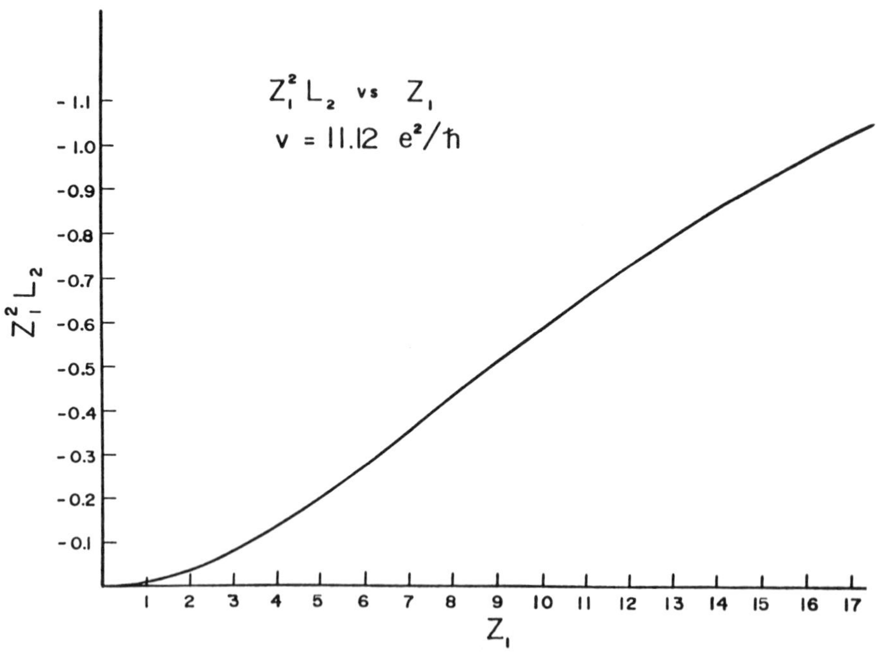

Figure 26. Bloch correction Z_1 dependence for $v = 11.12\, e^2/h$.

experiments. Since the algebraic signs of the Barkas and Bloch corrections are opposite, it is possible for these terms to cancel.

Summarizing, we may use Equation (22) if we let

$$L_0 \Rightarrow \alpha L_0 \tag{27}$$

$$L_1 \Rightarrow \alpha L_1 \tag{28}$$

$$L_2 \Rightarrow \frac{\rho_{\text{LOC}}}{Z_2 N} L_2 \tag{29}$$

where ρ_{LOC} is the local average electron density on the channeled particle path, which for our considerations is at the center of the channel. If we further divide by Z_1^2 and the target density we obtain the quantity S/Z_1^2.

One thing is immediately clear. The constant value of S/Z_1^2 tells us that either (a) both the Bloch and Barkas effects are so small they have not been detected or (b) both may be large but they have virtually canceled each other out.

Reasonable theoretical estimates of the magnitude of the Z_1^3 effects in the data require knowledge of ρ_{LOC} and $\hbar\omega$. It is not uncommon in channeling energy-loss theories to assume for ρ_{LOC} in silicon channels a value obtained by using only valence electrons whose density is assumed uniform over the whole crystal (De 75, Ap 67). This gives $\rho_{\text{LOC}} = 0.20$ electrons/Å^3. Alternatively one may reconstruct charge densities from x-ray form factors such as those given by Doyle and Turner (Do 68). For the 110 channel in silicon one obtains $\rho_{\text{LOC}} = 0.0672$ electrons/Å^3. The Bloch correction is included for both of these charge densities in the plot in Figure 25. The Barkas effect must clearly be included in the description if reasonable agreement between theory and experiment is to be obtained.

As mentioned earlier the contribution to L_1 from distant collisions has been caculated by Ashley *et al.* (As 72). Using this work and assuming with Lindhard (Li 76) that a nearly equal contribution also comes from close collisions and using Equation (28) for channeled particles we find $\alpha L_1 = 1.53 \times 10^{-2}$. For α we used the experimental value obtained as the ratio of our result, $S/Z_1^2 = 21.2\ (1 \pm 0.05)$ keV/mg/cm^2, to the random published experimental value of S/Z_1^2 (Es 78a) for He^{++} in silicon. Use of the latter quantity is required since random S/Z_1^2 values cannot be obtained for the totally stripped ions in these experiments due to charge state fluctuations. Figure 25 shows both the first-order result for S/Z_1^2 incremented by only the resulting Barkas contribution and incremented by both Barkas and Bloch contributions (in curve labeled "total theory"). The Bloch contribution in the latter is taken using the 0.067-electron/Å^3 center channel charge density. The resulting theoretical prediction is seen to be in poor agreement with the experimental points in the figure. We require substantially better cancellation of the Barkas and Bloch terms to improve the agreement. This can be achieved either by increasing ρ_{LOC} or reducing L_1. Stimulated by this state of affairs we have sought further theoretical justification for the low value of ρ_{LOC} adopted in the final comparison (Figure 25). Self-consistent-field pseudopotential calculations performed by Dr. M. Schluter of Bell Laboratories indicate a value of $\rho_{LOC} = 0.043$ electron/Å^3 with an estimated error of 25%. Rather than adopt this, or any other number, as a concrete value for ρ_{LOC} we merely use it as further justification for discarding the uniform valence gas picture (with $\rho_{LOC} = 0.02$ electron/Å^3) mentioned earlier. It therefore appears that the Barkas contribution to the data is grossly overestimated by the current theory of the effect.

It should be realized that Equation (24) for L_1 assumes that the excitation of the passing ion serves as a perturbation on the bound target electron motion. In particular, the recoil of the electron during the course of the collision should be small compared to its atomic orbit size for the calculational scheme used in the derivation of L_1 to work. When this is the case the role of the target binding force is manifest through the appearance of $\bar{\bar{\omega}}$ in Equation (24). The Z_1^3 effect is therefore a measure of the importance of binding forces in modifying the recoil of atomic electrons during the course of the collision. For the very large Z_1 and hence electronic recoils in these experiments target binding must play a diminished role in describing these recoils during a substantial part of the collision. Indeed, during these times it may well be more accurate to describe the evolution of the electronic states as being basically Coulombic in the field of the Z_1 ion with the binding force of the target nucleus acting as a perturbation. Insofar as contributions to the Barkas effect are concerned, the perturbation plays a smaller role the higher Z_1 becomes and hence the Barkas effect disappears for high Z_1. It does not appear that the present structure of the theory of the Barkas effect can accommodate this possibility. It is suggested, however, that the explanation for the reduced Z_1^3 contribution from polarization effects in the experiments lies in the above explanation.

The near absence of higher-order Z_1 effects for well-channeled ions is intriguing. Perhaps the result is a simple cancellation of what should be two large effects. This conclusion would be completely changed if the minimum $\langle 110 \rangle$ charge density were significantly below 0.067 Å^{-3}; in that case the conclusion would be that both the Bloch and Barkas corrections were beyond the reach of the experiments. In either case, the lack of visible higher-order Z_1 effects in planar channeling in Au (Da 77) as well as in $\langle 110 \rangle$ Si suggest that some long-cherished notions need further attention.

4. Charge Changing Collisions

As outlined in the introduction, the charge state distribution of an energetic ion beam emerging from a solid will be determined by electron capture, loss, and excitation processes occurring in the solid and may be modified by exit surface effects and relaxation processes following emergence. For ions in the energy region discussed in this chapter, charge state equilibrium is attained in even the thinnest foils used. For target atoms of $Z_2 \gg 2$ the charge changing cross sections for charge states near equilibrium are 10^{-16} to 10^{-17} cm^2; most solids have densities of $\sim 5 \times 10^{22}$ atoms/cm^3. Thus, mean free paths for charge exchange are in the order of 20 to 200 Å in a random medium where the ions are free to interact with all of the target electrons.

Under channeling conditions, however, the ions are constrained to collisions with only valence and/or conduction electrons with velocities in the order of v_0. Thus for ions with $v_i \gg v_0$ we have simulated a high-density, almost static electron gas.

4.1. Capture and Loss under Channeling Conditions

A relatively straightforward example of the effect of channeling on charge state distributions (CSD) is seen in Figure 27 for 60-MeV iodine ions channeled in Au (Lu 70). First, it can be noted that the equilibrium charge state distribution for channeled ions is lower here than for randomly directed particles. This is due to the suppression of multiply ionizing collisions which are small impact parameter events. Second, it can be seen that the CSD changes with target thickness and that equilibrium for channeled ions does not take place until the target thickness is $\sim$1000 Å because of the reduced cross sections for both capture and loss events. The suppression of the capture cross section occurs because these ions are traveling at velocities equivalent to 250-eV electrons and can interact most directly with the $5d$ electrons of Au which are bound with only $\sim$10 eV.

The effect is different and more startling for higher-velocity almost totally stripped ions as can be seen in Figure 28 (Da 72). Here we show the emergent charge state distributions for 40-MeV oxygen ions channeled in the [110] axis of an Au crystal 0.33 μm (3300 Å) thick as a function of the input charge state of the oxygen ion together with the CSD obtained from injection in a "random" direction. The random distribution (dashed curve) is representative of charge-state equilibrium since it is independent of a pathlength (0.143 to 0.663 μm) in the crystal and independent of input charge state (6+, 7+, or 8+). The channeled-ion charge state distributions are clearly nonequilibrium cases. Since the 8+ fraction is enhanced for all initial charge states, it is immediately obvious that an equilibrium charge state distribution for channeled ions will contain a higher fraction of higher charge states than the random case. Moreover, it is clear that some of the charge-changing cross sections for the channeled ions in attaining

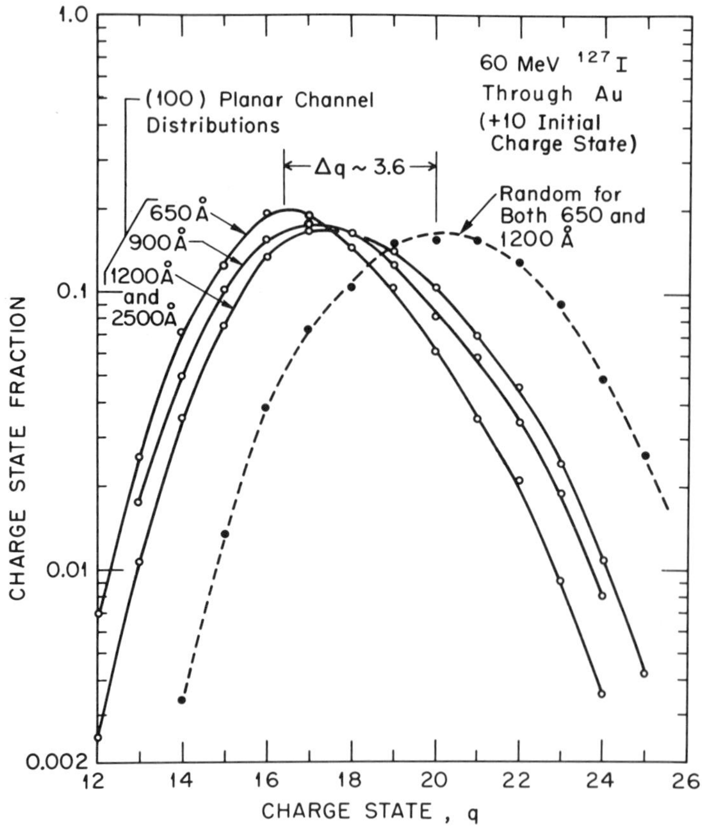

Figure 27. Emergent charge state distributions of $E_0 = 60$ MeV I ions emerging from Au in a random crystal direction and in a (100) planar channeling direction as a function of crystal thickness.

equilibrium must be quite small, i.e., if we assume that equilibrium is attained in approximately three mean-free paths for charge exchange, then charge-changing cross sections of $\ll \sim 5 \times 10^{-18}$ cm^2 must be involved if equilibrium is not achieved in 0.5 μm with target densities of $\sim 5 \times 10^{22}$ atoms cm^{-3}.

These nonequilibrium effects persist at lower energies and are observed in planar as well as axial directions. The tendency toward higher charge states diminishes with decreased channel dimension and approaches the random distribution.

Since the emergent charge state distributions for these channeled ions

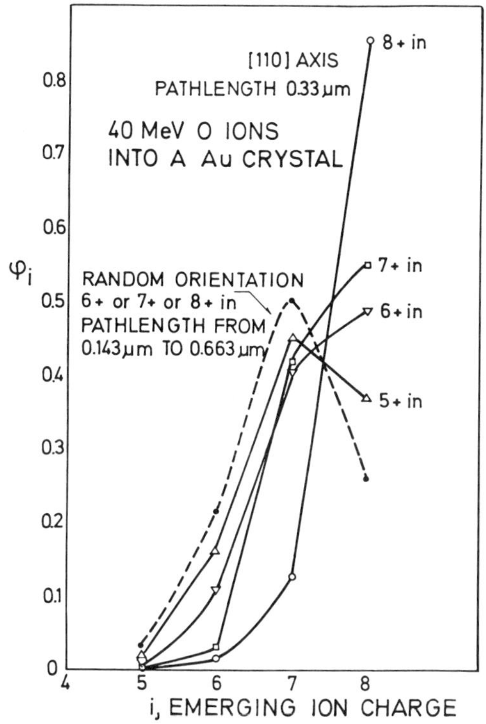

Figure 28. Emergent charge state distributions for $E_0 = 40$ MeV oxygen ions from Au in a random direction (independent of incident charge or thickness) and from a [110] axial channel as a function of incident charge state.

are markedly far removed from equilibrium, it should be possible to adduce the relevant charge-changing cross sections from variations in charge state distribution with change in pathlength and, in some cases, from the additional information obtained from different input charge states. Thus, for example, a capture cross section from 8+ to 7+, $\sigma_{8,7}$, was determined to be 2.5×10^{-19} cm^2 for 40 MeV O^{8+} channeled in an Au {100} planar channel and the loss cross section 7+ to 8+, $\sigma_{7,8}$, was 4.2×10^{-19} cm^2. The ratio $\sigma_{7,8}/\sigma_{8,7}$ causes the observed predominance of 8+ in the channel and the small magnitude of the cross section leads to the large distances required for equilibrium.

An interesting analogy can be made between the channeled and random charge-exchange cross sections and those for 40-MeV O ions in He and in Ar. In He $\sigma_{7,8} = 2.5 \times 10^{-19}$ and $\sigma_{8,7} = 9 \times 10^{-20}$ while in Ar $\sigma_{7,8} = 3 \times 10^{-18}$ and $\sigma_{8,7} = 1.5 \times 10^{-17}$ (Ma 65). In the case of He, there are no electrons with sufficient orbital velocity to meet the matching criterion, and the capture cross section is a factor of 100 lower than that in Ar, where such electrons are available for transfer.

Collisions in He may thus be compared with collisions in channels both in magnitude and in capture to loss ratio $-\sigma_{7,8}/\sigma_{8,7} = 2.8$ for He and 3.2 for Au {100}. When the fraction of O^{6+} is neglected, these ratios also represent the equilibrium charge state fraction σ_8/σ_7. Although, because of their larger magnitude, it is not possible to measure these cross sections in random Au, it can be noted that the ratio ϕ_8/ϕ_7 emerging from random Au is 0.5, which is comparable to the value 0.2 obtained in Ar gas targets.

In an experiment of the type discussed in Section 2.1 the energy-loss spectrum of a given exit charge state was measured for a set of different input charges. The stability of fixed charge states for 27.5-MeV oxygen (6^+, 7^+, and 8^+) ions was measured as a function of oscillation amplitude in the {111} planar channel of Au. It was found that for the A_0 (minimum amplitude) energy loss group both the magnitude of the energy loss (Section 3.1) and the charge fraction were strongly dependent on input charge but for the A_1 energy-loss group no input charge dependence was observed. These results imply that for amplitudes which carry the ion to within ~0.6 Å the only electrons available are the conduction $5d^{10}$ $6s^1$ electrons, which are too slow to be effectively captured; at shorter distances higher-velocity bound electrons become available for capture. The consequences of this effect on "convoy" electron production are discussed in Section 4.3.

4.2. Radiative Electron Capture

The phenomenon of radiative electron capture (REC), whereby an ion captures an electron into one of its shells and emits a photon, has been known for some time from studies in astrophysics and in the physics of laboratory plasmas. The process has been described theoretically by Bethe and Salpeter (Be 57). The first observation of REC by energetic heavy ions in solids was reported only recently, however, by Schnopper *et al.* (Sc 72). They found a new feature in the x-ray spectrum which appeared as a broad band well above the highest-energy characteristic x-ray lines of the projectile ion and attributed it to REC. The REC phenomenon becomes most probable when the ion is totally stripped and no outer-shell electrons are available to fill K vacancies. Then a free or weakly bound electron of the target atom can be captured directly into the K shell of the moving ion and be stablized by the emission of a photon. The energy E_p of the REC photon resulting from a capture of a free electron by an ion of charge Z_1, mass M, and energy E is

$$E_p = Z_1^2 \mathrm{Ry} + m_e E/M_1 \tag{30}$$

where Ry is the Rydberg energy and m_e is the electron mass. The first term is the binding energy of the K electron in a hydrogenlike ion and the second represents the kinetic energy of the electron in the rest frame of the ion (e.g., for an ion moving at 1 MeV/amu encountering an electron at rest the second term would be ~500 eV).

As noted by Schnopper, the width of the observed REC line is affected by a number of things. In measurements utilizing heavy-ion beams incident on amorphous solids, the heavy ions will have a distribution of charge states inside the solids and each charge state can result in a REC photon of a different energy as can be seen from Equation (30). In most vases the differences in energy are the order of the resolution of the detecting systems, thus the line can be shifted in energy as well as broadened. The intrinsic width of a REC line arising from a single charge state should reflect the velocity distribution of the target electrons captured by the moving ion. Even the small energies of the free electrons of the target (~5 eV) contribute significantly to the REC linewidth when transformed to the rest frame of the moving projectile.

In the work reported by Appleton *et al.* (Ap 79) REC was studied using 17–40 MeV oxygen ions channeled in Ag and Si. The experiment consisted of passing an O ion beam in a selected charge state through thin crystals in a channeling or random direction and observing the spectrum of x rays emanating from the target with a Si(Li) detector. The potential advantages of studying REC phenomena with channeled beams are manifold and a number of these advantages are demonstrated in Figure 29, which shows an x-ray spectrum taken under channeling [011], and random incidence in Ag. (The peak at ~3 keV is from Ag L x rays, and the anticipated REC x-ray energy for capture of an electron at rest in the laboratory frame is 1.8 keV.)

First, the channeled O ions remain predominantly in a fixed charge state (Section 4.1). In contrast, heavy ions incident in polycrystalline targets have a distribution of charge states, and the energies and widths of the REC radiation for the different charge states will interfere and complicate interpretation. A vertically expanded version of the random spectrum of Figure 29 shows (Ap 79) a poorly resolved broad peak at about 1.3 keV strongly asymmetric to lower energies, which is likely due to various oxygen charge states.

Second, the channeling effect greatly reduces close encounters between the ions and bound target electrons without diminishing the ion interactions with the free or conduction electrons. This renders the REC of free electrons a much more prominent feature relative to the characteristic target x-ray background. This is clearly demonstrated in Figure 29. The normalization here is to the Ag L x-ray peak. Under channeling conditions this peak is reduced by a factor of ~25.

Finally, in the channeling configuration most of the incident ions appear to be in a single charge state interacting primarily with conduction electrons. This imposes well-defined boundary conditions on the REC effect which

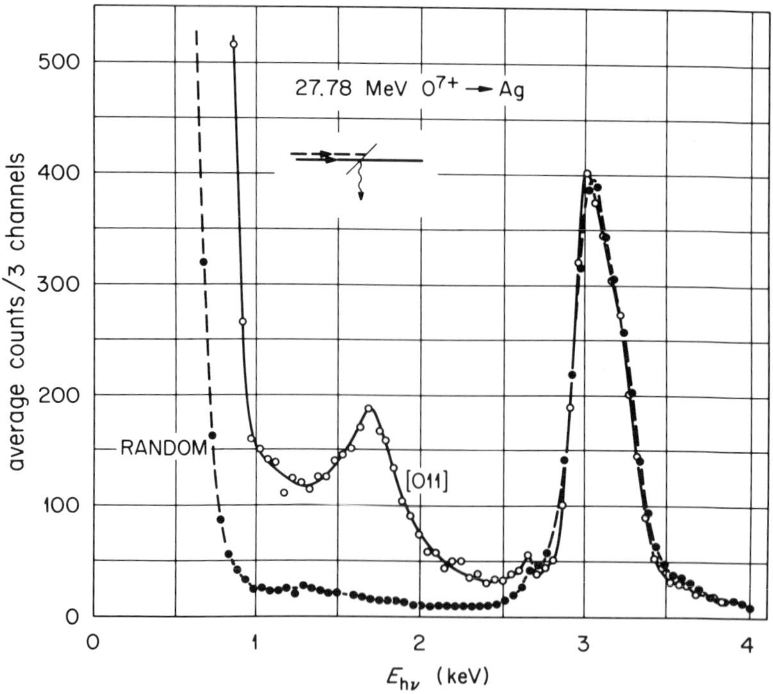

Figure 29. Photon spectra (normalized at the 3-keV Ag L x-ray peak) measured for 27.78 MeV O^{7+} ions transmitted parallel to the [011] direction and in a random direction of a 0.62-μm Ag single crystal.

should be more amenable to theoretical calculations (i.e., a hydrogenlike ion in a free-electron gas).

The simplest comparison that can be made with theory is the REC peak energy. For all injection energies ranging from 17.5 to 40 MeV, the measured peak energies for channeled oxygen ions agreed well with Equation (30) with $Z_1 = 8$.

More complex to analyze are the REC peak shapes and REC cross sections. To do this analysis requires assumptions concerning the electron environment through which the ion moves and a consideration of other processes which lead to radiation in the same energy region as REC. To include all possible trajectories of the channeled ions the assumption was made that the ion samples, on straight line trajectories, all portions of the Wigner–Seitz sphere corresponding to impact parameters b greater than b_{min} (Ap 79). This minimum value effectively excludes the small impact parameters suppressed by channeling and in general depends on v, Z_1, and the lattice potentials.

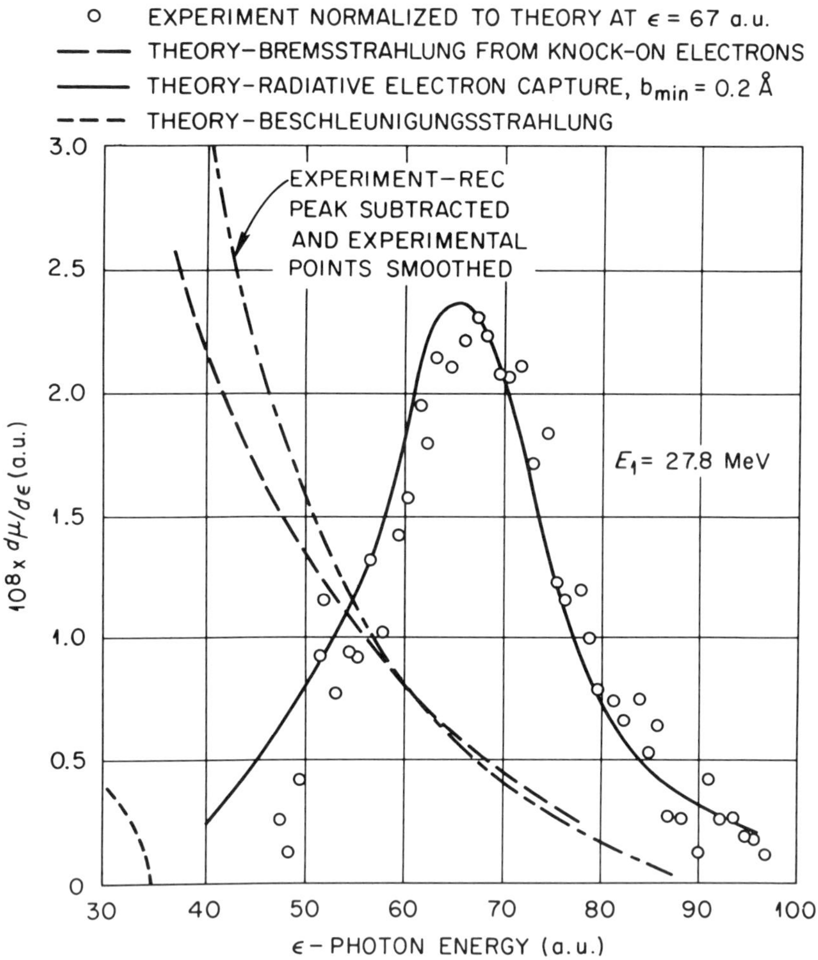

Figure 30. Comparison of measured REC peak and subtracted background with theoretical calculations detailed in (Ap 9).

A comparison of the theoretical peak shape with the data is shown in Figure 30. Using $b_{\min} = 0.2$ Å the fit is reasonably good and the background attributable to bremsstrahlung arising from knock-on electrons is well matched. The energy dependence of the width is not as satisfactory (see Figure 13 of Ap 79). The predicted energy dependence of the cross section for REC is in qualitative disagreement with the observations (Figure 31). The general decline of the cross section with increasing energy would be expected for any theory based upon considerations similar to those of Bethe and Salpeter (Be 57). This is illustrated in a comparison of the data obtained

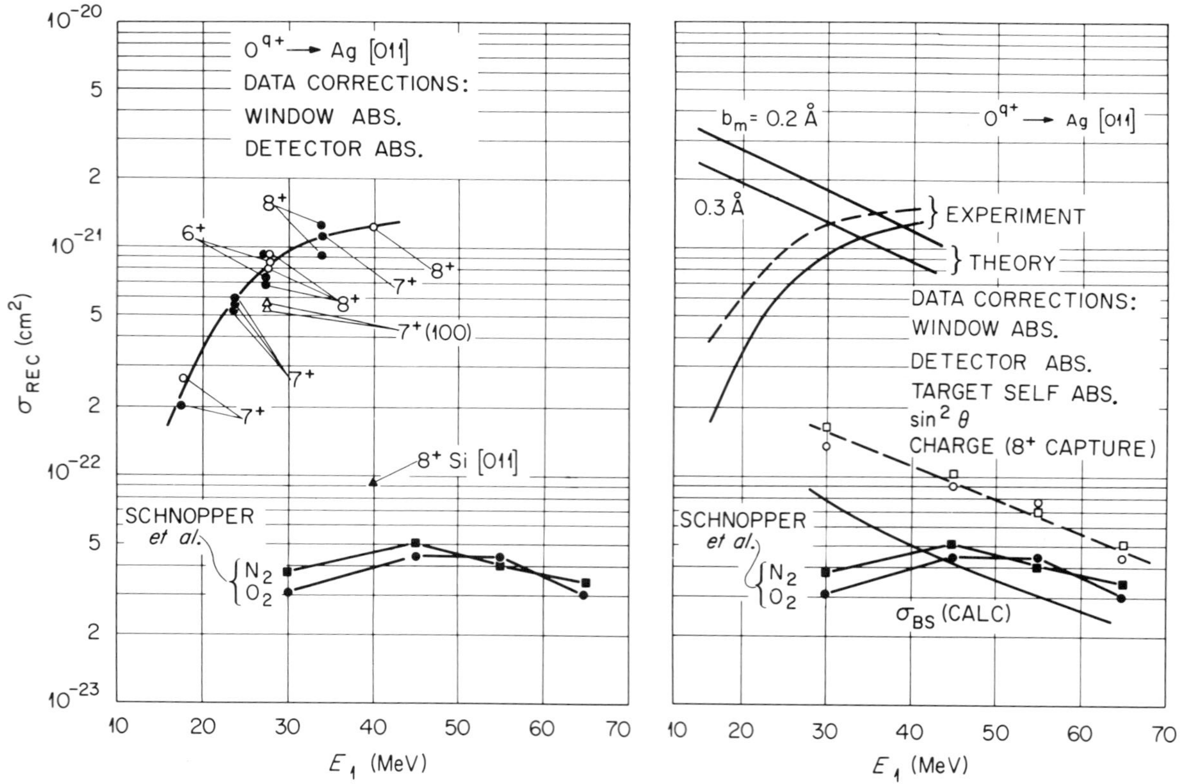

Figure 31. Measured REC cross sections versus oxygen ion energy. Left portion shows data for channeled ions (Ap 79) and gas target results (Sc 74) corrected for window and detector absorbtion. Right-hand portion shows data further corrected for angular anisotropy of emission ($\sin^2 \theta$) and for fraction of the beam in charge 8+ (bare ion) together with calculation from Bethe and Salpeter (σ_{BS}) for the gas target case and that of Appleton *et al.* (Ap 79) for the channeled ion case.

by Schnopper *et al.* (Sc 74) using O ions on gaseous targets of N_2 and O_2 in the lower portion of the figure (closed points) which when corrected for fractional charge state, etc. (open points) can be compared with the Bethe–Salpeter calculation. The rise in cross section observed by Appleton *et al.* is well documented; however, it is not anticipated and is, at this writing, still unexplained.

4.3. Electron Capture and Loss to Continuum States (Convoy Electron Production)

If an energy analysis is made of the electrons ejected in the forward direction following a collision of a swift ion with either a gaseous or solid target, a sharp cusp will appear in the spectrum at an electron velocity $\mathbf{v}_e$ which matches the ion velocity $\mathbf{v}_i$ (Ha 70, 71; Cr 70; La 80) (see Figure 32). In ion–atom collision (Va 79), the electrons originate from capture to the continuum (ECC) for fast bare or nearly bare projectiles, and from loss to the continuum (ELC) when loosely bound projectile electrons are available. The ECC cusps are strongly skewed toward lower velocities and exhibit a full width at half-maximum roughly proportional to v. A close examination of ELC data shows that ELC cusps are nearly symmetric, with widths independent of v in the velocity range 6–18 a.u., a result unpredicted by recent theory.

In contrast, the cusps characterizing "convoy"-electron production in

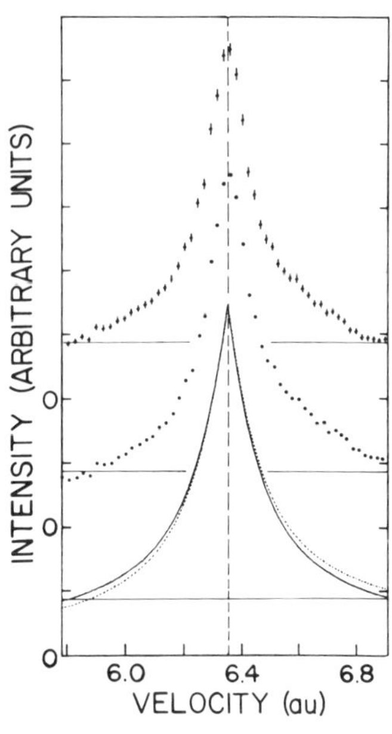

Figure 32. Spectrum of convoy electrons emergent near 0° from 16 MeV O^{3+} ions traversing a 30-μg/cm^2 carbon foil. The upper points are obtained from the raw spectrum (lower points) through a correction factor to account for the electron-escape-depth velocity dependence. The lowest curves represent fitted cusp shapes which show the degree of symmetrization produced with the above correction (El 81).

ion–solid collisions are skewed toward high electron velocities, but exhibit velocity-independent widths that seem to be very similar to ELC widths. While the shape of the convoy peaks is independent of projectile Z_1 and of target material, the yields in polycrystalline targets exhibit a strong dependence on projectile Z and velocity. Numerous attempts have been made to link convey-electron production to binary ECC or ELC processes, usually at the exit surface. As an alternative, a solid-state "wake-riding" model has been proposed (Br 77). Measured dependences of shape and yield on projectile charge state and energy are inconsistent with the predictions of either theory (De 74, Dr 76).

To aid in unraveling this puzzling array of similarities and differences, coincidence experiments were carried out to investigate the dependence of shape and yield of the convoy-electron spectrum on the charge q_e of the associated emergent ion, for ions traversing polycrystalline targets as well as for well-channeled ions traversing a gold single crystal (El 81). For the best channeled ions, the fact that their charge often does not change during their entire passage (Section 4.1) through the crystal makes such channeled ions useful probes.

For the random solid targets used (30 μg/cm^2 C, 50 μg/cm^2 Al, 100 μg/cm^2 Au) as well as for the Au single crystal (~300 μg/cm^2) oriented in a random (nonchanneling) direction, it was found that the convoy-electron yield per projectile ion is independent of emergent projectile charge state, and simply mirrors the statistical fraction of the corresponding projectile-charge-state distribution. Since in both the ion–atom and the ion–solid collision case there is rather strong projectile Z dependence, this observation is a surprise, and argues strongly against a surface-layer origin for convoy electrons.

For 2.4-MeV/u oxygen beams traversing the Au single crystal, Table 6 displays the convoy-electron yield $Y(\langle hkl \rangle)$ per emergent ion for a particular initial and final charge state. The measured fraction $[\mathrm{CF}(q, q_e)]$ of emergent ions with this final charge state is also displayed in parentheses. $Y(\langle hkl \rangle)$ and Y(rand) refer to particular channeling and random directions, respectively. For the random direction the yield is found to be independent of the final projectile-charge state, q_e. Hence the fraction of convoy electrons coincident with a particular q_e simply mirrors the unweighted fraction of projectiles having that q_e. It is also evident that convoy-electron production for channeled ions is much suppressed, with the greatest suppression arising in the most open channel ($\langle 110 \rangle$) and for the most compact projectiles (O^{8+}, O^{7+}).

In a channel, a large fraction, F_A, of ions is confined to collisions with Au conduction electrons ($5d^{10}6s^1$), for which the capture and loss cross sections are sharply reduced; the remaining, smaller fraction, $F_B = 1 - F_A$, experiences collisions with more strongly bound electrons. Ions in group B

Table 6. Percentage Convoy-Electron Yield per Emergent Ion, for O^{q+} Incident at 2.4 MeV/u on Au in the 110, 100, and Random Directions[a]

q_{in}	q_e^{out}	8+	7+	7+
8+	$Y(\langle 110\rangle)$	21 (68)	39 (28)	82 (4)
	$Y(\langle 100\rangle)$	37 (59)	58 (35)	79 (6)
	Y(rand)	100 (26)	100 (59)	100 (15)
7+	$Y(\langle 110\rangle)$	29 (42)	24 (51)	58 (7)
	$Y(\langle 100\rangle)$	37 (52)	47(42)	71 (6)
	Y(rand)	100 (25)	100 (60)	100 (15)
6+	$Y(\langle 110\rangle)$	37 (31)	29 (42)	21 (27)
	$Y(\langle 100\rangle)$	39 (49)	45 (42)	47 (9)
	Y(rand)	100 (27)	100 (57)	100 (16)

[a] The yield is normalized to the measured random yield of $\sim 3.8 \times 10^{-4}$ e/ion. The number in parentheses is the percentage of emergent ions in state q_e.

approach charge-state equilibrium rapidly, and the corresponding convey-electron yield Y_B should be well approximated by Y(rand).

For ions traversing the most open channel ($\langle 110\rangle$) near channel center, where the low electron density suppresses both capture and loss, and low electron momentum further suppresses capture, the lowest convey-electron yields are to be expected. Assuming, therefore, that for group-A ions the convey-electron yield $Y_A(8^+, \text{in}; 6^+, \text{out})$ is very small, the fractions $F_B(q_e)$ of class-B ions that emerge with charge q_e from measured yields and charge fractions for each channel may be inferred. Knowing the F_B's, $Y_A(q, q_e)$ for all q and q_e were calculated for a consistency check and confirm that all the Y_A's are small compared to Y_B (i.e., close to zero). The fraction of ions belonging to group B is given by $F_B = \sum_{q_e} F_B(q_e)$ and $F_A = 1 - F_B$. It was found that $F_A \simeq 0.78$ for the $\langle 110\rangle$ channel and $F_A \simeq 0.68$ for the $\langle 100\rangle$ channel. Using these F_A values and crystallographic values for the geometric area of each channel, the effective area available to group-B ions and unavailable to group-A ions is 1.3 Å^2 out of 5.83 Å^2 for $\langle 110\rangle$, and also 1.3 Å^2 out of 4.14 Å^2 for $\langle 100\rangle$. A unique value $r_{crit} = 0.65$ Å, corresponding to $\pi r_{crit}^2 = 1.3$ Å^2, can be deduced. Thus, only ions that have impact parameters <0.65 Å (group B) produce convoy electrons, with an efficiency approximating the random value.

Based on this work, it was possible to construct a simple model which reproduced most of the observed results. Convoy-electron production is initiated in close collisions throughout the bulk while the final ion-charge state is determined at the exit surface. A close relationship between ELC and convoy-electron production is suggested by the fact that the widths of the convoy-electron and ELC cusps are equal within uncertainty and are independent of projectile nuclear charge, projectile velocity, and target. Convoy-electron production is then initiated by single- or multiple-electron capture events having cross sections $\sim 10^{-17}$ cm^2 (Ma 65), predominantly to excited states $\gtrsim 90\%$ of the time (Br 74) followed immediately by ELC. Subsequent electron scattering (elastic and inelastic) leads mainly to scattering into a wide range of angles, effectively extinguishing the convoy-electron population. (Some unknown degree of repopulation is possible by secondary elastic scattering.) The net production of several convoy electrons per emergent ion is depleted by electron scattering to $\sim 10^{-3}$–10^{-2} observable electrons per ion. In Au atoms the 6*s* and 5*d* electrons have kinetic energies <10 eV. But the 5*p* and 4*f* electrons—which have binding energies of ~250–460 eV and are therefore far more efficient at contributing to capture according to the Bohr $v_e \sim v$ matching criterion—have mean radii ranging from 0.60 to 0.28 Å. Therefore, the "magic" distance of 0.65 Å can be assigned a plausible physical interpretation.

The interpretation of the data is that ions must approach an atomic string within the well-defined distance of ~0.65 Å to generate convoy electrons, independent of the particular channel chosen. This value is very suggestive of electron capture in the bulk, because of the availability of more strongly bound electrons (5*p* and 4*f*) at radii less than or equal to this value (but unavailable at greater radii).

The skewness of the electron velocity distributions from solids toward electron velocities $v_e > v$ can be qualitatively and quantitatively explained by considering the velocity dependence, $\sim v_e^{1.6\pm0.1}$, of the mean free path of free electrons in bulk solids (As 79). The observed cusp shape may be by a velocity-dependent factor reflecting the exponential attenuation due to electron scattering within the bulk. The result of this procedure is shown in Figure 32 for the typical case of 16-MeV O^{3+} traversing C. The resultant symmetric, peak-normalized curve is closely similar to experimental ELC cusps from C^{q+}, O^{q+} traversing Ne and Ar. Since no fitting procedure beyond overall peak normalization is used, the ability to symmetrize quantitatively the skewed peak is viewed as support for the bulk production of convoy electrons, a small fraction of which escape through the surface.

A curious and unexplained enigma remains unresolved by the otherwise very successful model. Three facts need to be reconciled. The free-electron-scattering data suggest that all of the observed convoy electrons—though they are produced throughout the bulk—originate with the final ~20 Å of passage through the target (otherwise they scatter out). Yet the mean free path for projectile–ion charge changing under our conditions is ~200 Å, so that any ion traversing the final 20 Å of target has little likelihood of changing charge. The fundamental question posed is as follows: How can the correlation between emergent-ion charge and convoy-electron yield be broken in a distance of ~20 Å? Unless the correlation is broken, it is very difficult to understant why the convoy-electron yield is strongly dependent on projectile nuclear charge ($Z_1^{\sim 2.7}$), yet is independent of the emergent-ion charge (Z_1 screened by zero to two tightly bound K electrons).

5. Resonant Coherent Excitation

From the work discussed in Section 4.1 on charge exchange, and in Section 3.1 on the effects of bound electrons on stopping, it is clear that the concept of states of certain ions moving in channels is a maningful one, i.e., they are not collision broadened to the extent that discrete eigenstates are not useful concepts in their description. In this case it is also meaningful to discuss transitions between eigenstates (i.e., to perform a type of spectroscopy to better describe these states). In the work described in this section the lattice periodicity supplies the requisite exciting frequencies and the excitation event is signaled by an increased ionization probability for the excited ion passing through a channel (Da 78, Mo 79).

Although the potential controlling the detailed trajectory of a channeled ion can be well represented as a continuum, the ion passing between ordered rows of atoms with velocity v_i experiences a coherent periodic perturbation of frequencies [$\nu = K(v/d)$, $K = 1, 2, 3, \ldots$, where d is the distance between atoms in the row (Figure 2)]. When one of these frequencies coincides with $\nu_r = E_{ij}/h$, where ΔE_{ij} is the energy difference between states i and j of the ion, a resonant coherent excitation might occur. The excited state, having a higher cross section for ionization by electrons present in the channel, would be less likely to survive its passage in its initial (input) charge state. To carry out these experiments an ion beam of a given selected charge state is passed through the crystal and the emergent charge state distribution is measured; in some cases the individual emergent charge is energy analyzed to assess the fraction in a given energy loss group (see Section 2.1).

In Figure 33 for incident hydrogenlike N^{6+} we display the charge fraction $[N^{6-}]/[N^{6+} + N^{7+}] = R$, surviving passage through and emerging from a $\langle 111 \rangle$ axial channel in an 850 Å thick Au crystal as a function of the incident energy. In the absence of resonances one would expect a monotonic downward trend. Instead, sharp decreases occur at certain velocities. The spacing along a $\langle 111 \rangle$ row is 7.074 Å. Resonances are seen for $n = 1$ to $n = 2$ transitions with $K = 4$, 5, and 6, and perhaps for $n = 1$ to $n = 3$, $K = 5$. The additional scales shown are given in terms of v/v_r, where v_r is the resonant velocity calculated for the $n = 1$ to $n = 2$ vacuum states of the ion.

The Oak Ridge group has observed more than 50 $n = 1$ to $n = 2$ resonances (axial and planar) with hydrogenlike ions of $Z = 5$–9 and heliumlike ions $Z = 7$–9 in the energy range $1 \leqslant E \leqslant 3.5$ MeV/amu. For all

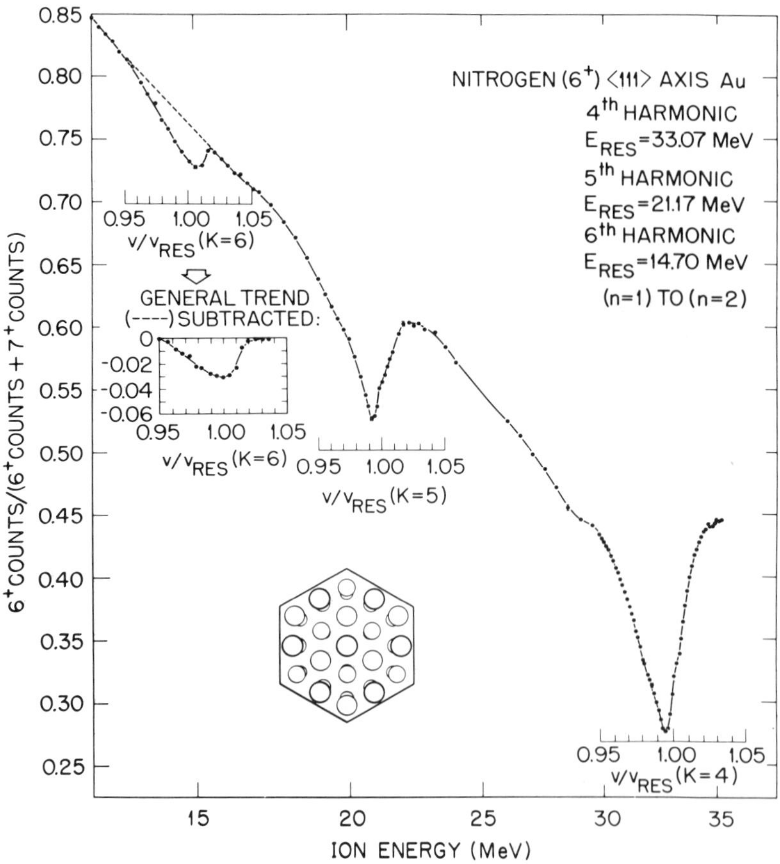

Figure 33. Ratio of N^{6+} counts to the sum of N^{6+} and N^{7+} counts as a function of initial energy for $\langle 111 \rangle$ axial channeling in Au. The incident beam is N^{6+}. The additional scales are in terms of the resonant velocity calculated for the vacuum states of N^{6+}. For $(n = 1)$ to $(n = 2)$ fourth harmonic, $E_r = 33.09$ MeV; fifth harmonic, $E_r = 21.19$ MeV; and sixth harmonic, $E_r = 14.72$ MeV.

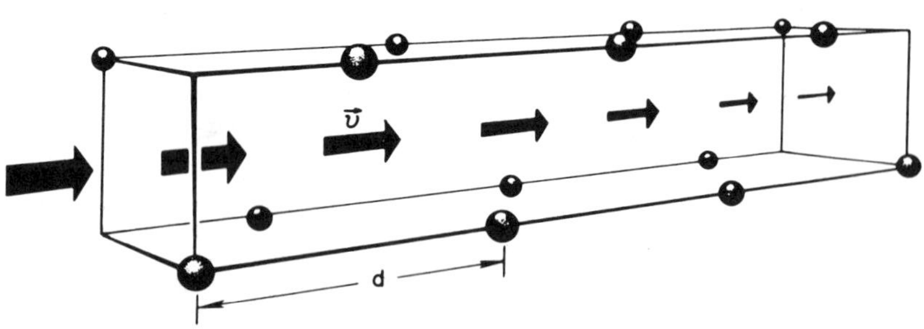

Figure 34. Schematic showing the influence of the symmetry of the perturbations on an ion penetrating in a $\langle 001 \rangle$ direction of a face-centered cubic lattice.

the simpler features of the resonances, some generalizations can be made: (1) narrower channels given stronger resonances, (2) higher harmonic resonances are weaker, (3) there are slight shifts in the resonance peaks to lower (v/v_r), and (4) the peaks are asymmetric towards lower (v/v_r).

To understand these and other features let us concentrate on a single channel, the $\langle 100 \rangle$ axial channel. The configuration of this channel, i.e., alternating pairs of atoms acting on the ion, is illustrated in Figure 34. The electric poential in this axis can be shown to have the following form:

$$V = \sum_{klm}^{\text{even}} V_{klm} \cos(2\pi kz/a) \cos(2\pi lx/a) \cos(2\pi my/a)$$
$$+ \sum_{klm}^{\text{odd}} V_{klm} \sin(2\pi kz/a) \sin(2\pi lx/a) \sin(2\pi my/a) \qquad (31)$$

where the axis is centered on the direction of motion $\langle 100 \rangle$ and x and y are centered on the othogonal $\langle 010 \rangle$ and $\langle 001 \rangle$ channels. The term a denotes the unit cell length for the $\langle 100 \rangle$ axis, $a = d$.

Using a Molière potential to describe the Au atom, the Fourier components of the electric field can be calculated and are shown for $K = 2$, 3, and 4 in Figure 35. The value at the centerline is for those ions which pass exactly down the center of the channel. The A and B directions denote the field experienced along paths that take the ion toward atomic rows and in a direction between atomic rows, respectively. The explanations of two of the features mentioned above are immediately obvious as higher harmonics have weaker Fourier components and trajectories constrained closer to atomic rows have stronger Fourier components.

More significantly a qualitative difference between even and odd harmonics is demonstrated. For centerline trajectories even harmonics only the parallel component is present, where in odd harmonics only the perpendicular component is present. Thus, for example, in the $1s$ to $2p$ transition only $1s \to 2p_z$ excitation is possible in even harmonic.

Other features appear if one considers injecting an ion at a slight angle to the centerline (i.e., the ion has transverse velocity in the B direction of Figure 35). Thus in addition to the high frequency (v_i/d) a lower-frequency component corresponding to the rate at which atomic rows are crossed is introduced and here again differences between odd and even harmonics enter.

For odd harmonics the field contains the frequencies $(v/a)(k \sin\theta \pm l \cos\theta)$; $l = 1, 3, 5, \ldots$. The $2p_x$ and $2p_y$ (degenerate) states are the strongest, and only the sideband frequencies $v_0 \pm v_0/K \tan\theta$ appear. The central frequency v_0 is missing because for odd K the field changes one cycle as the trajectory moves along direction B a distance a. For even harmonics the field

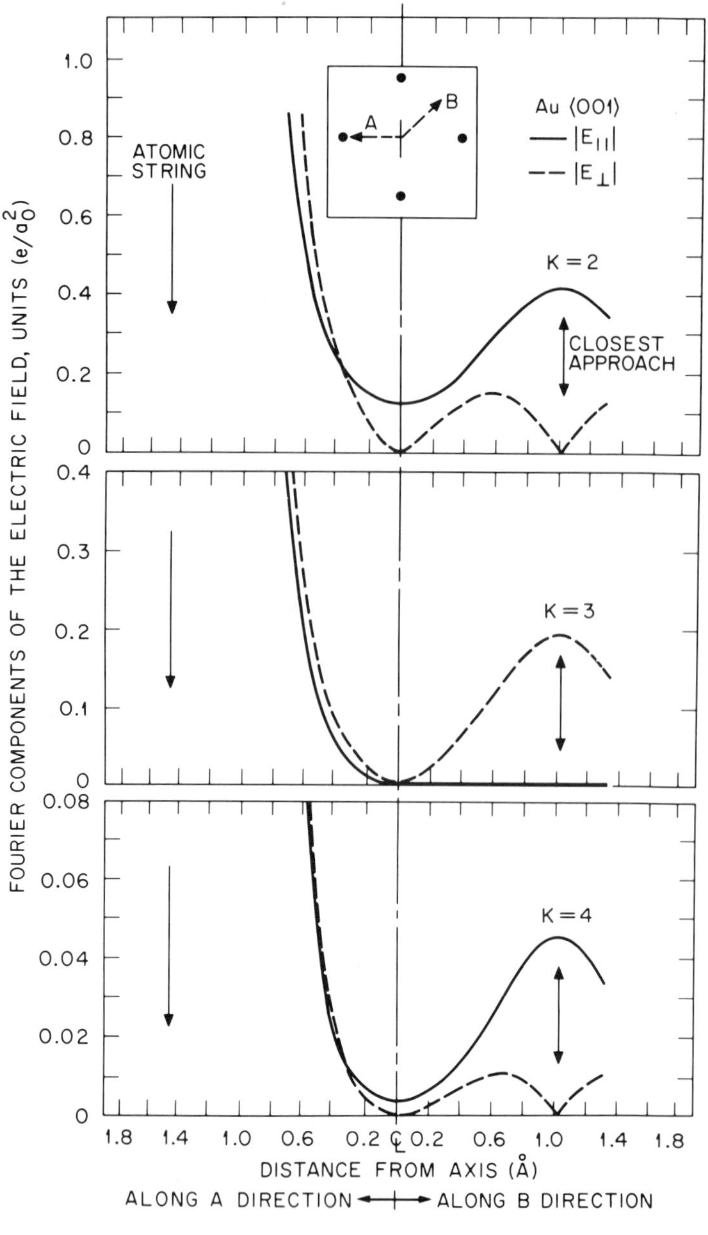

Figure 35. Fourier components of the longitudinal and transverse parts of the electric field for trajectories parallel to the ⟨001⟩ axis of Au versus displacement from the channel center line; toward an atomic row (A) and between atomic rows (B). Solid line shows longitudinal component; dashed line shows transverse component.

contains frequencies $(v/a) \times (k \cos\theta \pm l \sin\theta)$, $l = 0, 2, 4,\ldots$. The central component is mainly in the z direction so that the central frequency v_0 is maintained.

The predictions of this model are amply borne out in Figures 36 and 37. Figure 36 shows an odd ($K = 3$) harmonic resonance for O^{7+} ions. For

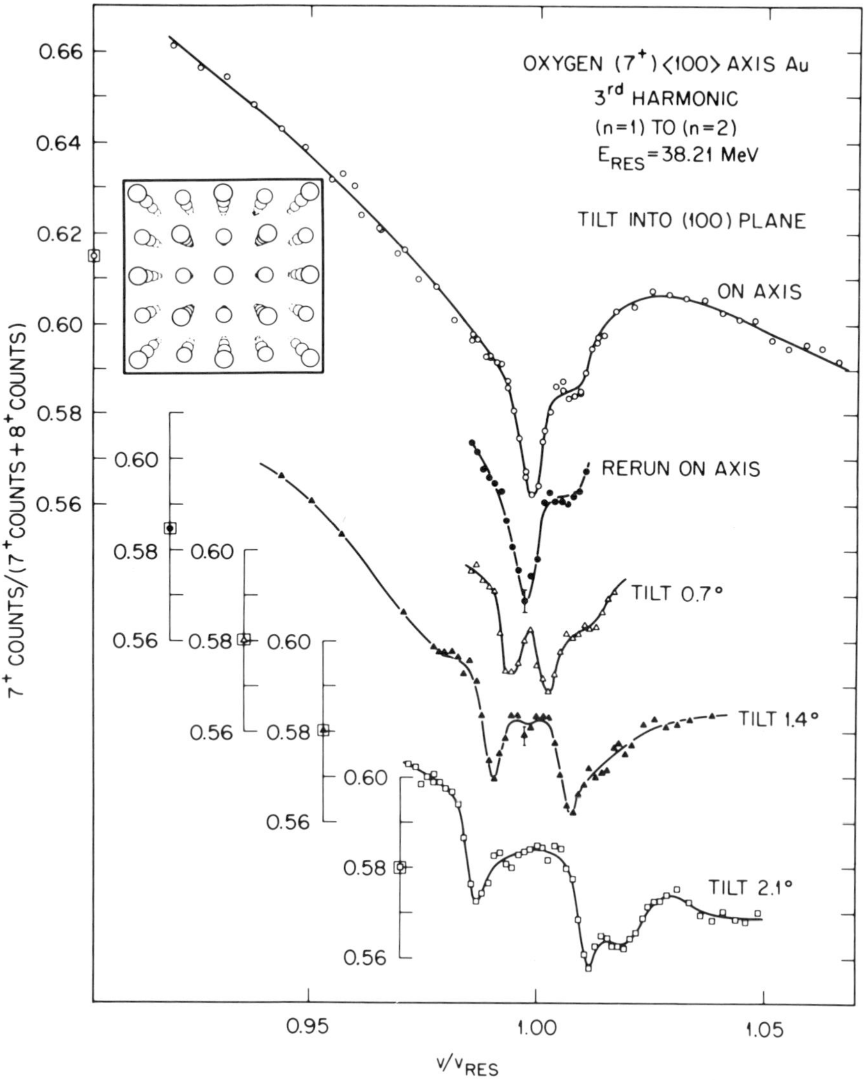

Figure 36. Effect of a small tilt in the B direction of Figure 34 [into a (100) plane] upon an (odd) third harmonic resonance: O^{7+} incident in ⟨100⟩ axial channeling in Au. For ($n = 1$) to ($n = 2$) the third harmonic resonant velocity v_r corresponds to an energy of 38.22 MeV.

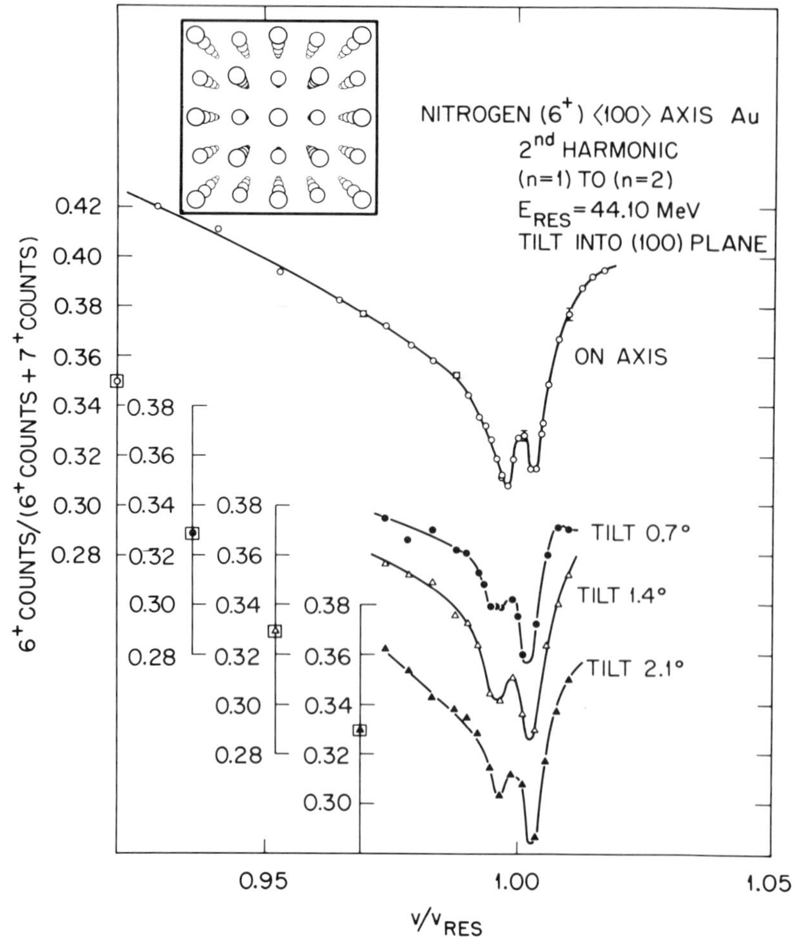

Figure 37. Effect of a small tilt in the (B) direction of Figure 34 upon an (even) second harmonic resonance. N^{6+} $\langle 100 \rangle$ axial channeling in Au. For $(n = 1)$ to $(n = 2)$ the ion energy required for second harmonic resonance is 44.10 MeV.

tilts in the B direction the central frequency disappears and only sidebands are seen. For example, a tilt of 0.7° causes an ion to step over a distance a once every 80 atoms ot its passage in the Z direction (thus once per 240 cycles of the third harmonic) giving sidebands 0.4% above and below the frequency $\nu = K(v/a)(\pm 0.4\%$ in $v/v_r)$. Figure 37 shows an equivalent tilting experiment but this time for an even $(K = 2)$ harmonic for N^{6+} ions. The central feature is seen to be maintained, but note that the central feature here is a doublet.

At this point we have demonstrated the mechanics of the excitation process well enough. The central question now is what can be learned about the crystal field and the detailed state of the ion in its electronic milieu.

Two principal effects occur which perturb the energy levels of an ion moving in a crystal channel compared to its vacuum state. First is that the ion is contained in a potential gradient (crystal field) formed by bounding atomic rows or planes. Simply put, the potential becomes more positive moving away from the center of the channel so that larger orbits on the channeled ion are more weakly bound than inner ones compared to the vacuum state. The result is, e.g., a reduction in the $n = 1$ to $n = 2$ spacing and accounts for the general observation of lowered resonant velocities. Ions moving on paths that are off the center of the channel experience sharper field gradients and probably account for the asymmetric broadening to lower velocities.

The second effect is caused by the presence of an electron "wake"; electrons scattered from the ion ($v_i \gg v_0$) create an enhanced electron density wave which follows the ion; the distance to the first node in the wave being $\sim 2\pi h v_i / \omega_p$, where ω_p is the plasmon frequency and the integrated enhanced charge density in this part of the wake is equal to $-q$, the charge on the moving ion. Thus the ion experiences a velocity-dependent dc field which can act to Stark-mix the ionic states. Since the field acts always in the z direction the mixing occurs between the $2p_z$ and $2s$ states and the doublets observed in even harmonics are due to these splittings.

It follows that the separation between the two minima in Figure 37 is a measure of the wake field in the vicinity of the moving ion. If we assume a uniform field ξ, the first-order Stark splitting is $6e\xi a_0/Z$. The data of Figure 37 give a value of $0.10e^2/a_0$ for this splitting, from which we deduce $\xi = 0.12e/a_0^2$, which is of the expected order of magnitude.

Crawford and Ritchie (Cr 79) have calculated the details of the anticipated effects of these perturbations using Hartree–Fock relativistic wave functions for the Au atoms with Wigner–Seitz boundary conditions. The results for O^{7+} ions in $\langle 100 \rangle$ Au at $v_i = 10v_0$ is shown in Figure 38. Since the $2p_{x,y}$ orbitals extend toward the bounding atomic rows they are more strongly shifted by the static potential than the $2p_z$ or $2s$. The mixing of the $2s$ and $2p$ by the polarization from the wake field is indicated on the right side of the figure. Transitions exciting $2p_z$ (even) populate the top and bottom states while odd harmonics excite the central line. As the velocity is increased, the length of the wake field is extended, the Stark field is decreased, and the resultant shifts from the vacuum level are decreased. Figure 39 shows the results of a detailed calculation (Cr 9) and a comparison with measurement for the $K = 2$ and 3 harmonics for N^{6+} in $\langle 100 \rangle$ Au. The most notable disagreement occurs for the lower line for $K = 2$ which appears near the $2p_{x,y}$ level rather than the $2sp_z$. This offset occurs for all the ions used and may possibly be explained as follows: the $2sp_z$ should

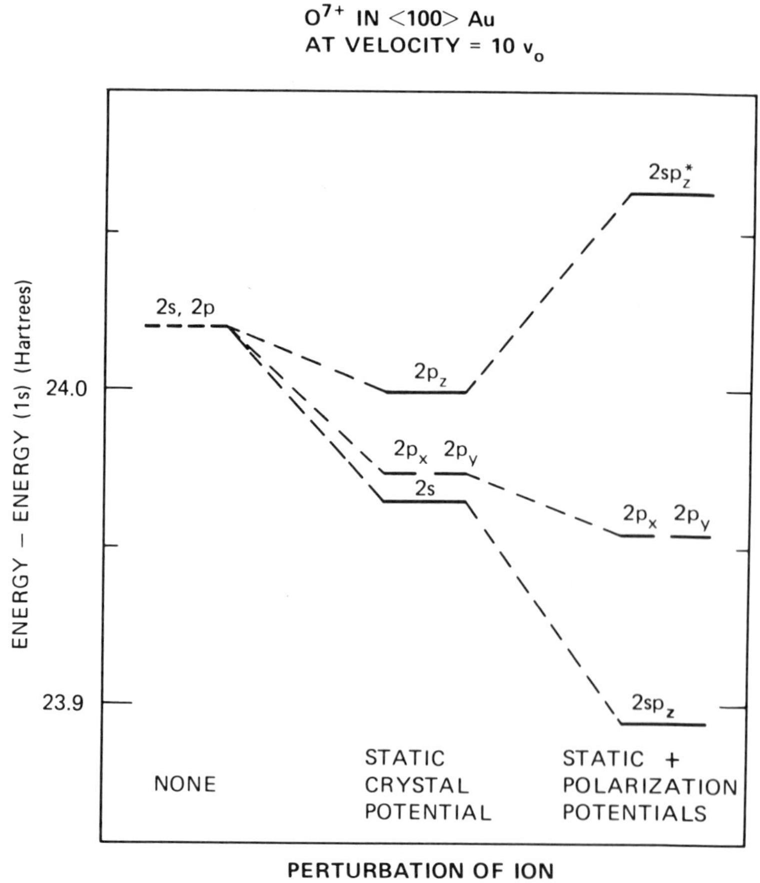

Figure 38. Effect of static crystal field and polarization (wake) field upon the energy levels of $^{16}O^{7+}$ in ⟨100⟩ axial channeling in Au at a velocity of $10\,v_0$.

only have ~1/3 the strength of the $2sp_z^*$. On the other hand, for trajectories with are not exactly on the center line, the $2p_x$ component can become quite strong for the even harmonics and may contribute more strongly than the $2sp_z$ to the total intensity.

To this point we have been discussing only axial channeling and slight tilts into the (100) plane (i.e., $\cos\theta \simeq 1$); in fact resonance coherent excitation (RCE) should occur everywhere in the plane and should be observable as long as collisional processes do not destroy the integrity of the initial and final states (Da 80). Here, however, we need two indices to describe the harmonic $K(k, l)$ in question, i.e., in equation (31) (Stark-split) harmonics should appear at any value of θ for k and l both even. Odd

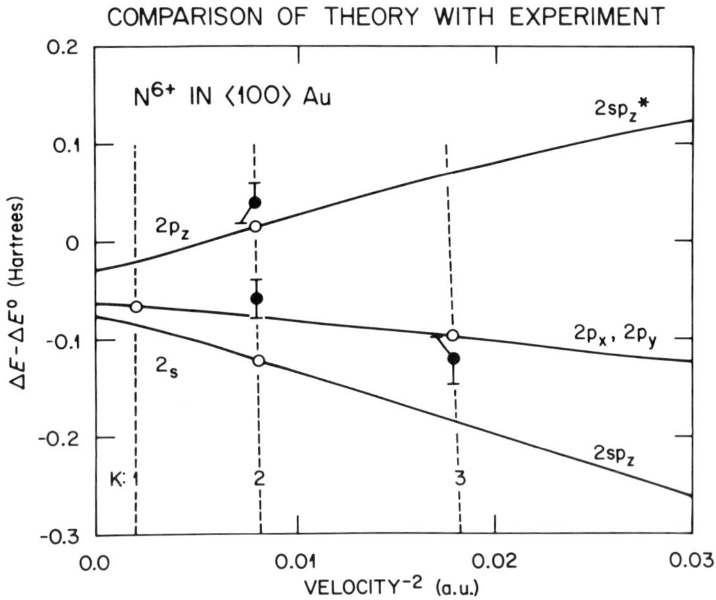

Figure 39. Calculated shift of $n = 1$ to $n = 2$ transition energies from vacuum level for N^{6+} as a function of ion velocity. Open points are calculated. Closed points are measured at the $K = 2$ and $K = 3$ resonances.

harmonics should appear at k and l both odd. [The field strengths should vary $\sim(k^2 + l^2)^{-1}$.] These expectations are demonstrated in Figure 40 for N^{6+} ions in the (100) plane of Au. The ⟨100⟩ axial channel is at $\theta = 0°$, the ⟨110⟩ at $\theta = 45°$, and other higher-order axial channels appear in between. The ordinate gives the energies in MeV/amu, where resonances should appear. Axial ⟨100⟩ resonances for $K = 2$ and 3 have already been shown (Figures 36 and 37), and the slight θ tilts for $K = 2$ (Figure 37) can now be identified as a (2, 0) resonance and for $K = 3$ (Figure 36) the "side bands" referred to above are really the (3, 1) and 3, −1) planar resonances.

In Figure 41 we show energy scans taken at $\theta = 6°$ and $\theta = 38°$ in the (100) plane for N^{6+}. The 6° tilt demonstrates the continued splitting of the (3, 1) and 3, −1) resonances indicated in Figure 36. The 38° tilt, however, introduces a Stark-split (2, 2) resonance into the picture (cf. Figure 40). Aside from a further demonstration of the generality of RCE, planar measurements provide other parameters which can be used better to determine the crystal and ionic parameters. The Stark splitting of the line depends to a large extent upon the strength of the wake field at the ionic core, which should be velocity dependent. Thus measuring the splitting of a (2, 2) resonance from, e.g., 3 MeV/amu at $\theta = 3°$ to 1.6 MeV/amu at $\theta = 38°$

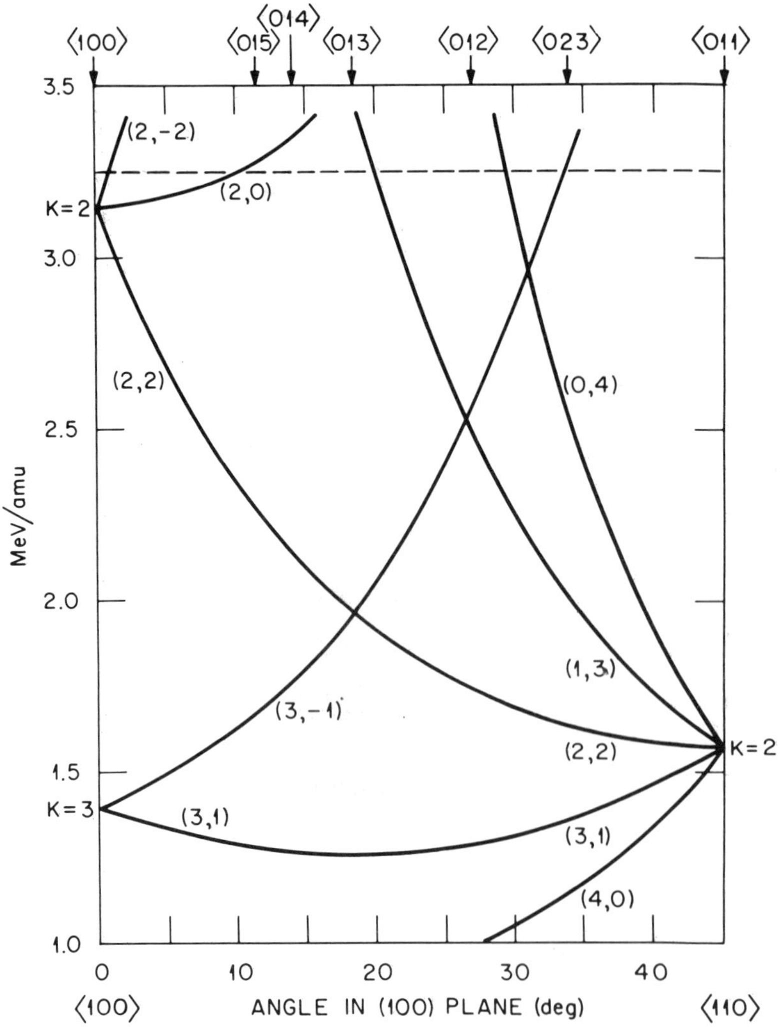

Figure 40. Calculated energy in MeV/amu for coherent $n = 1$ to $n = 2$ excitation of N^{6+} in the (100) plane of Au as a function of angle between the ⟨100⟩ and ⟨110⟩ axes.

provides a means for determining the velocity dependence in the same harmonic. Tilting into a plane also involves a change in the projected path length through the crystal so that changes in resonance depths for the same harmonic at different tilt angles can yield information on coherence lengths; e.g., in Figure 41 the (3, 1) harmonic appears at 1.35 MeV/amu at 3° and again at 38°. The resonance depth should also depend on interplanar spacing (i.e., stronger field components accompany narrower spacings) so that the strengths of transverse gradients can be tested.

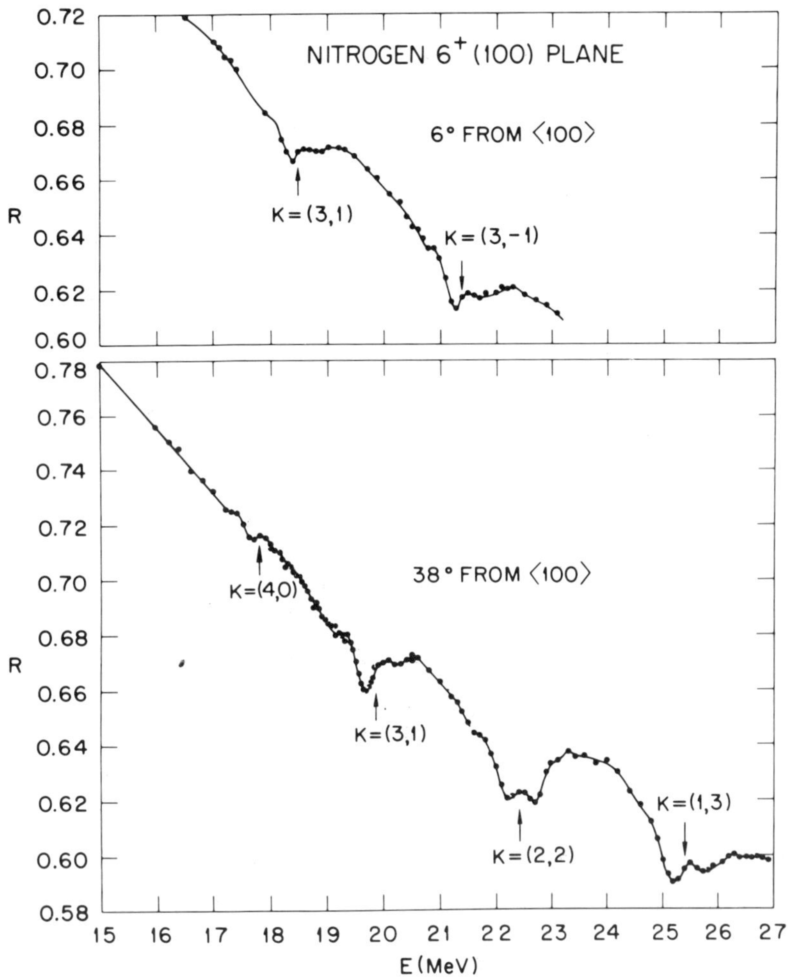

Figure 41. Demonstration of resonances for N^{6+} in the (100) plane. Note Stark split (2, 2) resonance.

Resonances in the heliumlike ions have also been investigated, but here the situation is not as clear as for the hydrogenlike cases. Because of the nondegeneracy of the 2*s* and *sp* levels in He, Stark mixing is not expected and indeed no doublet is observed for even harmonics in ⟨100⟩ channeling. A doublet structure is, however, observed in the odd harmonics, and is not understood.

Although many well-channeled particles will retain charge stability, it is also true that, even for very thin crystals, some of these ions will change charge. A pure beam of one charge state will thus produce well-channeled

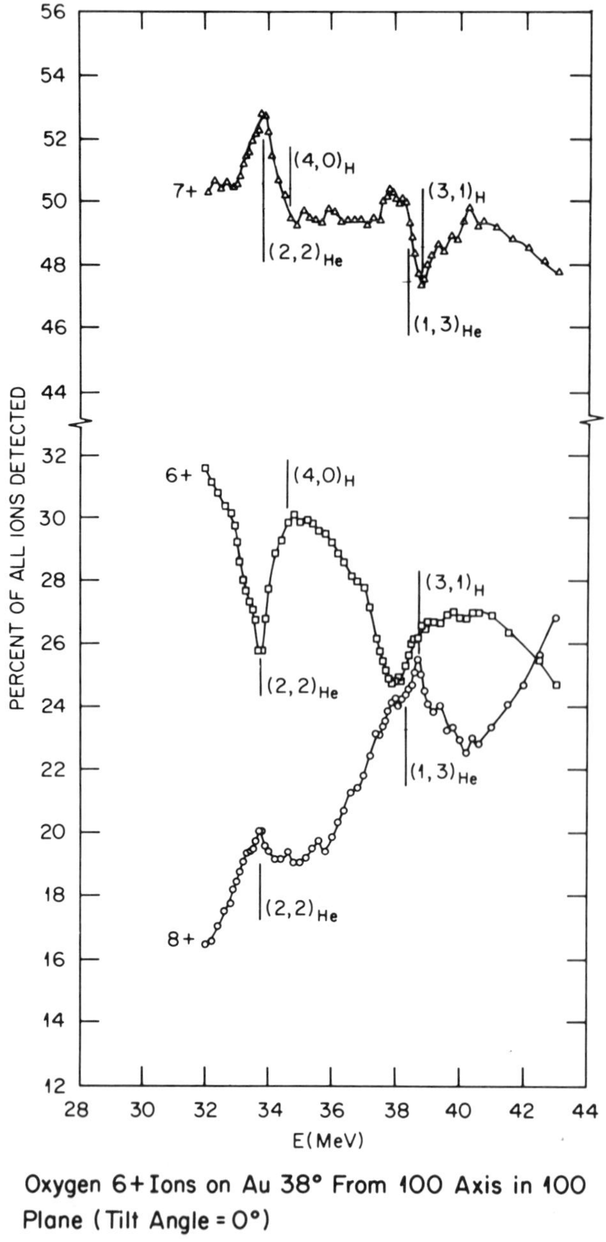

Figure 42. Emergent charge fractions from incident O^{6+} ions channeled in the (100) plane of Au 38° from the ⟨100⟩ axis.

particles of other charge states even in the absence of resonances; thus it is expected that both heliumlike and hydrogenlike resonances will occur together in the crystal since particles of both kinds will be available. Heliumlike resonances will depress the two-electron population and enhance the one-electron population; hydrogenic resonances will depress both the one-electron and two-electron population, the latter because the feed material to the two-electron ions by electron capture by one-electron ions is reduced. These effects are illustrated in Figure 42 for {100} planar channeled O^{6+} ions (Mo 81). (In this work populations for all emergent charge states were measured.) The single peaked (2, 2) resonance for heliumlike O^{6+} depresses the O^{6+} population, adds to the O^{7+} population which then by collisional ionization feeds the O^{8+}. At ~38 MeV a heliumlike (1, 3) resonance occurs which depresses O^{6+}, but an almost coincidental (3, 1) hydrogenlike resonance occurs at ~38.4 MeV, depleting the population of O^{7+} to form O^{8+}. Data of this type can establish relative resonance strengths.

At this writing it may be said that the RCE effect is qualitatively understood but that many of the quantitative details need further clarification. With this clarification should come a much clearer picture of the states of channeled ions and the nature of their interactions with crystal fields, both static and dynamic.

References

(An 77) H. H. Andersen, J. F. Bak, H. Knudsen, P. Møller Petersen, and B. R. Nielsen, *Nucl. Instrum. Methods* **140**:537 (1977); see also H. H. Andersen, J. F. Bak, H. Knudsen, and B. R. Nielsen, *Phys. Rev. A* **16**:1929 (1977).

(Ap 67) B. R. Appleton, C. Erginsoy, and W. M. Gibson, *Phys. Rev.* **161**:330 (1967).

(Ap 72a) B. R. Appleton, C. D. Moak, T. S. Noggle, and J. H. Barrett, *Phys. Rev.* **28**:1307 (1972).

(Ap 72b) B. R. Appleton, C. D. Moak, T. S. Noggle, and J. H. Barrett, *Radiat. Eeff.* **13**:171 (1972).

(Ap 9) B. R. Appleton, R. H. Ritchie, J. A. Biggerstaff, T. S. Noggle, S. Datz, C. D. Moak, H. Verbeek, and V. N. Neelavathi, *Phys. Rev. B* **19**:4347 (1979).

(As 72) J. C. Ashley, R. H. Ritchie, and W. Brandt, *Phys. Rev. B* **5**:2393 (1972). S. H. Morgan and C. C. Sung, *Phys. Rev. A* **20**:818 (1979).

(As 73) J. C. Ashley, R. H. Ritchic, and W. Brandt, *Phys. Rev. A* **8**:2402 (1973).

(Ba 63) W. H. Barkas, N. J. Dyer, and H. H. Heckman, *Phys. Rev. Lett.* **11**:26 (1963); see also H. H. Heckman and P. J. Lindstrom, *Phys. Rev. Lett.* **22**:871 (1969).

(Ba 71) J. H. Barrett, *Phys. Rev. B* **3**:1527 (1971).

(Ba 75) J. H. Barrett, B. R. Appleton, T. S. Noggle, C. D. Moak, S. Datz, J. A. Biggerstaff, and R. Behrisch, *Atomic Collisions in Solids* (S. Datz, B. R. Appleton, and C. D. Moak, eds.), Plenum Press, New York (1975), pp. 645–668.

(Ba 62) B. W. Batterman and D. R. Chipman, *Phys. Rev.* **127**:690 (1962).

(Be 30) H. A. Bethe, *Ann. Phys. (Leipzig)* **5**:325 (1930).

(Be 57) H. A. Bethe and E. E. Salpeter, *Quantum Mechanics of One- and Two-Electron Atoms*, Academic, New York (1957), p. 320.

(Bl 33) F. Bloch, *Ann. Phys. (Leipzig)* **16**:285 (1933).

(Bo 13) N. Bohr, *Phil. Mag.* **25**:10 (1913).

(Bo 40) N. Bohr, *Phys. Rev.* **58**:654 (1940).

(Bo 41) N. Bohr, *Phys. Rev.* **59**:270 (1941).

(Bo 67) E. Bonderup, *K. Dan. Vidensk. Selsk. Mat.-Fys. Medd.* **35**: No. 17 (1967).

(Bo 70) F. Bonsignori and A. Desalvo, *J. Phys. Chem. Soc.* **31**:2191 (1970).

(Br 77) W. Brandt and R. H. Ritchie, *Phys. Lett.* **62A**:374 (1977).

(Br 82) M. Breinig, S. B. Elston, S. Huldt, L. Liljeby, C. R. Vane, S. D. Berry, G. A. Glass, M. Schauer, I. A. Sellin, G. D. Alton, S. Datz, S. H. Overbury, R. Laubert, and M. Suter, *Phys. Rev. A* **25**:3015 (1982).

(Br 2) M. D. Brown and C. D. Moak, *Phys. Rev. B* **6**:90 (1972).

(Br 41) J. H. M. Brunings, J. Knipp, and E. Teller, *Phys. Rev.* **60**:657 (1941).

(Bu 81) Yu. V. Bulgakov and A. F. Tulinov, eds., *Atomic Collisions in Solids*, Moscow State University Press, Moscow (1981), Vol. 8, p. 668.

(Cr 79) O. H. Crawford and R. H. Ritchie, *Phys. Rev. A* **20**:1948 (1979).

(Cr 70) G. B. Crooks and M. E. Rudd, *Phys. Rev. Lett.* **25**:1599 (1970).

(Da 69) S. Datz, C. D. Moak, T. S. Noggle, B. R. Appleton, and H. O. Lutz, *Phys. Rev.* **179**:315 (1969).

(Da 70) S. Datz, C. D. Moak, B. R. Appleton, M. T. Robinson, and O. S. Oen, *Atomic Collision Phenomena in Solids* (D. W. Palmer, M. W. Thompson, and P. D. Townsend, eds.) North-Holland, Amsterdam (1970), p. 374.

(Da 72) S. Datz, F. W. Martin, C. D. Moak, B. R. Appleton, and L. B. Bridwell, *Radiat. Eff.* **12**:163 (1972).

(Da 75a) S. Datz, B. R. Appleton, and C. D. Moak, eds., *Atomic Collisions in Solids*, Plenum Press, New York (1975), Vol. 2, p. 972.

(Da 75b) S. Datz, B. R. Appleton, J. A. Biggerstaff, M. D. Brown, H. F. Krause, C. D. Moak, and T. S. Noggle, *Atomic Collisions in Solids* (S. Datz, B. R. Appleton, and C. D. Moak, eds.), Plenum Press, New York (1975), p. 63.

(Da 77) S. Datz, J. Gomez del Campo, P. F. Dittner, P. D. Miller, and J. A. Biggerstaff, *Phys. Rev. Lett.* **38**:1145 (1977).

(Da 78) S. Datz, C. D. Moak, O. H. Crawford, H. F. Krause, P. F. Dittner, J. Gomez del Campo, J. A. Biggerstaff, P. D. Miller, P. Hvelplund, and H. Knudsen, *Phys. Rev. Lett.* **40**:943 (1978).

(Da 80) S. Datz, C. D. Moak, O. H. Crawfod, H. F. Krause, P. D. Miller, P. F. Dittner, J. Gomez del Campo, J. A. Biggerstaff, H. Knudsen, and P. Hvelplund, *Nucl. Instrum. Methods* **170**:15 (1980).

(De 74) K. Dettmann, K. G. Harrison, and M. W. Lucas, *J. Phys. B* **7**:269 (1974).

(De 75) K. Dettmann, *Z. Phys. A* **272**:227 (1975).

(Do 68) P. A. Doyle and P. S. Turner, *Acta Crystallogr. Sect. A* **24**:390 (1968).

(Dr 76) F. Drepper and J. S. Briggs, *J. Phys. B* **9**:2063 (1976).

(Ei 71) F. H. Eisen and M. T. Robinson, *Phys. Rev. B* **4**:1457 (1971).

(El 81) S. B. Elston, I. A. Sellin, M. Breinig, S. Huldt, L. Liljeby, R. S. Thoe, S. Datz, S. Overbury, and R. Laubert, *Phys. Rev. Lett.* **46**:321 (1981).

(Es 78a) H. Esbensen and J. A. Golovchenko, *Nucl. Phys.* **A298**:382 (1978).

(Es 78b) H. Esbensen, O. Fich, J. A. Golovchenko, S. Madsen, H. Nielsen, H. Schiøtt, E. Eggerhoj, C. Vraast-Thomsen, G. Charpak, S. Majewski, G. Odniec, G. Petersen, F. Sauli, J. P. Ponpon, and P. Siffert, *Phys. Rev. B* **18**:1039 (1978).

(Fa 63) U. Fano, *Ann. Rev. Nucl. Sci.* **13**:1 (1963).

(Fi 59) O. B. Firsov, *Zh. Eksp. Teor. Fiz.* **36**:1517 (1959). [*Sov. Phys. JETP* **9**:1076 (1959)].

(Ge 74) D. S. Gemmell, *Rev. Mod. Phys.* **46**:129 (1974).

(Go 81) J. A. Golovchenko, A. N. Goland, J. S. Rosner, C. E. Thorn, H. E. Wegner, H. Knudsen, and C. D. Moak, *Phys. Rev. B* **23**:957 (1981).

(Ha 70) K. G. Harrison and M. Lucas, *Phys. Lett.* **33A**:142 (1970).

(Ha 71) K. G. Harrison and M. Lucas, *Phys. Lett.* **35A**:402 (1971).

(Hi 74) K. W. Hill and E. Merzbacher, *Phys. Rev. A* **9**:156 (1974).

(Ja 80) D. P. Jackson, J. E. Robinson, and D. A. Thompson, *Nucl. Instrum. Method* **170**:604 (1980).

(Ja 72) J. D. Jackson and R. L. McCarthy, *Phys. Rev. B* **6**:4131 (1972).

(Kn 41) J. Knipp and E. Teller, *Phys. Rev.* **59**:659 (1941).

(La 40) W. E. Lamb, Jr., *Phys. Rev.* **58**:696 (1940).

(La 80) R. Laubert, I. A. Sellin, C. R. Vane, M. Suter, S. B. Elston, G. D. Alton, and R. S. Thoe, *Nucl. Instrum. Methods* **170**:557 (1980), and references therein.

(Li 63) J. Lindhard, M. Scharff, and H. E. Schiøtt, *K. Dan. Vidensk. Selsk., Mat.-Fys. Medd.* **33**:14 (1963).

(Li 65) J. Lindhard, *K. Dan. Vidensk. Selsk. Mat.-Fys. Medd.* **34**: No. 14 (1965).

(Li 76) J. Lindhard, *Nucl. Instrum. Methods* **132**:1 (1976).

(Lu 71) C. C. Lu, T. A. Carlson, F. B. Malik, T. C. Tucker, and C. W. Nestor, Jr., *At. Data* **3**:1 (1971).

(Lu 63) H. Lutz and R. Sizmann, *Phys. Lett.* **5**:113 (1963).

(Lu 66) H. O. Lutz, S. Datz, C. D. Moak, and T. S. Noggle, *Phys. Rev. Lett.* **17**:285 (1966).

(Lu 70) H. O. Lutz, S. Datz, C. D. Moak, and T. S. Noggle, *Phys. Rev. Lett.* **33A**:309 (1970).

(Ma 71) J. R. Macdonald and F. W. Martin, *Phys. Rev. A* **4**:1965 (1971).

(Mo 66) C. D. Moak, J. W. T. Dabbs, and W. W. Walker, *Revs. Sci. Instrum.* **37**:1131 (1966).

(Mo 68) C. D. Moak, H. O. Lutz, L. B. Bridwell, L. C. Northcliffe, and S. Datz, *Phys. Rev.* **176**:427 (1968).

(Mo 74) C. D. Moak, S. Datz, B. R. Appleton, M. D. Brown, H. F. Krause, and T. S. Noggle, *Phys. Rev. B* **10**:2681 (1974).

(Mo 76) C. D. Moak, B. R. Appleton, J. A. Biggerstaff, M. D. Brown, S. Datz, T. S. Noggle, and H. Verbeek, *Nucl. Instrum. Methods* **132**:95 (1976).

(Mo 79) C. D. Moak, S. Datz, O. H. Crawford, H. F. Krause, P. F. Dittner, J. Gomez del Campo, J. A. Biggerstaff, P. D. Miller, P. Hvelplund, and H. Knudsen, *Phys. Rev. A* **19**:977 (1979).

(Mo 81) C. D. Moak, J. Gomez del Campo, J. A. Biggerstaff, S. Datz, P. F. Dittner, H. F. Krause, and P. D. Miller, *Phys. Rev. B* **25**:4406 (1982).

(Mo 47) G. Molière, *Z. Naturforsch.* **2a**:133 (1947).

(Ne 61) R. S. Nelson and M. W. Thompson, *Proc. R. Soc. London Ser. A* **259**:458 (1961).

(Pi 63) G. R. Piercey, F. Brown, J. A. Davies, and M. McCargo, *Phys. Rev. Lett.* **10**:399 (1963).

(Ro 63) M. T. Robinson and O. S. Oen, *Appl. Phys. Lett.* **2**:30 (1963); *Phys. Rev.* **132**:2385 (1963).

(Ro 69) M. T. Robinson, *Phys. Rev.* **179**:327 (1969).

(Ro 71a) M. T. Robinson, *Phys. Rev. B* **4**:1461 (1971).

(Ro 71b) M. R. Robinson, *Proc. Battelle Colloquium on Interatomic Potentials and Simulation of Lattice Defects*, July 14–19 (1971).

(Sc 72) H. W. Schnopper, H. D. Betz, J. P. Delvaille, K. Kalata, A. R. Sohval, K. W. Jones, and H. E. Wegner, *Phys. Rev. Lett.* **29**:898 (1972).

(Sc 74) H. W. Schnopper, J. P. Delvaille, K. Kalata, A. R. Sohval, M. Abdulwahab, K. W. Jones, and H. E. Wegner, *Phys. Rev. Lett.* **47**:61 (1974).

(St 12) J. Stark and G. Wendt, *Ann. Phys. (Leipzig)* **38**:921 (1912); J. Stark, *Phys. Z.* **13**:973 (1912).

(Va 79) C. R. Vane, *IEEE Trans. Nucl. Sci.* **26**:1078 (1979).

(Zi 77) J. P. Ziegler, ed., *The Stopping and Ranges of Ions in Matter*, Pergamon Press, New York (1977), Vols. 1–5, and references therein.

6

DONALD S. GEMMELL (top) is Director of the Physics Division and a Senior Physicist at the Argonne National Laboratory. He received his undergraduate education at the University of Adelaide and his Ph.D. from the Australian National University in 1959. After three years as a Research Fellow at AERE, Harwell, he joined the staff of the Argonne National Laboratory.

ZEEV VAGER (bottom) is a Professor of Physics at the Weizmann Institute of Science in Rehovoth, and a Senior Physicist in the Physics Division at the Argonne National Laboratory. He received his M.Sc. from the Hebrew University and his Ph.D. in physics from the Weizmann Institute in 1963. During 1969–1970, he was a Visiting Professor at the State University of New York at Stony Brook and during 1975 and 1976 he was a Visiting Scientist at the Argonne National Laboratory.

The Electronic Polarization Induced in Solids Traversed by Fast Ions

DONALD S. GEMMELL AND ZEEV VAGER

1. Introduction

In this chapter we discuss the response of the electrons in a solid target to the passage of a fast positively charged ionic projectile. "Fast," in this context, means projectile velocities, v, large compared with the Fermi velocities of the target electrons, but not so large that a relativistic treatment is required. That is, $v_0 \ll v \ll c$, where v_0 is the Bohr velocity, $e^2/\hbar$. This corresponds to projectile energies lying in the range roughly between 0.1 and 10.0 MeV/nucleon. The restriction here to nonrelativistic velocities is just for convenience; in general the extension of the discussion to relativistic projectile velocities is straightforward.

We shall be primarily concerned with the polarization "wake" that is induced in a solid during its traversal by a fast ion. It is well known [see, for example (Fr 54)] that if a stationary positive test charge (e.g., a positron) is embedded in a solid, the electrons of the solid respond by "heaping up" on the charge in such a way as to neutralize it and to screen out its field at distances large compared with the Debye length. The Debye length is proportional to the average electron velocity in the solid. If now the positively charged "impurity" is moved at a velocity large compared with the target electron velocities, the induced polarization charge lags behind the projectile and the screening distance becomes primarily determined by the

DONALD S. GEMMELL • Physics Division, Argonne National Laboratory, Argonne, Illinois 60439. ZEEV VAGER • Department of Nuclear Physics, The Weizmann Institute of Science, Rehovoth, Israel. The submitted manuscript has been authored by a contractor of the U. S. Government under contract No. W-31-109-ENG-38. Accordingly, the U. S. Government retains a nonexclusive, royalty-free license to publish or reproduce the published form of this contribution, or allow others to do so, for U. S. Government purposes.

velocity of the target electrons relative to the projectile. Under steady-state conditions this polarization wake moves with the projectile as though attached to it. In this respect it is analogous to the wake trailing a boat moving through water. Although individual electrons in the solid respond in various ways to the passage of the projectile, their time-averaged charge distribution travels along with the projectile. The induced wake charge acts as a brake on the projectile. The wake concept therefore provides a method for describing the stopping power of the solid medium.

Although the idea of a polarization wake has been recognized for many years [see, for example, the articles by Bohr (Bo 48), Pines and Bohm (Pi 52), and Neufeld and Ritchie (Ne 55)], it is only comparatively recently with the advent of experiments involving the slowing down (Br 74) and dissociation (Ge 75) of fast molecular-ion beams in foils, that the concept has begun to receive increased attention. When a fast diatomic molecular-ion projectile strikes a foil, it rapidly gives rise to a pair of ions (separated typically by about one angstrom) which then begin to move apart in a "Coulomb explosion." Inside the foil the motion of each fragment, in addition to being influenced by its own wake, is modified by the wake of its partner fragment. A detailed analysis of the momenta of these fragments after they emerge from the foil can thus yield precise information about the charge distributions associated with the wakes of the projectiles during their passage through the solid.

In the following sections we review theoretical models for the wake, we describe some recent experiments with molecular-ion beams, and we illustrate the type of information about wakes that is obtainable in such experiments.

The wake concepts discussed here have implications in several aspects of heavy-ion science. Some of the more obvious examples are to be found in descriptions of stopping power, the transient-field effect (Be 80), the resonant coherent excitation of swift ions traversing crystals (Da 78), the emission of electrons from solids traversed by fast ions (La 80, Sc 78), hot-atom chemistry, the determination of molecular-ion structures (Ge 80b), and the effects of dynamic screening on atomic states of fast ions (Be 76). For related discussions, see Chapter 6 of Volume 3, by Goldring and Hass; Chapter 1 of Volume 5, by Reinhardt and Greiner; Chapter 3 of Volume 5, by Martinson; and in this volume Chapter 2, by Cacace; Chapter 3, by Ziegler and Biersack; and Chapter 5, by Datz and Moak.

2. The Wake

In developing theoretical descriptions of the wake, one is forced, because of the complexity of the problem, to make several simplifying assumptions. Common to all theories thus far is the treatment of the projectile as being a point charge. The influence of the electronic structure of the projectile is thereby neglected. The projectile is usually assumed to have an effective charge $Z_{\text{eff}}e$ that fits the stopping power. This charge is an equilibrium value and is quickly established after the projectile strikes the foil (typicaly within a few angstroms of penetration). It is an average value that reflects the balance of the rapid processes of electron capture and loss occurring as the ion traverses the target [see Chapter 3, by Ziegler and Biersack; see also (Be 72)].

A further common approximation is to treat the target as an isotropic homogeneous electron gas embedded in a uniformly distributed positive background (the "jellium" model). In this approximation the detailed electronic structure of the target is not taken into account. (For example, the polarization contributions of tightly bound inner shells of the target atoms are neglected.)

2.1. Bohr's Model

In his treatise in 1948 on "The Penetration of Atomic Particles Through Matter" (Bo 48), Niels Bohr introduced the term "wake." From an elementary calculation of the charge accumulated in the wake of an ion traversing a solid, Bohr derived an expression for the stopping power. Figure 1 is a reproduction from Bohr's paper. The projectile traveling with velocity v is indicated at the foot of the arrow on line C. The circles represent the target atoms which, for simplicity, are assumed to contain only a single electron with orbital angular frequency ω. For a target electron located a distance r from the projectile trajectory, the collision time is

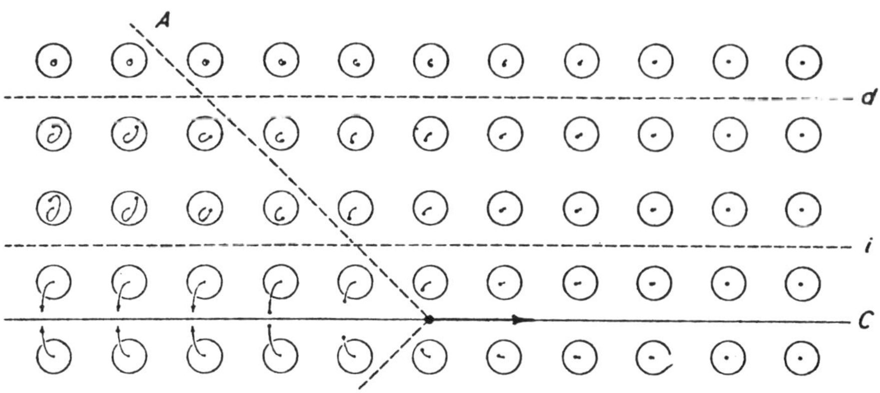

Figure 1. Reproduction of a figure from Bohr's 1948 paper (Bo 48) illustrating the production of a polarization wake behind a fast ion traversing a solid.

approximately $2r/v$. Thus, for those electrons lying in a conical wake (cone half-angle $\sim\pi/4$) trailing behind the projectile, collisions with the projectile are essentially completed. Using the momentum approximation, Bohr estimated the net inward displacement of each wake electron to be $\delta = 2Ze^2/mv^2$, where Ze is the projectile charge and m is the electron mass. Target electrons at distances from the projectile trajectory greater than the adiabatic limit $d = v/\omega$ make no net contribution to the induced polarization.

Since the displacement δ is independent of the electron distance from the line C, the electronic density remains unchanged except in a small region of width δ on the line C. The excess electronic charge induced at a distance x behind the projectile is $2\pi ne\delta x\, dx$, where n is the number of target electrons per unit volume. The braking force that this charge exerts on the projectile is then

$$dF = 2\pi nZe^2\delta \frac{dx}{x} \tag{1}$$

and the stopping power is given by

$$-\frac{dE}{dx} = \int dF = \int_{r_{\min}}^{r_{\max}} 2\pi nZe^2\delta \frac{dx}{x} \tag{2}$$

If, for the integration limits, we use $r_{\max} = d$, and $r_{\min} = \hbar/mv$, representing the quantum-mechanical uncertainty in the minimum impact parameter, then Eq. (2) yields

$$-\frac{dE}{dx} = \frac{4\pi nZ^2e^4}{mv^2} \ln\left(\frac{mv^2}{\hbar\omega}\right) \tag{3}$$

For comparison, the general formula derived by Bethe (Be 30) is

$$-\frac{dE}{dx} = \frac{4\pi nZ^2e^4}{mv^2} \ln\left(\frac{2mv^2}{\hbar\langle\omega\rangle}\right) \tag{4}$$

where $\hbar\langle\omega\rangle$ is an average excitation energy characteristic of the target material.

Thus the Bohr picture for the wake is a line of excess electronic charge behind the projectile. The linear charge density is $(4\pi ne^2/m)(Zex/v^2)$ and it tapers off at about $x_{\max} = d = v/\omega$.

2.2. The Electron-Gas Model

It is convenient to adopt a simplified model of the solid in which the effect of the ionic fields of the target atoms is replaced by a uniform background of positive charge. This electron-gas model has proved extremely valuable in describing the properties of metals [see, for example, Bohm and Pines (Bo 53)].

A classical free electron gas with a density of n electrons per unit volume can sustain plasma oscillations with an angular frequency given by

$$\omega_p = (4\pi n e^2/m)^{1/2} \tag{5}$$

In a quantum-mechanical description, volume plasmons with an energy $\hbar\omega_p$ can be excited. Plasmons have energies that typically lie in the range 5–30 eV and so are not readily excited thermally. However, they are excited by the passage of charged particles as has been elegantly proven in experiments on the energy losses of keV electron beams traversing thin metallic foils [see, for example, the review article by Raether (Ra 65)].

In Equation (5) it is sometimes necessary to use an effective mass rather than the actual electron mass. If, however, one uses the actual electron mass and expresses n in units of electrons per cubic angstrom, then the plasmon energy becomes

$$\hbar\omega_p = 37.1 n^{1/2} \text{ eV} \tag{6}$$

Thus, if one assumes for carbon four electrons per atom as contributing to the electron gas and assumes the density of a carbon foil to be $\rho = 2.0 \text{ g/cm}^3$, one calculates from Equation (6) $\hbar\omega_p = 23.5$ eV [the measured value (Bu 68) is 25.9 eV]. Similarly, one calculates for aluminum (three electrons per atom, $\rho = 2.7 \text{ g/cm}^3$) $\hbar\omega_p = 15.8$ eV [the measured value (Ra 65) is 15.0 eV].

Plasmon excitations are damped at rates that depend upon various factors such as the electronic structure of the solid and the size of the microcrystallites in the solid. For carbon foils the damping is rapid (the classical oscillation is extinguished in a few plasma periods), but in aluminum the damping is usually slow and the classical oscillations persist over many plasma periods (Ra 65).

2.2.1. Dielectric Descriptions of the Wake

Almost all theories of stopping power are based on first-order perturbation theory [see, for example (Fa 63)]. For such theories the response of a medium to a perturbing projectile is linear. Thus, if the perturbing projectile has an effective charge Ze then the polarization wake charge and potential are proportional to Z. The stopping power is then proportional to Z^2. The so-called Z^2 stopping-power theories reproduce the experimental data quite well and therefore one might expect that such theories should estimate correctly electrical forces not only at the location of the projectile but also in its vicinity (see Sections 2.2.2 and 3.3.3 for a discussion of this point). The linear response of the medium can be characterized completely by a dielectric function $\varepsilon(\mathbf{k}, \omega)$ where $\mathbf{k}$ is the wave vector and ω is the frequency of an electromagnetic disturbance in the medium. In electrostatics the dielectric constant is actually a tensor that when multiplied by the electric field $\mathbf{E}$ gives the displacement vector $\mathbf{D}$. Here we assume an isotropic homogeneous medium where the dielectric tensor can be reduced to a scaler. The Fourier components of the polarization and projectile charge-density distributions are related by

$$\rho_{\text{pol}}(\mathbf{k}, \omega) = \rho_{\text{proj}}(\mathbf{k}, \omega)[1/\varepsilon(\mathbf{k}, \omega) - 1] \tag{7}$$

Let us assume a point charge Ze to be moving with a velocity $\mathbf{v}$ along the z' axis of a Cartesian coordinate frame $(x', y', z') \equiv \mathbf{r}'$ fixed with respect to the solid medium. That is

$$\rho_{\text{proj}}(\mathbf{r}', t) = Ze\delta(x')\ \delta(y')\,\delta(z' - vt) \tag{8}$$

and the corresponding Fourier transform is

$$\rho_{\text{proj}}(\mathbf{k}, \omega) = Ze\,\frac{\delta(\omega - \mathbf{k}\cdot\mathbf{v})}{2\pi} \tag{9}$$

The Fourier components of the polarization (wake) charge distribution are then

$$\rho_{\text{pol}}(\mathbf{k}, \omega) = Ze\,\frac{\delta(\omega - \mathbf{k}\cdot\mathbf{v})}{2\pi}\left[\frac{1}{\varepsilon(\mathbf{k}, \omega)} - 1\right] \tag{10}$$

and this transforms back to give

$$\rho_{\text{pol}}(\mathbf{r}) = \frac{Ze}{(2\pi)^3}\int\left[\frac{1}{\varepsilon(\mathbf{k}, \mathbf{k}\cdot\mathbf{v})} - 1\right]e^{i\mathbf{k}\cdot\mathbf{r}}\,d^3k \tag{11}$$

where $\mathbf{r} = \mathbf{r}' - \mathbf{v}t$ is a coodinate frame that moves with the projectile. This is the fundamental equation relating the wake polarization charge to the dielectric properties of the medium. Note that the time coordinate is eliminated through the transformation to the frame of the moving projectile. If the projectile does not change its velocity appreciably, there is established a stationary (in the projectile frame) polarization charge distribution (wake) in the vicinity of the projectile.

A further reduction of Eq. (11) is frequently found in the literature. The cylindrical symmetry around the z axis is used for an integration over the azimuthal angle of $\mathbf{k}$. The result is

$$\rho_{\text{pol}}(r_\perp z) = \frac{Ze}{(2\pi)^2}\int_0^\infty dk_\perp\,k_\perp J_0(k_\perp r_\perp)\int_{-\infty}^\infty dk_z\left[\frac{1}{\varepsilon(\mathbf{k}, k_z v)} - 1\right]e^{ik_z z} \tag{11$'$}$$

where $r_\perp$ and $k_\perp$ are the components of $\mathbf{r}$ and $\mathbf{k}$ perpendicular to the z axis.

The wake potential and forces are easily derived from Equation (11) with the aid of the Poisson equation. For example, the wake potential is given by

$$\phi(\mathbf{r}) = \frac{Ze}{2\pi^2} \int \left[\frac{1}{\varepsilon(\mathbf{k}, \mathbf{k} \cdot \mathbf{v})} - 1 \right] \frac{e^{i\mathbf{k} \cdot \mathbf{r}}}{k^2} d^3k \tag{12}$$

2.2.1a. The Simple Plasma Wake. This model (Va 76a) was developed in order to account for the dissociation patterns observed in high-resolution experiments with fast molecular-ion beams (Section 3). The dielectric function was approximated with the high-frequency form [see, for example (Ja 75)]

$$\varepsilon(\omega) = 1 - \frac{\omega_p^2}{(\omega + i\gamma)^2} \tag{13}$$

where γ is a damping parameter.

Inserting Equation (13) into Equation (11) and choosing the limit $\gamma \to 0$, we find

$$\rho_{\text{pol}}(\mathbf{r}) = \frac{Ze}{a} \sin\left(\frac{z}{a}\right) \delta(x)\, \delta(y)\, \theta(-z) \tag{14}$$

where $a = v/\omega_p$ and $\theta(x) = 0$ for $x < 0$, $\theta(x) = 1$ for $x \geqslant 0$. This spatially oscillatory, linear charge-density distribution extends only behind the projectile, not in front of it. The wavelength, $\lambda = 2\pi a$, is typically a few tens of angstroms and is simply the distance traveled by the projectile in one plasma oscillation period.

It is easy to visualize the origin of the above polarization wake. A point charge Ze embedded in a medium with a dielectric constant ε acts like an effective charge of Ze/ε. Thus a polarization charge of magnitude $Ze(1/\varepsilon - 1)$ is induced at the location of the foreign charge. If now the charge moves, then as it passes by each point in the medium it starts a local oscillation of polarization charges. The full polarization is achieved only after a quarter of the period time. By that time the particle has already moved forwards a distance of $\lambda/4$.

Equation (14) describes the polarization wake in terms very similar to the Bohr picture (Section 2.1). The first term of an expansion of the sine function in Equation (14) results in an expression identical with Bohr's for the linear charge density. The tapering-off at v/ω in Bohr's picture corresponds approximately to the turning point of the sine function [Equation (14)] at $\lambda/4$.

The width of the distribution of polarization charges cannot be indefinitely small. A limit is set by the quantum-mechanical minimum impact parameter for electrons, $\hbar/mv$, where m is the mass of the electron [see, for example (Ja 75)]. This is taken into account when evaluating the potential distribution corresponding to the polarization wake, by substituting $R^2 + (\hbar/mv)^2$ for R^2 (where $R^2 = x^2 + y^2$). The expression for the potential is then

$$\phi(R, z) = -\frac{Ze}{a}\int_0^\infty \sin\frac{\zeta}{a}\,[R^2 + (\hbar/mv)^2 + (\zeta + z)^2]^{-1/2}\,d\zeta \tag{15}$$

an example of which is shown plotted in Figure 2.

The derivative of this potential evaluated at the projectile position yields the stopping power ($-dE/dz$). From Equation (15) we find

$$-\frac{dE}{dz} = Ze\left(\frac{\partial\phi}{\partial z}\right)_{z=0,R=0} \approx \frac{4\pi n Z^2 e^4}{mv^2}\ln\left(\frac{1.123mv^2}{\hbar\omega_p}\right) \tag{16}$$

which is a close approximation to the usual Bethe formula that applies in this velocity range [Eq. (4)].

This model can easily be adapted to take into account plasmon damping.

2.2.1b. The Wake Using Lindhard's Dielectric Function. In 1954 Lindhard (Li 54) published a pioneering paper "On the Properties of a Gas

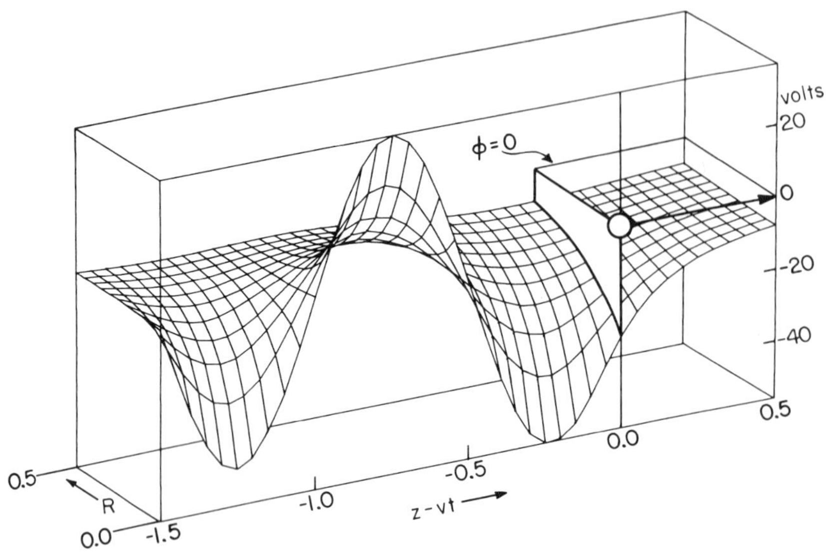

Figure 2. Potential distribution [Equation (15)] associated with the polarization wake of a 400-keV proton traversing carbon ($\hbar\omega_p = 25.0$ eV). Distances are shown in units of $\lambda = 2\pi a = 14.5$ Å.

of Charged Particles." A general formula for the dielectric function was derived using a quantum-mechanical first-order treatment for a free-electron gas:

$$\varepsilon(\mathbf{k}, \omega) = 1 + \frac{2m^2\omega_p^2}{\hbar^2 k^2} \sum_n \frac{f(E_n)}{N} \left[\frac{1}{k^2 + 2\mathbf{k} \cdot \mathbf{k}_n - (2m/\hbar)(\omega + i\gamma)} + \frac{1}{k^2 - 2\mathbf{k} \cdot \mathbf{k}_n + (2m/\hbar)(\omega + i\gamma)} \right] \tag{17}$$

where E_n is the energy and $\mathbf{k}_n$ is the wave vector of the nth electronic state in the medium. The distribution function $f(E_n)$ for the occupation of the states is normalized so that $N = \sum f(E_n)$ is the total number of electrons.

Equation (17) is used extensively in the literature for describing stopping-power mechanisms, for the derivation of sum rules (e.g., Li 64), and for the treatment of screening inside solids (for example Ki 63, Zi 72). Equation (17) is also the basis for many approximate dielectric functions appearing in the literature. If we rewrite Equation (17) in a slightly different form:

$$\varepsilon(\mathbf{k}, \omega) = 1 + \sum_n \frac{f(E_n)}{N} \times \frac{\omega_p^2}{\dfrac{\hbar^2 k^4}{4m^2} + \dfrac{2\hbar}{m}(\omega + i\gamma)\mathbf{k} \cdot \mathbf{k}_n - \left[\dfrac{\hbar}{m}(\mathbf{k} \cdot \mathbf{k}_n)\right]^2 - (\omega + i\gamma)^2} \tag{17'}$$

then we easily recognize the high-frequency approximation [Equation (13), Section 2.2.1a]. Each member of the sum represents the contribution of a particular electronic state to the dielectric function. If one replaces the summation in Equation (17′) with an effective parameter that is independent of the electronic state n, then one obtains pole approximations to the dielectric function (see, for example, Ec 79).

Equation (17) is quite general and can be used for solids where a knowledge of the function $f(E_n)$ is available. The sensitivity of the dielectric function to $f(E_n)$ is reduced appreciably if the velocity of the projectile is much greater than the electronic orbital velocities inside the solid target. This is the regime where plasma models are applicable. In this regime we may expect metals, insulators, and semiconductors to behave similarly in response to the passage of a fast ion. The requirement for high velocity can be easily achieved with respect to the valence electrons of solids and thus wake phenomena depend more on the density of those electrons that meet this requirement rather than on the detailed phenomenological characteristics (like conductivity) of a specific solid. For a related discussion see (Ra 65).

An important application of Equation (17) is for the case of a degenerate free electron gas with Fermi velocity v_F (Li 54)

$$\varepsilon^l(u, z) = 1 + \frac{\chi^2}{z^2}\left\{\frac{1}{2} + \frac{1}{8z}[1-(z-u)^2]\ln\frac{z-u+1}{z-u-1} + \frac{1}{8z}[1-(z+u)^2]\ln\frac{z+u+1}{z+u-1}\right\} \tag{18}$$

where

$$\chi^2 = \frac{e^2}{\pi\hbar v_F}, \qquad z = \frac{k}{2k_F}, \qquad k_F = \frac{mv_F}{\hbar}, \quad \text{and} \quad u = \frac{\omega + i\gamma}{kv_F}$$

When $\varepsilon(k, \omega) = \varepsilon^l(u, z)$ is introduced into Equations (11) and (12) the three-dimensional integration can be carried out to produce the Lindhard wake potential and a wake force field. The numerical integration (Fa 80) is carried out in cylindrical coordinates. The angle integration is straightforward and poles in the k_z integration are treated analytically.

2.2.1c. The Plasmon-Pole Approximation to the Dielectric Function. Echenique *et al.* (Ec 79) have discussed the wake polarization and potential in terms of various approximations of Lindhard's dielectric function. One approximation that has been employed for some time by Ritchie and collaborators is the so-called plasmon-pole dielectric function

$$\varepsilon(k, \omega) = 1 + \frac{\omega_p^2}{\dfrac{\hbar^2 k^4}{4m^2} + \beta^2 k^2 + \omega_g^2 - \omega(\omega + i\gamma)} \tag{19}$$

The plasmon dispersion parameter is given by $\beta^2 = \frac{3}{5}v_F^2$ (Ra 65) and ω_g is an effective parameter for solids that have a band gap (e.g., semiconductors). Although the expression in Eq. (19) is mathematically simpler than that in Equation (18), evaluation of the wake potential and forces [Equations (11), (12)] has to be done numerically. Furthermore, the integration procedure is no easier to perform than that described in Section 2.2.1b for the Lindhard function [Eq. (18)].

2.2.2. *The Coulomb Wake*

In this section we discuss limitations on the use of linear response theories for the wake. It will be seen that there exists a wide variety of experimental conditions where there is a need for a nonlinear theory of the electronic wake forces. An approximate nonlinear theory (Fa 80) that makes use of the scattered Coulomb wave function for the target electrons will be described.

When a heavy nucleus of charge Ze with velocity v passes through a solid, it captures and loses electrons around it so that it rapidly achieves an equilibrium effective charge. Bohr (Bo 48) made the assumption that an ion traveling through condensed matter sheds those electrons that have orbital velocities smaller than the ion velocity. Using the approximation that the effective charge is a point charge, the orbital velocity of those electrons that are just barely bound to it is approximately $Z_{\text{eff}}e^2/\hbar$. Using Bohr's criterion, an estimate of the effective charge is then given roughly by

$$\begin{array}{lll} \text{Case I:} & Z_{\text{eff}} \approx \hbar v/e^2 & \text{for } Ze^2/\hbar v > 1 \\ \text{Case II:} & Z_{\text{eff}} \approx Z & \text{for } Ze^2/\hbar v < 1 \end{array} \tag{20}$$

Thus, for Case I to apply, it is sufficient to have an ion velocity $v < Zv_0$; that is, $-\eta = |Z_{\text{eff}}e^2/\hbar v| \approx 1$. But the Born approximation (upon which linear theory relies) holds only when $-\eta \ll 1$. However, it is well known that the stopping power is not very sensitive to inaccuracies inherent in the Born approximation description (the Born approximation reproduces Rutherford scattering even for regions where $-\eta \geqslant 1$). But for a description of the electronic polarization near a moving ion, it is not enough to have the correct asymptotic behavior (e.g., cross sections). For a wake description we need to obtain the electronic wave function in the vicinity of the moving ion, and the Born approximation will not be valid for this when the ion velocity corresponds to Case I [Equation (20)]. The correct electronic wave function near a point charge projectile is given by the Coulomb scattering wave function ψ_c (see, for example, Sc 49), so that

$$\rho_{\text{pol}}(\mathbf{r}) = ne[|\psi_c(\eta, \mathbf{r})|^2 - 1] \tag{21}$$

A simple comparison can be made for the ratio of polarization charge-density induced on the projectile as given by Equation (21) and by the linear theories (Lindhard or plasmon-pole approximation)

$$\frac{\rho_{\text{pol}}(0)(\text{Coulomb})}{\rho_{\text{pol}}(0)(\text{linear})} \sim 2 - \frac{1}{\pi|\eta|} \to 1.7 \qquad (\text{for } |\eta| = 1) \tag{22}$$

There is thus almost a factor of 2 more charge density on top of the projectile for velocities in the region pertaining to Case I [Equation (20)]. One may therefore conclude that for the description of wakes at such velocities there is a need for a better theory than the linear theories.

When a nonlinear theory is needed, other complications are introduced. For example, when two ions are traveling near each other the combined wake poential is no longer the sum of the individual potentials of each ion. Furthermore, the effective charge is clearly not a point charge. A theory which deals successfully with all of these problems does not yet exist.

The Coulomb-wake charge induced by the motion of a heavy ion with an effective point charge Ze can be calculated as follows. In the frame of the moving projectile, electrons with initial velocities $\mathbf{v} + \mathbf{u}$ are scattered from a point charge, where $\mathbf{u}$ is distributed in a Fermi sphere with a radius v_F. Near the projectile the electron–electron interaction can be ignored. Further away from the projectile at about a radius of v/ω_p the influence of the projectile is screened. Thus the excess density of the electrons (relative to the homogeneous background) is given by

$$\rho_{\text{pol}}(\mathbf{r}) = ne[\langle|\psi_c(\eta, \mathbf{r})|^2\rangle_u - 1] \tag{23}$$

where $\langle\ \rangle_u$ is an average over the Fermi velocity distribution. The density $\rho_{\text{pol}}(\mathbf{r})$ is then reduced to zero for radii greater than v/ω_p from the projectile.

2.3. Fluctuations in the Wake

The wake potential represents a statistical average of a quantity that is subject to fluctuations. Echenique *et al.* (Ec 79) have analyzed the extent to which these fluctuations can be expected to modify the energy straggling suffered by an ion that is moving in the wake trailing behind a swift partner ion. These authors concluded that the straggling is not very different from the value it would have if it were an independent (monatomic) projectile. The arguments used were similar to the statistical arguments of Bohr (Bo 48) and in addition invoked application of the plasmon-pole aproximation for the wake fields. In the c.m. frame, the motion of the fragments arising from a molecular ion dissociating in a solid is slow compared with the motion of the surrounding target electrons. These considerations indicate that only relatively small errors will be introduced in calculations if the straggling of the ionic fragments is taken to be the same as for independent ions.

3. Experiments with Fast Molecular-Ion Beams

As noted above, the motions of the dissociation fragments produced when fast molecular ions strike a solid target are influenced not only by their own wakes, but also by the wakes of their partner fragments.

3.1. Stopping-Power Effects

The mutual wake interaction within the cluster of dissociation fragments arising from a fast molecule was first recognized in measurements of Brandt *et al.* (Br 74) who compared the slowing down in foils of beams of H_2^+ and H_3^+ ions with that for proton beams of the same velocity (Figure 3). As the

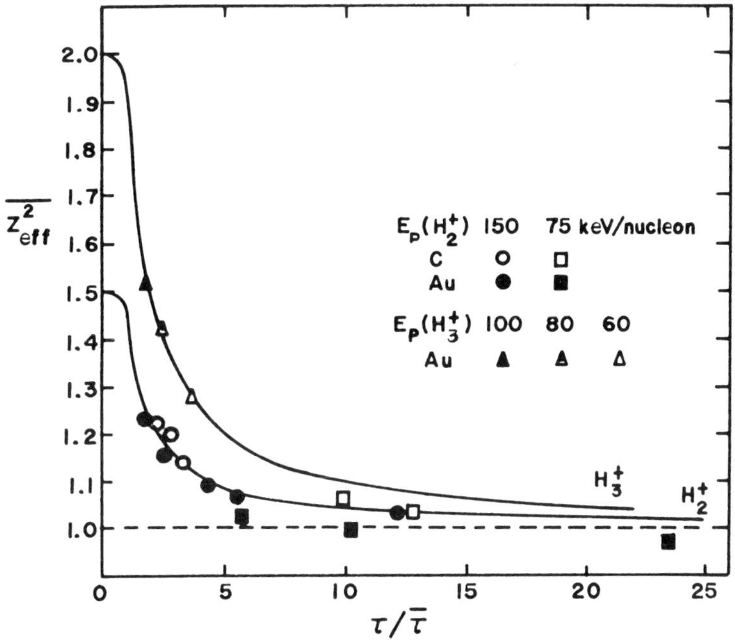

Figure 3. Measured energy-loss ratios $\Delta E(H_2^+)/2\Delta E(H^+)$ and $\Delta E(H_3)^+/3\Delta E(H^+)$ in carbon and gold foils as a function of the dwell time in the foils in units of $\bar{\tau}$ ($\bar{\tau}$ is the average time needed for the internuclear separation to become equal to v/ω_p). The solid lines are calculated on the basis of a model of interfering wakes. From Brandt *et al.* (Br 74).

foil dwell times increase, the slowing down for the molecular ions approaches equality with that for protons of the same velocity. This is simply because the molecular clusters Coulomb-explode apart and for large enough separations (i.e., large compared with v/ω_p) the individual proton fragments are not influenced by the wakes of their partners. However, the situation is quite different at short dwell times. When the interproton separation within a cluster is small (compared with v/ω_p), the cluster acts somewhat like a point charge $[Z_c^2 \approx (\sum Z)^2]$ as far as "distant" collisions with target electrons are concerned, while it acts like a collection of separate protons $[Z_c^2 \approx \sum (Z^2)]$ as far as energy transfer in close collisions is concerned (here Z_c is an effective charge for the cluster). The solid lines in Figure 3 are calculated on the basis of the interference between the wakes that each fragment induces in the target. The good agreement between the data and the calculations can be interpreted as experimental confirmation of the equipartition of stopping power into contributions from close and distant collisions (Bo 48). Thus as $\tau/\bar{\tau} \to 0$, $z_{\text{eff}}^2 \to \frac{1}{2}\{\sum Z)^2\}/N$, where the summation extends over the N protons in the cluster. Related measurements with swift light molecular ions have been performed by other workers (Ew 62, Ma 67, De 74, Ec 78, Ny 78, and La 79). Tape *et al.* (Ta 76) observed increases of 10% to 60% in the slowing down of oxygen fragments from 3.2–5.8-MeV O_2^- projectiles when compared with that for monatomic oxygen projectiles of the same velocity. Theoretical treatments of these stopping-power effects have also been given by other authors (e.g., Ar 75, Br 76, Ar 78, and St 78).

One expects that the stopping-power modifications will depend on the spatial orientation of the projectile relative to the beam direction. Recent measurements by Steuer *et al.* (St 81) in which the slowing down of nitrogen fragments from 1—3.6-MeV N_2^+ projectiles has been determined, show a reduction of 10%–20% for projectiles in which the internuclear axis is aligned close to the beam direction. The opposite sign of the molecular effect (compared with that found in most previous measurements) is a consequence of the fact that at lower velocities a significant fraction of the leading ion's wake charge lies between the separating fragments.

3.2. High-Resolution Measurements of Fragment Momenta

The stopping-power measurements outlined above are not as sensitive to the detailed form of the wake as are high-resolution measurements on both the longitudinal and transverse momentum components of fragments arising from the dissociation of molecular-ion projectiles in foils. We therefore devote the major part of Section 3 to this latter type of measurement.

3.2.1. Collision-Induced Dissociation at High Velocities

Until recently, most studies on the collision-induced dissociation of molecular ions were performed using gaseous targets and ion-beam energies in the keV range [see, for example, (Co 78) for a review]. At energies in the MeV range, however, accurate measurements with foil targets become feasible. This is mainly because in foils high fragment charge states are attained giving rise to vigorous Coulomb explosions, while at the same time multiple scattering of the fragments in the foil can be reduced to acceptable levels at the higher velocities.

3.2.2. Coulomb Explosions

In this section we explore the implications of a simplified model for the Coulomb explosion of a fast molecular ion in a foil. We take as an example a tightly collimated beam of 2-MeV HeH^+ ions incident on a 100-Å carbon foil (Figure 4). The model can later be refined and extended to other molecular projectiles.

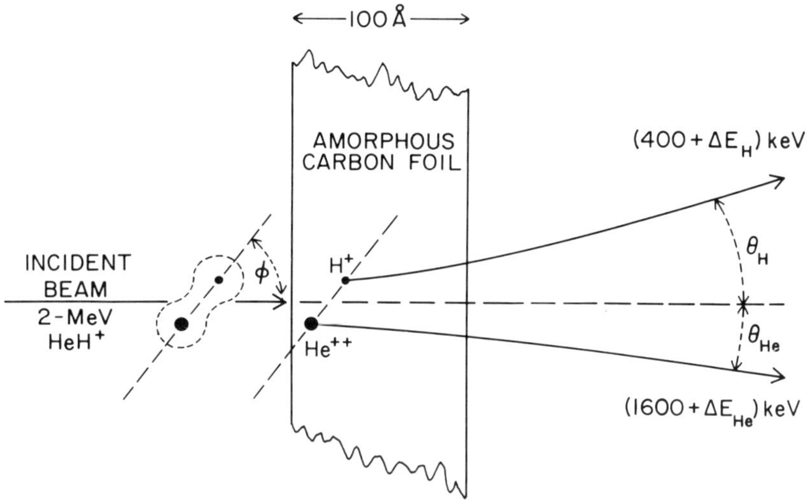

Figure 4. A (very) schematic picture of a 2-MeV HeH^+ projectile dissociating in a 100-Å-thick amorphous carbon foil.

For such a fast projectile the Coulomb explosion begins within about 10^{-17} s after the ion strikes the front surface of the foil. This is a consequence of the large cross sections for electron removal from the projectile in collisions with target electrons [typically $\sim 10^{-16}$ cm^2—see, for example, (Bo 48)].

Since the mean time to strip the two electrons from each projectile is short compared with the dwell time in the foil (1.14 fs at the beam velocity $v = 8.8 \times 10^8$ cm/s) and with the times for molecular vibration ($\sim 10^{-14}$ s) and rotation ($\sim 10^{-12}$ s) of the projectile, we may make the approximation that the electron stripping and the start of the consequent Coulomb explosion both occur instantaneously at the front surface of the foil. Let the internuclear vector from the α particle to the proton in a given HeH^+ projectile have a length r_0 at this instant and be inclined at an angle ϕ relative to the beam direction. Let the values of r_0 have a probability distribution $D(r_0)$ that depends upon the population of the various vibronic states of the projectile.

If all the HeH^+ projectiles were in the ground electronic, vibrational, and rotational state, the most probable value of r_0 would be expected to be 0.79 Å (based, for example, on the potentials calculated by Kolos and Peek (Ko 76)]. Let us consider a projectile having this particular value of r_0 at the moment of entry into the foil. Electron stripping then produces an α-particle–proton pair with a mutual Coulomb energy $\varepsilon = 2e^2/r_0 = 36.5$ eV. This potential energy is converted into kinetic energy in the c.m. as the Coulomb explosion develops. Since this Coulomb energy is much larger than the energies that the fragments have due to vibration and rotation in the incoming projectile, we temporarily make the further approximation that at the instant of stripping ($t=0$) the α particle and the proton are at rest in the c.m. frame.

If, for the moment, we ignore further interactions of the projectile nuclei with the target foil, then the internuclear separation as a function of time, $r(t)$, is found by solving

$$\mu\ddot{r}=Z_1Z_2e^2/r^2 \tag{24}$$

with the initial conditions

$$r(0)=r_0, \qquad \dot{r}(0)=0 \tag{25}$$

where $\mu=M_1M_2/(M_1+M_2)$ is the reduced mass of the two fragments and Z_1e and Z_2e are their charges. From this, one finds that the time for the internuclear separation to grow to r is given by

$$t(r/r_0)=t_0 f(r/r_0) \tag{26}$$

where

$$t_0=[\mu r_0^3/(2Z_1Z_2e^2)]^{1/2} \tag{27}$$

and

$$f(x)=x^{1/2}(x-1)^{1/2}+\ln[x^{1/2}+(x-1)^{1/2}] \tag{28}$$

If μ is expressed in a.m.u. and r_0 is in angstroms, then

$$t_0=1.90[\mu r_0^3/Z_1Z_2]^{1/2}\ \text{fs} \tag{29}$$

In a time t_0, 19% of the initial Coulomb potential energy is converted into kinetic energy of the separating fragments. The final asymptotic ($t \gg t_0$) relative velocity of recession is

$$\dot{r}(\infty)=[2Z_1Z_2e^2/\mu r_0]^{1/2} \tag{30}$$

For our example of 2-MeV HeH^+ ($Z_1=1$, $Z_2=2$), $t_0=0.84$ fs and $\dot{r}(\infty)=9.4\times10^6$ cm/s (about 1% of the beam velocity). The internuclear separation as a function of time is shown for this example in Figure 5. From Figure 5 one also sees that with our dwell time in the target of 1.14 fs, a significant portion of the Coulomb explosion occurs after the fragments emerge from the foil (the internuclear separation at exit is 1.11 Å).

The final energies and angles of the fragments in the laboratory (LAB) frame are easily obtained by referring to the vector diagram in Figure 6. Let

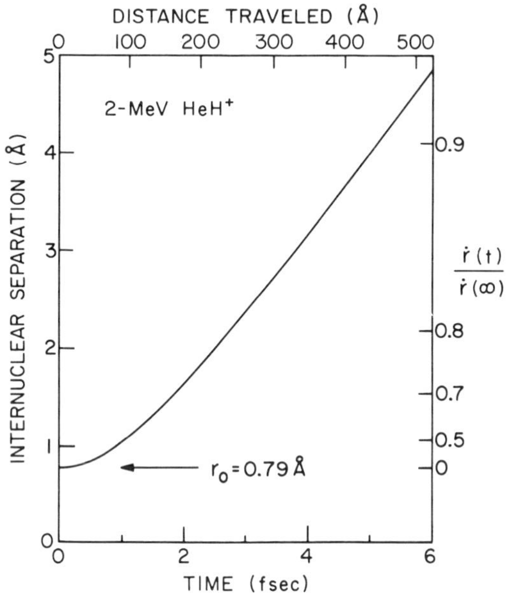

Figure 5. The dependence upon time of the internuclear separation between the α particle and the proton of a HeH^+ ion whose internuclear separation at $t=0$ is 0.79 Å. The nuclei are assumed to be stationary at $t=0$ in the c.m. frame. The right-hand ordinate gives the fractional growth of the Coulomb-explosion velocity as a function of time. The scale at the top shows the distance traveled in the LAB frame by the projectile assuming the beam energy to be 2 MeV.

u be the asymptotic c.m. velocity acquired by the fragment of mass M_1 as a result of the Coulomb explosion. That is,

$$u = (\mu/M_1)\,\dot{r}(\infty) \tag{31}$$

The asymptotic LAB velocity of this fragment is then $\mathbf{v}+\mathbf{u}$. The shift in LAB angle is

$$\theta = (u \sin \phi)/v = u_\perp/v \tag{32}$$

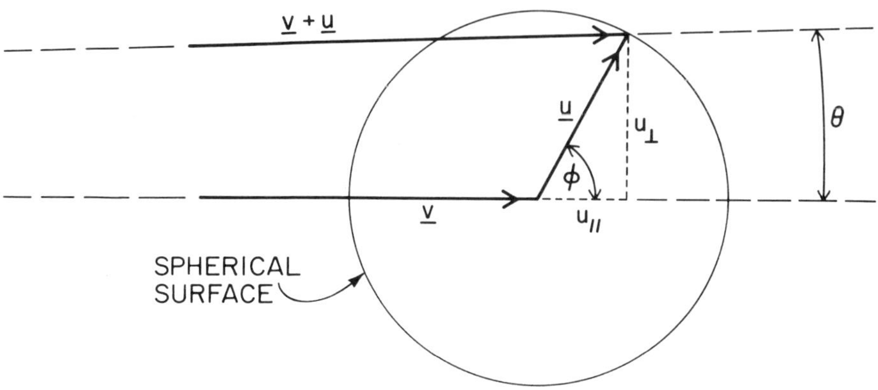

Figure 6. Vector diagram illustrating the relationship between c.m. and LAB coordinates for a fragment emerging from a Coulomb-exploded molecular-ion projectile.

and the shift in LAB energy is

$$\begin{aligned} \Delta E &= \tfrac{1}{2}M_1 |\mathbf{v} + \mathbf{u}|^2 - \tfrac{1}{2}M_1 v^2 \\ &\approx M_1 vu \cos\phi = M_1 vu_{||} = 2E_1(u_{||}/v) \end{aligned} \tag{33}$$

where

$$E_1 = \tfrac{1}{2}M_1 v^2 \tag{34}$$

(We neglect terms with relative magnitudes on the order of u/v.)

From this, one sees that for each fragment species there is a maximum LAB angle $\theta_{\max}$ such that for $\theta > \theta_{\max}$ none of these fragments will be observed. For $\theta < \theta_{\max}$ two classes of this fragment species will be observed: one with an upward shift in energy and one with an equal shift downwards in energy. If $\theta = \theta_{\max}$, then $\phi \approx 90°$; i.e., at $t = 0$ the internuclear vector of the incident projectile lies normal to the beam direction. For the fragment of mass M_1

$$\theta_{\max} = u/v = (\varepsilon_1/E_1)^{1/2} \tag{35}$$

where

$$\varepsilon_1 = (\mu/M_1)(Z_1 Z_2 e^2/r_0) \tag{36}$$

In the LAB frame the fragments emerging along the beam direction ($\theta = 0$) correspond to the other extreme orientation of the incoming projectile ($\phi = 0$ or $\phi = 180°$). Such fragments although suffering no angular shift in the LAB, are observed with the maximal energy shift $\pm\Delta E_{\max}$. The sign of the shift depends on whether the fragments are leading (+) or trailing (–) their partners. For the fragment of mass M_1

$$\Delta E_{\max} = M_1 vu = 2(\varepsilon_1 E_1)^{1/2} \tag{37}$$

In our example of 2-MeV HeH^+ with $r_0 = 0.79$ Å, Equations (35) and (37) give for the protons, $\theta_{\max} = 8.5$ mrad (0.49°) and $\Delta E_{\max} = 6.8$ keV. For the α particles, $\theta_{\max}$ is a factor of 4 smaller and $\Delta E_{\max}$ is the same.

It is interesting to note the "amplifying" effect of the cross-term in Equations (33) and (37). A c.m. energy on the order of eV gives rise to a LAB energy shift on the order of keV when MeV projectiles are employed.

If the spatial orientations of the incident projectiles are isotropically distributed, we may think of the tip of the vector **u** in Figure 6 as populating with uniform probability the surface of a spherical shell of radius u. Thus, for a given projectile orientation, there is a $(4\pi u^2)^{-1}$ weighting factor for the values of u that arise from an initial distribution $D(r_0)$ of internuclear spacings. The distribution function for u is

$$G(u) = (4\pi u^2)^{-1} D(r_0) \, |dr_0/du| \tag{38}$$

where $G(u)$ and $D(r_0)$ are normalized by

$$\int_0^\infty D(r_0) \, dr_0 = 1, \qquad 4\pi \int_0^\infty G(u) \, u^2 \, du = 1 \tag{39}$$

For a simple Coulomb explosion we use Equations (30) and (31) to obtain

$$G(u) \propto u^{-5} D(r_0(u)) \tag{40}$$

From the foregoing it is apparent that measurement of the LAB variables θ and ΔE for one of the fragment species (M_1, say) is equivalent to a measurement of $u_\perp$ and $u_{\parallel}$ in the c.m. system. Knowledge of the Coulomb dissociation potential then permits determination of $D(r_0)$. Actually just one of two measurements suffices: either the distribution of θ for $\Delta E = 0$, or the distribution of ΔE for $\theta = 0$.

A two-parameter measurement $(\theta, \Delta E)$ of the yield of fragments of mass M_1 would be expected *a priori* to give a uniform "ring pattern" since it

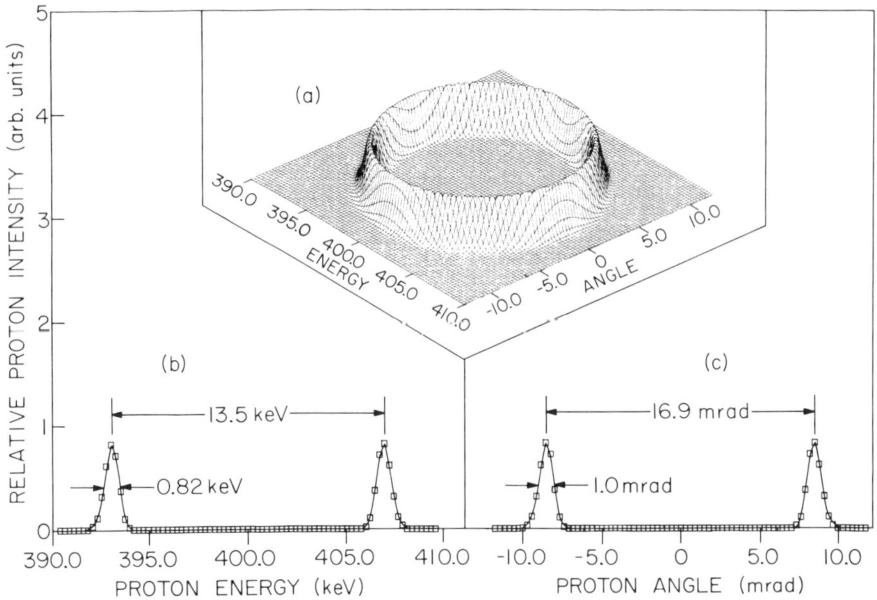

Figure 7. (a) "Ring pattern" calculated for the joint energy-angle distribution for protons arising from the simple Coulomb explosion of 2-MeV HeH^+. (b) Energy distribution for $\theta = 0$. (c) Angular distribution for $\Delta E = 0$.

corresponds to a cut across the center of the sphere indicated in Figure 6. In Figure 7 we show such a ring pattern calculated for the protons from our 2-MeV HeH^+ example. For the calculation we have assumed that all projectiles are in the ground vibronic state and we have used $D(r_0)$ obtained from the potential given by Kolos and Peek (Ko 76). Energy and angle distributions such as those in Figures 7b and 7c are sometimes called a "cross" since they represent orthogonal diametric cuts across the ring pattern.

We have thus far ignored the influence of the medium apart from its role in initiating the Coulomb explosion. In practice, of course, several phenomena occurring as the ion fragments traverse the foil will influence their momenta. The most important of these are multiple scattering, energy loss, energy straggling, wake effects, and charge exchange. Later in this section we discuss the influence of all of these phenomena.

3.2.3. Apparatus for High-Resolution Measurements

To explore the details of ring patterns having "radii" of a few mrad and a few keV, experimental resolution widths at least an order of magnitude smaller than these values are desirable. This poses some technical problems, particularly in the angular coordinate. We describe here the apparatus in use at Argonne and at the Weizmann Institute.

Figure 8 shows the experimental arrangement in use at Argonne's 4-MV Dynamitron accelerator.*

Magnetically analyzed molecular-ion beams are collimated to have a maximum angular divergence of ± 0.09 mrad and a spot size of 1 mm at the target position. A set of "predeflector" plates permits electrostatic deflection of the beam incident on the target. Similarly a set of "postdeflectors" is used to deflect charged particles emerging from the target. The pre- and postdeflectors are used in combination so as to avoid detecting particles that arise from spurious incident beams (e.g., fragments arising from dissociation of the primary beam along the long flight path between the beam collimators). A 25-degree electrostatic analyzer having a relative energy resolution of $\pm 3 \times 10^{-4}$ is located several meters downstream from the target. An aperture placed ahead of the analyzer accepts a 1-mm-diam group of trajectories originating at the target position.

Distributions in energy and angle are made for particles emerging from the target by varying the voltages on the horizontal predeflectors and/or the

* The Argonne apparatus has evolved over several years. Major contributions to its development have been made by P. J. Cooney, D. S. Gemmell, E. P. Kanter, W. J. Pietsch, I. Plesser, A. J. Ratkowski, M. J. Steuer, Z. Vager, and B. J. Zabransky.

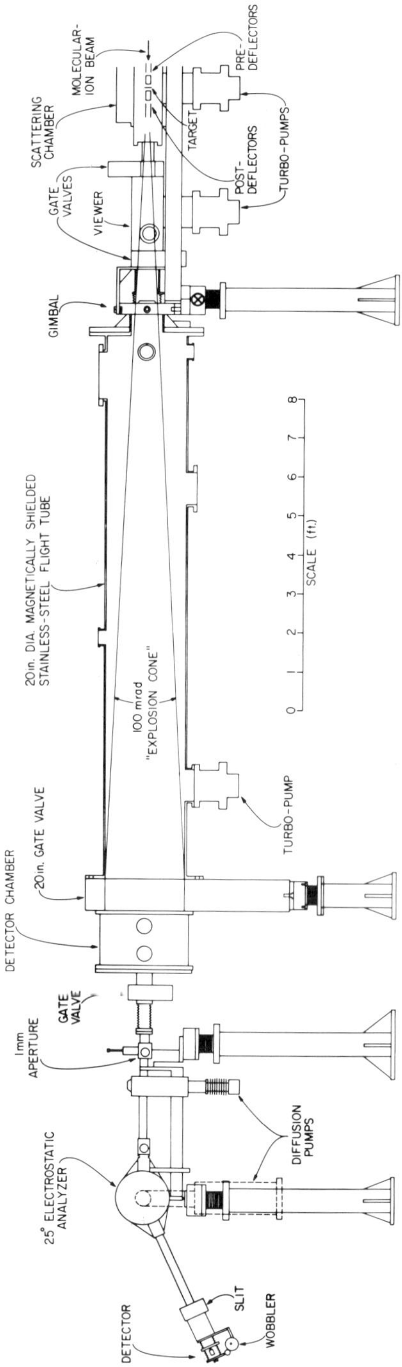

Figure 8. Schematic diagram of the experimental arrangement at Argonne's 4-MV Dynamitron accelerator.

postdeflectors in conjunction with that on the electrostatic analyzer. [The vertical ("Y") deflectors are normally used only for "trimming" purposes.] The overall angular resolution is ± 0.15 mrad.

The detector chamber shown in Figure 8 is a recent addition to the beam line (Ge 80a,b). It permits the coincident detection of dissociation fragments. The chamber houses two detectors that can positioned to an accuracy of about 0.001 in. anywhere on a 20-in.-diam circular area subtending an angle of 100 mrad at the target.

With the apparatus resolutions

$$\delta\theta = \pm 1.5 \times 10^{-4}\ \text{rad}, \qquad \delta E/E_1 = \pm 3 \times 10^{-4} \tag{41}$$

one can measure $u_\perp$ and $u_{||}$ with equal accuracy. From Equations (32) and (33) we find

$$\delta u_\perp = \delta u_{||} = \pm 1.5 \times 10^{-4} v \qquad \text{(Argonne)} \tag{42}$$

A different approach is used at the Weizmann Institute's 14UD Pelletron accelerator. Molecular-ion beams are produced by initially accelerating negative molecular ions and converting them into positive molecular ions in a gas stripper at the accelerator terminal. Subsequent magnetic analysis and angular collimation produces a beam with ± 0.07 mrad angular divergence and 0.3 mm diameter at the target position. A set of pre- and postdeflectors is used in a manner similar to that at Argonne. The experimental arrangement is shown in Figure 9. Fragments (e.g., protons as shown in Figure 9) from Coulomb explosions in the target are restricted by a variable entrance slit to a horizontal "ring" in velocity space. A horizontal magnetic field (spectrometer) disperses the velocity components $u_{||}$ parallel to the beam while leaving unaffected the velocity components $u_\perp$ in the horizontal plane. The fragments are recorded in a detector (Br 80) which is composed of two independent parts mounted in a common vessel: firstly, a drift chamber which provides a bidimensional measure of the position of the incident particle, and secondly, a parallel-plate avalanche counter that provides a fast timing signal. The drift time of electrons produced by an incident particle gives a measure of its position along the spectrometer focal plane, and hence of its energy (E). The orthogonal position, obtained by extracting induced signals from a delay-line, is proportional to the angle (θ) of the particle trajectory in the plane defined by the entrance slits to the spectrometer. The resolution of the drift chamber is about 0.5 mm FWHM in both dimensions. The overall angle and energy resolutions are

$$\delta\theta = \pm 1.8 \times 10^{-4}\ \text{rad}, \qquad \delta E/E = \pm 3.6 \times 10^{-4} \tag{43}$$

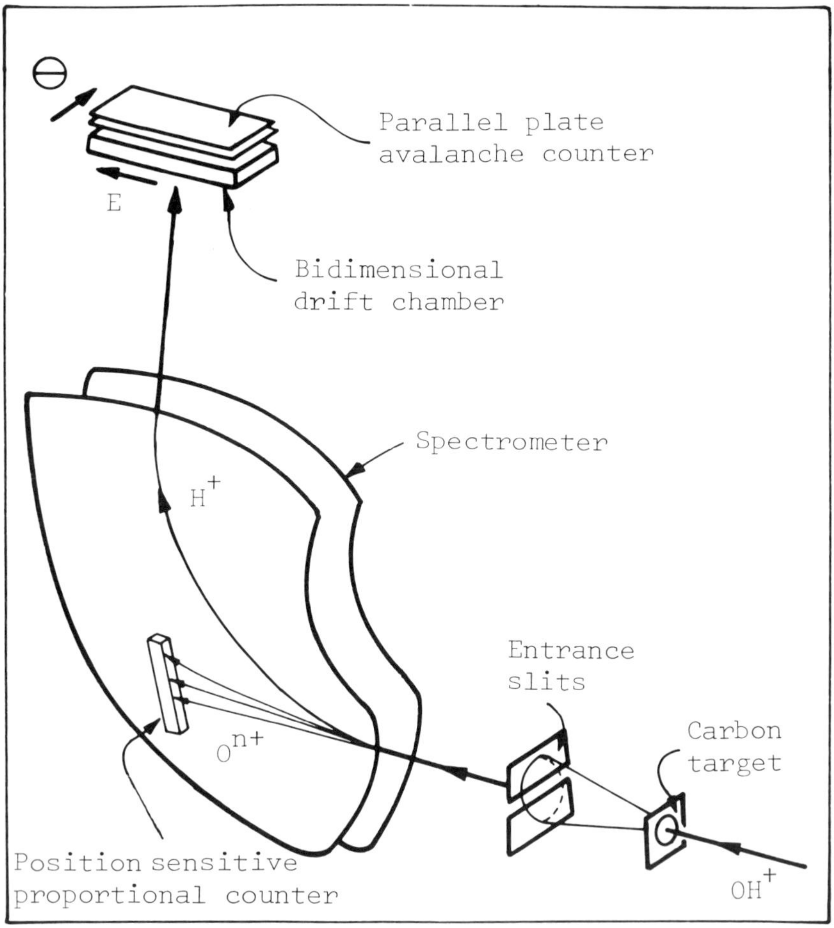

Figure 9. Experimental arrangement at the Weizmann Institute.

Also here we find that the perpendicular and the parallel components of the velocity **u** are measured with equal accuracy:

$$\delta u_{\perp} = \delta u_{\parallel} = \pm 1.8 \times 10^{-4} v \qquad \text{(Weizmann)} \tag{44}$$

Heavier ion partners (e.g., the oxygen ions shown in Figure 9) from the same molecular ions can be measured in coincidence with the aid of a position-sensitive proportional counter. This detector is inserted between the pole pieces of the spectrometer magnet. Different charge states of the heavier partner are incident on different positions on the detector. Thus, ring patterns of the lighter particles are sorted according to the charge state of the heavier partner.

3.2.4. Measurements on "Ring Patterns" for Diatomic Projectiles

Figure 10 shows a ring pattern and a cross measured for protons from 2-MeV HeH^+ incident upon a thick carbon foil (Va 76a). The ring has approximately the diameter predicted from our simple model (cf. Figure 7). The most obvious feature not predicted from the model is the nonuniform distribution of proton intensity around the ring. There is an enhanced intensity for trailing protons and (to a much lesser degree) for leading protons. This redistribution of particle flux is a consequence of the interaction of the fragments with the wakes that their partners induce in the foil. It is observed for both heteronuclear and homonuclear projectiles.

The widths of the peaks in the measured cross are larger than expected for the HeH^+ ground state alone (Figure 7). This is because $D(r_0)$ is a broader distribution (excited states of the projectile contribute) and because of multiple scattering. The asymmetry in the peak widths in the $\theta = 0$ energy spectrum is caused by wake effects.

Wake effects are also responsible for the slightly asymmetric location of the peaks in the energy spectrum with respect to the expected center of the

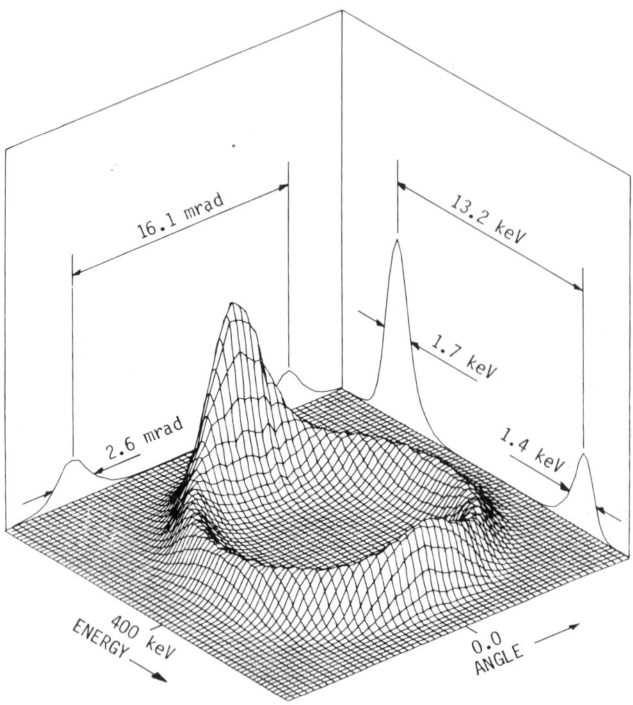

Figure 10. "Ring pattern" and "cross" for protons from 2.0-MeV HeH^+ dissociating in an 85-Å-thick carbon foil (Va 76a).

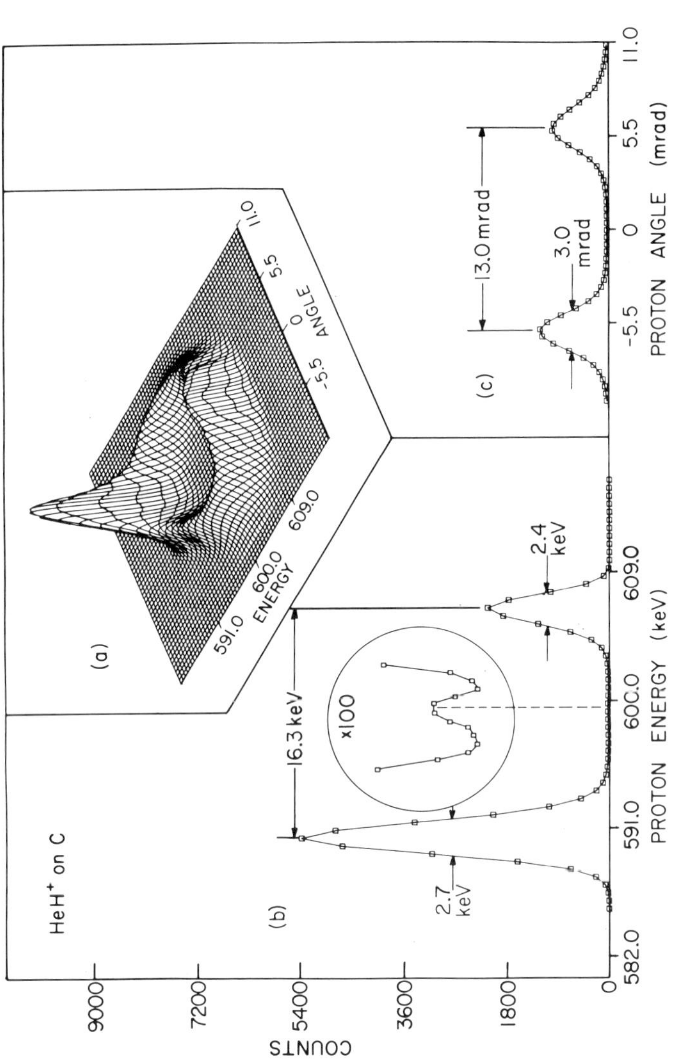

Figure 11. (a) "Ring pattern" and (b), (c) "cross" for protons from 3.0-MeV HeH^+ dissociating in a 195-Å-thick carbon foil (Ka 79).

ring. At this central position one observes a very small peak. This can be seen clearly in the data shown (Figure 11) for 3-MeV HeH^+. The central peak arises from dissociations producing He^0 and a proton (Va 76b, Ka 79, Ge 80c). Considering the weighting factors given in Equation (38), it is evident that the fraction of dissociations resulting in He^0–H^+ pairs is very small. These weighting factors coupled with the breadth of $D(r_0)$ cause a slight reduction in the ring diameter as compared with the values given in Figure 7 and by Equations (35) and (37).

Figures 12–15 show similar results for the foil-induced dissociation of beams of H_2^+, $^3He_2^+$, N_2^+, and OH^+. (For further examples of data of this type, see Go 74, Va 76b, Ge 77, Ga 78, La 78, Br 79a, b, Ge 79a, b, Ka 79, Br 80, Ge 80a, c).

When fast molecular ions dissociate in a dilute gas target, wake effects, screening, and multiple scattering no longer play a role. Figure 16 shows two examples. One sees immediately that the rings are uniformly populated. The diameters of the rings correspond to the dissociations $H_2^+ \to H^+ + H^+$ and $HeH^+ \to He^+ + H^+$ [there is a very much weaker outer ring in Figure 16b

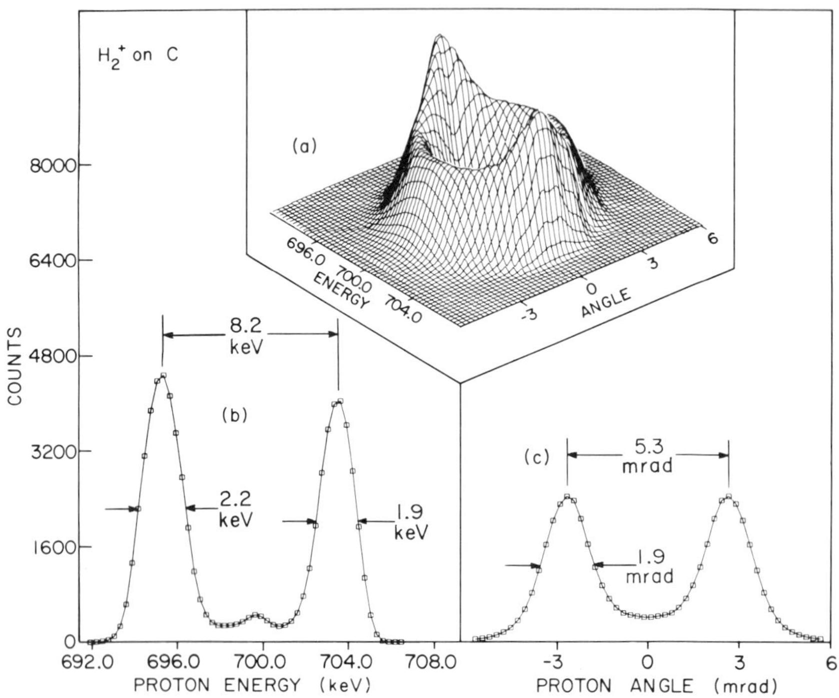

Figure 12. (a) "Ring pattern" and (b), (c) "cross" for protons from 1.4-MeV H_2^+ dissociating in an 88-Å thick carbon foil (Ka 79).

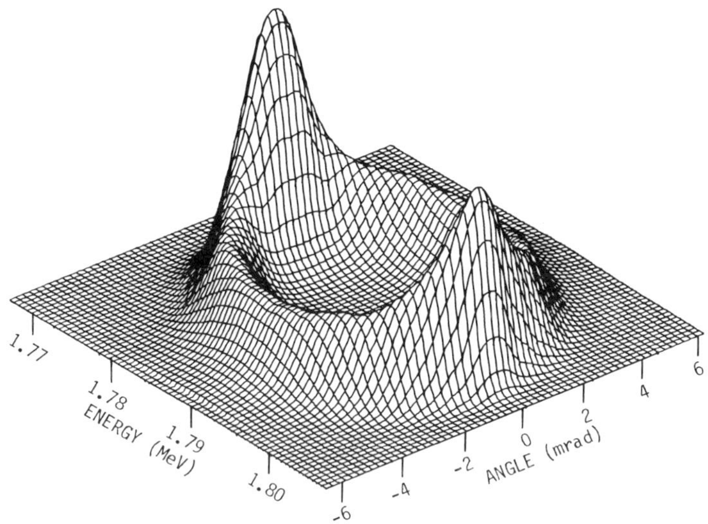

Figure 13. "Ring pattern" for $^3He^{++}$ from 3.6-MeV $^3He_2^+$ dissociating in a 118-Å-thick carbon foil (Ge 79a).

corresponding to $HeH^+ \rightarrow He^{++} + H^+$]. The patterns display dominant central peaks. The central regions can be quantitatively accounted for (Ka 79) in terms of collision-induced transitions of the projectiles to excited electronic states which then decay dissociatively yielding one neutral and one charged fragment. Gentle collisions of this type are possible in gases (unlike solids) where large-impact-parameter collisions are favored. For such excitations the resultant c.m. energy acquired by the fragments can be small (⟨0.1 eV, say) if the initial internuclear separation is large [i.e., on the tail of the distribution $D(r_0)$]. These fragments, although arising in only a small fraction of the dissociations, are detected for all initial projectile orientations and thus give rise to the large central peak.

3.3. Understanding the Ring Patterns

In the preceding discussion we have seen that the gross features of the ring patterns (e.g., their diameters and the shapes of their rims) can be reasonably well understood in terms of the Coulomb-explosion mechanism and taking into account the initial distributions $D(r_0)$. In this section we examine the detailed shapes of the measured rings and discuss the physical processes that determine them. Of particular interest is the very obvious redistribution of particle intensity around the ring. All of the rings obtained with foil targets display an enhancement of intensity along the zero-angle

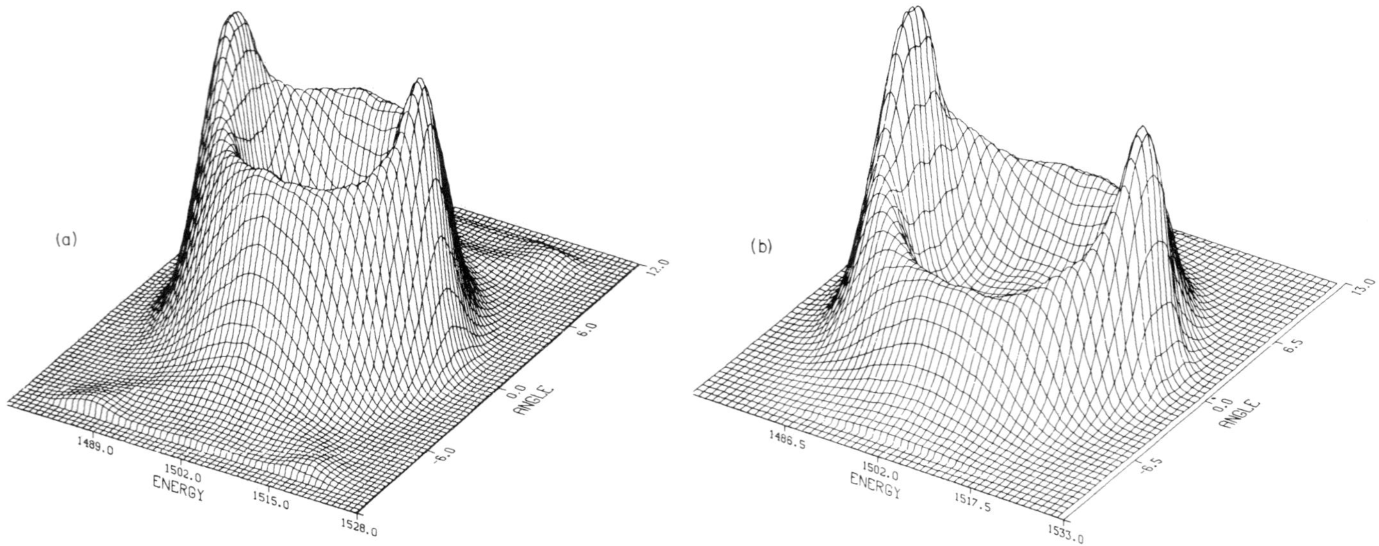

Figure 14. "Ring patterns" for (a) N^+ and (b) N^{4+} fragments arising from the dissociation of 3-MeV N_2^+ in a 75-Å-thick carbon foil (Ge 81).

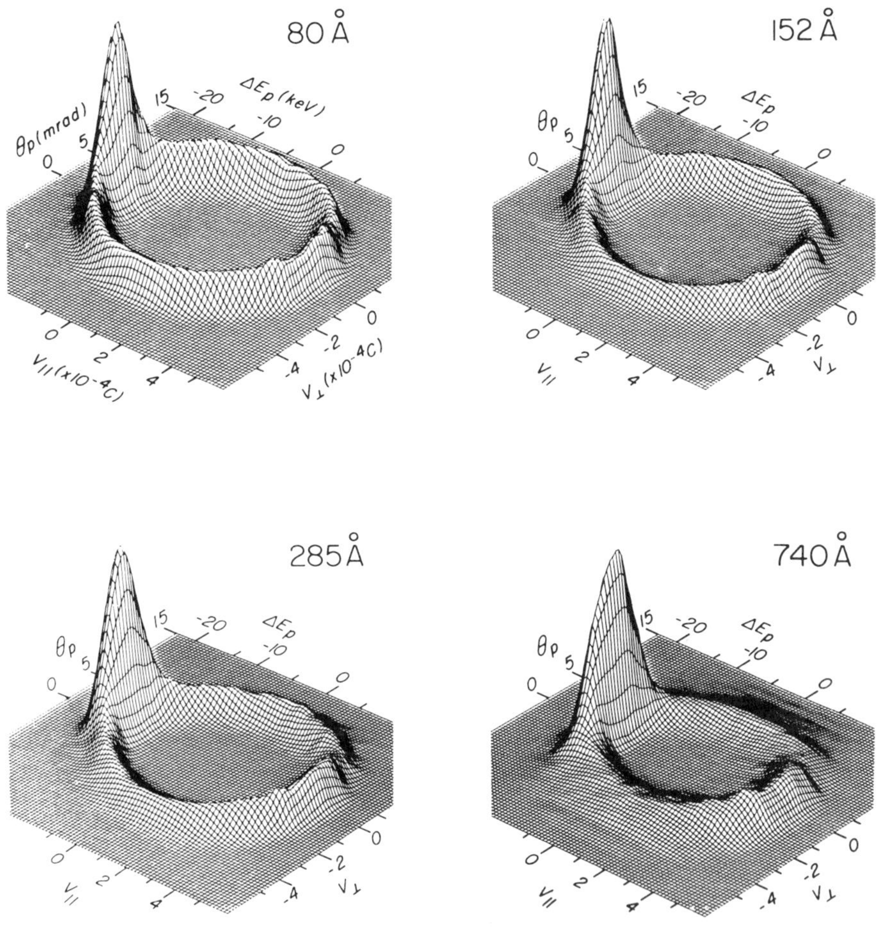

Figure 15. "Ring patterns" measured for protons in coincidence with O^{6+} ions from 11.2-MeV OH^+ projectiles. The data shown are for four different thicknesses of carbon target (Br 80).

axis, and along this axis there is a "backwards–forwards" asymmetry—there are more trailing fragments than leading fragments.

3.3.1. Physical Phenomena Related to Ring Patterns

In the motion of the molecular fragments in the solid target one can distinguish between two regimes of time scales, namely, the electronic time scale and the Coulomb-explosion time scale. As a consequence of the large electron-ion cross sections and the high projectile velocities, the time scale

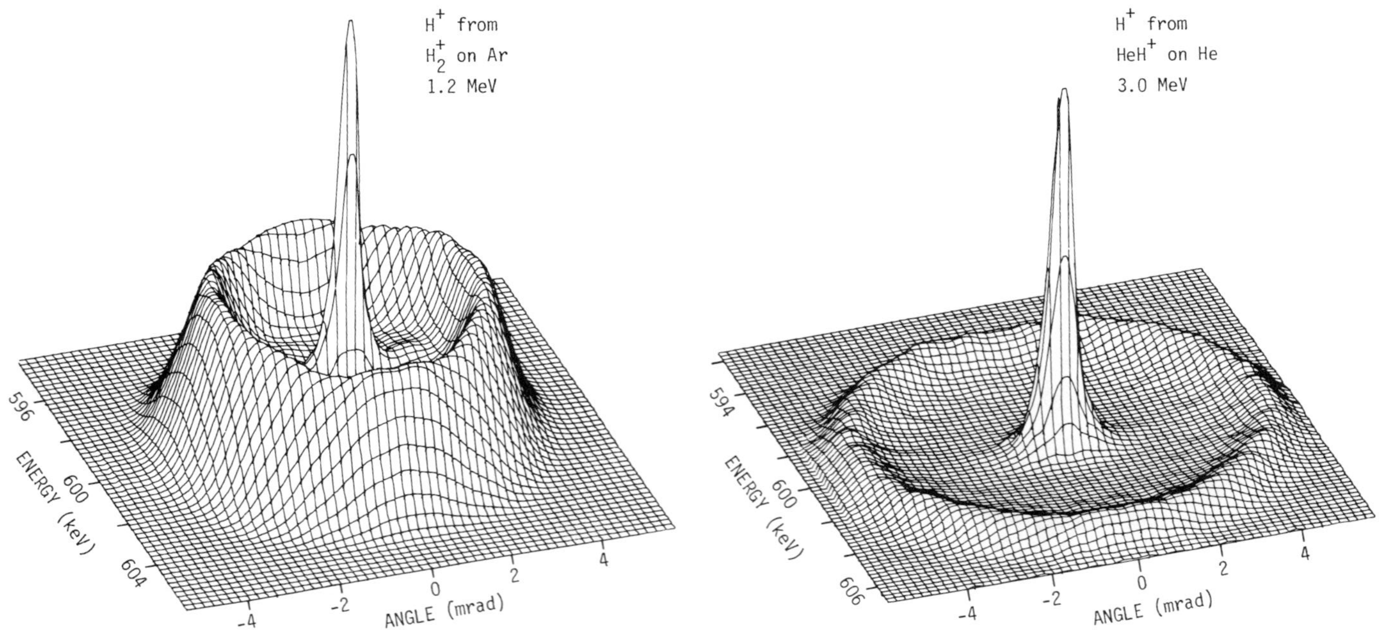

Figure 16. Joint energy-angle distributions for protons from 1.2-MeV H_2^+ and 3.0-MeV HeH^+ dissociating in gaseous Ar and He, respectively (Ka 79).

for the electronic motion is around 10^{-17} s. For our example of 2-MeV HeH^+ ions incident on a 100-Å carbon foil (Figure 4), both the dwell time in the target (1.14 fs) and the characteristic Coulomb-explosion time (0.84 fs) are of the order of 1 fs. We may therefore conclude that as a first approximation we can neglect all the statistical fluctuations connected with the electronic motion and consider only the average field due to the electrons at any given relative position of the exploding fragments. In this approximation statistical processes like energy straggling and charge exchange can be ignored. On the other hand, the statistical process of multiple scattering is of crucial importance since it adds significantly to the separation of the fragments during their dwell time in the target. We are left then to deal with the average field of the target electrons (wake), the Coulomb explosion, and the multiple scattering.

Since the parameter $Z_1Z_2e^2/\hbar\dot{r}(\infty)$ is always much greater than unity for all molecular ions [see Equation (30)], classical trajectories (Bo 48) can be used in calculating the c.m. motion of the dissociation fragments. This statement is true as long as Coulomb forces dominate [see (Ka 79) for a discussion of the validity of classically calculated trajectories for other dissociation potentials].

3.3.2. The Simulation of Ring Patterns, First Version

The first work on the analysis of a 2-MeV HeH^+ ring pattern (Va 76a) used a simulation procedure with the following properties. The α-particle and proton motions were computed under the assumption that they are influenced only by (a) the direct Coulomb repulsion, and (b) wake forces corresponding to potentials of the form given in Equation (15). The initial internuclear separations were taken to be distributed as a Gaussian with mean value $\bar{r} =$ 0.79 Å and standard deviation 0.15 Å. The calculated results were smeared in the angular coordinate by the measured multiple scattering. The plasmon energy was obtained by matching Equation (16) to the measured stopping power of carbon for 400-keV protons. This gave $\hbar\omega_p = 25.0$ eV. The result of this calculation is shown in Figure 17. Comparing Figures 10 and 17, it can be seen that the calculation succeeds in reproducing the backward peak and the small forward peak as well as the other gross features of the ring. It is also clear that many details, such as ratios of the side peaks to the front and back peaks, are not very well reproduced.

3.3.3. Comparison of 11.2-MeV OH^+ Rings with Theory

Breskin *et al.* (Br 80) have studied the proton ring patterns arising from the dissociation of 11.2-MeV OH^+ in several carbon targets. The protons were detected in coincidence with emergent oxygen ions in 4^+, 5^+, 6^+, and

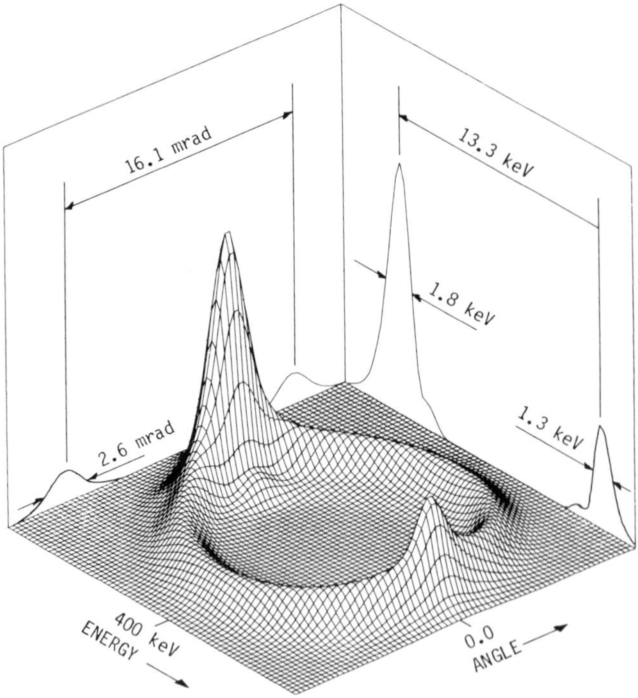

Figure 17. "Ring pattern" calculated on the basis of a simple plasma wake for protons from 2-MeV HeH^+ dissociating in an 85-Å-thick carbon foil (Va 76a).

7^+ charge states. Some examples of the data are shown in Figure 15. The proton velocity distributions display two prominent features. First, to a good approximation, all protons are distributed on a circular ring of radius u and width δu. This indicates that the radial forces between the oxygen and the proton are almost angle independent. The magnitudes of u and δu depend on the initial internuclear separations, which are characterized here by a Gaussian centered ar $r_0 = 1.05$ Å and a standard deviation $\sigma = 0.17$ Å. The second prominent feature is the very large backward peak.

We now describe a study of the patterns of the proton ring patterns focussing our attention on the shapes of the backward peaks and comparing them with simulations of the experiment of the type described in Section 3.3.2.

Three wake models are examined, the simple plasma wake (Section 2.2.1a), the Lindhard wake (Section 2.2.1b), and the Coulomb wake (Section 2.2.2). The wake potentials for the three different models are shown in Figure 18. For each model the simulation program described in Section 3.3.2

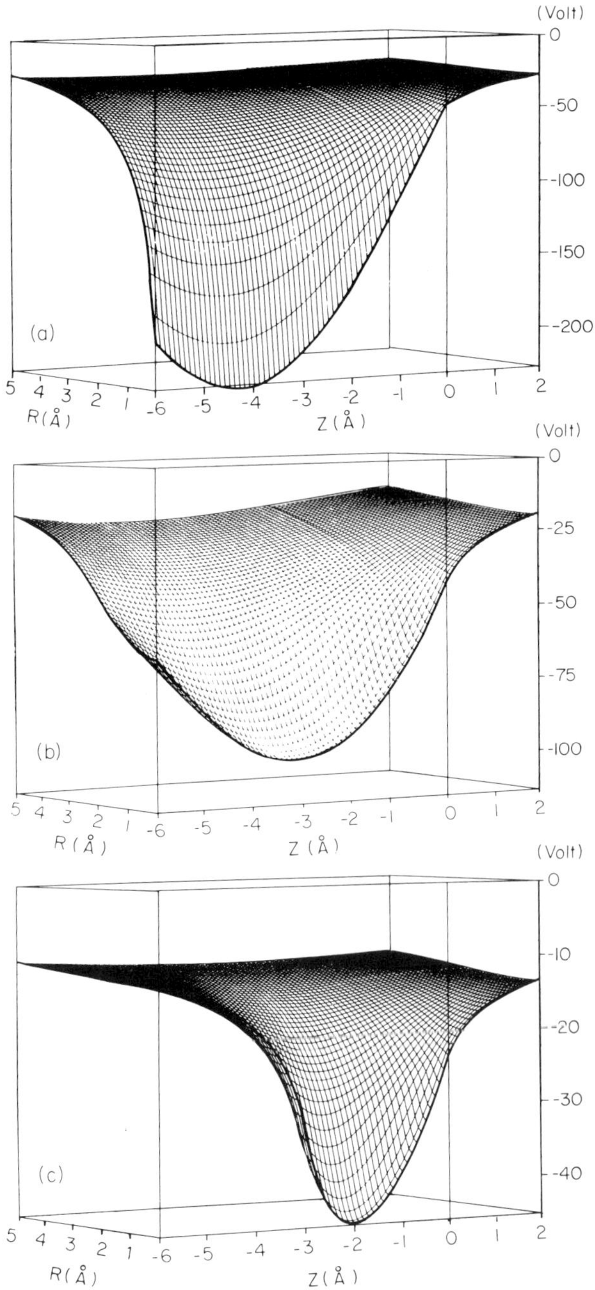

Figure 18. Potentials near a moving oxygen ion computed with (a) classical plasma polarization wake, (b) Lindhard dielectric function, and (c) Coulomb scattering of electrons. The ion track is along the Z axis and R is the radial distance (Fa 80).

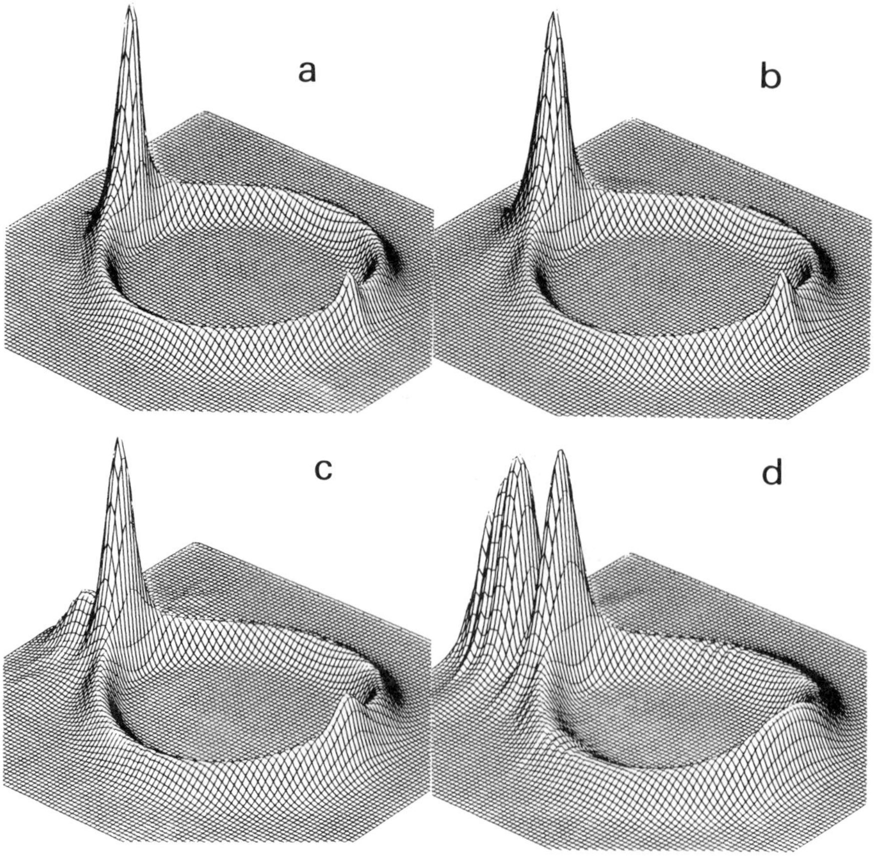

Figure 19. Computer simulation of proton velocity distribution arising from OH^+ dissociation in carbon foils of thickness (a) 80 Å, (b) 152 Å, (c) 285 Å, and (d) 740 Å. The protons are in coincidence with the 6^+ final charge state of the outgoing oxygen ions. Wake forces are derived from the plasma wake model (Figure 18a).

produced calculated rings (Figures 19–21) for the four different target thicknesses appearing in the rings of Figure 15. It is clear that the differences in the predictions of the three models show mainly in the rings obtained for thicker targets. Indeed, inspection of Figures 19–21 shows that the simple wake model fails to reproduce the data. Between the Lindhard-wake model prediction and the Coulomb-wake model prediction one would probably choose the latter one. But it appears that none of the above calculations really reproduces the overall features of the observed data, especially not the backward-peak to side-peak ratio.

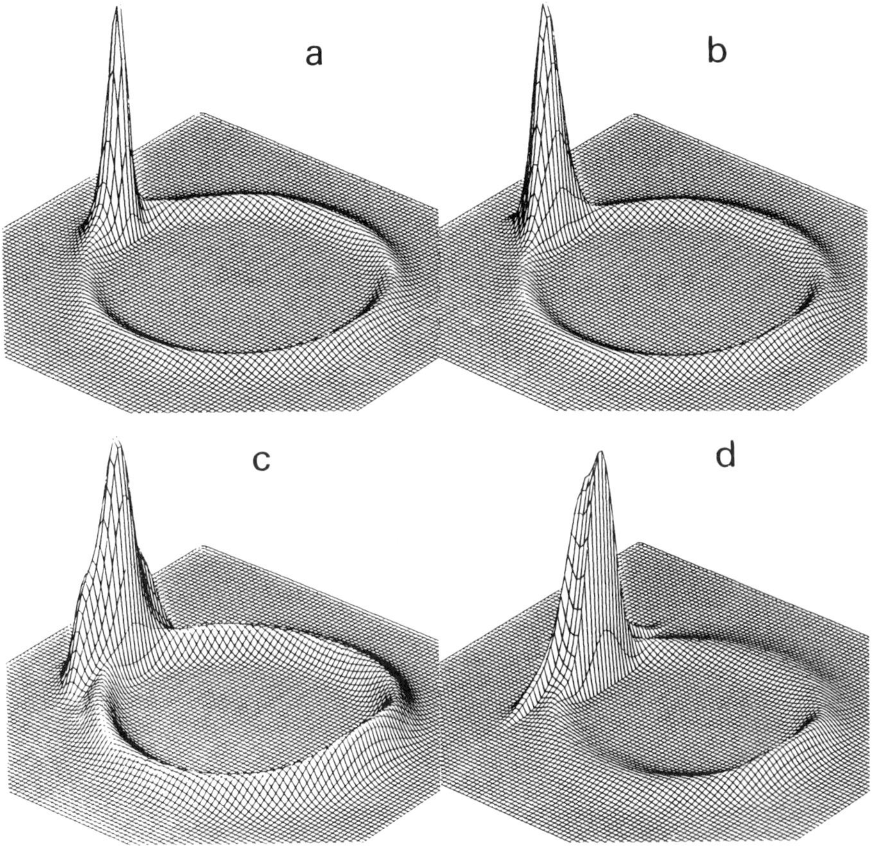

Figure 20. Same as Figure 19 except that wake forces are derived from the Lindhard dielectric function (see Figure 18b).

In the simulations with the Coulomb wake (Section 2.2.2) it was found that averaging of the wake polarization over the different target-electron velocities below the Fermi velocity v_F is of crucial importance. If one uses only a single representative velocity v, the wake forces show spatial structure exhibiting a wavelength $2\pi\hbar/mv$. The end result is then a very much distorted ring that obviously does not resemble the one determined experimentally. We conclude that for dielectric functions which include single-electron collisions (Lindhard, plasmon-pole, Coulomb) it is important to include the Fermi averaging. This averaging is not done in the plasmon-pole approximation.

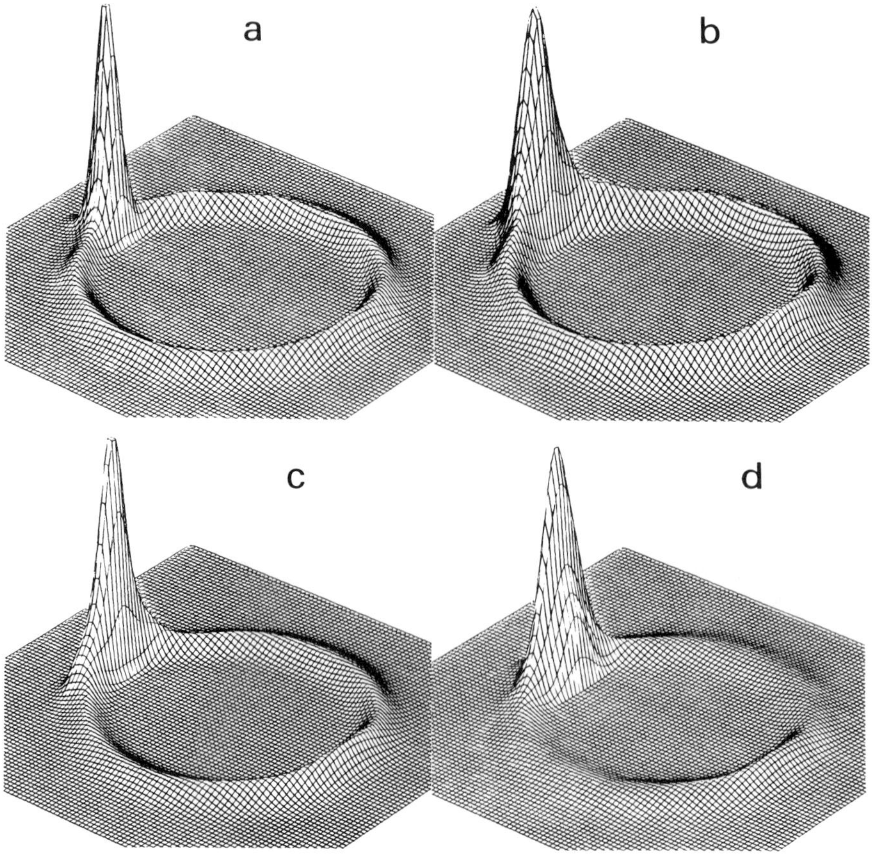

Figure 21. Same as Figure 19 except that wake forces are derived from polarization density due to Coulomb scattering of electrons (see Figure 18c).

3.3.4. An Improved Analysis for Molecular-Ion Dissociation in Foils

A new approach to the analysis of molecular-ion dissociation data was developed recently (P1 80). To see the advantages of this approach one has to summarize first the disadvantages of the old approach.

The main disadvantages of the old approach are as follows:

(1) Initial rotational velocities were assumed to be unimportant and therefore ignored. Yet, for an angular resolution of 1.5×10^{-4} rad, an initial rotational energy of the order of 0.01 eV corresponds to about one unit of resolution. One can therefore expect that the shape of the backward peak in particular should reflect the initial rotational states of the projectiles. A

possible remedy for the above difficulty would be to perform the calculation also for a given initial rotational velocity distribution. But computing initial velocity and initial orientation for many values of initial radii is not practical.

(2) The influence of multiple scattering is simulated in the old approach by smearing the final direction of the outgoing ions by an amount that is appropriate for the multiple scattering of a single ion. As will be shown later, this procedure can lead to significant errors, especially when the multiple scattering angles become large. Multiple scattering should be taken into account continuously together with the influences of the wake and the Coulomb forces.

In the new approach, we consider the time evolution of the density function characterizing a molecular-ion beam. In a typical Coulomb-explosion experiment, a beam of molecular ions is prepared in an ion surce where a population of different rotational and vibrational levels is established. The ions are then accelerated and the beam is magnetically analyzed and collimated with a specific accuracy. In the target the molecules dissociate and the final vectorial velocities of the fragments are observed. We now address ourselves to the following problem. If the density function describing the beam characteristics is given and if the interaction with the target is specified, what is the density matrix for the final products? Once the density matrix for the final products is given one can predict quite simply the outcome of any measurement made with a particular experimental setup.

In practice, the experimental setup and the experimental results are given and the goal is to extract the distribution of internal degrees of freedom in the incident molecular ion, or the interaction with the target, or both.

If the time development of the initial beam density function $\sigma(x_1, \ldots, x_n, v_1, \ldots, v_n, t)$ has no statistical behavior (e.g., multiple scattering, straggling, etc.) then by Liouville's theorem

$$\frac{d\sigma}{dt} = \frac{\partial\sigma}{\partial t} + \sum \frac{\partial\sigma}{\partial x_k} v_k + \sum \frac{\partial\sigma}{\partial v_k} a_k = 0 \tag{45}$$

where a_k is the acceleration on the coordinate x_k.

If the velocity of the particles does not change appreciably, then one can correct Equation (45) for statistical forces by a simple additional term:

$$\frac{\partial s}{\partial t} + \sum \frac{\partial\sigma}{\partial x_k} v_k + \sum \frac{\partial\sigma}{\partial c_k} a_k + \frac{1}{2} \sum_{kl} M_{kl} \frac{\partial^2\sigma}{\partial v_k \, \partial v_l} = 0 \tag{46}$$

This is an equation of the Fokker–Planck type (Wa 45). The matrix M_{kl} is a constant and is usually diagonal (for conventional multiple scattering and straggling).

Once the initial conditions $\sigma(x_1,\ldots,x_n,\ v_1,\ldots,v_n,0)$ given by the preparation of the beam are set, then the solution of Equation (46) for $t \to \infty$ provides us with the expected outcome of an experiment.

A method is introduced (Pl 80) for solving Equation (46) in a reasonable approximation so as to overcome all the criticisms listed above. The method is based on a special analytical solution of Equation (46) for the case where the potential function $V(x_1,\ldots,x_n)$ is quadratic. The solution then has the following properties:

(1) If the initial distribution has Gaussian form

$$\mu(\mathbf{z},0) = C(0)\exp\{-\tfrac{1}{2}[\mathbf{z}-\langle\mathbf{z}(0)\rangle]\,Q(0)^{-1}\widetilde{[\mathbf{z}-\langle z(0)\rangle]}\}, \qquad \mathbf{z} \equiv (\mathbf{x},\mathbf{v}) \tag{47}$$

then the development in time will retain the Gaussian form

$$\mu(\mathbf{z},t) = C(t)\exp\{-\tfrac{1}{2}[\mathbf{z}-\langle\mathbf{z}(t)\rangle]\,Q(t)^{-1}\widetilde{[\mathbf{z}-\langle\mathbf{z}(t)\rangle]}\} \tag{48}$$

(2) The average $\langle\mathbf{z}\rangle$ obeys Newton's law

$$\langle\dot{\mathbf{x}}\rangle = \langle\mathbf{v}\rangle, \qquad m_\alpha\langle\dot{v}\rangle_\alpha = -\left.\frac{\partial V}{\partial x_\alpha}\right|_{\langle\mathbf{x}\rangle} \tag{49}$$

In other words, the average coordinates form a simple trajectory.

(3) Each such trajectory is accompanied by a matrix of Gaussian standard squared deviations—the covariance matrix $Q(t)$.

The development in time of the covariance matrix can be shown to be given by the following formal solution:

$$Q(t) = e^{Bt}\,Q(0)\,e^{\tilde{B}t} + \int_0^t e^{B\tau}Ge^{\tilde{B}\tau}\,d\tau \tag{50}$$

where

$$B = \left(\begin{array}{c|c} 0 & I \\ \hline -A & 0 \end{array}\right), \qquad A_{\alpha\beta} = \frac{1}{m_\alpha}\frac{\partial^2 V}{\partial x_\alpha\,\partial x_\beta}$$

and

$$G = \left(\begin{array}{c|c} 0 & 0 \\ \hline 0 & M \end{array}\right)$$ [M is the straggling-multiple scattering matrix—see Equation (46)]

(4) When there is only one particle involved or when there are several particles and $A = 0$, then

$$Q(t) = Q(0) + \left(\begin{array}{c|c} \tfrac{1}{3}Mt^3 & \tfrac{1}{2}Mt^2 \\ \hline \tfrac{1}{2}Mt^2 & Mt \end{array}\right) \tag{51}$$

For *all* other cases the multiple scattering problem will depend on the matrix A and will therefore behave differently from Equation (51). This important point has been ignored in previous analyses of Coulomb-explosion data.

The way the above solution helps in the case of a general initial distribution $\sigma(\mathbf{z}, 0)$ and a general potential $V(\mathbf{x})$ is as follows.

The general initial distribution is split into a sum of Gaussians:

$$\sigma(\mathbf{z}, 0) = \sum_m \mu_m(\mathbf{z}, 0) \tag{52}$$

The criterion for the widths of the individual distributions is such that the general potential within each individual Gaussian is well approximated by a quadratic function.

The actual integration is carried out numerically making use of the formal solution [Equations (49) and (50)] for each step in time. The calculations are carried out in three-dimensional Cartesian coordinates.

This method differs from a simple Monte Carlo trajectory calculation by keeping track of the time development of the covariance matrix in all

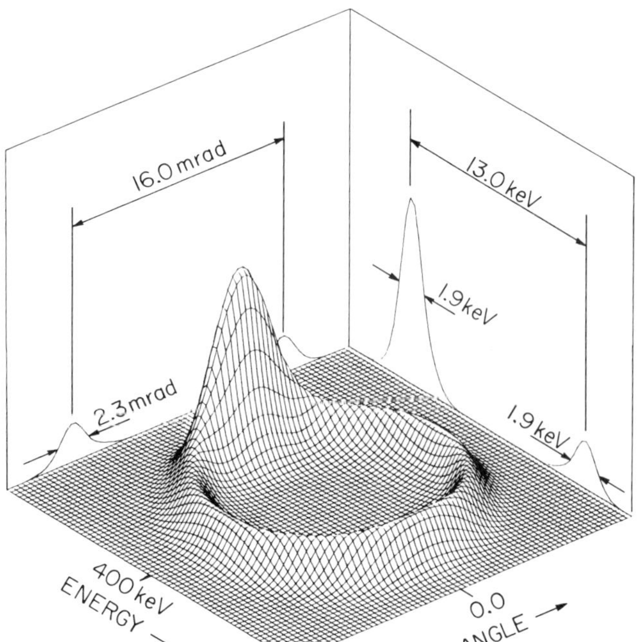

Figure 22. Calculated ring pattern for protons from 2.0-MeV HeH^+ dissociating in an 85-Å-thick carbon foil, using the new method. The same parameters were used as in the calculation of Figure 17, but the initial rotational velocities were those corresponding to 0.3 eV.

degrees of freedom. The advantage is a substantial reduction in the number of trajectories needed to cover all phase space even for diatomic molecular-ion dissociation. The method is particularly suitable when the input for the initial phase-space distribution corresponds to a molecule. This is true since in the rotating frame of a molecule (in its zero-point vibrational motion) the phase space of the nuclei is well approximated by a multivariant Gaussian distribution.

As yet, very few calculations have been carried out using the new approach. Here we introduce one ring calculation just to demonstrate the power of the method. A lot of work is still needed until all of the features appearing in the experimental data can be interpreted.

Figure 22 shows a ring calculated for the proton distribution from 2-MeV HeH^+ dissociating in an 85-Å-thick carbon foil. Comparison with Figures 10 and 17 shows that the new simulation method comes much closer to reality than the older one. An initial rotational velocity distribution was used with standard deviation corresponding to an energy of 0.3 eV. This initial rotational velocity distribution was needed in order to obtain a realistic proton distribution. Although the Lindhard dielectric function [Eq. (18)] was used, the real test of the validity of the different wake models has to await calculations for thicker targets.

Acknowledgments

The work described here has been supported in part by the Office of Basic Energy Sciences (Division of Chemical Sciences), U. S. Department of Energy and by the United States–Israel Binational Science Foundation.

References

(Ar 75) N. R. Arista and V. H. Ponce, *J. Phys. C* **8**:L188 (1975).
(Ar 78) N. R. Arista, *Phys. Rev. B* **18**:1 (1978).
(Be 30) H. A. Bethe, *Ann. Phys. (Leipzig)* **5**:325 (1930).
(Be 72) H. D. Betz, *Rev. Mod. Phys.* **44**:465 (1972).
(Be 76) F. Bell, H. D. Betz, H. Panke, and W. Stehling, *J. Phys. B* **9**:L443 (1976).
(Be 80) N. Benczer-Koller, M. Hass, and J. Sak, *Ann. Rev. Nucl. Part. Sci.* **30**:53 (1980).
(Bo 48) N. Bohr, *K. Dan. Vidensk. Selsk. Mat.-Fys. Medd* **18**(8) (1948).
(Bo 49) D. Bohm and E. P. Gross, *Phys. Rev.* **75**:1851 (1949).
(Bo 53) D. Bohm and D. Pines, *Phys. Rev.* **92**:609 (1953).
(Br 74) W. Brandt, A. Ratkowski, and R. H. Ritchie, *Phys. Rev. Lett.* **33**:1329 (1974).
(Br 76) W. Brandt and R. H. Ritchie, *Nucl. Instrum. Methods* **132**:43 (1976).
(Br 79a) A. Breskin, A. Faibis, G. Goldring, M. Hass, R. Kaim, Z. Vager, and N. Zwang, *Phys. Rev. Lett.* **42**:369 (1979).

(Br 79b) A. Breskin, A. Faibis, G. Goldring, M. Hass, R. Kaim, I. Plesser, Z. Vager, and Z. Zwang, p. 1 of (Ge 79b).

(Br 80) A. Breskin, A. Faibis, G. Goldring, M. Hass, R. Kaim, Z. Vager, and N. Zwang, *Nucl. Instrum. Methods* **170**:93 (1980).

(Bu 68) R. E. Burge and D. L. Mosell, *Philos. Mag.* **18**:251 (1968).

(Co 78) R. G. Gooks, ed., *Collision Spectroscopy*, Plenum Press, New York (1978).

(Da 75) M. H. Day, *Phys. Rev. B* **12**:514 (1975).

(Da 78) S. Datz, C. D. Moak, O. H. Crawford, H. I. Krause, P. F. Dittner, J. Gomez del Campo, J. A. Biggerstaff, P. D. Miller, P. Hvelplund, and H. Knudsen, *Phys. Rev. Lett.* **40**:843 (1978).

(De 74) K. Dettmann, K. G. Harrison, and M. W. Lucas, *J. Phys. B* **7**:269 (1974).

(Ec 78) J. C. Eckardt, G. Lanschner, N. R. Arista, and R. A. Baragiola, *J. Phys. C* **11**:L851 (1978).

(Ec 79) P. M. Echenique, R. H. Ritchie, and W. Brandt, *Phys. Rev. B* **20**:2567 (1979).

(Ew 62) R. I. Ewing, *IRE Trans. Nucl. Sci.* **9**:207 (1962).

(Fa 63) U. Fano, *Ann. Rev. Nucl. Sci.* **13**:1 (1963).

(Fa 80) A. Faibis, R. Kaim, I. Plesser, and Z. Vager, *Nucl. Instrum. Methods* **170**:99 (1980).

(Fr 54) J. Friedel, *Adv. Phys.* **3**:446 (1954).

(Ga 78) M. J. Gaillard, J.-C. Poizat, and J. Remillieux, *Phys. Rev. Lett.* **41**:159 (1978).

(Ge 75) D. S. Gemmell, H. Remillieux, J.-C. Poizat, M. J. Gaillard, R. E. Holland, and Z. Vager, *Phys. Rev. Lett.* **34**:1420 (1975); *Nucl. Instrum. Methods* **132**:61 (1976).

(Ge 77) D. S. Gemmell, P. J. Cooney, W. J. Pietsch, A. J. Ratkowski, Z. Vager, B. J. Zabransky, A. Faibis, G. Goldring, and I. Levine, Priceedings of the 7th International Conference on Atomic Collisions in Solids, Moscow (1977), Vol. I, p. 74, Moscow State University Publishing House (1981).

(Ge 79a) D. S. Gemmell, Proceedings of the 6th International Congress of Radiation Research, Tokyo, Japan (1979), pp. 132–144.

(Ge 79b) D. S. Gemmell, ed., Proceedings of the Workshop on Physics with Fast Molecular-Ion Beams, Argonne, IL (1979) Argonne National Laboratory Physics Division Informal Report ANL/PHY-79-3, August 1979.

(Ge 80a) D. S. Gemmell, P. J. Cooney, and E. P. Kanter, *Nucl. Instrum. Methods* **170**:81 (1980).

(Ge 80b) D. S. Gemmell, *Chem. Rev.* **80**:301 (1980).

(Ge 80c) D. S. Gemmell, *Nucl. Instrum. Methods* **170**:41 (1980).

(Ge 81) D. S. Gemmell, I. Plesser, and B. J. Zabransky (to be published).

(Go 74) J. Golovchenko and E. Laegsgaard, *Phys. Rev. A* **9**:1215 (1974).

(Ja 75) J. D. Jackson, *Classical Electrodynamics*, (2nd ed.) Wiley, New York (1975).

(Ka 79) E. P. Kanter, P. J. Cooney, D. S. Gemmell, K.-O. Groeneveld, W. J. Pietsch, A. J. Ratkowski, Z. Vager, and B. J. Zabransky, *Phys. Rev. A* **20**:834 (1979).

(Ki 63) C. Kittel, *Quantum Theory of Solids*, Wiley, New York (1963).

(Ki 77) M. Kitagawa and Y. M. Ohtsuki, *Phys. Rev. B* **16**:5321 (1977).

(Ko 76) W. Kolos and J. M. Peek, *Chem. Phys.* **12**:381 (1976).

(La 78) R. Laubert and F. K. Chen, *Phys. Rev. Lett.* **40**:174 (1978).

(La 79) R. Laubert, *IEEE Trans. Nucl. Sci.* **NS-26**:1020 (1979).

(La 80) R. Laubert, I. A. Sellin, C. R. Vane, M. Suter, S. B. Elston, G. D. Alton, and R. S. Thoe, *Nucl. Instrum. Methods* **170**:557 (1980).

(Li 54) J. Lindhard, *K. Dan. Vidensk. Selsk. Mat.-Fys. Medd.* **28**(8) (1954).

(Li 64) J. Lindhard and A. Winther, *K. Dan. Vidensk. Selsk. Mat. Fys. Medd.* **34**(4) (1964).

(Ma 67) V. V. Makarov and N. N. Petrov, *Sov. Phys. Solid State* **8**:2993 (1967).

(Ne 55) J. Neufeld and R. H. Ritchie, *Phys. Rev.* **98**:1632 (1955); J. Neufeld and R. H. Ritchie, *Phys. Rev.* **99**:1125 (1955).

(Ny 78) A. R. Nyaiesh, W. Steckelmacher, and M. W. Lucas, *J. Phys. C* **11**:2917 (1978).

(Pi 52) D. Pines and D. Bohm, *Phys. Rev.* **85**:338 (1952).

(Pl 80) I. Plesser and Z. Vager, Proceeding of the 38th Bat-Sheva Seminar, Annals of the Israel Physical Society, Vol. 4 (1980).

(Ra 65) H. Raether, *Springer Tracts Mod. Phys.* **38**:85 (1965).

(Ri 76) R. H. Ritchie, W. Brandt, and P. M. Echenique, *Phys. Rev. B* **14**:4804 (1976).

(Sc 49) L. I. Schiff, *Quantum Mechanics*, McGraw-Hill, New York (1949).

(Sc 78) W. Schaefer, H. Stoecker, B. Mueller, and W. Greiner, *Z. Phys.* **A288**:349 (1978).

(St 78) J. Steinbeck and K. Dettmann, *J. Phys. C* **11**:2907 (1978).

(St 81) M. F. Steuer, D. S. Gemmell, E. A. Johnson, E. P. Kanter, and B. J. Zabransky, *IEEE Trans. Nucl. Sci.* NS-30, 1069 (1983).

(Ta 76) J. W. Tape, W. M. Gibson, J. Remillieux, R. Laubert, and H. E. Wegner, *Nucl. Instrum. Methods* **132**:75 (1976).

(Va 76a) Z. Vager and D. S. Gemmell, *Phys. Rev. Lett.* **37**:1352 (1976).

(Va 76b) Z. Vager, D. S. Gemmell, and B. J. Zabransky, *Phys. Rev. A* **14**:638 (1976).

(Wa 45) M. C. Wang and G. E. Uhlenbeck, *Rev. Mod. Phys.* **17**:323 (1945).

(Zi 72) J. M. Ziman, *Principles of the Theory of Solids*, Cambridge University Press, London (1972)

7

PETER K. HAFF is a Senior Research Associate at the California Institute of Technology. He received his undergraduate education at Harvard University and his graduate education at the University of Virginia leading to the award of the Ph.D. in 1970. He was a Research Fellow at the University of Washington during 1970–1972, at Caltech during 1972–1974, and a Senior Research Fellow at Caltech during 1974–1975. Following a year as a Visiting Scientist at the Niels Bohr Institute in Copenhagen, he served as an Assistant Professor of Physics at Yale University from 1976 through 1980, when he returned to Caltech.

Erosion of Surfaces by Fast Heavy Ions

PETER K. HAFF

1. Introduction

When an ion beam strikes a surface, the beam–target interaction can lead to the ejection of target atoms from the surface. This phenomenon of target erosion on an atomic scale is known under the general rubric of "sputtering." The word "sputtering" has been applied to a variety of circumstances where particulate objects strike a surface with the subsequent emission of target atoms, although the ejection mechanism may differ from case to case. Thus photons, electrons, and light and heavy ions at both high and low energy, have all been shown under appropriate circumstances to lead to ejection of target atoms. In the general sense of target erosion, sputtering has been known for more than one hundred years; A. W. Wright, to whom the first Ph.D. degree in science in the United States was awarded in 1861, performed early experiments on surface erosion using discharge tubes (Wr 77).

Today the phenomenon of sputtering, in one or another of its several forms, is studied by many researchers around the world. Ion accelerators, often using heavy ions, are an important tool in this study. It is a fact that in any experiment in which an ion beam strikes a target, sputtering occurs. Sometimes the study or use of this process is the purpose of the experiment. Often sputtering is only an unwanted, but perhaps negligible, side effect, as in many nuclear physics experiments. Sometimes sputtering causes

---

PETER K. HAFF • Division of Physics, Mathematics, and Astronomy, California Institute of Technology, Pasadena, California 91125. Work supported in part by NSF[PHY79-23638], NASA[NGR-05-002-333], and the Caltech President's Fund.

deleterious effects which are difficult to avoid, for example at fusion-reactor containment walls (Vo 75).

The most widely known and discussed sputtering phenomena at the present time involve the processes surrounding the bombardment of selected surfaces with ions in the keV/amu range. Sputtering at these energies has seen detailed study because the required ion beams are easily obtained, and the effects are large. If the sputtering yield S is defined as the average number of target atoms ejected per incident ion, then keV/amu ion beams typically produce yields on the order of unity or larger. Reviews of ion-induced sputtering in the keV/amu, or low-energy, regime can be found in (Si 72, To 76).

The topic of the present review involves ion-induced sputtering, but primarily in the MeV/amu range. We emphasize the role of heavy ions, but not to the exclusion of all light ion sputtering. Systematic experiments at high energies are rather few in number, and most of them have been performed in the past five years or so. The results of these experiments are evidently not explicable in terms of the theoretical picture which has successfully described and correlated many low-energy experiments. In the following sections we will first outline briefly, for purposes of comparison and contrast, the nature of the sputtering process at low energies, and then proceed to a discussion of experiments at high energies. This will involve a summary of the motivations for initially undertaking experiments at such energies, and a short description of the experimental techniques themselves. Few new calculations are presented in the present paper, but it was thought it would be useful to draw together into one place some of the results which have been obtained by a number of workers in the area.

Experimental work in the MeV/amu regime has recently been carried out by several groups. At Caltech (Gr 79, Gr 80, Se 80a,b) heavy-ion beams from a tandem accelerator are being used to investigate high-energy sputtering of a variety of materials. Preliminary investigations have revealed unusually high sputtering rates in Al_2O_3, B_2O_3, UO_2, $LiNbO_3$, H_2O, and UF_4. Most of this work is in progress and has not yet been published. However, results from an extensive series of investigations into the sputtering properties of UF_4 are available, and we will focus upon them here. This uranium-containing compound has been of much interest because of the sensitivity with which sputtered ^{235}U can be detected. At Bell Laboratories (Br 78, Br 80a,b), important work on high-energy sputtering of water ice and other volatile targets [e.g., SO_2 (Hi 81)] has been carried out. With the exception of several data points for C and O bombardment of H_2O ice, the published work of these authors has centered mainly on sputtering with H and He beams. Researchers at Chalk river (Bø 80, Ol 80) have reported interesting experiments involving moderate-energy bombardment of frozen Xe and Kr, and experiments utilizing both fission fragments and heavy ions have been carried out by a number of workers [e.g., (Dü 80, Ma 76a,b)] to study the release of large adsorbed organic molecules.

All of this work would rightfully fit, in principle, under the title of the present chapter. However, for reasons of length it was felt best to limit the discussion primarily to the work of the Caltech and the Bell Laboratories groups. The Chalk River work raises very interesting questions, but the beam energies are generally not so high that strict separation of the usual low-energy results from the new phenomena expected to be found at higher energies is always possible. The work on biological molecules is likewise important, but it is not clear that the ejection of such large objects from the surface is describable in the same way as is the ejection of smaller molecules. To keep the present discussion within limits, a review of this work has not been included here.

2. Sputtering at Low Energies

Before proceeding to erosion phenomena at high energies, it is first of interest to discuss briefly the more familiar sputtering events in the keV/amu range. At 1 keV/amu, the velocity of an ion is approximately $v_i = 4 \times 10^7$ cm s^{-1}; this may be compared with the velocity of an electron in the outermost orbital of a target atom, $v_e \approx \alpha c \approx 2 \times 10^8$ cm s^{-1}. Thus keV/amu ions generally have velocities which are smaller than the smallest electron velocities, and therefore ionization processes are not an important mode of energy loss at these energies. Instead, energy is lost mainly in elastic collisions between the projectile ion and target atoms, and this (mean) energy loss per unit length of projectile path is the nuclear stopping power $(dE/dx)_n$.

The description of sputtering at these energies is based upon the concept of the collision cascade, Figure 1. When the projectile undergoes a collision with a target atom, that atom recoils with some kinetic energy. Subsequent collisions of this primary recoil and its daughters with stationary target atoms lead to the formation of the collision cascade. If the primary recoil was generated in the vicinity of the target surface, some members of the cascade may be able to escape through that surface to form part of the flux of sputtered particles.

There is an energy cost U imposed on any particle leaving the surface. The value of U may be sufficiently approximated by identifying it with the sublimation energy of the target. If ΔE is the average energy transferred in a collision between the projectile ion and a target atom, then an estimate of the total number of target particles which reach some smaller energy ε is $\approx \Delta E/\varepsilon$. The target sputtering yield would accordingly be proportional to $\Delta E/U$. It is useful at this point to define two length scales λ and d. λ we take to be the

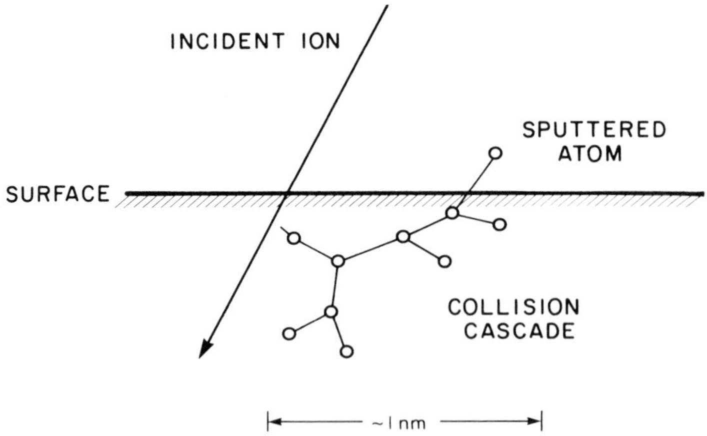

Figure 1. Schematic illustration of collision cascade.

mean free path between projectile and target-particle collisions, and d the mean free path between collisions of recoiling atoms. If $\lambda \gg d$, then the probability of a recoiling atom created within λ of the surface actually escaping as a sputtered particle is $\approx d/\lambda$. Consequently, the sputtering yield S would be of the form

$$S \approx \frac{\Delta E\, d}{\lambda U} \tag{1}$$

Identifying $\Delta E/\lambda$ with the nuclear stopping power $(dE/dx)_n$, we recover the standard expression for the yield (Si 69)

$$S = \text{const}\, \frac{1}{U}\, (dE/dx)_n \tag{2}$$

Moreover, from the relation $S \approx \Delta E/\varepsilon$, for arbitrary energy $>U$, we obtain the differential form

$$dS = \frac{d\varepsilon}{\varepsilon^2} \tag{3}$$

which gives the energy spectrum. Actually, the arguments leading to (3) apply inside the target, and do not account for the loss of energy U suffered upon crossing the surface, hence the divergence as $\varepsilon \to 0$. The sputtered flux is of course finite, and treating the surface as a refracting barrier leads to an exterior differential sputtering flux of the form

$$dS = \frac{\varepsilon\, d\varepsilon}{(\varepsilon + U)^3} \tag{4}$$

which reduces to (3) if $\varepsilon \gg U$. Equation (2) remains valid, and has been amply verified for many beam–target combinations (Si 69). A more rigorous derivation, based upon the Boltzmann transport equation, may be found in (Si 69). The energy spectrum, Equation (3), has been verified by a number of authors (Th 68, We 78a,b). Because of the general nature of the arguments used to arrive at Equation (2), it must be expected that *any* ion beam will produce collision cascade sputtering from *any* target. It is this circumstance which makes sputtering so important. However, as the beam energy increases beyond a few keV/amu, $(dE/dx)_n$ begins to decrease in magnitude. This occurs because the scattering cross section σ, which determines λ, is a decreasing function of energy in this energy region ($\sigma =$ the Rutherford cross section at high energies). At energies on the order of 1 MeV/amu, the collision cascade sputtering yield prediction is usually very small ($S \approx 0.01$ for 20 MeV F on U metal).

3. Sputtering at High Energies

3.1. General Remarks

We turn, therefore, to the erosion of solid surfaces when the beam energy lies in the region where the electronic stopping power $(dE/dx)_e$ is significant. A 1 MeV/amu nucleus has a velocity $v_i \approx 1.4 \times 10^9\ \mathrm{cm\ s^{-1}}$, so that ionization processes are efficient. Energy transfer through ionization is the principal mode of energy loss, and the electronic stopping power $(dE/dx)_e$ is large. Figure 2 shows the relative values of the nuclear and the electronic stopping powers for a moderately massive ion (Cl) in a solid Al target. At the lower energies, as discussed in the preceding section, the interaction between beam ion and target atom is manifested primarily in the recoil of the target atoms a whole, with little ionization. As the velocity of the incident particle approaches that of the atomic electrons, in both relative and absolute terms the energy deposited as kinetic atomic recoil energy decreases, so that the collision cascade contribution to sputtering becomes smaller and smaller. In some materials, however, it has been discovered that as the beam energy moves into the electronic stopping region, the sputtering yield actually increases. It is the purpose of this section to give a brief account of some of the experimental results which have been obtained in the last several years at these higher beam energies, and to outline some models which have been advanced to explain them.

3.2. Possible Connections with Track Formation

In 1976 Haff pointed out (Ha 76) that the "ion explosion" mechanism which had been advanced by Fleischer, Price, and Walker (Fl 65) to explain

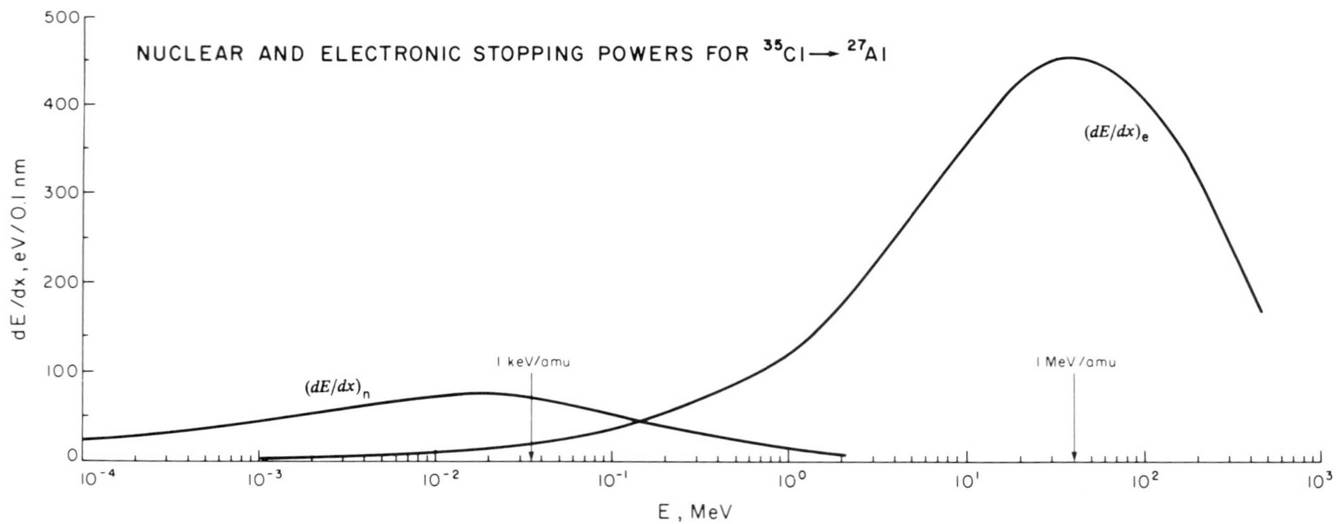

Figure 2. Nuclear and electronic stopping powers versus energy for ^{37}Cl bombardment of an ^{27}Al target. The nuclear stopping is dominant in the few keV/amu range, where collision cascade sputtering is important. In the MeV/amu range, high-energy sputtering is the dominant erosion mechanism for certain dielectrics. The electronic stopping power controls sputtering in this region.

the production of nuclear tracks by ionizing particles in dielectric media would also predict the existence of a substantial sputtering yield at high energies. Nuclear tracks are trails of radiation damage formed by the passage of a fast ion in certain dielectric materials. Fleischer *et al.* (Fl 65) had suggested that neighboring ion pairs were produced by the passage of the fast projectile and that the ions subsequently recoiled under their mutual Coulomb repulsion. The result would be displacement of atoms from their unperturbed crystal lattice positions, with the resulting formation of a strained and chemically reactive region around the ion path. The activated volume could then be attacked with suitable chemical etchants to make the track visible in a microscope.

The model suggests that particle erosion from the surface ought to accompany track production. In this picture the recoil due to the mutual Coulomb repulsion of neighboring ions leads to the generation of low-energy daughter recoils, and some of these recoils near the surface may escape as sputtered atoms. The key feature here is that the energy dependence of the sputtering yield would be associated with the electronic rather than the nuclear stopping power. From general considerations we can try to predict some details of sputtering yields generated by the ion explosion mechanism (Ha 76). Following similar arguments used in connection with our discussion of collisional sputtering, we anticipate that the electronic sputtering yield S_e will reflect the number of recoiling secondaries. This in turn will be proportional to the total energy released during the ion repulsion step of the model, i.e., proportional to the square of the average number of electrons removed per atom along the projectile trajectory. The appropriate function describing this process is the primary ionization rate dJ/dx. This quantity is related to the electronic energy loss $(dE/dx)_e$, but it does not weight the electron δ rays by energy. This model would therefore predict

$$S_e \approx (dJ/dx)^2 \tag{5}$$

Because it is not weighted by electron energy, dJ/dx falls off more quickly at high energies than $(dE/dx)_e$. Figure 3 shows how the shape of $(dJ/dx)^2$ compares with that of $[(dE/dx)_e]^2$. The primary ionization rate dJ/dx has indeed been found to be more closely associated with the production of tracks than is the quantity $(dE/dx)_e$ (Fl 75).

Fleischer (Fl 80) has recently pointed out that suggestive evidence for particle ejection associated with track production was available as early as 1964. Knorr (Kn 64) had observed tracks produced by fission fragments in LiF crystals. An occasional feature seen was a long groove on the crystal surface which gradually transformed into a ridge. Evidently fission fragments incident at very oblique angles had scribed a groove, and as the fragment penetrated deeper and slowly buried itself, the recoil of atoms along the path

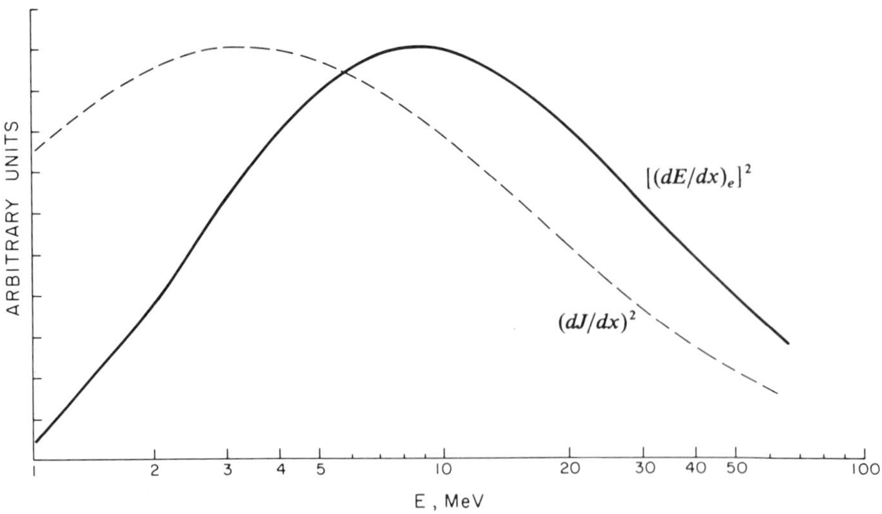

Figure 3. Calculation of $[(dE/dx)_e]^2$ and $(dJ/dx)^2$ for F incident on a heavy target like UF_4. The primary ionization peaks at a lower energy than the stopping power, and falls off more quickly with increasing energy.

caused the surface of the crystal above the shallow track to bulge outward. Although the experiment provides no direct proof of particle ejection, it does show that substantial rearrangement can accompany track formation. This suggests that atomic rearrangement is occurring when tracks are produced, and not just electronic rearrangement. The latter might be sufficient to produce an etchable track, but the former is certainly required for sputtering to occur.

3.3. Role of Target Strength

There are further predictions of a general nature which may be made from this picture of high-energy sputtering. Projectiles of given mass and energy will register tracks in some target materials, but not in others. If dJ/dx drops below a certain "threshold" value for a specified target, then track registration does not occur. For a given ionization rate dJ/dx, the controlling factor, according to the picture of Fleischer *et al.* (Fl 65, Fl 75), is essentially the theoretical mechanical tensile strength of the target material. Targets which are mechanically weak are more easily disrupted along the projectile path by the electrostatic stresses accompanying the ionization. A classification of target materials according to their Young's modulus is thus also a classification according to track registration properties. The above picture, which associates sputtering with track formation, therefore predicts that this same classification scheme will, at least approximately, rank target materials according to their sputtering yields in the electronic stopping region.

The threshold ionization levels need not be identical for sputtering and track registration; for track formation a more or less continuous trail of damage is necessary in order to allow a chemical etchant to produce a visible track feature. Sputtering, however, could still arise as the result of fluctuations in the ionization. A local region of ionization higher than the average could lead to the emission of sputtered particles if such a fluctuation occurred at the surface, even though the average ionization rate for that particular projectile were below the threshold for track production. There should, however, be a general correlation between track production and high-energy sputtering, according to this picture, and if the track registration threshold falls below the value suggested by dJ/dx for a particular beam–target combination, then sputtering should occur.

3.4. Role of Electrical Conductivity

There is another criterion in addition to mechanical strength which must be invoked in order to satisfactorily divide track registering from non-track-registering materials. For the mechanical strength criterion to have meaning, the local ionization must persist for a sufficient length of time that substantial atomic readjustment can occur. Since, as a general rule, tracks do not form in electrical conductors, it is proposed that in those substances the flow of electrons back to the positively charged core of the track occurs so fast that the ion explosion never has a chance to develop. We therefore expect high-energy sputtering to be associated principally with materials of low electrical conductivity. Although few experiments have been performed, there are no examples of sputtering of bulk metal targets at high energy where the yield exceeds significantly the magnitude predicted by the collision cascade theory; e.g., Griffith (Gr 79) has shown for a metallic uranium target that no anomalous erosion occurs under bombardment by ions in the electronic stopping regime. All the target materials mentioned in Section 1 that show substantial erosion at high energy are poor electrical conductors. Quantitative estimates of how high the target resistivity must be in order for a specified ion to induce electronic sputtering are unavailable. Some idea of the order of magnitude involved, however, can be gotten by noting the values of resistivity for track forming materials (Fl 75) (silicate minerals, 10^6–10^{20} Ω cm; MoS_2, 3,000–25,000 Ω cm; and V_2O_5 glass, 2,000–25,000 Ω cm) versus the resistivity for non-track-forming materials (Ge and Si, 10–2,000 Ω cm; metals, 10^{-6}–10^{-4} Ω cm).

3.5. Experiments with UF_4 Targets

3.5.1. Total Yield Measurements

The first experiments designed explicitly to test the ion explosion picture of sputtering were undertaken by Griffith (Gr 79) and collaborators (Gr 80). They used a tandem accelerator to bombard a UF_4 target with ^{19}F and other ions in the MeV/amu range. The choice of UF_4 as a target was based on several considerations. The physical and chemical properties of this substance have been well studied. It is chemically stable, and its vapor pressure is sufficiently low that an ultrahigh vacuum can be maintained even at elevated temperatures. These features make it possible to maintain a clean target surface for extended periods of time, even at 10^{-8} torr, by the simple method of heating the target. The electrical resistivity of this material is also high, which is evidently required if we are searching for a phenomenon associated with the ion explosion model of track formation. Finally, a target containing ^{235}U was chosen because of the extremely high sensitivity to sputtered ^{235}U atoms which recently developed techniques afford (Gr 78).

Thin films of UF_4 (200 to 500 nm thick) were evaporated onto a Cu block, and baked at 150–250°C in ultrahigh vacuum in order to remove any contaminants incorporated into the target or adsorbed on the target surface. Collection of sputtered atoms following bombardment by high-energy ^{19}F ions was done in two different ways. In the first, an Al foil was positioned to catch atoms and molecules ejected from the target. In order to avoid unknown effects due to change of surface composition following prolonged bombardment or to the presence of implanted ^{19}F atoms, beam fluences were kept sufficiently low that typically less than a monolayer of target material was removed during any one run. The detection of sputtered ^{235}U atoms trapped on the catcher foil was carried out by exploiting the track formation properties of fission fragments (Gr 78). The catcher foil was placed in contact with a sheet of mica, and then this assembly was exposed to a neutron flux in order to fission a certain fraction of the ^{235}U atoms on the foil. The fragments recoiling into the mica leave damage trails which can be subsequently etched to form tracks, and then counted to determine a sputtering yield. By counting tracks as a function of position on the foil, an angular distribution can be obtained. Figure 4 shows a typical yield (Gr 80) in arbitrary units as a function of ϑ, the ejection angle measured from the target normal. The solid curve is proportional to $\cos \vartheta$.

This is the distribution expected when sampling that part of an isotropic flux which passes through a small element of surface area. In low-energy sputtering calculations the collision cascade is usually assumed to be isotropic (except for the highest-energy particles), and this flux is then refracted through the surface barrier U before reaching the detector. Passage

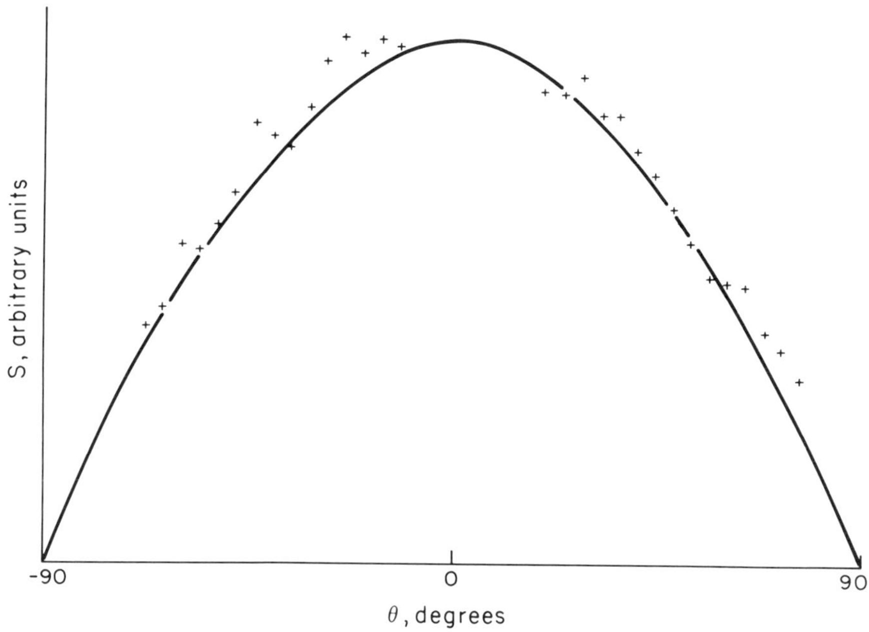

Figure 4. Angular distribution of ^{235}U sputtered from UF_4 by high-energy F. The crosses are experimental points (Gr 80), and the curve is proportional to cos ϑ. $\vartheta = 0$ corresponds to the target normal.

through the surface generallly changes the direction of particle motion. However, it can be shown that the cos ϑ shape is preserved (Th 68). Similar arguments for high-energy sputtering would also lead to a cos ϑ prediction for the angular distribution if the subsurface flux were isotropic.

Of greater interest is the value S of the sputtering yield itself. The yield as a function of ^{19}F energy is depicted in Figure 5. The numbers beside the experimental points indicate the charge state of the incident ^{19}F ion. The maximum yield is on the order of $\approx$6 U atoms for each incident F ion. The yields shown are two to three orders of magnitude larger than predicted on the basis of the collision cascade mechanism discussed above, and must be due to an entirely different cause. The ion explosion mechanism is one possibility, and will be discussed again subsequently in relation to these experiments. A second point to notice is that there is evidently some charge state dependence of the yield, with projectiles of higher incident charge states leading to marginally larger yields. Some charge state dependence was suggested in the original discussion of the ion explosion picture (Ha 76). The theoretical curves in Figure 5 are discussed below in connection with Equation (15).

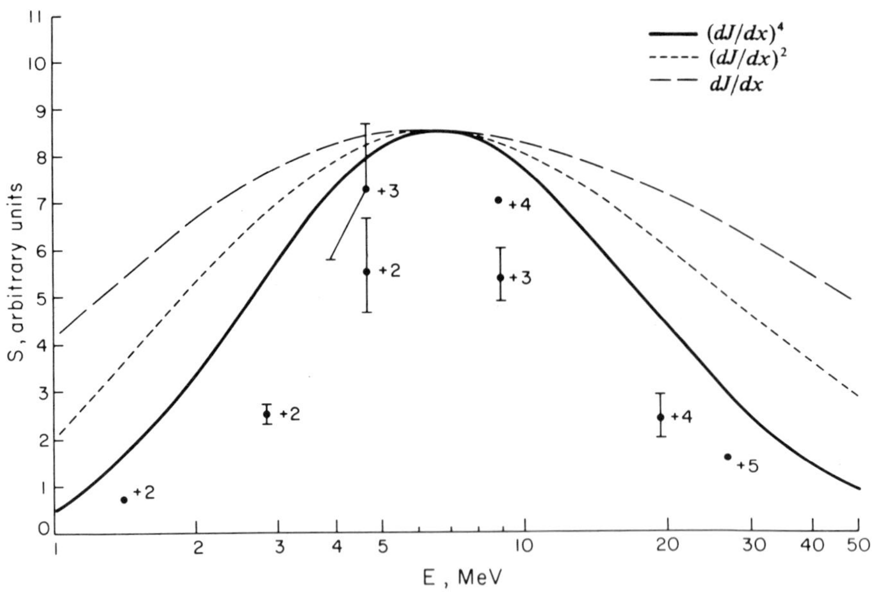

Figure 5. Sputtering yield as a function of incident ^{19}F energy. The points are experimental values (Gr 80, Se 80a). The upper curve is proportional to dJ/dx, the middle curve is proportional to $(dJ/dx)^2$, and the lower curve is proportional to $(dJ/dx)^4$. The numbers beside the experimental points correspond to the charge state of the incident F ions.

3.5.2. Energy Spectrum Measurements

A second type of experiment utilizing a UF_4 target has turned out to provide information crucial to the development of what is presently the most successful picture of high-energy sputtering by heavy ions in insulating targets. Utilizing a mechanical spectrometer developed by Weller and Tombrello (We 78a), Griffith *et al.* (Gr 80) measured the energy spectrum of sputtered U atoms arising from fast ^{19}F bombardment. Sputtered particles collected on a rotating wheel were counted using the track techniques described above. Both neutral atoms and ions sputtered from the target can be counted in this way. By imposing a target bias, the two components can be separated. In this way the ratio of sputtered ions to neutrals for 4.74-MeV ^{19}F bombardment was found to be ≈ 0.2 (Se 80b). The electrostatic acceleration of the charged ions necessary to separate them from the neutrals has so far made it impossible to determine their energies. For this reason, and because the charged population is not dominant, we focus attention on the ejected neutral atoms. The energy spectrum of neutral particles obtained by this for bombardment by 4.7-MeV ^{19}F is illustrated in

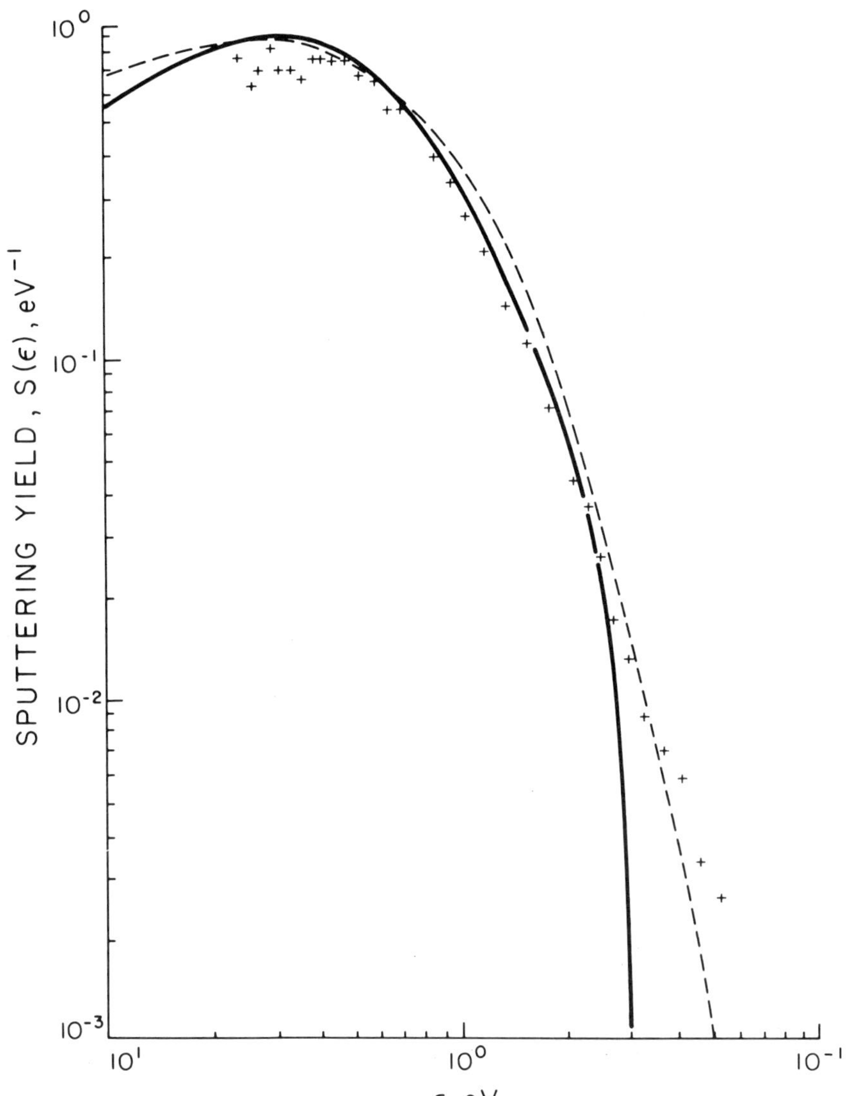

Figure 6. Energy spectrum of neutral particles sputtered from UF_4 by 4.74-MeV ^{19}F ions. The crosses are experimental points (Gr 80) and the solid curve is proportional to the flux $\varphi(\varepsilon) = \varepsilon \exp(-\varepsilon/kT)\, d\varepsilon$ with $T = 3500$ K, equation (10). For an explanation of the dashed curve, see text following equation (22) and see Figure 13.

Figure 6. A fit to the data using the general functional form (not shown in the figure)

$$S \approx \frac{E}{(E+U)^n} \tag{6}$$

where U and n are constants, yielded $U = 1.2$ eV and $n = 6.1$. In the discussion of low-energy sputtering where the collision cascade dominates the sputtering yield, both theory and experiment indicate a value of U on the order of a few electron volts, and a value of $n \approx 3$. The measured value $n = 6.1$ for fast ^{19}F bombardment is a significant departure from the low-energy behavior, the sputtered flux in the high energy case falling off much faster with increasing energy. Low-energy bombardment by 80-keV ^{20}Ne produced a spectrum which could be fit by the form (6) with $n = 2.64$ and $U = 0.71$ eV.

3.6. Thermal Model of High-Energy Sputtering

3.6.1. Energy Spectrum

The parameters appearing in Equation (6) have a physical interpretation only in terms of the collision cascade process ($n \approx 3$). Otherwise n and U must be regarded simply as fitting parameters which characterize a convenient functional form. We are therefore not obliged to retain the form (6) for a description of high-energy sputtering, where electronic processes are certainly important and where the collision cascade has yet to be shown to play a role. Following this philosophy, Seiberling and her collaborators (Se 80a,b) analyzed further, high-resolution, experiments on ^{19}F bombardment of UF_4 in terms of a Maxwell–Boltzmann velocity distribution. The solid curve in Figure 6 has the Maxwell–Boltzmann shape for particles of mass $M = 235$ amu at a temperature of $T = 3500$ K. The normalization is adjusted to fit the data.

The agreement between the data and the Maxwell–Boltzmann distribution is good at energies below about 2 eV, but not so good at higher energies. The thermal distribution (at a single temperature) predicts far fewer particles than are actually observed in this energy range. A possible theoretical explanation of this discrepancy is presented in Section 3.8 below. However, the fluxes at high energies are very small, and the data here are therefore less reliable than at the lower energies (Gr 81). At the present time we therefore feel it necessary to exercise caution in interpreting details of the spectrum in this energy range. The fit to the Maxwell–Boltzmann shape involves only two parameters (normalization and T), while in the case of Equation (6) three were required (normalization, U and n). Moreover, Equation (6) does not appear to have a well-determined physical basis when applied outside the cascade regime. The thermal model is therefore an economical one which deserves further investigation.

We note that a result relevant to this discussion has recently been obtained by Weller and Weller (We 81), who studied the relaxation toward thermal equilibrium of selected primary recoil distributions using the Boltzmann transport equation in velocity space. These authors found that an energy distribution quite similar to a Maxwell–Boltzmann form could be obtained from a variety of initial recoil distributions, as long as the peaks of those distributions were at very low energy ($\leqslant 1$ eV). For distributions peaked at higher energies, the differential sputtering yields began to resemble more strongly the expected collision cascade power law form.

An interesting idea due to Seiberling (Se 80a,b) was to connect the observed "thermal" sputtering distribution with a microscopic picture of the excitation processes occurring along the projectile track. Her idea is based upon a modification of the direct ion explosion mechanism. When an ion in the MeV/amu range enters a solid, essentially all the energy loss is taken up by ionization and electron excitation. For sputtering to occur, atoms near the surface must acquire some kinetic energy. One way to acquire energy would be via direct collisions with electrons energized by the passage of the beam ion. An upper limit on the rate of energy transfer in such a process is found by considering the effect of a head-on collision every lattice spacing. The collision time is $\tau_{ea} \approx 7 \times 10^{-16}$ s for 1-eV electrons. The fraction of the electron energy transferred in each collision is equal to four times the ratio of the electron mass m to the uranium atomic mass M. Thus the time for an electron to transfer directly ≈ 1 eV to an atom is certainly no less than

$$t_{ea} \approx 4 \frac{m}{M} \tau_{ea} \tag{7}$$

or $t_{ea} \approx 7.5 \times 10^{-11}$ s.

Let us assume for the moment that energy is pumped into random atomic motion along the path at a rate determined by the time constant t_{ea} in the above equation. In order for sputtering to occur, the rate of energy loss from atomic recoils along the projectile track to sinks other than sputtering itself ought to be less than 1 eV/t_{ea}. The most important mode of energy loss is by heat conduction to the relatively undisturbed lattice surrounding the track. If K is the thermal diffusivity of the target at the ambient temperature, then the time constant for heat conduction is given in order of magnitude by

$$t_{hc} \approx \frac{r_0^2}{K} \tag{8}$$

where r_0 is the radius of the region within which most of the atomic motion occurs. The requirement of a limitation on the rate of heat conduction is an entirely new criterion not envisioned in the original ion explosion model, either as applied to track formation or to sputtering. It is interesting to note, however, that Sigrist and Balzer (Si 77a,b) had proposed a relation between the thermal properties of certain minerals and their track registration thresholds. Figure 7 shows the thermal diffusivity of these targets plotted versus the smallest specific energy loss which still leads to track formation. The higher the thermal diffusivity, the greater the ionization required in order to produce tracks. This relationship is of the same general kind which Seiberling *et al.* (Se 80a,b) suggested ought to hold in a comparison of Equations (7) and (8). If K is too large, then $t_{hc} < t_{ea}$, and the energy supplied to the atoms by the electrons is conducted away too fast for sputtering or track formation to occur. This argument is independent of the mechanism which supplies energy to the atoms, as long as the above inequality applies, and is therefore of more general significance than it would be if restricted to the case of direct collisional energy supply.

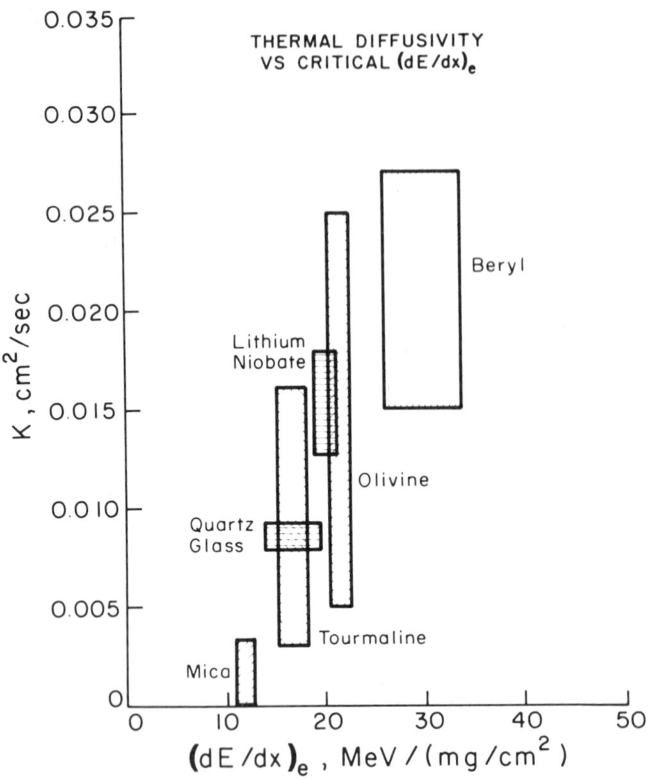

Figure 7. Thermal diffisuvity K of track-registering materials plotted versus the minimum stopping power required to produce tracks. Data adapted from Sigrist and Balzer (Si 77a).

For UF_4 at 60°C, $K = 8 \times 10^{-3}\ \text{cm}^2\ \text{s}^{-1}$. Taking $r_0 = 2$ nm, a choice discussed below, we have $t_{hc} \approx 5.2 \times 10^{-12}$ s. This is about ten times smaller than the estimate of t_{ea} above, which was surely a lower limit on the collision time scale. Therefore, in UF_4 it appears that electron collisions cannot supply energy fast enough to U atoms to lead to any significant particle emission.

Another method of heat transfer into the lattice is via the ion explosion mechanism described above. Adjacent ions will experience a force on the order of e^2/a_0^2, which has a value equal to a few electron volts per 10^{-8} cm. The time required for a U atom accelerated in this way to 1 eV to collide with an adjacent atom is, in UF_4, $t_{aa} \approx 5 \times 10^{-13}$ s. This is an order of magnitude faster than the heat conduction time in UF_4, and therefore Seiberling *et al.* (Se 80a,b) have argued that the several collisions allowed by the difference between t_{hc} and t_{aa} lead to a "thermalized" core of atoms from which sputtering occurs before the flux is quenched by heat conduction.

Before quenching occurs, the atoms along the projectile track are envisioned to reach a Maxwell–Boltzmann velocity distribution

$$F(\bar{v})\,d\bar{v} = n(M/2\pi kT)^{3/2}\, 4\pi\bar{v}^2 \exp(-M\bar{v}^2/2kT)\, d\bar{v} \tag{9}$$

where n is the atomic number density of the target particles of mass M. If it costs an amount of energy U for an atom to leave the surface of the hot cylinder and escape as a sputtered particle, then the flux of atoms sputtered into solid angle $d\Omega$ at ϑ with speed v is

$$\begin{aligned}\varphi(v, \Omega)\, dv\, d\Omega = n(M/2\pi kT)^{3/2} \exp(-U/kT)\\ \times \exp(-Mv^2/2kT) \cos\vartheta\, v^3\, dv\, d\Omega\end{aligned} \tag{10}$$

The angular distribution is proportional to $\cos\vartheta$, which is in agreement with the data of Griffith *et al.* (Gr 80), Figure 4. Equation (10) is the basis for the solid curve in Figure 6.

The differential sputtering yield itself is related to the flux by the equation

$$S(v, \Omega)\, dv\, d\Omega = \varphi(v, \Omega)\, At\, dv\, d\Omega \tag{11}$$

and the total yield is then

$$S = \int S(v, \Omega)\, dv\, d\Omega \tag{12}$$

In Equation (11), A is the area of the heated cylinder that is exposed at the surface of the target; it is taken to be πr_0^2. The quantity t is the duration of the spike, which is given by the thermal quenching time, $t \approx t_{hc} \approx r_0^2/K$. With these criteria, the total sputtering yield is

$$S = n(\pi r_0^4/K)(kT/2\pi M)^{1/2}\, e^{-U/kT} \tag{13}$$

For sputtering of UF_4 the value of r_0 has been deduced using Equation (13) together with measured values of S, T, and U. At 4.74 MeV, $^{19}F^{+2}$ ions give a yield of $S = 5.5$. A value of U was determined by fitting Equation (6) to an energy spectrum obtained with low-energy ^{20}Ne bombardment, yielding $U = 0.71$ eV. Finally, a fit of Equation (10) to the data in Figure 6 for $M = 235$ amu, gave $T = 3500$ K, about three times the melting temperature. This number leads to the value $r_0 \approx 2.4$ nm, comparable to the radii of latent (unetched) tracks as observed by TEM techniques.

3.6.2. *Dependence on Projectile Energy: Evidence from UF_4 Sputtering*

The preceding expressions do not yet allow one to make any direct connection with the ionizing properties of the incident beam. Evidently, some of the parameters in Equations (10) and (13) must be expressible in terms of the velocity and the nuclear charge of the incident beam. With such a relation, we would then have a direct linkage between beam characteristics and sputtering yields.

One way to do this is to assume that the average kinetic energy E_0 per atom within the cylinder of radius r_0, once equilibrium has been obtained, is proportional to the interaction energy of ionized atoms per unit volume along the ion track; i.e., $E_0 \approx (Ne)^2/a_0\, dr_0^2$, where N is the number of ions produced per lattice spacing d. Identifying kT with E_0, for low ambient target temperatures, and setting $N = (dJ/dx)d$, we have

$$kT = D\left(\frac{dJ/dx}{r_0}\right)^2 \tag{14}$$

where D is a constant. If it is permissible to consider kT to be approximately constant over that part of the sputtering event during which the majority of the particles from a given spike are emitted, as is assumed implicitly in the derivation of Equation (13), then r_0 is proportional to dJ/dx. Thus by Equation (13)

$$S \approx (dJ/dx)^4 \tag{15}$$

Figure 5 compares the excitation function data (Gr 80, Se80a,b) for 4.74-MeV ^{19}F on UF_4 with the calculated (and normalized) yield based on Equation (15). Curves proportional to dJ/dx and to $(dJ/dx)^2$ are also shown. The fourth power law provides a considerably better fit to the data than does the square law.

We note that additional information on the energy dependence of the sputtering mechanism is provided by experiments (Gr 80) in which UF_4 was bombarded by ^{16}O and ^{20}Ne. The sputtering yields, at a common velocity of 0.25 MeV/amu, are in the ratio 1:1.47:3.16 for $^{16}O^{+2}$, $^{19}F^{+2}$, and $^{20}Ne^{+2}$, respectively. Since dJ/dx, like $(dE/dx)_e$, is proportional to the square of the effective projectile charge Z_{eff}, Equation (15) implies that $S \approx Z_{\text{eff}}^8$. Using an approximate expression derived from Heckman *et al.* (He 63).

$$Z_{\text{eff}} = Z(1 - 10^{-45.667\beta/Z^{0.55}}) \tag{16}$$

where $\beta = v/c$ and Z is the nuclear charge of the projectile, we find the results plotted in Figure 8. The rate of increase in the yield S as a function of Z is significantly faster than a Z_{eff}^4 law based on the relation $S \approx (dJ/dx)^2$ would predict. It is true that high powers of Z_{eff} are involved, so that small changes in the effective charge expression will be come greatly magnified in the final result. However, we have eliminated some of the uncertainty this introduces by considering ratios only.

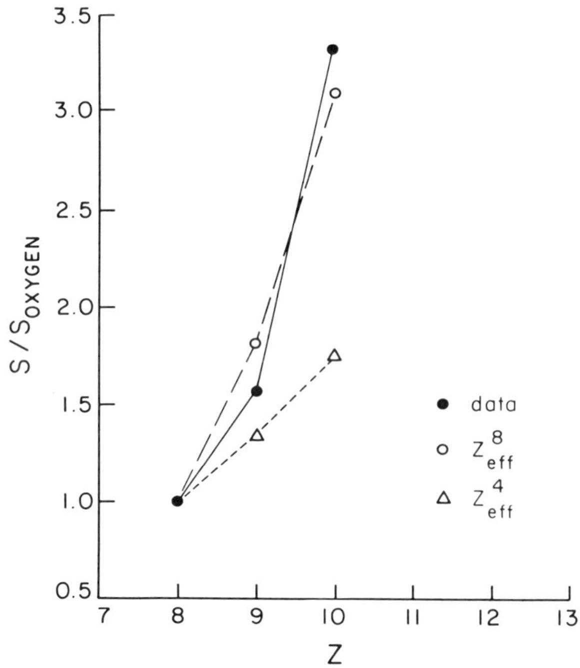

Figure 8. Sputtering yield for heavy-ion bombardment of UF_4 as a function of nuclear charge of the incident ion. Solid circles are data (Se 80a), relative to ^{16}O, for $^{16}O^{+2}$, $^{19}F^{+2}$, and $^{20}Ne^{+2}$ bombardment at 0.25-MeV/amu. The open circles are predictions based upon a Z_{eff}^8 law, which would be expected for a $(dJ/dx)^4$ dependence of the sputtering yield, equation (15). The triangles are the prediction of a Z_{eff}^4 law.

The proportionality of S to $(dJ/dx)^4$ is not what would be expected by the arguments leading to Equation (5). There we found $S \approx (dJ/dx)^2$. In Equation (15) the four powers of dJ/dx arise, under the stated assumption $r_0 \approx dJ/dx$, from the result of multiplying the flux φ by the spike area πr_0^2 and spike lifetime $r_0^2/4K$. Equations (5) and (15) are fundamentally different because of the assumptions about equilibrium which are implicit in the two formulas. In Equation (5), there is no equilibrium, in the sense that the energy released in the initial ion explosion is thought of as being inexorably divided among more and more particles each with smaller and smaller energy. At the same time the pulse spreads out over space. The proportionality to $(dJ/dx)^2$ arises from counting the number of daughter atoms which have energies sufficient to escape the target surface energy U. After all target recoil atoms have had their energy decreased below this value, it is of no use to wait for a longer period of time, as no more atoms will be sputtered, for the particular pulse in question.

In Equation (15), the energy is envisioned as shared, for the duration of the pulse, among a finite number of atoms. At any instant in time, there is a flux of atoms impinging on the surface area πr_0^2, and this same flux continues for the duration of the pulse $r_0^2/4K$. Even when all the atoms in a thin layer near the surface with sufficient energy to escape have done so, the model allows still more to escape if $t < r_0^2/4K$. The reason is that, by the assumption of equilibrium, atoms are scattered to higher energy states as well as to lower energy states. Thus there is the additional implicit assumption that the depletion of energetic atoms caused by loss to sputtering can be quickly restored by collisions occurring just a little deeper in the target. This point has not been investigate quantitatively.

What is the evidence that the atoms along the track are actually thermalized? Thermalization can occur only through the repeated mutual collisions of atoms in the volume which is to be thermalized. In the case of the ion explosion model one may imagine that, by simultaneously setting into motion a collection of neighboring atoms with energies so low that they cannot move far away from one another, the requirements for producing a thermalized flux are met. However, the point has not been investigated in detail. The main arguments in favor of a thermal model seem to be (1) that Equation (6) with $n = 6.1$, although fitting the data very well, is entirely *ad hoc* while the spectrum according to the Maxwell–Boltzmann form, which also fits the data reasonably well, except at high energy as noted earlier, can be tied to a definite picture of the atomic motion; (2) that the low recoil energies and consequently confined recoil volume suggested by the ion explosion model are at least consistent with the multiple (nonlinear) collision processes required for thermalization to occur; (3) that the temperature of the spike as deduced from the measured spectrum is on the order of, or somewhat higher than, the target melting temperature, which would certainly lead to considerable atomic motion; and (4) that the thermal conduction properties of the target appear to play a role in the formation of nuclear tracks, a process conjectured to be related to sputtering at high ion energy. We caution, however, that experiments with light ions can give different results for the excitation curve $S(E)$, as discussed below in Section 3.6.3.

3.6.2.1. Role of Target Thermal Properties. It may be that the most significant aspect of the model as developed by Seiberling *et al.* (Se 80a) is its emphasis on target thermal properties. If we consider the phenomenon of track formation for the moment, we see that there are several very broad cuts which may be made against the class of all possible solid targets in order to divide the track registering materials from those which do not register tracks, Figure 9. Fleischer *et al.* (Fl 65, Fl 75) applied the criterion of mechanical strength and electrical conductivity. Good electrical conductors do not register tracks in bulk, while materials of low mechanical strength form tracks at lower ionization levels than do intrinsically strong materials. The prediction of the Seiberling model is that the best candidates for track formation, and for enhanced sputtering, are mechanically weak solids which are good electrical *and* thermal insulators. Some evidence for this point of view, in terms of registration thresholds, has been obtained independently by Sigrist and Balzer (Si 77a,b). The high-energy sputtering results on UF_4, on U metal (Gr 79) where no enhanced sputtering was seen, and on water ice targets (to be discussed below), are all consistent with this point of view. However, at the present time there are only rather sparse experimental results which deal directly with the sputtering yields at high energy, and much more evidence will be required before it is known whether sputtering yields in the ionizing regime actually correlate, like track registration, in the way suggested by Figure 9.

3.6.3. Dependence on Projectile Energy: Evidence from H_2O Sputtering

We recall for a moment the discussion leading to Equation (5). The associated arguments were based upon the supposition that the ion explosion process produces a standard collision cascade, leading to the prediction that $S \approx (dJ/dx)^2$. However, we found in the last section that a crude thermalized-ion explosion model led to a $(dJ/dx)^4$ dependence. Below, in Section 3.8, we will find that a more detailed treatment of the thermal model leads to the prediction $S \approx \hat{\varepsilon}^2$ [Equation (19)], where $\hat{\varepsilon}$ is the linear energy-transfer density associated with the spike. Interpreting $\hat{\varepsilon}$ as the energy introduced per unit length into the spike due to the Coulomb repulsion of neighboring ions, we have $\hat{\varepsilon} \approx (dJ/dx)^2$. Consequently, the sputtering yield is again found to have the dependence $S \approx (dJ/dx)^4$, as in Equation (15). [The strict proportionality to the fourth power of the ionization rate actually holds only for a

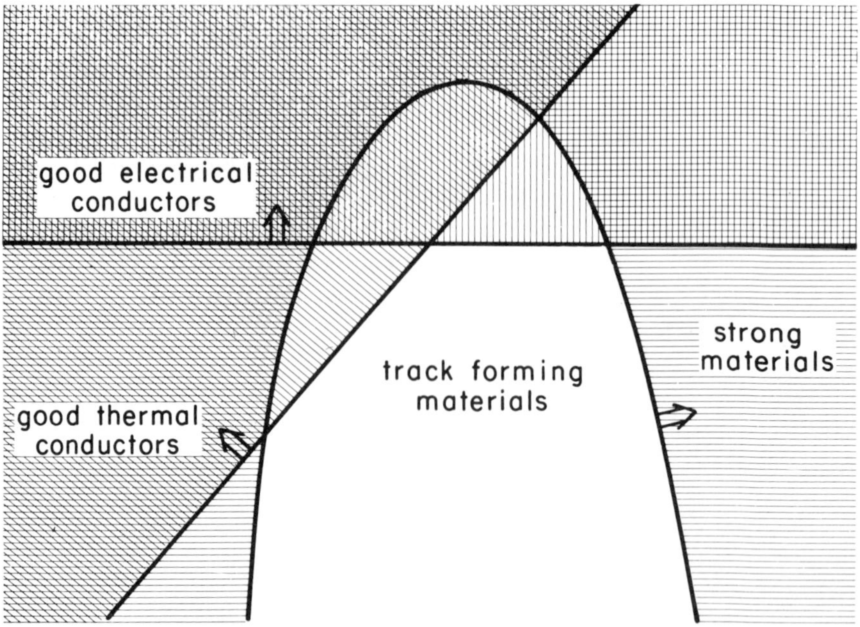

Figure 9. Proposed filter for distinguishing track-registering from non-track-registering materials. Strong materials (horizontal lines), good thermal conductors (left-leaning lines), and good electrical conductors (vertical lines) are hypothesized to require greater ionization to produce tracks than are the materials located in the unlined region of the figure. For a given bombarding ion and fixed energy, every target material falls within a definite region of the figure. In general, a change of beam ion or energy will require a redistribution of targets, since track production is a function of beam characteristics.

spike of zero width at time $t = 0$. For a finite spike at $t = 0$, the yield is somewhat more sensitive still to dJ/dx. See Equation (21) below.]

A different energy dependence has been discovered for H and He bombardment of H_2O ice by the Bell Laboratories group (Br 80a,b). These workers have advanced evidence for the proportionality of S to $[(dE/dx)_e]^2$, for which they suggest an ion explosion mechanism is responsible. {Given existing uncertainties in the actual values of dJ/dx and $(dE/dx)_e$, it would be difficult to distinguish between a $(dJ/dx)^2$ and a $[(dE/dx)_e]^2$ dependence.} These measurements, although comparable to the UF_4 work, are carried out with entirely different techniques. The UF_4 experiments utilize the presence of ^{235}U to detect individual sputtered atoms. It is for this reason that it is possible to measure the energy spectrum at low beam fluences as described above. In the ice experiments, the Rutherford backscattering technique is used to determine the thickness of material removed from the target in a given sputtering run. In order for surface erosion to be detected by this method, a large number of monolayers must be removed. By studying the erosion rate as a function of the integrated beam flux, it has been possible to assure that potential nonlinear processes, due, for instance, to target contamination with projectile ions, are unimportant. Although the direct identification of sputtered species has not yet been made (i.e., whether sputtered particles are atoms, molecules, molecular fragments, or molecular clusters), it has nevertheless been shown that the stoichiometry of the sputtered material is the same as that of the target.

In their studies of H_2O sputtering Brown *et al.* (Br 80b) showed that there was a low-temperature region below about 100 K, where the yield S was independent of the ambient temperature T {and proportional to $[(dE/dx)_e]^2$}. It is the sputtering in this region which we may compare more or less directly with the UF_4 work (where S was also found to be insensitive to the ambient temperature between 70 and 170°C). At the higher temperatures studied, the yield of H_2O depended linearly on the electronic stopping power, and arguments based on radiation chemistry were advanced (Br 80b) to explain the energy dependence. Strong temperature dependences of the sputtering yields of other dielectric materials have also been reported. Biersack and Santner (Bi 76) noticed such a dependence for alkali halide targets, and Bøttiger *et al.* (Bø 80) have reported extensive work on frozen Xe showing strong temperature effects. In the latter two cases it is not clear whether at sufficiently low temperatures a temperature-independent region exists or not. These temperature dependences have not as yet been incorporated into even a semiquantitative description of the sputtering process. To the extent that F-center formation, or similar processes, are concerned, the mechanisms fall outside the scope of this paper, and we shall therefore continue to concern ourselves solely with sputtering phenomena where the ambient target temperature has been shown to exert little effect on the yield.

Results for the sputtering of H_2O ice by protons (Br 80a) are shown in Figure 10, along with a (normalized) curve proportional to $[(dE/dx)_e]^2$ for comparison. The electronic stopping power is taken from the compilation of Andersen and Ziegler (An 77), with the use of Bragg's rule for combining stopping powers of different elements. The square-law curve evidently provides a good fit to the data.

In Figure 11, the ice sputtering data of (Br 80a) for H, He, and (at 1.5 MeV energy only) C and O are collected together. The solid line is proportional to $[(dE/dx)_e]^2$. This curve seems to correlate the H data, but the He, C, and O data points lie considerably above the curve. According to (Br 80a) the deviations apparent for He are qualitatively consistent with the fact that the equilibrium He charge state is not obtained in the thin surface layer from which the sputtering presumably occurs. There remains a problem, however, with the C and O points.

Seiberling has argued (Se 80c) that it may be misleading to compare

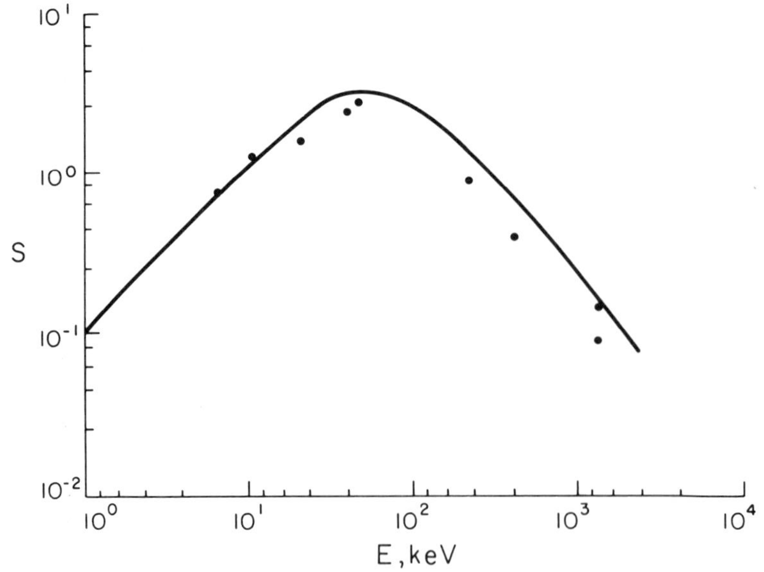

Figure 10. Sputtering yield for proton bombardment of H_2O ice. The points are taken from the data of (Br 80a), and the curve is proportional to $[(dE/dx)_e]^2$.

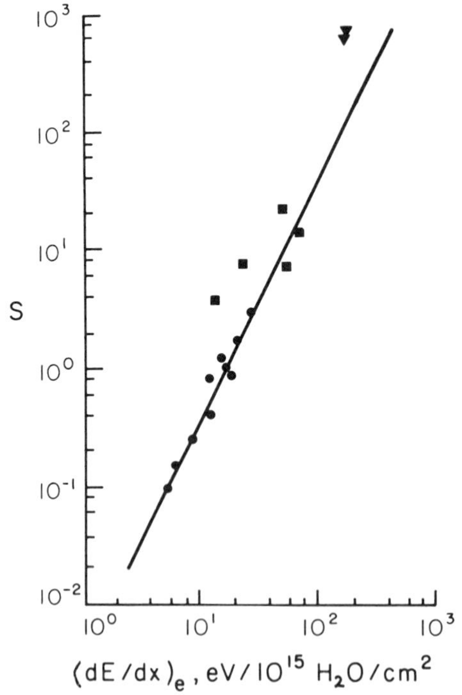

Figure 11. Proton, alpha particle, C, and O sputtering yields for H_2O ice plotted versus the square of the electronic stopping power [adapted from (Br 80a)]. The solid circles correspond to proton bombardment, the squares to alpha particle bombardment, and the triangles to the C and O data.

heavy-ion and light-ion sputtering yields on the same theoretical footing. Whereas O electronic energy loss in water at ≈1.5 MeV is on the order of 1000 eV/nm, H energy loss, in the ≈100-keV range where $(dE/dx)_e$ is a maximum for this projectile, is only ≈100 eV/nm. Thus a heavy ion like O can create a continuous trail of significant ionization, of the sort which is implicitly assumed in the formulation of the ion explosion model. But the energy loss for light ions is much smaller and it may therefore not be the mean energy loss, but the fluctuations, which are the dominant factor in determining the sputtering yield in that case.

Brown *et al.* (Br 80a) have pointed out that the relative stopping powers of heavy versus light ions are not accurately known, so that the fact that C and O fall above the $[(dE/dx)_e]^2$ curve may be due simply to ignorance of the true $(dE/dx)_e$ values for these ions in water. This illustrates a general problem in comparing high-energy sputtering data with theoretical calculations. Essentially all theoretical results are necessarily cast into forms which contain either the electronic stopping power $(dE/dx)_e$ or a related function such as dJ/dx. However, these functions are often not accurately known, especially dJ/dx. One is therefore never entirely certain whether to ascribe a discrepancy between theory and experiment to the theoretical arguments which invoke $(dE/dx)_e$, or to the independently derived values of $(dE/dx)_e$. Thus the C and O abscissae of Figure 11 were derived (Br 80a) from the tables of Northcliffe and Schilling (No 70), while the H and He points were assigned stopping power values based upon the work of Andersen and Ziegler (An 77) and Ziegler (Zi 77). For these light ions there exist substantial discrepancies between the two compilations, and therefore one's confidence in the relative assignment of abscissa values for the stopping powers of C and O in Figure 11 is diminished.

3.7. Alternative Mechanisms

3.7.1. Collision Cascade via Ion Explosion

Another point of view which can be used to model the experimental results again starts with an ion explosion mechanism, but explicitly describes sputtering as arising from unequilibrated collisions, as in the standard low-energy sputtering theories. The tacit assumption made in using a spectrum of the form $\varepsilon/(\varepsilon + U)^3$ is that the initial injection of energy into the cascade, via the recoil of an atom struck by the projectile, occurs at an energy $\varepsilon = E_0$ which is much greater than U. For typical low-energy experiments $E_0 \approx$ several hundred eV, and $U \approx$ few eV, so that the criterion is well satisfied. However, in the case of high-energy sputtering, the initial ion recoils due to Coulomb repulsion are expected to have energies $\leqslant 10$ eV. Therefore, the energy spectrum expected would not be of the form

$\varepsilon/(\varepsilon + U)^3$, even if the collision cascade process were dominant, although the general arguments used to arrive at Equation (1) are still valid, and therefore the total yield will remain proportional to $(dJ/dx)^2$.

It can be shown in the collision cascade picture (Ha 80a), for example, that if the initial distribution Φ of recoiling ions in the ion explosion is proportional to ε^{-5}, then the sputtering yield is given by the form $\varepsilon/(\varepsilon + U)^6$, where U is the surface binding energy. This is essentially the form found by Griffith *et al.* (Gr 80) empirically. The high power of ε^{-1} is consistent with the condition that the primary ion recoils arising from mutual Coulomb repulsion are essentially all of low energy ($\leqslant 10$ eV, say). The divergence as $\varepsilon \to 0$ is of no consequence since recoils of energy $<U$ cannot contribute to sputtering. There is no explicit justification for the form ε^{-5} from first principles, except that it assures that only low-energy particles can be involved in the sputtering process—a fact which is assured by the ion explosion model anyway. The thermal model has the advantage that no assumption is required about the shape of Φ, because the collisions occurring prior to sputtering have erased all memory of the starting conditions.

3.7.2. Pulsed Laser Analog

A fundamental difficulty with the ion explosion picture as developed so far is the requirement that somehow a positive atomic core can be maintained long enough for the atoms to gain the required kinetic energy before the electrons "come back" and neutralize the spike. No convincing argument has been advanced to explain how this can occur. Recently, C. C. Watson (Wa 81) has proposed a mechanism which avoids this difficulty. His idea is based on an analogy with an approach which has been useful in the study of pulsed laser annealing processes [see (Yo 80), for example]; here, as in high-energy sputtering, energy is first deposited into electronic motion.

Watson's approach has been to treat the electronic excitations, for sufficiently high energy densities, as a fermi gas at an elevated temperature $T > 0$. The gas fills a volume which is determined by the minimum electron excitation energy possible in the undisturbed medium. Atomic kinetic energy derives from the change in average interatomic potentials, which are a function of the temperature of the electron gas. No charge separation is required in this picture.

Although numerical results are not yet available, the idea seems promising. Much of the philosophy of the ion explosion model is retained—for example, the (lattice) thermal diffusivity will presumably remain an important parameter, as discussed in Sections 3.6.1 and 3.6.2.1. The principal difference is the way in which the electronic and atomic motions are coupled. The ion explosion picture can be viewed as an oversimplification of the model presented by Watson. In the ion explosion scheme, the electronic screening length for a few ion cores in the immediate vicinity of the track is arbitrarily taken to infinity, while in the heated fermi gas model this increase in the screening length is more moderate, consistent with the temperature and density of the gas, but involves a much greater number of atoms.

Until this model is developed further, however, we will stick with an ion explosion description of the sputtering process.

3.8. Elaboration of the Zeroth-Order Thermal Model

The result given in Equation (13) for the total sputtering yield was found by multiplying the flux φ at a fixed temperature T by the exposed surface area of the spike and by the appropriate time scale. A more detailed calculation of S would involve an integration of φ over the surface area dr^2 and over time t, thus taking into account the diffusion of the heat spike. The temperature profile resulting from a delta-function input of energy at $r = 0$ and $t = 0$ is

$$T(r, t) = \frac{\hat{\varepsilon}}{4\pi K t} \exp(-Cr^2/4Kt) \tag{17}$$

where $\hat{\varepsilon}$ is the constant energy per unit length in the spike, and C is the heat capacity per unit volume. Following the technique of Vineyard (Vi 76), we find that the integration

$$S(E, \Omega) = \iint \varphi(\varepsilon, \Omega, T(r, t))\, 2\pi r\, dr\, dt \tag{18}$$

yields

$$S(\varepsilon) = \frac{15nk^2\hat{\varepsilon}^2\varepsilon}{32\sqrt{2}\,\pi M^{1/2}KC(U+\varepsilon)^{7/2}} \tag{19}$$

This energy spectrum clearly cannot fit the data of Figure 6, since we know that a function of the form $\varepsilon/(\varepsilon + U)^{6.1}$ is required. Equation (19) predicts far too many particles at high energies. The reason for this is that near time $t = 0$ the temperature of the spike is extremely high (by assumption). In the calculation of Seiberling *et al.* (Se 80a,b) the temperature is assumed to be constant for the duration of the spike.

An infinitely high temperature at $t = 0$ is physically impossible. A thermal description is certainly not valid until at least several collisions have occurred, so that the initial spike is not of zero width. Furthermore, the electrons scattered in the ionization process spread out over a finite distance, thus providing a means for producing secondary ionization several lattice spacings away. A better calculation of the yield S therefore might begin with an extended spike of radius r_i and a finite temperature T.

Although we could directly integrate Equation (18), it is somewhat more transparent, although less accurate, to use a calculational scheme due to Seitz and Koehler (Se 56). Here we imagine that the heat spike is always of a rectangular shape in space, as illustrated in Figure 12, and that the width evolves in time according to $r = (4Kt/C)^{1/2}$, since a diffusion process is involved. At a given radius r, the height of the spike T is related to the deposited linear energy density according to

$$kT = \frac{2\hat{\varepsilon}C}{3\pi n 4Kt} \tag{20}$$

which just expresses energy conservation. The double integral in Equation (18) is now performed with the restriction $r < (4Kt/C)^{1/2}$, and $t > t_i =$

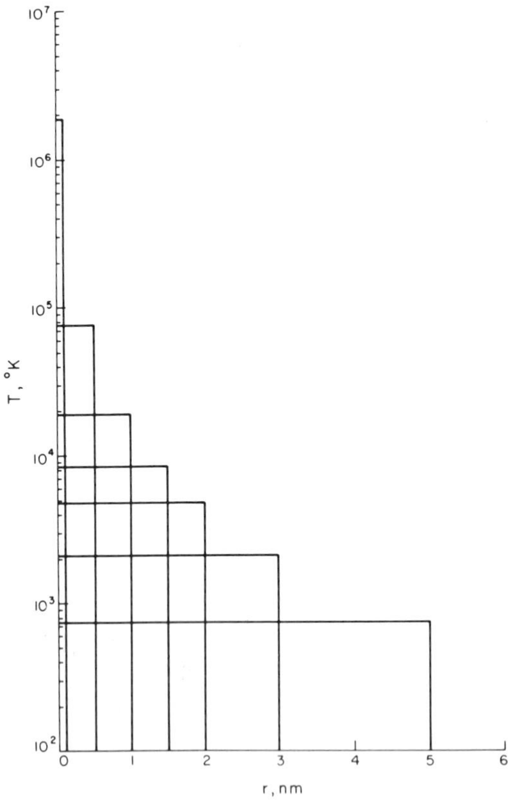

Figure 12. Temperature versus spike radius in the Seitz–Koehler model (Se 56), for an initial linear energy density $\hat{\varepsilon} \approx 100$ eV/nm.

$r_i^2 C/4K$. This procedure yields

$$S(\varepsilon) = \frac{nk^2\hat{\varepsilon}^2\varepsilon}{2^{1/2}\pi^{3/2}M^{1/2}KC(\varepsilon + U)^{7/2}} \int_{\gamma t_i}^{\infty} x^{5/2} e^{-x}\, dx \tag{21}$$

where

$$\gamma = \frac{2\pi K(U + \varepsilon)}{k\hat{\varepsilon}} \tag{22}$$

The result (21) is essentially the same as (19) except for an extra factor involving an incomplete gamma function. This factor leads to a suppression of particles at high energies [relative to Equation (19)] through the dependence of γ on energy. The spectrum is illustrated in Figure 13 and compared with a Maxwell–Boltzmann spectrum at a fixed temperature. An extra dimensionless factor α has been used to multiply the lower limit of integration γt_i, i.e., $\gamma t_i \to \alpha\gamma t_i$, and the values of α are shown on the figure. For the solid lines we have taken $\hat{\varepsilon} = 3$ eV, $r_i = 1$ nm, and $U = 3$ eV. For $\alpha = 1.5$ (heavy line), the curve based on (21) is indistinguishable, at the resolution of the figure, from a Maxwell–Boltzmann spectrum at $T = 3500$ K. Actually, a value of $U = 3$ eV will not lead to a good value for the total yield, since we already found that $U = 0.71$ is required for the zeroth-order model. Choosing this latter value of U, setting $\alpha = 0.7$, and leaving other parameters unchanged, leads to the dashed curve in Figure 13.

The dashed curve is not quite as good a fit as the line with $\alpha = 1.5$, but a change of normalization at $\varepsilon = 0.3$ eV would improve it greatly. With any reasonable normalization at 0.3 eV, the dashed curve still has the desirable property of predicting substantially more particles at high energies than the Maxwell–Boltzmann distribution. This is illustrated by the dashed curve plotted on the logarithmic scale of Figure 6, from which it is clear that the agreement at the high-energy end of the spectrum is substantially improved. However, recall from our earlier discussion that the high-energy data points are the most difficult to measure, and therefore we cannot yet attach as much significance to agreement or disagreement with theory in this energy range as we can at lower energies.

It is necessary to refrain from pursuing these developments further since a complete investigation would require more attention to the calculation of $\hat{\varepsilon}$, to the temperature dependence of the thermal diffusivity (see, e.g., Si 80), and to the modifications which may be required to account for evaporative cooling of the spike. However, the above developments do suggest that a more elaborate thermal model may be able to retain the essence of the results obtained initially by Seiberling *et al.* (Se 80a), at the same time dispensing with some of the uncertainties raised by the approximations in their treatment.

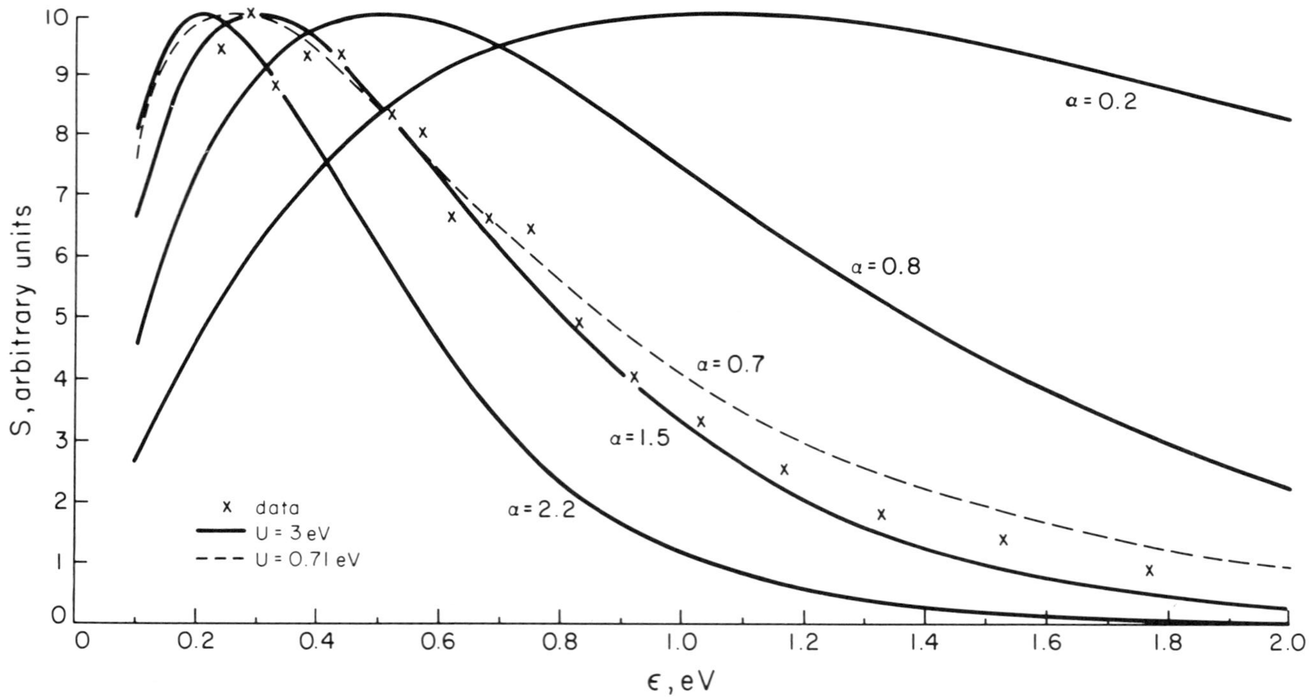

Figure 13. The solid curves are the prediction of equation (21) for 4.74-MeV ^{19}F sputtering of UF_4 with $U = 3$ eV. The role of α is discussed in the text. For $\alpha = 1.5$, the calculated curve is indistinguishable, at the resolution of the figure, from a Maxwell–Boltzmann distribution at 3500 K. The dashed curve is taken from equation (21) with $U = 0.71$ eV, $T = 3500$ K, and $\alpha = 0.7$. The crosses are some of the data points of (Gr 80). For energies greater than 2 eV, refer to the dashed curve of Figure 6, which is the same as the dashed curve of this figure, except for a normalization factor.

4. Applications

Investigations into the phenomenon of high-energy sputtering were begun only in the middle of the last decade, and the subject is still far from completely understood. Nevertheless, some applications have been suggested, and it now appears that this erosion process occurs in certain natural environments. We will discuss briefly some possible directions in which the work discussed in this chapter may be applied in the future.

In unpublished notes, Seiberling (Se 80d) has pointed out that high-energy sputtering may become useful as a tool for surface analysis. The technique of secondary ion mass spectrometry (SIMS), using low-energy particles to sputter surface ions, which are then analyzed in a mass spectrometer, is well known. However, while neutral particle sputtering in this regime is well understood, the process of ion emission is not. The method relies heavily on the use of standards, and the spectra are sensitive to surface contamination. Lack of a quantitative theory for ion emission during low-energy ion bombardment limits the usefulness of this method. The investigation of high-energy sputtering has just begun and no comprehensive experimental or theoretical studies have yet been done. However, there is evidence that in some cases atoms may be evaporating from a region in the target that is in local thermodynamic equilibrium. If this is the case for dielectric targets in general, then a quantitative theoretical description of ion emission processes may be possible. This would also mean that one would be less dependent on having standard samples available that were nearly identical to the samples under investigation. If the secondary ions produced by high-energy sputtering can be correlated in a simple way with the composition of the surface being sputtered, then the use of such heavy-ion beams could lead to the development of a new method of surface analysis.

Another area of application, which has already drawn the attention of some of those working in the field of high-energy sputtering, centers around erosion processes on planetary bodies in the solar system. Fast protons and heavy ions are known to be associated with the magnetosphere of Jupiter. Lanzerotti *et al.* (La 78) first pointed out the possible implication for surface erosion by proton fluxes on the icy Jovian satellites Ganymede, Europa, and Callisto. Sputtering of those surfaces can potentially lead to erasure of meteorite impact craters (Ha 79), to modification of the satellite albedo (Ha 80b), to the generation of a thin atmosphere (La 78, Wa 80), and may provide an important source for heavy ions in the Jovian magnetosphere with concomitant implications for magnetospheric dynamics. Conventional low-energy sputtering can produce similar results, although sputtering yields for high-energy bombardment can be orders of magnitude greater. An important distinction between the effects of the two processes lies in the expected difference in the energy spectra of the sputtered atoms. Atoms sputtered by low-energy impact typically have energies on the order of a few electron volts. The energy spectrum of atoms sputtered by fast ions has been measured only for the UF_4 target, but in that case the distribution of emitted atoms peaks at 0.3 eV, corresponding to a spike temperature of a few thousand degrees. On the large satellites of Jupiter the kinetic energy that an O atom, for example, needs to escape the satellite gravity is about 0.6 eV. Therefore low-energy sputtering can easily remove material from the planetary body altogether, while, to the extent that the results suggested by UF_4 apply to H_2O ice, the effect of sputtering by fast ions may be mainly to rearrange matter upon the surface.

The kinds of experimental investigations that are required to resolve and clarify these questions are clear—direct measurements of energy spectra for H_2O ice, for instance. A picture of the true influence of natural ion fluxes on solar system surfaces, then, can only come in time as the problems raised in this chapter are subjected to continued experimental and theoretical scrutiny.

5. Summary and Conclusions

In the past five years or so an accumulating body of experimental results and theoretical developments have been brought to bear on the phenomenon of high-energy sputtering. It is clear now that there is a class of materials which exhibit the effect. UF_4 and water ice have received the most attention, but various frozen gases and a number of minerals have also been examined and shown to erode at anomalously high rates. The collision cascade theory so successful at lower energies fails at high energy. It is clear that the dominating role of the nuclear stopping power in the keV/amu range yields to the electronic stopping power as beam energies are increased.

A certain amount of circumstantial evidence, as well as semiquantitative arguments, support the notion of a thermalized core around the ion trajectory, for the case of heavy-ion bombardment. The most impressive piece of evidence is the direct measurement of the sputtered atom energy spectrum, which can be fit reasonably well with a Maxwell–Boltzmann distribution function at a definite temperature T. However, a more careful calculation of the spectrum and of the total yield, which would include the details of spike coupling to electrons and of evaporative cooling, has not yet been done.

There appear to be differences in the dependence of the sputtering yield on incident beam energy as derived from the UF_4 and H_2O experiments. In the former case, the yield function is fit by a form $(dJ/dx)^4$, while in the ice experiments a $[(dE/dx)_e]^2$ [or $(dJ/dx)^2$] dependence is indicated. Theoretical considerations can produce either dependence, depending on what mechanism is invoked for ultimately translating the initial ionization disturbance along the projectile path into kinetic energy of target atoms. One possibly important difference between the two sets of experiments is that the ice sputtering yield was performed principally with light ions, and the UF_4 runs with heavy ions. The ionization rates are thus greatly different, and the low $(dE/dx)_e$ for protons and alpha particles may diminish the applicability of the thermalized ion explosion model in this case.

An important problem arising in the analysis of any high-energy sputtering experiment was mentioned in the text but deserves reiteration here. Namely, experimental results have been analyzed chiefly in terms of the electronic stopping power, or some related function. However, $(dE/dx)_e$ is generally not known with precision, especially near its maximum value, which one expects to coincide with the maximum of the sputtering yield. There exists, then, the dilemma of deciding whether disagreement between experiment and theory is due to failings in the underlying theoretical formulation, or in the details of calculating (or measuring) $(dE/dx)_e$ and dJ/dx. This introduces an uncertainty into the problem of unraveling the origins of high-energy sputtering which has not yet been dealt with adequately.

The application of high-energy sputtering to surface analysis is very speculative, and we will not comment upon it further here. However, the study of sputtering in its relation to planetary satellites and ring systems is already an area of endeavor which has seen much activity in the past few years. The role of high-energy sputtering in this environment is coming to be more fully appreciated. In the 1970s the return of lunar samples sparked a research effort which established firmly the role of ion accelerators in experiments designed to untangle the history of rocks subjected to eons of bombardment by the solar wind and by cosmic rays. In a similar way the Voyager spacecraft missions to Jupiter, Saturn, and beyond are providing a flood of data rich in information about the nature of both light- and heavy-ion interactions with a diverse range of planetary surfaces. Already fundamental investigations of high-energy sputtering have found application in this regard, and it is clear that ion accelerators can continue to make a substantial contribution to our expanding knowledge of the solar system.

Acknowledgments

I would like to thank my colleagues at Caltech for numerous discussions of the topics covered in this chapter. Special acknowledgment is due to J. E. Griffith, L. E. Seiberling, T. A. Tombrello, and C. C. Watson.

References

(An 77) H. H. Andersen and J. F. Ziegler, in *The Stopping and Ranges of Ions in Matter*, Vol. 3 (J. F. Ziegler, ed.), Pergamon Press, New York (1977).

(Bi 76) J. P. Biersack and E. Santner, *Nucl. Instrum. Methods* **132**:229 (1976).

(Bø 80) J. Bøttiger, J. A. Davies, J. L. Ecuyer, N. Matsunami, and R. Ollerhead, *Rad. Effects* **49**:119 (1980).

(Br 78) W. L. Brown, L. J. Lanzerotti, J. M. Poate, and W. M. Augustyniak, *Phys. Rev. Lett.* **40**:1027 (1978).

(Br 80a) W. L. Brown, W. M. Augustyniak, E. Brody, B. Cooper, L. J. Lanzerotti, A. Ramirez, R. Evatt, and R. E. Johnson, *Nucl. Instrum. Methods* **170**:321 (1980a).

(Br 80b) W. L. Brown, W. M. Augustyniak, L. J. Lanzerotti, R. E. Johnson, and R. Evatt, *Phys. Rev. Lett.* **45**:1632 (1980b).

(Dü 80) P. Dück, W. Treu, W. Galster, H. Frölich, and H. Voit, *Surf. Sci.* **95**:603 (1980).

(Fl 65) R. L. Fleischer, P. B. Price, and R. M. Walker, *J. Appl. Phys.* **36**:3645 (1965).

(Fl 75) R. L. Fleischer, P. B. Price, and R. M. Walker, *Nuclear Tracks in Solids*, Univ. of California Press, Berkeley (1975).

(Fl 80) R. L. Fleischer, *Nuclear Track Production in Solids*, Technical Information Services, No. 80 CRD128, General Electric Co., Schenectady, New York (1980).

(Gr 78) R. Gregg and T. A. Tombrello, *Rad. Effects* **35**:243 (1978).

(Gr 79) J. E. Griffith, Sputtering of Insulators in the Electronic Stopping Region, Ph.D. Thesis, California Institute of Technology (1979).

(Gr 80) J. E. Griffith, R. A. Weller, L. E. Seiberling, and T. A. Tombrello, *Rad. Effects* **51**:223 (1980).

(Gr 81) J. E. Griffith, private communication (1981).

(Ha 76) P. K. Haff, *Appl. Phys. Lett.* **29**:473 (1976).

(Ha 79) P. K. Haff, C. C. Watson, and T. A. Tombrello, *Geochimica et Cosmochimica Acta* Suppl. 11 (Proc. of Lunar and Planet. Sci. Conf. Tenth, Houston, Texas), Pergamon Press, New York (1979), p. 1685.

(Ha 80a) P. K. Haff, unpublished calculations (1980a).

(Ha 80b) P. K. Haff, in *Proc. of the Symposium on Thin Film Interfaces and Interactions* (J. E. E. Baglin and J. M. Poate, eds.), The Electrochemical Society, Princeton (1980b), p. 21.

(He 63) H. H. Heckman, E. L. Hubbard, and W. G. Simon, *Phys. Rev.* **129**:1240 (1963).

(Hi 81) T. W. Hill, EOS **62**:25 (1981).

(Jo 80) R. E. Johnson and R. Evatt, *Rad. Effects* **52**:187 (1980).

(Kn 64) T. G. Knorr, *J. Appl. Phys.* **35**:2753 (1964).

(La 78) L. J. Lanzerotti, W. L. Brown, J. M. Poate, and W. M. Augustyniak, *Geophys. Res. Lett.* **5**:155 (1978).

(Ma 76a) R. D. Macfarlane and D. F. Torgerson, *Phys. Rev. Lett.* **36**:486 (1976a).

(Ma 76b) R. D. Macfarlane and D. F. Torgerson, *Intl. J. Mass Spectrom. Ion Phys.* **21**:81 (1976b).

(No 70) L. C. Northcliffe and R. F. Schilling, *Nucl. Data Tables* **A7**:233 (1970).

(Ol 80) R. W. Ollerhead, J. Bøttiger, J. A. Davies, J. L'Ecuyer, H. K. Haugen, and N. Matsunami, *Rad. Effects* **49**:203 (1980).

(Se 56) F. Seitz and J. S. Koehler, *Solid State Physics* **2**:251 (1956).

(Se 80a) L. E. Seiberling, J. E. Griffith, and T. A. Tombrello, *Rad. Effects* **52**:201 (1980a).

(Se 80b) L. E. Seiberling, A Thermalized Ion Explosion Model for High Energy Sputtering and Track Registration, Ph.D. Thesis, California Institute of Technology (1980b).

(Se 80c) L. E. Seiberling, private communication (1980c).

(Se 80d) L. E. Seiberling, unpublished notes (1980d).

(Si 69) P. Sigmund, *Phys. Rev.* **184**:383 (1969).

(Si 72) P. Sigmund, *Rev. Roum. Phys.* **17**:969 (1972).

(Si 77a) A. Sigrist and R. Balzer, *Helv. Phys. Acta* **50**:49 (1977a).

(Si 77b) A. Sigrist and R. Balzer, *Rad. Effects* **34**:75 (1977b).

(Si 80) P. Sigmund and C. Claussen, in *Proceedings of the Symposium on Sputtering*, Perchtoldsdorf/Wien, Austria (P. Varga, G. Betz and F. P. Viehböch, eds.) (1980).

(Th 68) M. W. Thompson, *Phil. Mag.* **18**:377 (1968).

(To 76) P. D. Townsend, J. C. Kelly, and N. E. W. Hartley, *Ion Implantation, Sputtering and their Applications*, Academic Press, New York (1976).

(Vi 76) G. H. Vineyard, *Rad. Effects* **29**:245 (1976).

(Vo 75) F. L. Vook, H. R. Birnbaum, T. H. Blewitt, W. L. Brown, J. W. Corbett, J. H. Crawford, Jr., A. N. Goland, G. L. Kulcinski, M. T. Robinson, D. N. Seidman, and F. W. Young, Jr., *Rev. Mod. Phys.* **47** Suppl. 3 (1975).

(Wa 80) C. C. Watson, P. K. Haff, and T. A. Tombrello, *Geochimica et Cosmochimica Acta*, Suppl. 12 (Proc. of the Lunar and Planet. Sci. Conf. Eleventh, Houston, Texas) (1980), p. 2479.

(Wa 81) C. C. Watson, unpublished calculations (1981).

(We 78a) R. A. Weller and T. A. Tombrello, *Rad. Effects* **37**:83 (1978a).

(We 78b) R. A. Weller, Jr., The Energy Spectra of Uranium Atoms Sputtered from Uranium Metal and Uranium Dioxide Targets, Ph.D. Thesis, California Institute of Technology (1978b).

(We 81) R. A. Weller and M. R. Weller, to be published (1981).

(Wr 77) A. W. Wright, *Am. J. Sci.* **13**:49 (1877).

(Yo 80) E. J. Yoffa, *J. Phys. (Paris) Col. C4*, suppl. to **41**:C4 (1980).

(Zi 77) J. F. Ziegler, *The Stopping and Ranges of Ions in Matter*, Vol. 4 (J. F. Ziegler, ed.), Pergamon Press, New York (1977).

8

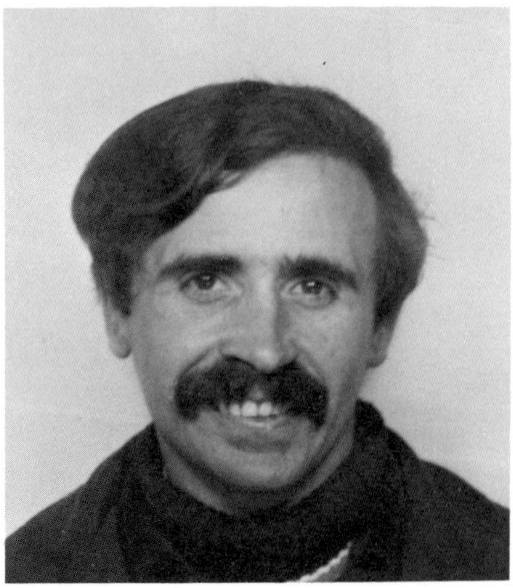

COLIN A. ENGLISH (top) is a Senior Scientific Officer at the Atomic Energy Research Establishment (AERE) at Harwell, England. Educated in England, he holds both B.Sc. and D.Phil. degrees.

MICHAEL L. JENKINS (bottom) is a staff member in the Department of Metallurgy and Science of Materials at Oxford University, and a Research Fellow of Wolfson College at Oxford. He, too, holds the B.Sc. and D.Phil. degrees.

Heavy-Ion Damage in Solids

C. A. English and M. L. Jenkins

1. Introduction

Much of the impetus for studying irradiation damage in solids stems from the demands of nuclear technology. The radiation fields encountered in all critical areas of present and projected reactors present an intensely hostile environment. At typical reactor operating temperatures a whole range of largely irreversible and deleterious effects may occur in reactor components—creep, swelling, embrittlement, sputtering etc.—which lead to serious degredation in properties and impose severe limitations on reactor design. Perhaps the best-known example is radiation-induced void swelling, originally discovered in studies of metal fuel cladding by Cawthorne and Fulton (Ca 67), and subsequently found to be a major potential problem in structural components, which has had enormous implications in metallurgical design criteria for fast reactors. It has been estimated that the cost benefit of designing void-resistant low-swelling alloys for use in reactors into the next century is more than $10 billion!

The problem of radiation damage in solids has been approached in two ways. The first, which might be called an engineering approach, is to fully characterize a material for a particular set of conditions—dose, dose rate, temperature, prior history etc.—in terms of macroscopic measurements (swelling rate, fracture toughness, etc.). This approach gives little direct information on the atomistic processes responsible for the property changes, and has not altogether been successful in predicting behavior under other irradiation conditions. Even rather small changes in service conditions can

C. A. English • Metallurgy Division, Atomic Energy Research Establishment, Harwell, Didcot, Oxfordshire OX11 0RA, England. M. L. Jenkins • Department of Metallurgy and Science of Materials, University of Oxford, Parks Road, Oxford, England.

sometimes lead to quite unexpected results. For these reasons a second approach, that of characterizing the microstructure and identifying atomistic mechanisms, has become common. Several different techniques for gaining information on fundamental processes are available—x-ray, neutron and particle scattering, or in some cases defect spectrometry—but the most direct are the "imaging" techniques of transmission electron microscopy and field-ion microscopy.

Heavy-ion irradiation, particularly in conjunction with these microscopy techniques, has made major contributions to the understanding of fundamental radiation damage mechanisms and in the simulation of fast neutron damage. Advantages of the use of beams of heavy ions are that the energy, dose, and dose rate of the ion beam and the irradiation temperature can be controlled to fine limits. Since dose rates of charged particles can be much higher than of reactor fast neutrons, large doses can be built up in convenient times.

Experiments can be designed such that basic processes such as displacement cascade production can be studied, or the recoil spectrum made similar to the fast neutron case. Such experiments, and the information on damage mechanisms which they deliver, are the subject of most of this chapter. The importance of understanding damage processes and obtaining data on the performance of in-reactor components has meant that metal systems have received vastly more attention than ionic or covalent solids. Our review will thus be biased towards metals, although the growing interest in heavy-ion damage in nonmetals and semiconductors will be reflected where appropriate.

In considering heavy-ion damage we first discuss qualitatively the passage of a heavy ion through a solid. The heavy-ion loses energy through interacting with both the electrons and the atoms of the solid. The rate of energy deposition depends on the ion velocity as this governs the ratio of electronic energy loss to nuclear energy loss. For ions with initial energies of a few MeV electronic losses predominate, but as the ion slows down nuclear energy loss becomes increasingly important and is dominant for most ion–target combinations at energies below a few hundred keV. If the energy transferred to a lattice atom is greater than some threshold value E_d (typically $\approx$40 eV) then the atom will be displaced from its site, creating a Frenkel defect, i.e., a vacancy (empty lattice site) and an interstitial (an atom occupying a nonlattice site). If the transferred energy is much greater than a few keV the knock-on atom is able to displace many atoms, which may in turn go on to produce further displacements. This sequence of events is shown schematically in Figure 1, and is termed a displacement cascade. The knock-on atoms rapidly enter regimes where collisions become increasingly closely spaced. The central cascade regions may then contain several hundred Frenkel defects within a volume of typical dimension $\approx$10 nm. For

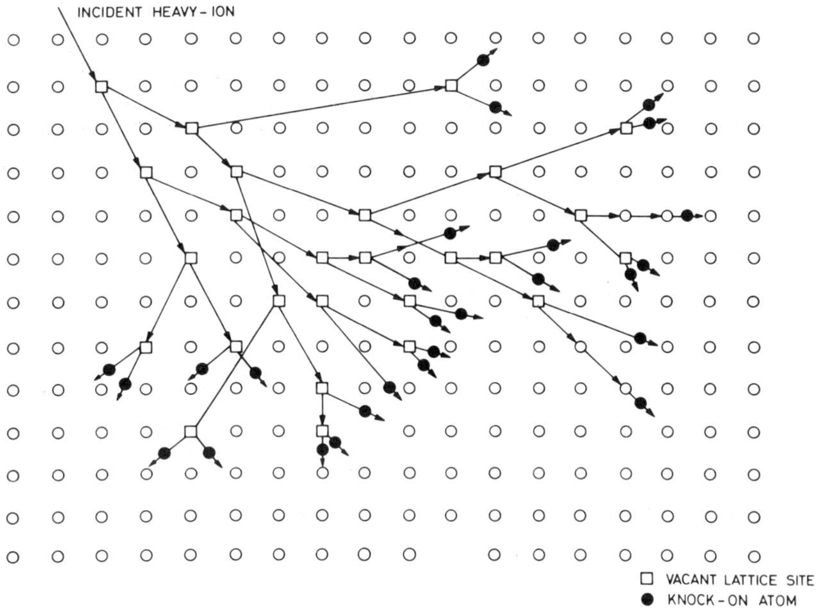

Figure 1. Schematic diagram of collision cascade.

the purpose of discussion it is convenient to consider separately three different regimes of incident ion energy E:

a. Low energy, $E < 5$ keV;
b. Medium energy, 5 keV $< E <$ 200 keV;
c. High energy, $E >$ 200 keV.

At energies $E < 5$ keV individual impacting ions have only sufficient energy to produce isolated Frenkel pairs or low-energy displacement cascades in a region very close (1–2 nm) to the target surface. Experiments in this energy regime are designed to gain information on low-energy transport mechanisms such as focused replacement sequences which may inject interstitials over long distances (tens of nm) into the specimen (Section 2).

At energies 5–200 keV the damage is again produced in a narrow layer within a few tens of nanometers of the target surface. The ions have sufficient energy to generate violent collision cascades as described above. The production of cascades is the major primary damage mechanism in fast-neutron irradiation, when typically over 95% of displacements occur in cascades of energy 5 keV $< E <$ 200 keV. In low-dose ($<10^{16}/m^2$) heavy-ion experiments the incident ions are used to simulate the primary knock-ons produced under neutron irradiation, so that individual cascade regions can be studied. Because of the fundamental importance of the displacement cascade in the understanding of neutron damage, it has received considerable theoretical attention, with both analytical approaches and various computer simulation treatments. These are reviewed in Section 3, together with the experimental techniques which have been developed to allow direct comparison between theory and experiment. An important phenomenon associated with cascades in this energy range is the development at the cascade core of a vacancy rich "depleted zone," which may collapse to form a vacancy aggregate visible in the electron microscope. Such vacancy clustering has now been studied in a wide range of pure materials and alloys and the results and their implications are discussed in Section 3. Other phenomena which occur in this energy range—interstitial clustering, and radiation enhanced solute redistribution—are also treated.

At energies $E > 200$ keV, and especially at energies $E > 1$ MeV, the incident ions may penetrate far into the target, to produce dense collision cascades only near the end of their paths when the energy falls to $\lesssim$200 keV. The damage created under such conditions may be more representative of bulk material. Experiments in this energy range are often carried out at high doses and high temperatures to simulate high-dose reactor damage. Such experiments will probably become of increasing importance in the fusion reactor as it becomes necessary to test the radiation response not only of structural materials such as metal alloys but also components such as ceramic insulators. Some of the work carried out to date in this general area is reviewed in Section 4.

2. Low-Energy Irradiations (E < 5 keV)

As described in the Introduction, heavy ions in this energy range are able only to produce isolated Frenkel defects with the initial collision occurring close the surface. The threshold transferred energy for creating displacements, E_d, is dependent on the crystal lattice type (e.g., face-centered-cubic, fcc, or body-centered-cubic, bcc) and orientation of the axes with respect to the heavy-ion beam. Typical values for pure metals are in the range 20–40 eV. Silsbee (Si 57) pointed out that during irradiation, energy and momentum can be focused along low index crystallographic directions, and if the energy transferred in each of the collisions is greater than E_d then a sequence of replacement events may occur, resulting in a significant separation of the interstitials from their correspoding vacancies. This is an important process for low-energy ion irradiation of crystals, because although the incident particle can only penetrate a few atomic layers, interstitial point defects can be deposited at significantly greater depths. Two types of replacement collision sequences are envisaged (shown schematically

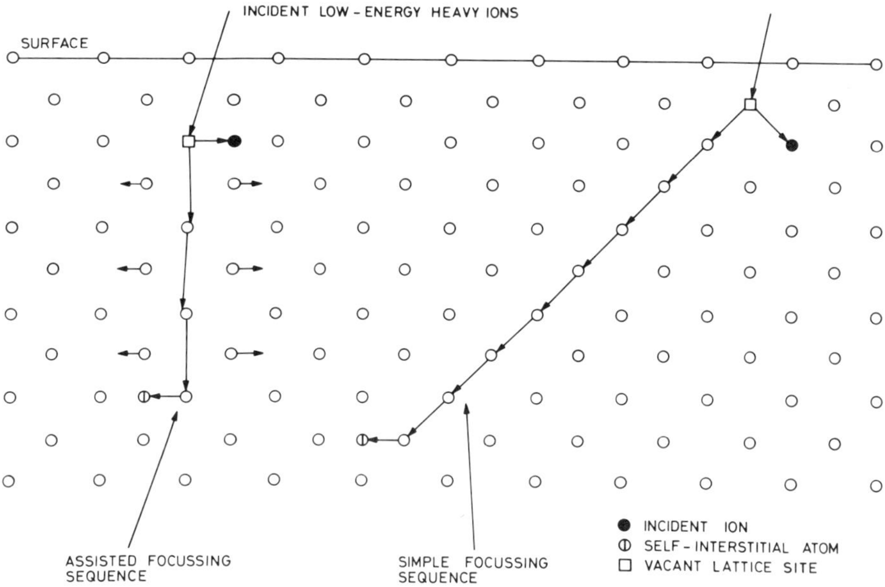

Figure 2. Schematic diagram of replacement collision sequences.

in Figure 2): simple focusing along close-packed rows in the lattice (such as $\langle 110 \rangle$ in fcc crystals and $\langle 111 \rangle$ in bcc crystals), and assisted focusing where atoms in adjoining rows assist the propogation of the sequence by helping to steer the moving atoms along the focusing direction. Assisted focusing occurs along low index directions where the distance of the adjacent atoms from the focusing axis is less than the atom separation along the axis (such as $\langle 100 \rangle$ rows in fcc crystals). Analytical treatments of the collision processes [see Thompson (Th 69a) for a review] show that focusing is essentially limited to low-energy events, typically <50–100 eV; for knock-ions above this energy focusing does not occur because of momentum transfer out of the row. The sequence length is limited by energy transfer to atoms outside the focusing direction and is still subject to considerable debate. The results of atomistic computer simulations (see Ro 74, Ho 75, and Sc 78) point to short replacement sequences, typically <30 replacements, although the precise results depend on the exact model used. The inclusion of thermal vibrations, for example, reduces the sequence length considerably (Te 78).

Experimental results from studies of the damage produced by low-energies ions have generally indicated that longer sequences occur. The most systematic studies have been those of Diehl and co-workers (Di 66, Di 67, Di 68, He 74, He 78, He 80). They employed (1–5)-keV argon ion irradiation of pure fcc and bcc metals and used transmission electron microscopy to observe the clusters formed by the highly mobile interstitials

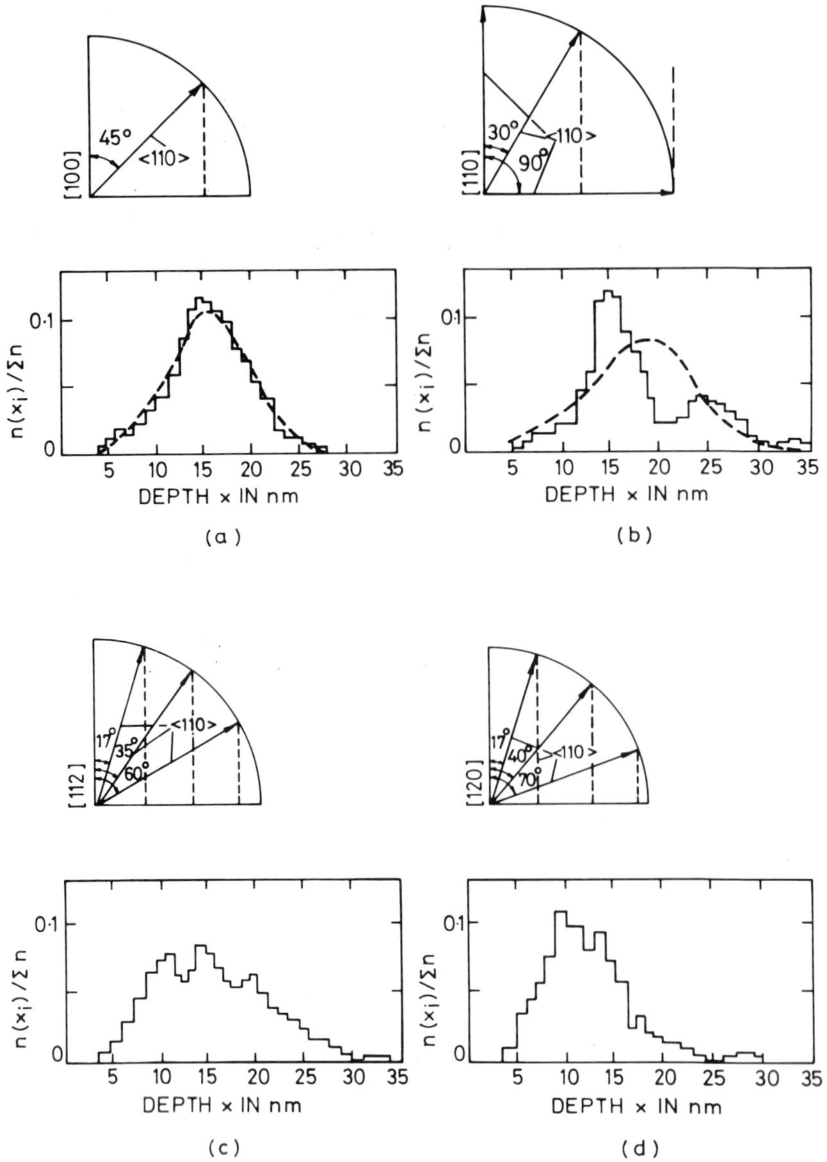

Figure 3. Depth distributions of interstitial agglomerates in Cu 1.3×10^{19} Ar^+/m^2, together with angles between the surface and all {110} directions, Surface orientations: (a) {100}, (b) {110}, (c) {112}, (d) {120}. (Courtesy of B. Hertel.)

deposited in the crystal. In the initial experiments on copper and gold irradiated at 130 and 300 K measurements of the depth of the interstitial clusters showed that the clusters were formed at depths considerably greater than the expected penetration of the Ar^+ ions. The form of the depth distribution was dependent on the crystallographic orientation of the foil surface, and was consistent with the injection of interstitials by the operation of replacement collision sequences along $\langle 110 \rangle$ directions, with cluster formation at the end of the sequence. Some of the implications and conclusions of the work were later questioned (Th 69b, Ve 70, Ec 74). Reservations were expressed on the role of the argon atoms in the clustering process, particularly as they themselves may penetrate into the crystal by channeling (Ec 74), the purity of the ion beam and the foils, and the possible role of contamination of the foil surface during ion bombardment. More recently Hertel and co-workers have repeated the initial study under more controlled conditions in an attempt to meet these criticisms. Typical results for copper are shown in Figure 3 for four orientations and the agreement is to be noted between the position of the measured peaks and the different projected $\langle 110 \rangle$ directions at each orientation. From the positions of the peaks the authors derived a value of 18 ± 1.5 nm for the average replacement collision sequence length in copper. The authors also observed bubbles adjacent to the surface which they ascribed to the clustering of the implanted argon ions. This and the presence of the double maximum in the depth distribution are powerful arguments in support of the original replacement collision sequence interpretation. It is interesting to note that at lower doses vacancy clusters were observed within $\sim$10 nm of the surface and that these were replaced by interstitial clusters at the higher doses. The authors also irradiated a bcc metal niobium and concluded that the results demonstrated that replacement collision sequences were propagated in bcc metals and that they possibly occurred along not only in the close-packed $\langle 111 \rangle$ direction but also in the $\langle 100 \rangle$ direction. Little is known about the propagation of replacement collision sequences in other structures.

3. Medium Energy Irradiation (5–200 keV)

In the Introduction it was described how heavy ions in this energy range dissipate their energy within $\sim$10–20 nm of the surface, creating a displacement cascade in coming to rest (shown schematically in Figure 1), and that this provides a unique opportunity for studying many aspects of cascade development. In Sections 3.2, 3.3, and 3.4 the contribution that the study of heavy-ion damage has made to the understanding of cascade processes, particularly in metals, is reviewed. Frequently, it will be necessary to make detailed reference to the conceptual and theoretical models of the cascade process, and for the sake of clarity we first review these concepts briefly in Section 3.1. In Section 3.5 the damage created by high doses of heavy ions is reviewed, and finally in Section 3.6 the use is discussed of heavy-ion damage to study solute redistribution under irradiation.

3.1. Models of Displacement Cascades

The entire collision process which brings the heavy ion to rest occurs in less than a lattice vibration period and within each cascade many simultaneously moving and interacting atoms are created. Many hundreds of vacancy and interstitial pairs may be created within a volume of average dimensions 10 nm. Individual cascades, even when created by incident ions of the same energy, may differ sharply between each other and the point defects created in the cascades may recombine and be annihilated or cluster to form point defect clusters or escape from the cascade region. Many attempts have been made to describe the stages of development of this complex process; in particular considerable attention has been given to the configuration and number of point defects created and the spatial distribution of the deposited energy, both within individual cascades and also averaged over a population of cascades.

The accepted picture of the primary state of damage, due to Seeger (Se 58), is of a vacancy rich region (or depleted zone) formed in the core of the cascade, surrounded by a shell of interstitial point defects ejected by mechanisms such as focused replacement sequences. The computer simulations of low-energy events by Vineyard and co-workers (Gi 60, Er 64, Er 65) confirmed the importance of replacement collision sequences in the development of the cascade. The "depleted zone" surrounded by a shell of interstitials was confirmed by Beeler and co-workers as a feature of high-energy cascades simulated in cubic materials using a computer code which followed the sequence of binary collisions suffered by the displaced atom (Be 63, Be 66). These computer simulations have revealed many other details of cascade development such as the dependance of E_d on crystallographic direction, the distance Frenkel pairs had to separated to survive spontaneous recombination, and the clustering of vacancies into small clusters. For a recent review see Agronavich and Kirsanov (Ag 76). Channeling of ions or knock-ons between rows or planes of atoms is important to the development of the cascade in crystalline materials because channeled ions create more diffuse cascades as they penetrate further into the crystal displacing fewer lattice atoms. Further, if in the creation of a cascade an atom is knocked onto a channeling direction with sufficient energy it may leave the region of primary damage and initiate a secondary branch of the cascade termed a subcascade. (For a discussion of channeling see the review by Datz and Moak, Chapter 5 in this volume.)

An important parameter which is widely used in all areas of radiation damage is the number of vacancies, ν, created in a cascade. The accepted formula is a modification of the expression originally derived by Kinchin and Pease (Ki 55):

$$\nu = \frac{kE_{\mathrm{DAM}}}{2E_d} \tag{1}$$

The damage energy E_{DAM} is that fraction of the bombarding ions energy that is available to produce displacements, and is calculated by subtracting the total inelastic energy loss summed over all collisions in the cascade [see Norgett *et al.* (No 72) for formulism]. The analytical theory employed in deriving the above predicts that for amorphous material the parameter k, the displacement efficiency, should be constant for a given scattering law and also be independant of energy. Torrens and Robinson (To 72a, To 72b) used computer simulation to verify that k was a constant independant of energy for crystalline solids as well as amorphous materials. They found that k was approximately equal to 0.8 for heavy metals such as copper, iron, and gold, and this is the value generally employed in equation (1).

The most complete description of the spatial distribution of the deposited energy created by heavy ions of this energy has come from the analytical treatment of the collision process due to Winterbon, Sigmund, and Sanders (Wi 70). From this theory many valuable parameters can be derived, such as the energy deposition profile and estimates of the cascade dimensions. It is important to note that the analytical theory gives a cumulative average over very many events. Under some conditions quantities derived from the theory may be quite atypical of the average individual cascade event. Some account may be taken of this by the introduction of correlation functions (Si 68, We 70).

A possibility that has received some attention in the literature is that a portion of the energy deposited by the ion may be transferred to the lattice in the form of lattice vibrations (Se 56). In this so-called "spike" a limited volume is created in which all the atoms are in motion, and in these regions the "temperature" may be raised to a high value for a short time until it is dissipated through atomic processes in the lattice. The importance of the spike phenomenon is that increased damage production, disordering, and annealing may occur over that predicted theoretically. For a discussion of spike lifetimes and the mean energy deposited per target atom see Sigmund (Si 74), and for the effect of spikes on recrystallization or amorphization of materials see (Na 70, Ca 78).

The above shows that a detailed conceptual picture, confirmed and developed by the analytical theories and computer simulations, has existed for some time for many aspects of cascades generated by heavy ions coming to rest in a solid. The one notable exception is that relatively little attention has centered on the formation of vacancy clusters from the vacancy-rich depleted zone at the cascade core (see Section 3.3). In the next section the results are discussed of direct observation of individual cascades created by heavy ions.

3.2. Direct Observation of Individual Displacement Cascades

Two largely complementary experimental techniques have been developed which allow the relatively direct observation of displacement cascades. The first, field-ion microscopy, allows the detailed reconstruction of the point-defect structure of a cascade. The second, transmission electron microscopy of ordered alloys, allows regions of local disordering, produced by replacement events in the cascade, to be imaged. Both techniques yield results which allow direct comparison with the analytical theories or cascade models derived by computer simulation. Each technique has its drawbacks. Field ion microscopy is applicable to only a limited number of materials, mostly refractory metals; defect configurations may be affected by the high mechanical stresses produced by the electric field. The transmission electron microscope technique is applicable only to order–disorder alloys, and good contrast is obtained only in very thin regions of the foil ($t < 30$ nm) where under some irradiation conditions the cascades may be truncated by the foil surface. Despite these limitations, the two techniques have been able to confirm directly and dramatically many of the central ideas of cascade theory.

3.2.1. Field-Ion Microscopy

The field-ion microscopy is the only currently available instrument with truly atomic resolution, which with the use of field evaporation of the specimen can be used to probe the specimen in three dimensions. Initial experiments on tungsten, platinum, and iridium employed a relatively coarse field evaporation technique and because of this concentrated on large vacancy clusters. The vacancy portion of the damage could usually be classified in one of three states: depleted zones, voids, and dislocation loops. The morphology observed was found to be dependent on irradiation temperatures, ion energy and mass, and the metal irradiated; see for example Buswell (Bu 70) and Hudson and Ralph (Hu 72). An important aspect of the work of Seidman and co-workers at Cornell University is the use of pulsed field evaporation to dissect specimens on an *atom-by-atom* basis. In this way a complete picture of both the interstitial and vacancy defect distributions within cascades can be built up. We summarize some of their observations below.

The vacancy structure of a depleted zone in tungsten created by a 30-keV Mo^+ ion is shown in Figure 4 (from We 81a). This compact cascade contains 169 vacancies, 124 of which are part of three large clusters. Within this depleted zone the average local vacancy supersaturation is close to 9 at. %. The stability of the zone with respect to collapse is a remarkable and unexpected feature, not confined only to tungsten. The primary state of the *interstitial* component has been investigated by experiments on tungsten at cryogenic temperatures (<15 K) when vacancies as well as interstitials are immobile. This has provided evidence for the existence of replacement collision sequences (Section 2). Measurements of the lower bound $R_{\min}$ for the propagation distance of self-interstitial atoms away from the depleted zone yield a mean value $\langle R_{\min} \rangle = 16$ nm, with standard deviation 12 nm, averaged over measurements made after 20 keV W^+, 30 keV Cr^+, and 18 keV Au^+ ion irradiations (We 81b). $R_{\min}$ appeared to vary little from ion to ion. Although this value of 16 nm may be an overestimate of the mean range of replacement collision sequences in tungsten, the result is direct experimental evidence for the efficient separation of vacancies and interstitials in the cascade confirming the model discussed in Section 3.1. It is generally believed that no other mechanism can account satisfactorily for such large propagation distances at this temperature.

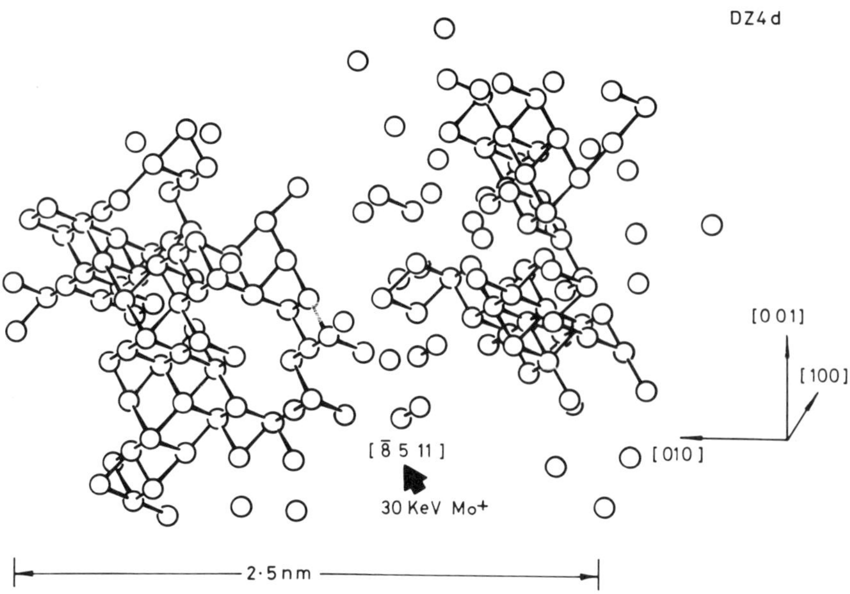

Figure 4. Visualization of depleted zone, DZ4d, which was created by a single 30-keV Mo^+ ion. The figure was constructed from experimental data with the aid of the OR TEP program; see Appendix to Wei *et al.* (We 81) for details. (Courtesy of D. N. Siedman.)

With the development of better automated systems for the evaluation of video films it has now become possible to investigate many more events—although even now the total is still limited—and hence to carry out more systematic investigations of the effects of the incident ion mass and energy. Some of the first such results for tungsten are as follows (We 81a):

(i) The number of vacancies ν contained within a depleted zone created by an ion of mass M_1 and energy E_1 was found to first order to be independent of M_1 and in good agreement with the modified Kinchin–Pease expression (Section 3.1). For both Kr^+ and W^+ ions the value of ν increased linearly as E_1 was increased from 15 to 70 keV.

(ii) The average diameter of depleted zones $\langle\lambda\rangle$ decreased with increasing M_1 at constant E_1 (30 keV) with a consequent increase in average vacancy concentration from ≈2 at. % for Ar^+ to ≈16 at. % for W^+ ions. This is in agreement with the trend predicted by theory. For both Kr^+ and W^+ ions with energies in the range 15–70 keV, however, the values of $\langle\lambda\rangle$ did not follow the trend of increasing $\langle\lambda\rangle$ with increasing E_1, predicted by linear cascade theory. This is probably a consequence of the theory treating cumulative averages, whilst experiments sample individual events.

(iii) The spatial arrangement of vacancies within depleted zones was a strong function of M_1 at constant E_1 (30 keV) with an increasing probability for subcascade formation with decreasing M_1. This was expected from computer simulations.

(iv) Evidence for nonlinear effects in dense cascades was found. The detailed results, briefly summarized in (i)–(iv) above, have been compared in detail with theory, and computer simulations. They have confirmed many expected features and revealed new features. The reader is referred to the original publications for details.

3.2.2. Transmission Electron Microscopy

Displacement cascade regions in order–disorder alloys can be studied using the technique developed by M. L. Jenkins and co-workers (Je 76a, Je 76b, Je 79). The method makes use of the fact that a displacement cascade in an ordered alloy induces local disordering: within a cascade region a "disordered zone" is created, embedded in the ordered matrix, which may be rendered visible in the electron microscope. [See (Je 76a) for details of the imaging technique.] An example of the disordered zones created in Cu_3Au by 100-keV Cu^+ ions is shown in Figure 5; the cascade regions are visible as dark dots in the thin areas of foil (for good contrast $t \lesssim 30$ nm). At this energy many cascades exhibit fine structure, with cascades developing extrusions or discrete subcascades. The method allows the numbers, sizes, and shapes of displacement cascades in ordered alloys to be studied in detail.

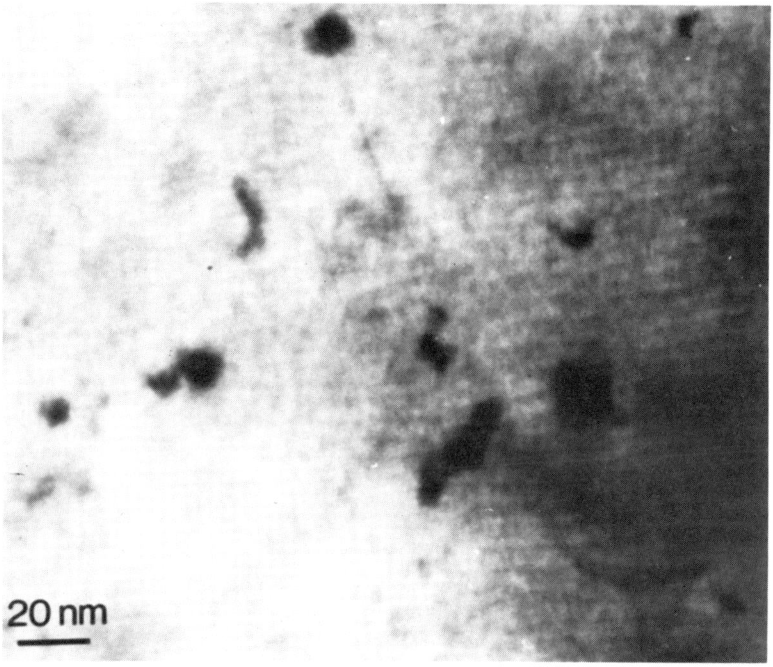

Figure 5. Micrographs of disordered zones created by 100-keV Cu^+ ions in ordered Cu_3Au, imaged in superlattice reflection $g = 110$.

So far most work has been done on the ordered alloy Cu_3Au. The chief results to emerge from studies of cascades in Cu_3Au created by Cu^+ ions of energy 5–200 keV were as follows:

(i) Each copper ion of energy $\geqslant$10 keV produces a visible disordered zone. Within some zones dislocation loops were observed (see Section 3.3).

(ii) The sizes of displacement cascades appear to be well described by the analytical theory (Section 3.1) for Cu^+ ion energies $\lesssim$50 keV. Above this energy quantities used to describe cascades sizes (e.g., the maximum cascade diameter) lay below the theoretical curve. This was attributed largely to the theoretical quantities relating to cumulative energy profiles, whilst the experiments sample single cascade events.

(iii) The development of extrusion and subcascade formation are observed at energies $\gtrsim$30 keV, and become more pronounced at higher energies. These are probably produced by quasichannelled knock-ons, although no clear crystallographic correlations could be established.

(iv) The large amount of disordering produced within cascades implies that many replacement events occur for every displacement, the ratio being at least ten and probably higher. The precise mechanism for this is not completely understood but may involve replacement collision sequences operating along ⟨110⟩ atom rows which contain both gold and copper atoms.

The method has now been developed into a powerful technique for investigating primary recoil spectra in different environments (En 81), when detailed comparison of the experimental data with theoretical damage models has generated valuable insight into damage mechanisms. Cascades have also been investigated in other alloy systems, with varying degrees of success: Zr_3Al, Nb_3Sn, Cu_3Pd, and Fe_3Al.

3.3. Point Defect Clustering in Individual Cascades

As outlined in Section 1, low-dose ($<10^{16}$ ions/m^2) heavy-ion damage gives a convenient method of studying some aspects of fast-neutron damage. In particular considerable attention has been given to point defect clustering in cascades in metals and their alloys. This is partly for the indirect evidence it gives on important cascade process such as vacancy production, and partly for information it yields on the clustering occurring in cascades produced by fast neutrons. The high generation rate of cascades by neutrons implies that clusters produced by cascade collapse will represent an important feature of the microstructure, and information on the clustering is important to theories of void swelling, irradiation creep, and growth. The use of the transmission electron microscope provides a very convenient analytical tool for examining the damage structures produced. The contrast arises from the strain field that the defect cluster sets up in the lattice and is now very well understood. From studies in a wide variety of pure metals a consistent picture of the damage structures at low doses has emerged. Under most conditions small vacancy loops are observed, produced heterogeneously at the sites of displacement cascades by the collapse of the vacancy-rich "depleted zones." It is thought that the interstitial component of the damage either forms submicroscopic clusters or is lost to the foil surface since interstitial loops are usually not observed at doses $<10^{16}$ ions/m^2 [but see Section 3.3.4 below—for detailed reviews see (Bl 73), (Wi 76a)].

Usually, the observed loop population is characterised in the following way. Two parameters, originally introduced by Merkle (Me 66) are used to describe the efficiency of vacancy clustering. The defect yield is defined as the fraction of displacement cascades which collapse to form dislocation loops. The cascade efficiency is a measure of the average fraction of the vacancies produced within those cascades which collapse that are retained in visible clusters (say, >1.5 nm). The latter is usually calculated as the ratio of the number of defect clusters of mean size to the theoretical number of vacancies generated in the displacement cascade. This theoretical number can be calculated from the standard formula of Norgett, Robinson, and Torrens (see Section 3.1). Established methods exist for identifying, sizing, and determining the vacancy or interstitial nature of a particular cluster. Methods for determining the precise defect geometry have been developed and both the Burgers vector and loop habit plane can usually be established. Frequently in this analysis use is made of the theoretically calculated or computer simulated images of known loops (see, for example, En 80).

In the remainder of this section we shall summarize observations made on point-defect clustering in cascades in metals. Most of the work has been carried out in fcc and bcc metals, and the results on pure materials are described in Section 3.3.1 and in alloys in Section 3.3.2.

3.3.1. Pure Metals

Heavy-ion damage at doses where cascade overlap does not occur has been studied in the pure fcc metals, copper, silver, gold, and platinum (e.g., Th 69a, Sc 70, Wl 71, Wi 72, Je 74, St 81a); in the pure bcc metals molybdenum, tungsten, and α-Fe (En 77, Ja 75, Je 78); and in one hexagonal metal cobalt (Fö 77). The observed loop population observed after self-ion irradiation of copper and molybdenum are illustrated in Figure 6. The black–white images arising from the small vacancy loops are clearly visible.

In each case the same general picture has emerged. The vacancies aggregate on the close-packed planes, and collapse leads to the formation of faulted loops. Subsequent faulting may occur, determined by loop size and stacking-fault energy. In fcc metals the Frank loops ($b = a/2\langle 111\rangle$) only unfault to form perfect loops in the high-stacking-fault energy metals such as nickel (Ro 81) while in the low-stacking-fault energy metals, e.g. silver, Frank loops dissociate on $\{111\}$ planes to form complicated geometries (Je 74). In bcc metals, because of the very high stacking fault energy, there is a strong tendency for loops to unfault to perfect loops ($b = a/2\langle 111\rangle$ or $b = a\langle 100\rangle$)—indeed only very few faulted loops have ever been identified. As perfect loops can glide through the lattice they may, if suitably oriented, be lost from the surface, so distorting apparent loop populations (Jä 75). This is particularly important in bcc metals.

Typical defect yields and cascade efficiencies measured after room temperature irradiations are presented in Table 1 for selected metals. The following trends can be seen from the results of self-ion irradiations:

(i) In fcc metals cascade collapse is very efficient with high values of defect yield and cascade efficiency.

(ii) In bcc metals the defect yield and cascade efficiency are on average lower than in fcc metals. α-Fe is exceptional; here, self-ion irradiation results in no visible damage.

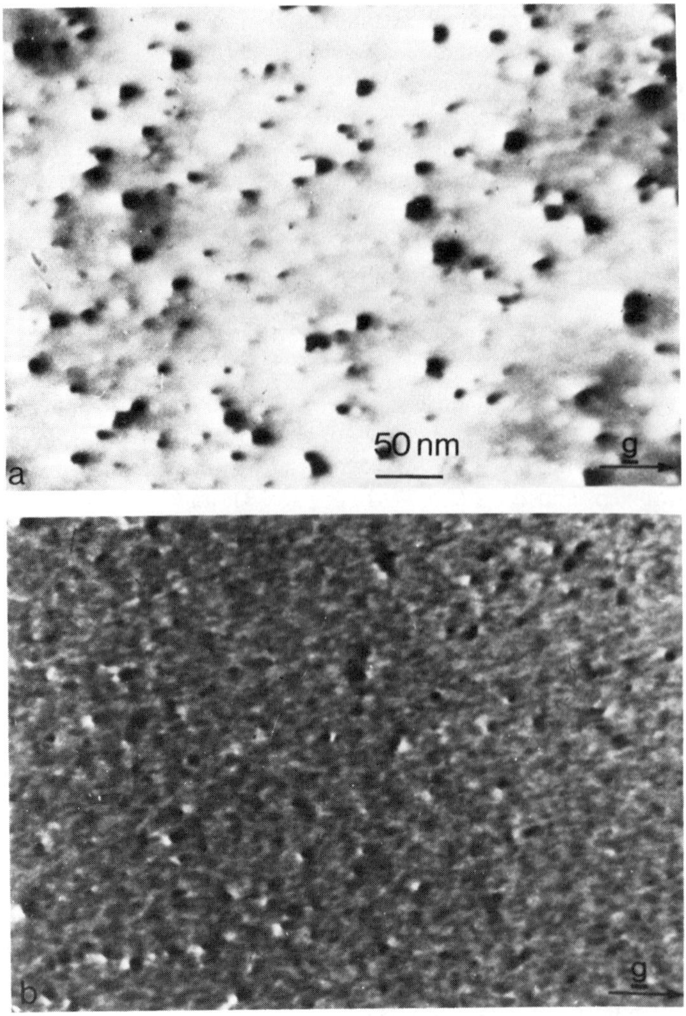

Figure 6. Micrographs showing vacancy cluster damage in (a) pure copper irradiated with 30-keV Cu^+ ions to 1.0×10^{16} ions/m^2 ($g = 200$); (b) pure molybdenum irradiated with 60-keV Mo^+ to 5×10^{16} ions/m^2 ($g = 110$).*

* Figures 6–9 are reproduced by permission of the United Kingdom Atomic Energy Authority and they remain UKAEA copyright.

(iii) In hcp cobalt the defect yield was much lower than in fcc metals (although this figure takes no account of possible loop loss to the surface). The cascade efficiency of 0.56 is comparable to fcc and bcc metals.

(iv) In both fcc and bcc metals the defect yield decreases as the ion energy increases. In fcc metals at high energies loops are often found to occur in bunches consistent with subcascade formation and nucleation of loops within individual subcascades.

(v) From Table 1 it is clear that increasing the ion mass at constant energy increases the defect yield. In nickel the yield is nearly doubled over the value obtained with self-ions when W^+ ions are used, while in α-Fe the defect yield after 80-keV ion irradiation increased from zero for self-ions to a true value close to unity for W^+ ions. Increasing the ion mass at constant energy causes the ion to deposit its energy in a smaller volume, i.e., both the deposited energy density and vacancy supersaturation are higher, and clearly these are important parameters in determining cascade collapse.

There is also evidence that the energy density may influence the loop geometry in bcc materials (α-Fe, Mo) with a higher probability for the formation of loops with energetically less favorable burgers vector ($\mathbf{b} = a\langle 100 \rangle$) in high energy-density cascades (Je 78, En 81).

Studies of cascade collapse at elevated temperatures have shown that the fcc materials studied (copper and nickel) (En 76, Ro 81) behave differently from the only bcc metal studied, molybdenum (En 77). In copper after 30-keV self-ion irradiations the defect yield decreased sharply for irradiation temperatures above 300°C. This was shown to be consistent with a model whereby cascade collapse occurs athermally, but where at elevated

Table 1. Vacancy Cluster Formation in Pure Metals

Metal	Orientation	Ion	Defect yield	Cascade efficiency	Ref.
Cu	[011]	30 keV Cu^+	0.5 ± 0.01	0.36 ± 0.02	(St 81)
(fcc)		30 keV W^+	0.6 ± 0.01	0.71 ± 0.02	
Ni	[011]	80 keV Ni	0.44 ± 0.05	0.25	(Ro 81)
(fcc)		80 keV W^+	0.81 ± 0.09	0.28	
Mo	[011]	60 keV Mo	0.12	0.36	(En 81b)
(bcc)		60 keV Xe^+	0.20	0.65	
α-Fe	[011]	80 keV Fe^+	Visible damage not observed		(Je 78)
(bcc)		80 keV W^+	0.18	—	
Co		60 keV Au	0.07	0.56	(Fi 77)
(hcp)					

temperatures loop shrinkage then occurs due to thermal emission of vacancies. The same mechanism accounts for the observations in nickel.

In molybdenum, however, the situation is quite different. Here the defect yield after 60-keV self-ion irradiation fell sharply at temperatures above 150–200°C, the temperature at which vacancies are able to undergo long-range migration in the lattice. This behavior is inconsistent with the above model as perfect loops in molybdenum in the observed size range will not shrink by thermal emission until 700–800°C. It would appear therefore that cascade collapse to form vacancy loops is strongly temperature dependent in molybdenum. This interpretation is given additional support by the observation of no significant decrease in vacancy loop numbers in molybdenum after 60-keV Xe^+ and W^+ ion irradiations for temperatures up to 425°C. These heavier ions are thought to create higher vacancy supersaturation leading to a greater driving force for collapse which offsets the loss of vacancies from the cascade region.

All these observations are in good accord with the accepted picture of cascade development outlined in Section 3.1 and with the direct studies of cascades described in Section 3.2. The high cascade efficiencies found for the loops formed from cascades which collapse points to an efficient separation of vacancies and interstitials in the cascade, consistent with the propagation of replacement collision sequences. In the more compact cascades the local vacancy supersaturation is increased, leading to a greater chemical driving force for collapse. The effect may be enhanced by reduced interstitial-vacancy recombination. In addition the local high-energy densities may lead to "spike" effects, which may facilitate collapse and the formation of energetically less favorable configurations. In comparing fcc and bcc metals the results in α-Fe at room temperature and molybdenum at elevated temperatures point to a difficulty in cascade collapse in bcc metals which is not found in fcc metals. This has important implications for the damage structures developed under fast neutron irradiation. It should be stressed in considering cascade collapse that the aggregation of vacancies in the cascade center can occur when the vacancies in the lattice are virtually immobile; for example cascade collapse occurs in molybdenum at room temperature whereas vacancies only undergo long-range migration at about 200°C. Indeed in the common fcc metals it has been established that cascade collapse does occur down to temperatures as low as 10 K (Ho 66). The reasons for this enhanced vacancy mobility are not properly understood.

3.3.2. Vacancy Clustering in Alloys and Technological Materials

Much of the interest in heavy-ion damage stems from its relevance in understanding neutron damage effects in reactors (see Section 1). Theoretical studies have established that collapse to vacancy loops has important implications in the understanding of void swelling in the fast reactor (Bu 75). For this reason there is considerable interest in cascade collapse in materials of possible technological importance. Since these materials are often alloys of considerable complexity attempts have also been made to isolate alloying effects in much simpler materials, e.g., copper binary alloys. These experiments will be described before results in austenitic stainless steels.

3.3.2a. Effects of Substitutional Alloying Elements. By far the most complete range of experiments has been carried out by Stathopoulos *et al.* (St 81b). A comparative study was made of the effect of alloying by carrying out irradiations with 30-keV Cu^+ and W^+ ions on a series of high-purity Cu–Al and Cu–Ge and a selection of Cu–Si, Cu–Ni, Cu–Zn, and Cu–Be alloys at room temperature. The main effect of the alloying was that, in all cases, the defect yields were higher than in pure copper, and as illustrated in Figure 7 increased linearly with solute content in the two alloy series. No simple explanation was possible to account for this behavior. However, a correlation was found with the modulus of the solute-atom size factor, suggesting that the dominant mechanism accounting for the increased yields was dechaneling of ions and planarly channeled knock-ons. Substitutional impurities had also been found to be important in bcc metals (En 77), although no systematic studies have been undertaken.

3.3.2b. Effects of Interstitial Impurities. The only systematic study made to date of the effect of interstitial elements was carried out by English *et al.* (En 77) on molybdenum doped with nitrogen. A very small concentration of nitrogen—17 appm—was found to have a dramatic effect on the damage observed after 60-keV self-ion irradiation. Compared with undoped material the defect yield fell by a factor of 5 and the cascade efficiency by a factor of 2. The nitrogen atoms are clearly having a major influence on the collapse process, although the mechanism responsible for this is not yet known.

Possible effects of interstitial impurities have been alluded to in other studies, although not studied systematically. Robinson and Jenkins (Ro 81), for example, considered that the drop in yield between heavy-ion irradiated nichrome alloys and pure nickel was probably due to the interstitial oxygen content rather than the changes in major alloying elements.

3.3.2c. Results in Stainless Steels. Vacancy cluster damage has been reported in two stainless steels, types 316 and 321, (En 75, Wi 76b, Ro 81) following Cr/W^+ and Ni^+/W^+ ion irradiations, respectively. Defect yields and cascade efficiencies were very much lower than in pure fcc metals. Robinson and Jenkins (Ro 81) reported that the defect yield decreased significantly in 321 steel compared to a ternary alloy with the same Fe–Ni–Cr composition, suggesting that minor alloying or impurity elements—particularly interstitial elements—were influencing the collapse process. Williams (private communication) found a similar insensitivity to

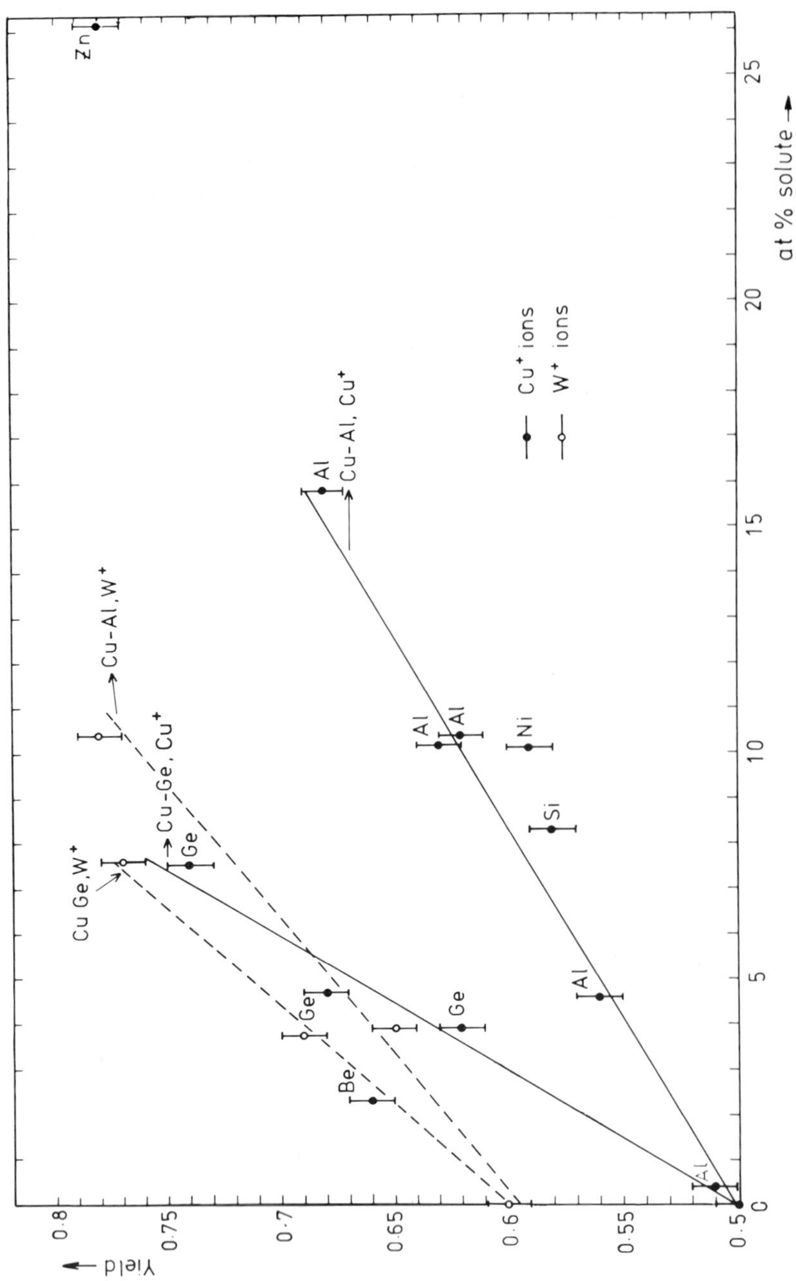

Figure 7. Defect cluster yields as a function of at. % alloying content for 30-keV Cu^+ and W^+ irradiations of Cu(5N) and copper alloys. (Courtesy of A. Stathopoulos).

base composition in a series of ternary Fe–Ni–Cr alloys, but a strong dependence on the amount of carbon in solution.

3.3.3. Interstitial Clustering in Cascades

In most studies of low-dose heavy-ion damage interstitial clustering has not been observed. However, Ruault *et al.* (Ru 80) have reported the observation of interstitial loops in gold believed to have been produced heterogenously at cascade sites, under conditions when the mean cascade energy density was low. Such is the case, for example, when the incident ion mass is low and its energy high. Ruault *et al.* have suggested that this is evidence for a "spike threshold," arguing that at maximum deposited energy densities $\gtrsim 2$ eV/atom a transition occurs to spike behavior, and lattice collapse leads to the formation of vacancy loops. At lower energy densities interstitial cluster formation may occur due to the interstitial mobility whilst vacancies are immobile. There is still controversy, however, as to the mechanism whereby visible interstitial loops can be produced in single cascade events, and the interpretation of these results is still under question. Interstitial loops are of course observed in high-dose experiments at $>10^{16}$ ions/m^2 where they are produced by homogeneous nucleation and growth.

3.4. Studies of Cascades in Nonmetals

Although ion implantation is a widely used method for doping elemental and III–V semiconductors, relatively little is known about the structure of displacement cascades in these semiconductors; work has been carried out on Si, Ge, GaP, and GaAs. Under some conditions—particularly when the deposited energy is high—cascade structures in these materials are thought to be essentially amorphous, but their detailed structure may be more complicated. For example in a transmission electron microscope study of radiation damage in silicon Mazey *et al.* (Ma 68) reported zones of approximately 5 nm in diameter after 80-keV Ne^+ bombardment. After bombardment with 10^{17} ions/m^2 the electron diffraction pattern indicated that the material had become amorphous (see Section 3.5). Dislocation loops at cascade sites have not been reported, although Matthews (Ma 71) used transmission electron microscopy to observe that the discrete amorphous zones present after implantation disappear after annealing to 600°C, but give rise to small dislocation loops after annealing to 700°C.

Rutherford backscattering has been used to study the lattice disorder and has provided valuable insight into the relative disordering of atomic and molecular ion implants. For example Moore *et al.* (Mo 75), who injected

40-keV As_2^+ and 20-keV As^+ into GaAs crystals at 20°C, found that the damage created per arsenic atom was about 50% greater in the diatomic ion case. Mitchell *et al.* (Mi 74) found a similar dependance with As^+, Sb^+, Te^+, Bi^+ diatomic and atomic ions implanted into silicon and germanium. They argued that the observed amount of damage and the effective displacement energy in the highest density cascades was evidence for a thermal-spike type of model; see also (Th 78, Ah 78).

Very little appears to be known about the structure of individual cascades in other nonmetals, such as oxides, although in MgO Crawford and Dean Dragsdorf (Cr 73) found damage-related photoluminescence which was probably connected with displacement cascades.

3.5. High-Dose Irradiations

At high doses $>10^{16}$ ions/m^2 cascade overlap occurs, and the observed damage structures are not generally simply related to single cascade events. High-dose experiments in metals are often used to simulate later stages of the development of reactor damage, to study void swelling for example, but usually high energies (>200 keV) are employed as well in order to achieve good penetration and hence to simulate bulk effects. Work of this kind is summarized in Section 4 below. In high-dose medium-energy (<200 keV) experiments interstitial clustering has been observed, for example, the experiments of Masters in iron (Ma 65) and Johnson *et al.* in aluminum (Jo 73).

Doses $>10^{17}$ ions/m^2 are also typical of ion implantation in elemental and III–V semiconductors, and usually result in the creation of buried amorphous layers which must be recrystallized by heat treatments. The structures have been studied extensively, but since they have been considered in chapter 4 of this volume, they will not be considered further here. Oxides have also received attention in the literature, and again, high dose implantations can cause crystalline in amorphous transitions, although the dose at which the lattice disorder saturates may be considerably greater than in the elemental and III–V semiconductors (Na 73). Some oxides amorphize more readily than others and indeed amorphous ZrO_2 can be rendered crystalline by ion bombardment (Ke 70). Kelly and Naguib (Na 75) considered the data from 72 nonmetals and concluded that a thermal-spike model and a bond-type criterion are successful in explaining the observed pattern of behavior (see also Na 70). More recently Naguib and Kelly (Na 72) demonstrated from measurements of electrical conductivity and electron diffraction that under 40-keV Kr^+ ion bombardment initially crystalline MoO_3 and V_2O_5 first amorphize but then subsequently crystalize as lower oxides MoO_2 and V_2O_3. The authors concluded that bombardment-induced oxygen loss resulted in the formation of lower oxides which were not stable in the amorphous state at room temperature. They also noted that this behavior presented a severe limitation on the use of such oxides in a radiation environment. In WO_3 Lam and Kelly (La 72) measured very precisely the depth of the amorphization produced by (4–35)-keV Kr^+ ion bombardment by making use of interference colors, and deduced from this that the mean damage depth was in good agreement with the analytical theory of Winterbon *et al.* (Wi 70). In contrast to the production of amorphous regions under heavy-ion bombardment, Mazey and Barnes (Ma 66) in an early study of graphite demonstrated that high dose irradiation with argon ions produced intensive twinning.

3.6. Radiation-Enhanced Diffusion and Solute Redistribution

During irradiation large point defect supersaturations are produced and strong point defect fluxes may be induced in the neighborhood of point defect sinks. In a metal alloy target this can have a considerable effect on the spatial distribution of the different elements in the alloy. An important application of the damage created by heavy ions is in the study of this radiation enhanced solute redistribution. The interest in this area stems from the need to understand and predict the behavior of complex alloys used in advanced nuclear reactors, as solute redistribution may have a profound effect on the properties of these materials.

The attraction of heavy-ion damage is the very large concentration gradients of point defects set up both near the free surface and adjacent to the peak in the damage distribution, so that by using sputter profiling combined with a surface analysis technique the irradiation-induced compositional changes can be measured at increasing distances from the surface. The initial experiments concentrated on dilute alloys where the concentration of any solute was initially uniform throughout the bulk. Marwick and co-workers employed 75-keV Ni^+ ions to create the damage and secondary-ion mass spectrometry for the compositional analysis, while Okamoto and co-workers employed 3-MeV Ni^+ ions and Auger electron spectroscopy (for convenience the results from these higher-energy implants will be included here). The metal systems studied were chosen for their relevance to technological materials and have included austenitic stainless steels, Fe–Ni–Cr ternaries, and dilute nickel alloys. A detailed picture has been built up of the irradiation-induced redistribution of many different solute atoms. Okamato and Wiedersich (Ok 74) demonstrated that the surface of a stainless steel was enriched in silicon and nickel and depleted in chromium. Marwick and Piller (Ma 77) and Piller and Marwick (Pi 78) showed that in dilute nickel alloys after irradiation (500°C) oversize solute atoms (Mn, Al, Ti, Cr) were enriched in the peak of the distribution, which implies that they migrate against the flow of point defects, whereas the

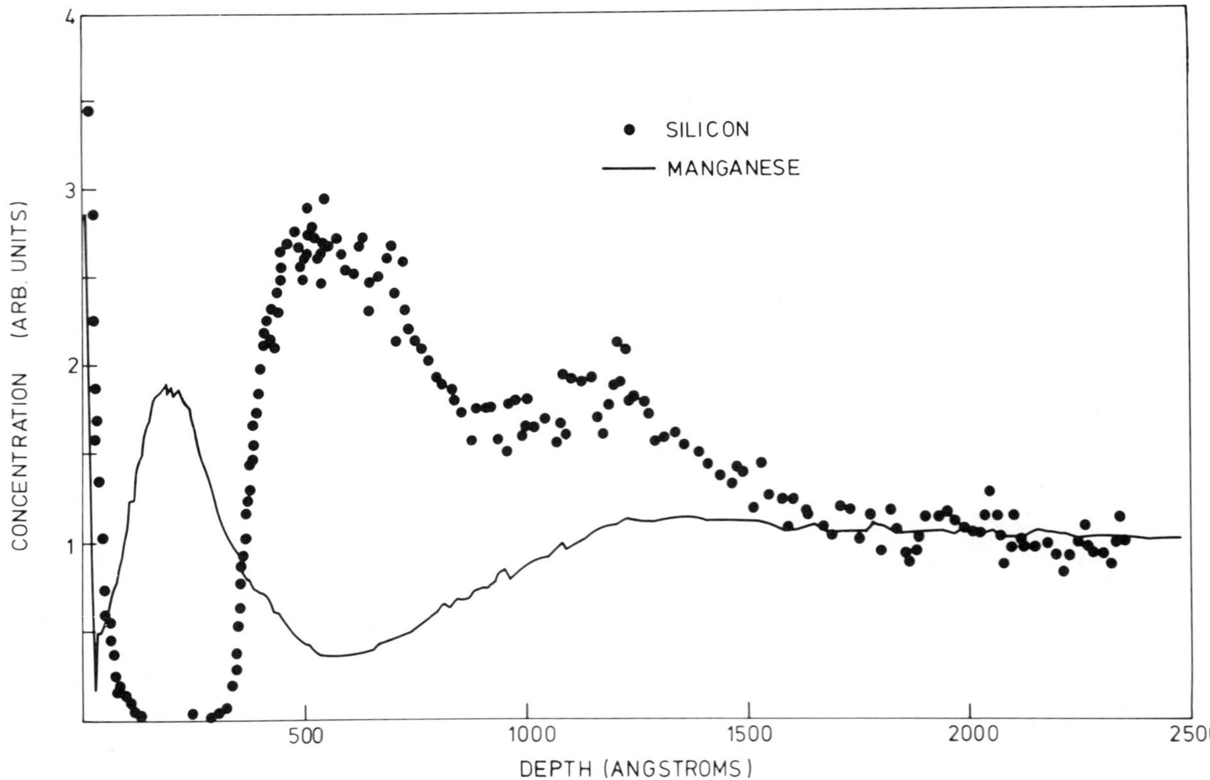

Figure 8. The depth distributions of manganese and silicon concentration as measured by secondary-ion mass spectrometry in a dilute nickel alloy after irradiation at 500°C. Manganese atoms have moved into, and silicon atoms out of, the defect-rich region centered on 20 nm depth. (Courtesy of A. Marwick.)

undersize silicon atoms were depleted in this region. Typical results for Si and Mn are shown in Figure 8. The temperature dependence of the segregation has been shown to be complex for elements such as silicon. In the studies by Piller and Marwick (Pi 78) irradiations at 15–25°C demonstrated that silicon atoms were depleted in the damage region (as at 500°C), and since vacancies are immobile at 15–25°C this implies the formation and migration of nickel interstitial–silicon complexes. At 500°C the migration of vacancy–silicon complexes cannot be ruled out. Rehn *et al.* (Re 78) conducted a detailed study of the temperature dependence of segregation to free surfaces during heavy-ion bombardment in a series of binary alloys and found that for silicon in Ni 1 at. % Si alloy surface segregation peaked at 560°C. Sethi and Okamoto (Se 80) found for silicon in Fe–20Cr–12Ni that a maximum in segregation occurred at 500°C followed by a further sharp increase in segregation at temperatures above 600°C. In contrast to this the segregation of Cr, Ni, and Mo showed a continuous increase in the temperature range studied.

Hobbs and Marwick (Ho 81) have studied the case of an initially nonuniform distribution of solute by initially implanting nickel with 30-keV Mn^+ ions and then following the Mn^+ implantation profile during subsequent 75-keV Ni^+ irradiation. The manganese is initially deposited closer to the surface than the damage distribution generated by the Ni^+ ions. During the Ni^+ irradiation the peak of the Mn distribution is driven to greater depths until a steady-state profile is established. The depth of the peak of this new distribution is independent of dose rate but surprisingly is further into the crystal than the calculated peak in the defect distribution created by the 75-keV Ni^+ ions. They also point out that in principle this type of experiment offers the real possibility of a quantitative comparison of the segregation rates of different solute species.

4. High-Energy Irradiation ($E > 200$ keV)

It is convenient to discuss the damage created by heavy ions in this energy range in two separate categories; first, the work which has concentrated on the primary damage created as the ion comes to rest; and second the large amount of effort into the behavior of materials, largely metals, which have suffered very high doses in which, on average, each individual atom of the lattice has been displaced many times. As stressed in the Introduction this latter area has been stimulated by the need to understand the behavior of in-core components in both fast breeder reactors and the proposed fusion reactors which will suffer very high displacement doses.

4.1. Study of the Primary Damage

High-energy ions lose energy through both electronic excitation and nuclear collisions. At the start of the range electronic range losses are expected to predominate but at the end of the range nuclear energy losses should be most important. Some attention has been given to experimental measurements of the damage produced in metals along the track of a heavy ion and comparison of this with theory. Noggle *et al.* (No 76) employed electrical resistivity to measure the damage along the path of 5-MeV Al^+ ions in evaporated aluminum. They found good agreement with the appropriate damage model (Li 63) although the peak damage rates were below the calculated value. Narayan *et al.* (Na 77), using an elegant specimen preparation technique, were able to study in the transmission electron microscope the number of point defects retained in dislocation loops along the path of Cu^+ ions in the energy range 5–38 MeV and 4- and 58-MeV Ni^+ ions at room temperature. Although the results were consistent with an energy-dependent electronic loss the proportionality constant deviated from the Lindhard value. The authors demonstrated that good agreement was obtained if the known deviations from the Lindhard theory were considered.

In metals electronic energy loss produces no net damage as the hole left when an electron is excited from below the Fermi surface to a continuum band is quickly filled by the decay of an electron into the empty hole. In nonmetals the situation is more complicated as electron–hole recombination may not readily take place. The lattice in the near vicinity may become unstable and atomic displacements may result, a process called radiolysis. For example in the alkali-halides a hole trapped by a halogen ion (e.g., Cl^-) results in a dumbbell molecule (e.g., Cl_2^-) which produces distortion in the $\langle 110 \rangle$ direction. In some ionic systems the relaxation along $\langle 110 \rangle$ when the hole recombines with the electron can result in an impulse sufficient to displace atoms (Po 66). Radiolysis is a very efficient mechanism in many systems (halides, hydrides, and organic solids); for example as little as 6 eV is required to produce a displacement in the halide sublattice of alkali halides (this should be compared to $E_d > 20$ eV for metals). Similarly, in the glass structure of fused silica electronic energy loss can create isolated ionized bonds which may relax to form permanent point defects (Ee 74). In other oxides such as MgO or Al_2O_3 atomic displacements are produced primarily by nuclear energy loss [see, for example, Evans *et al.* (Ev 72)], although the charge state of the defect may have important effects on the volume expansion and optical absorption of the material (Ar 74). If the energy loss through electronic excitation is very high then there is the possibility of ionization events on adjacent atoms along the ion track resulting in a cylinder of ions with positive charges (Fe 65). The strong mutual repulsion may create damage in a process called ion explosion damage. In many materials the trail of damage left by the heavy ion has a different etching rate than the undamaged material, and this has given rise to the widely used track detectors which have had widespread applications, particularly for detecting fission fragment tracks (see for example Ni 74).

The damage created in nonmetals by heavy ions of this energy range has not been as widely studied as in metals. It should be noted that as indicated above the production of defects in nonmetals is far more complex than in metals as displacements may occur on one or more sublattices; they may be charged and are often highly reactive [for a review of secondary defects in nonmetals see Ashbee and Hobbs (As 77)]. In contrast to metals it is frequently possible to study individual point defects as they may be optically active. Evans *et al.* (Ev 72) found that anion-sublattice-defect production (F centers) in MgO was remarkably inefficient, which led them to the conclusion noted above that displacements occur primarily through nuclear collisions. They found good agreement between the F-center production with depth and the predictions based on the LSS theory (Li 63). Diane and Gisclon (Di 78) have shown that the defects in silver chloride due to electronic losses and those resulting from nuclear energy losses have a very different optical behavior; the former are comparable to those produced by electromagnetic radiation while the latter were identified as silver colloids. Heavy-ion irradation of silica has received some attention partly through the possibility of using the radiation-induced density changes in silica for the production of integrated optic devices. In a study of ion-beam-induced luminescence Chandler, Jaque, and Townsend (Ch 79) found that their data could be understood if the luminescence centers are created in nuclear collisions but excited by the electronic energy losses. The possible use of silica based glasses for incorporating radioactive waste has stimulated interest in accelerated tests and postirradiation experiments on the structural and chemical stability of these materials. For example Antonini *et al.* (An 78) showed that a large amount of energy release occurred during postirradiation annealing of amorphous silica irradiated with 46.5-MeV Ni^+ ions and alpha particles to produce approximately 10% displaced atoms. The energy release occurred through the annealing of atomic defects, SiO_2 and strained or cleaved Si-O bands, and the authors also noted that effects due to nuclear and electronic energy loss seemed separable in the energy release spectrum.

4.2. Use of High-Energy Damage to Produce High Levels of Displacement Damage

The use of heavy ions to produce high levels of displacement damage in metals in relatively short times was greatly stimulated by the observation

(Ca 67) that voids were formed at elevated temperatures in the Dounreay fast reactor. High displacement rate experiments (10^{-4}–10^{-2} dpa/s) using heavy ions present a way of simulating in a few hours the damage that it would take months or years to accumulate in fast reactors where the displacement rates are typically 10^{-6}–10^{-8} displacements/atom/s. An example of the typical microstructure built up after a high dose irradiation is shown in Figures 9a and 9b for a stainless steel (FV548) irradiated in the Variable Energy Cyclotron at Harwell with 46.5-MeV Ni ions to 20 displacements per atom (dpa) and subsequently observed in the transmission electron microscope. In Figure 9a the irradiation induced voids are in contrast, where in Figure 9b, which shows the same field of view, the voids are not in contrast but the complex structure generated by interacting dislocation lines can be seen.

Such complex damage structures may arise in the following way. At elevated temperatures both interstitials and vacancies are mobile in the lattice. Some of the point defects created in cascades generated by the heavy ions will survive the clustering and recombination in the cascade and migrate out of the cascade region into the surrounding lattice. These point defects will either recombine with complementary defects, or be absorbed at sinks such as preexisting dislocation lines and grain boundaries or aggregate with like defects to form clusters, such as dislocation loops. Dislocations tend to attract interstitials rather more strongly than vacancies because of the greater elastic interaction with the larger dilational field of the interstitials. The bias

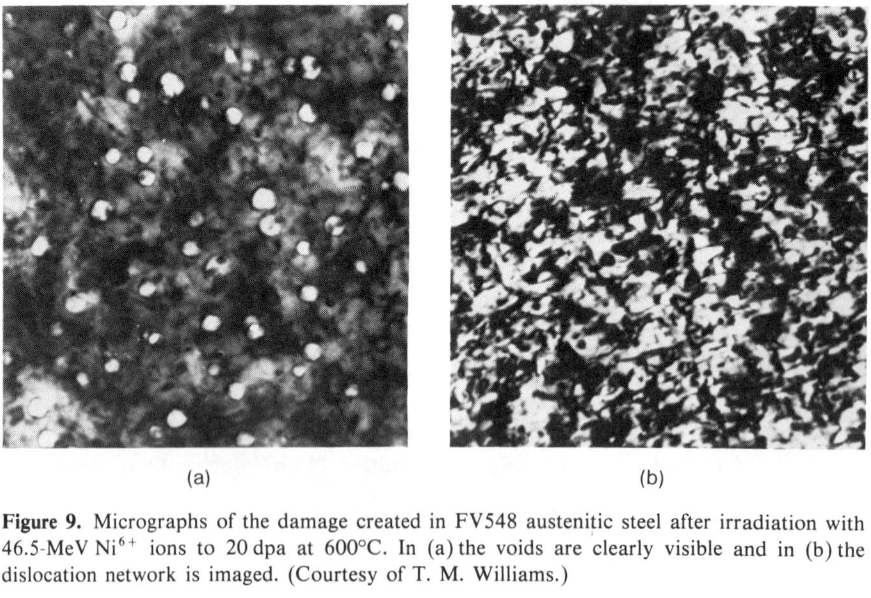

(a) (b)

Figure 9. Micrographs of the damage created in FV548 austenitic steel after irradiation with 46.5-MeV Ni^{6+} ions to 20 dpa at 600°C. In (a) the voids are clearly visible and in (b) the dislocation network is imaged. (Courtesy of T. M. Williams.)

causes rather more interstitials than vacancies to arrive at dislocations, so that line dislocations move (i.e., climb) through the lattice, while interstitial loops grow and vacancy loops shrink. An excess of vacancies will be built up in the lattice, and under these conditions any neutral sink in the system will grow. Voids are such neutral sinks, which can nucleate and grow when insoluble gas atoms such as helium are present to stabilize three-dimensional clusters of vacancies. Void growth is only favored in a relatively narrow temperature region. The lower limit is determined by the vacancy mobility. At about $0.3T_m$ (T_m = melting temperature) the vacancies are relatively immobile and void growth is suppressed, and recombination with the highly mobile interstitials dominates. A high-temperature cutoff occurs when the thermal emission of vacancies from voids exceeds the vacancy flow into voids. The precise temperature at which this occurs is determined by the vacancy diffusion coefficients and the dose rate, but is typically $0.5T_m$. In the reactor, helium produced by transmutation reactions induced by both thermal and fast neutrons is available for void stabilization. In heavy-ion experiments helium is often injected into the specimens either before or during the heavy-ion bombardment.

In alloy systems all of these basic processes occur, but the situation will be even more complicated. Point defect trapping by impurity atoms may modify diffusion and recombination. An initially uniform solute distribution may, as we have seen in Section 3.6, be affected by the strong point defect gradients set up near sinks, resulting in local enhancement or depletion of certain elements. This may in turn affect the properties of the sink, for example by altering dislocation climb. At the elevated temperatures of interest precipitation of second phases may occur, even in the absence of irradiation; this may be modified by irradiation with the precipitation of phases not expected from bulk equilibrium considerations. These second-phase particles may then represent an important sink for point defects or gas atoms, or may cause changes in the irradiation response by preferentially removing elements from the matrix. In general we expect a complex microstructure consisting of dislocations, second phase particles, and voids; the precise irradiation response will depend sensitively on the irradiation and material parameters. For example, for irradiations at 525C to 100 displacements per atom with 20-MeV C^+ ions, Nimonic PE16 gives 0.3% void swelling while pure nickel gives 10% swelling (Ne 71). It is to be noted that a sophisticated rate theory has been developed to model these processes (for reviews see Bu 77, Ma 78).

Heavy-ion simulation of void swelling has primarily been used as a rapid method of alloy screening, although it has also been used to study the basic mechanisms controlling void nucleation and growth in metals. The initial studies on void simulation were initiated in the UK by Nelson and co-workers (Ne 69, Ne 70, Ma 71) with work on pure nickel and AISI 316 stainless steel. The technique was rapidly taken up by other workers in the field, and this can be seen from the large number of papers on heavy-ion simulation of neutron damage in the proceedings of conferences on void swelling over the last decade (see for example Pu 71, Co 71, Ne 75, Bl 77, Po 80). It has been applied to many different metals, from pure fcc and bcc metals (see for example Gl 73, De 73, Sp 74), binary alloys (see for example Hu 75), to a wide range of alloys of technological interest such as austenitic stainless steels. Ion energies from 500 keV to 46 MeV have been employed, the advantage of the higher energies being that the damage is produced further away from the surface sink. The successes of this approach include the identification of possible low swelling alloys, such as a nickel-based alloy PE16 (Ne 69); and establishing the role of microstructural features in suppressing void swelling. Two examples of the latter are the importance in austenitic steels of the level of cold work (Ne 71b), and the presence of a fine distribution of NbC built up during irradiation (Wi 77, Wi 79a). A further important role played by simulation is in studying the effect of compositional variations on alloys' swelling characteristics, and establishing a "ranking order" of low-swelling alloys, although it has been observed that this "ranking order" is dependent on simulation technique (Ga 77). Some workers have attempted to correlate microstructural features after low-dose ion irradiation with the void swelling behavior at high doses, thus bridging the gap between basic and technological extremes of the dose range (Wi 79b, Wi 80). However, a recurrent theme underlying the more recent work is the problem of establishing a precise correlation with the void swelling induced by irradiation with fast neutrons. Not only must the effect of the great difference in dose rate be accounted for, but the effects of differences in recoil spectrum, and of ions injected into the metal in appreciable concentrations, and the influence of the free surface on the development of the damage must also be understood. Even with the aid of sophisticated rate theory models of void swelling such correlations are difficult to establish.

The use of heavy-ion simulation in studying the basic mechanisms of void swelling in pure metals is well illustrated by the work of Glowinski and co-workers on copper (Gl 73, Gl 76a, Gl 76b, Gl 76c). Here by irradiating pure copper with 500-keV Cu^+ ions they systematically studied the effect of dose rate, residual gas content, and implanted gas content (helium, hydrogen, carbon, and oxygen) on the development of the microstructure. They further demonstrated that the experimental curves for the variation of swelling were in good agreement with a computer simulation model.

Studies of the effects of alloy composition on void swelling behavior are exemplified by the work of Johnson and co-workers (Jo 78, Ba 77). A step height technique was employed to monitor the swelling induced by 5-MeV Ni^+ ion irradiation to doses >100 dpa of a wide range of Fe–Ni–Cr ternary alloys. The swelling was found to depend sensitively on Cr and Ni content. For example in the Fe–15% Cr series irradiated to 140 dpa at 675°C the swelling decreased with increasing Ni content to give very low swelling for Ni contents in the region 45%–60% Ni before rising slightly at higher Ni contents. The same technique was also used to study minor element effects. In a Fe–15Cr–20Ni alloy it was found that Si, Ti, Nb, and Zr reduced swelling while Mn, P, B, Al, C, and Mo increased the swelling slightly or had little beneficial effect. Most interestingly it was found that while Mo on its own slightly increased swelling, 1.5 wt. % Mo together with 0.7 wt. % Ti decreased the swelling by a factor of 3. Similarly the combination of Si and Ti additions together had a greater beneficial effect than the total reduction found by adding them individually. Minor elements have also been found to be effective in suppressing the void swelling in alloys with a different base composition (Le 77, Ma 80). The explanation for these compositional effects is still unclear, although several possibilities have been discussed in the literature. These include the effect of composition on diffusion, the position of a alloy in the complex Fe–Ni–Cr phase diagram and its propensity to form second phases.

The advent of microchemical analytical facilities on transmission electron microscopes has allowed investigation of the solute segregation and precipitation which occurs under irradiation in both neutron and heavy-ion irradiations. For example, Marwick *et al.* (Ma 78b) showed that under 46.5 MeV Ni^{6+} irradiation to 60 dpa at 590C and 650C of a PE16 matrix alloy that nickel was enriched at void surfaces with a resultant depletion of Fe and Cr. Enrichments of more than 50% were observed in a shell <60 nm thick about the voids. Similar effects had been observed in electron and neutron irradiated stainless steels (Ok 74, Ga 77). Radiation-induced precipitation effects have been studied in heavy-ion simulations, but the narrow damage layer presents experimental problems in precipitate identification, while the precise effect of the high dose rate on the precipitation is difficult to gauge.

The technique of heavy-ion simulation is also widely used for investigations into material response to the likely irradiation environment in proposed fusion reactors, where very high gas generation will occur, particularly helium and hydrogen. This environment can be simulated by helium (sometimes together with hydrogen) coimplantation or preinjection. Some differences in the damage evolution have been found, depending on whether the gas is introduced before or during the heavy-ion irradiation; dual beam irradiations with simultaneous injection of gas ions are likely to become increasingly important (for a review see Sp 80). A far wider range of materials, both metals and nonmetals, are candidate materials for the various components which will suffer irradiation damage. This has meant a corresponding increase in the range of materials studied in heavy-ion irradiation experiments. For example, titanium alloys have been extensively studied (Ag 80), as have refractory metal alloys; there is also an increasing interest in the damage created in ceramics (Re 77, Re 79).

References

(Ag 76) V. Agramovich and V. V. Kirsanor, *Sov. Phys. Usp.* **19**:1 (1976).

(Ag 80) S. C. Agarwal, G. Ayrault, D. I. Potter, A. Taylor, and F. V. Nolfi, *J. Nucl. Mater.* **85**, **86**:653 (1979).

(Ah 78) N. A. G. Ahmed, C. E. Christodoulides, and G. Carter, *Radiat. Eff.* **38**:221 (1975).

(An 78) M. Antoni, A. Manara, and P. Leusi, Proc. Intl. Topical Conference of Physics of SiO_2 and its Interfaces, Yorktown Heights, New York (1978).

(Ar 74) G. W. Arnold, G. B. Krefft, and C. B. Norris, *Phys. Lett.* **25**:540 (1974).

(As 76) K. H. G. Ashbee and L. W. Hobbs, Proceedings Conference on Point Defect Behaviour and Diffusional Processes, Bristol, U.K. (1976), p. 79.

(Ba 77) J. F. Bates and W. G. Johnston, Proceedings of the Conference on Radiation Effects in Breeder Reactor Structural Materials, Scotsdale, Arizona (1977), p. 625.

(Be 60) R. M. Berman, Bettis (U.S.A.) Report WAPD-BT-21, p. 33 (1960).

(Be 63) J. R. Beeler and D. G. Besco, *J. Appl. Phys.* **34**:287 (1963).

(Be 66) J. R. Beeler, *J. Appl. Phys.* **37**:3000 (1966).

(Bl 73) B. L. Eyre, *J. Phys. F* **3**:422 (1973).

(Bl 77) M. L. Bleiburg and J. W. Bennett, eds., Proceedings of the Conference on Radiation Effects in Breeder Reactor Structural Materials, Scotsdale, Arizona (1977).

(Bu 70) J. T. Buswell, *Phil. Mag.* **22**:787 (1970).

(Bu 75) R. Bullough, B. L. Eyre, and K. Krishan, *Proc. R. Soc. London* **A346**:81 (1975).

(Bu 77) R. Bullough and M. R. Hayns, Proceedings 2nd International Symposium on Continuum Models of Discrete Systems. Mt. Gabriel, Canada (1977).

(Ca 67) C. Cawthorne and E. J. Fulton, *Nature* **216**:575 (1967).

(Ca 78) G. Carter, D. G. Armour, S. E. Donnelly, and R. Webb, *Radiat. Eff.* **36**:1 (1978).

(Ch 79) P. J. Chandler, F. Jaque and P. D. Townsend, *Radiat. Eff.* **43**:45 (1979).

(Co 71) J. W. Corbett and L. C. Ianniello, eds., Proceedings Conference on Radiation-Induced Voids in Metals, Albany, New York (1971).

(Cr 73) J. R. Crawford and R. Dean Dragsdorf, *J. Appl. Phys.* **44**:385 (1973).

(De 73) J. Delaplace, N. Azam, L. Le Naour, M. Lott, and C. Fiche, *J. Nucl. Mater.* **47**:278 (1973).

(Di 66) H. Diepers and J. Diehl, *Phys. Stat. Sol.* **16**:K109 (1966).

(Di 67) J. Diehl, H. Diepers, and B. Hertel, *Can. J. Phys.* **46**:647 (1968).

(Di 68) H. Diepers, *Phys. Stat. Sol.* **24**:235 (1967).

(Di 78) C. H. Diane and J. L. Gisdon, Proc. IBBM Conference (1978).

(Ec 74) K. H. Ecker, *Radiat. Eff.* **23**:171 (1974).

(Ee 74) E. P. Eer Nisse, *J. Appl. Phys.* **45**:167 (1974).

(En 75) C. A. English, B. L. Eyre, K. Shoaib, and T. M. Williams, *J. Nucl. Mater.* **58**:220 (1975).

(En 76) C. A. English, B. L. Eyre, and J. Summers, *Phil. Mag.* **34**:603 (1976).

(En 77) C. A. English, B. L. Eyre, A. F. Bartlett, and H. N. G. Wadley, *Phil. Mag.* **35**:533 (1977).

(En 80) C. A. English, B. L. Eyre, and S. M. Holmes, *J. Phys. F* **10**:1065 (1980).

(En 81) C. A. English and M. L. Jenkins, *J. Nucl. Mater.* **96**:341 (1981).

(Er 64) C. Erginsoy, G. H. Vineyard, and A. Englert, *Phys. Rev.* **133**:A595 (1964).

(Er 65) C. Erginsoy, G. H. Vineyard, and A. Shinzu, *Phys. Rev.* **139**:A118 (1965).

(Ev 72) B. D. Evans, J. Comas, and P. R. Malinberg, *Phys. Rev. B* **6**:2453 (1972).

(Fe 65) R. L. Fleischer, P. B. Price, and R. M. Walker, *J. Appl. Phys.* **36**:3645 (1965).

(Fö 77) H. Föll and M. Wilkins, *Phys. Stat. Sol. (a)* **39**:561 (1977).

(Ga 77) F. A. Garner, R. W. Powell, S. Diamond, T. Lauritzen, A. F. Rowcliffe, J. A. Sprague, and D. Keefer, Proceedings International Conference on Radiation Effects in Breeder Reactor Structural Materials, Scotsdale, Arizona (1977), p. 543.

(Gi 60) J. B. Gibson, A. N. Goland, M. Milgram, and G. H. Vineyard, *Phys. Rev.* **120**:1229 (1960).

(Gl 73) L. D. Glowinski, C. Fiche, and M. Lott, *J. Nucl. Mater.* **47**:295 (1973).

(Gl 76a) L. D. Glowinski and C. Fiche, *J. Nucl. Mater.* **61**:22 (1976).

(Gl 76b) L. D. Glowinski and C. Fiche, *J. Nucl. Mater.* **61**:29 (1976).

(Gl 76c) L. D. Glowinski, J. M. Lanore, C. Fiche, and Y. Adda, *J. Nucl. Mater.* **61**:41 (1976).

(He 74) B. Hertel, J. Diehl, R. Gotthardt, and H. Sultze, in *Applications of Ion Beams to Metals*, Plenum Press, New York (1974), p. 507.

(He 80) B. Hertel, *Phil. Mag.* **40**:331 (1979).

(Ho 66) L. M. Howe, J. F. McGurn, and R. W. Gilbert, *Acta. Met.* **14**:801 (1966).

(Ho 75) D. K. Holmes and M. T. Robinson, *Atomic Collisions in Solids*, Plenum Press, New York (1975), p. 225.

(Ho 81) J. E. Hobbs and A. D. Marwick, *Radiat. Eff. Lett.* **58**:83 (1981).

(Hu 72) J. A. Hudson and B. Ralph, *Phil. Mag.* **25**:265 (1972).

(Hu 75) J. A. Hudson and S. J. Ashby, The Physics of Irradiation produced Voids Consultants Symposium, Harwell, England (1974), p. 140.

(Jä 75) W. Jäger and M. Wilkens, *Phys. Stat. Sol. (a)* **32**:89 (1975).

(Je 74) M. L. Jenkins, *Phil. Mag.* **29**:813 (1974).

(Je 76a) M. L. Jenkins and M. Wilkins, *Phil. Mag.* **34**:1155 (1976).

(Je 76b) M. L. Jenkins, K. H. Katerbau, and M. Wilkins, *Phil. Mag.* **34**:1141 (1976).

(Je 78) M. L. Jenkins, C. A. English, and B. L. Eyre, *Phil. Mag. A* **38**:97 (1978).

(Je 79) M. L. Jenkins, N. G. Norton, and C. A. English, *Phil. Mag. A* **40**:131 (1979).

(Jo 73) E. Johnson and J. A. Ytterhus, *Phil. Mag.* **28**:489 (1973).

(Jo 74) W. G. Johnston, T. Lauritzen, J. H. Rosolowski, and A. M. Turkalo, *J. Nucl. Mater.* **54**:24 (1974).

(Ke 70) R. Kelly and H. M. Naguib, *Proceedings International Conference on Atomic Collision Phenomena in Solids*, North-Holland, Amsterdam (1970), p. 172.

(Ki 55) G. H. Kinchin and R. S. Pease, *Rep. Prog. Phys.* **18**:1 (1955).

(La 72) N. Q. Lam and R. Kelly, *Can. J. Phys.* **50**:1887 (1972).

(Le 77) V. Levy, N. Azam, L. Nenaour, G. Didout, and J. Delaplace, Proceedings International Conference on Radiation Effects in Breeder Structural Materials, Scotsdale, Arizona (1977), p. 709.

(Li 63) J. Lindhard, M. Scharff, and H. E. Schiott, *K. Dan. Vid. Selsk. Mat. Fys. Medd* **33**:1 (1963).

(Lu 69) K. L. Van de Lugt, J. Comas, and E. A. Wolicki, *J. Phys. Chem. Sol.* **30**:2486 (1969).

(Ma 65) B. C. Masters, *Phil. Mag.* **11**:881 (1965).

(Ma 66) D. J. Mazey and R. S. Barnes, Proceedings Sixth International Congress for Electron Microscopy Kyoto, Japan (1966), p. 365.

(Ma 68) D. J. Mazey, R. S. Nelson, and R. S. Barnes, *Phil. Mag.* **17**:1145 (1968).

(Ma 71) D. J. Mazey, J. A. Hudson, and R. S. Nelson, *J. Nucl. Mater.* **41**:257 (1971).

(Ma 77a) M. D. Matthews, Harwell Report A.E.R.E., R6802 (1971).

(Ma 77b) A. D. Marwick and R. B. Piller, *Radiat. Eff.* **33**:245 (1977).

(Ma 78) L. A. Mansur, *Nucl. Technol.* **40**:5 (1978).

(Ma 78b) A. D. Marwick, W. A. D. Kennedy, D. J. Mazey, and J. A. Hudson, *Scr. Met.* **12**:1015 (1978).

(Ma 80) D. J. Mazey, D. R. Harries, and J. A. Hudson, *J. Nucl. Mater.* **89**:155 (1980).

(Me 66) K. L. Merkle, *Phys. Stat. Sol.* **18**:173 (1966).

(Mi 74) J. B. Mitchell, J. A. Davies, L. M. Howe, R. S. Walker, K. B. Winterbon, G. Foti, and J. A. Moore, Proceedings of the Fourth International Conference on Ion Implantation, Osaka, Japan (1974).

(Mo 75) J. A. Moore, G. Carter, and A. W. Tinsley, *Radiat. Eff.* **25**:49 (1975).

(Na 72) H. M. Naguib and R. Kelly, *J. Phys. Chem. Solids* **33**:1751 (1972).

(Na 73) H. M. Naguib, J. F. Singleton, W. A. Grant, and G. Carter, *J. Mater. Sci.* **78**:1633 (1973).

(Na 75a) H. M. Naguib and R. Kelly, *Radiat. Eff.* **25**:1 (1975).

(Na 75b) H. M. Naguib and R. Kelly, *Radiat. Eff.* **25**:79 (1975).

(Na 77) J. Narayan, O. S. Oen, and T. S. Noggle, *J. Nucl. Mater.* **71**:160 (1977).

(Ne 69) R. S. Nelson and D. J. Mazey, UKAEA Report, AERE-R 6046 (1969).

(Ne 70) R. S. Nelson, D. J. Mazey, and J. A. Hudson, *J. Nucl. Mater.* **41**:257 (1971).

(Ne 71a) R. S. Nelson, J. A. Hudson, D. J. Mazey, G. P. Walters, and T. M. Williams, Proceedings International Conference on Radiation Induced Voids in Metals. State Univ., Albany (1971).

(Ne 71b) R. S. Nelson, D. J. Mazey, and J. A. Hudson, Proceedings BNES Conference, Voids formed by Irradiation of Reactor Materials Reading, England (1971), p. 191.

(Ne 75) R. S. Nelson, The Physics of Irradiation Produced Voids, Consult Symp. Harwell, England (1974).

(Ni 74) V. A. Nikolaer and V. P. Perelygin, *Prib. Tekh. Eksp.* **2**:7 (1976).

(No 72) M. J. Norgett, M. T. Robinson, and I. M. Torrens, Harwell Report TP 494 (1972).

(No 76) T. S. Noggle, B. R. Appleton, J. M. Williams, O. S. Oen, J. A. Biggerstaff, and T. Inoata, *Scientific and Industrial Applications of Small Accelerators*, Inst. Electr. Electron Eng., New York (1976), p. 255.

(Ok 74) P. R. Okamoto and H. Wiedersich, *J. Nucl. Mater.* **53**:336 (1974).

(Pi 78) R. C. Piller and A. D. Marwick, *J. Nucl. Mater.* **71**:309 (1978).

(Po 66) D. Pooley, *Proc. Phys. Soc. (London)* **87**:257 (1966).

(Pu 71) S. F. Pugh, M. L. Loretto, and D. I. R. Norris, Proceedings Brit. Nucl. Energy Soc. Conference on Voids formed by the Irradiation of Reactor Materials, Reading, England (1971).

(Re 77) M. D. Rechtin and H. Wiedersich, *Radiat. Eff.* **31**:181 (1977).

(Re 78) L. E. Rehn, P. R. Okamoto, D. I. Potter, and H. Wiedersich, *J. Nucl. Mater.* **74**:242 (1978).

(Re 79) M. D. Rechtin, *Radiat. Eff.* **42**:129 (1979).

(Ro 74) M. T. Robinson and I. Torrens, *Phys. Rev.* **39**:5008 (1974).

(Ro 81) T. M. Robinson and M. L. Jenkins, *Phil. Mag.* **43**:999 (1981).

(Ru 79) M. O. Ruault, H. Bernas, and J. Chaumont, *Phil. Mag.* **39A**:757 (1979).

(Sc 57) J. O. Schiffgens, D. W. Schwarz, R. G. Ariyasu, and S. E. Cascadden, *Radiat. Eff.* **39**:221 (1978).
(Sc 70) T. Schober, *Phys. Stat. Sol. (a)* **1**:307 (1970).
(Se 56) F. Sietz and J. A. Koehler, *Solid State Phys.* **2**:307 (1956).
(Se 80) V. K. Sethi and P. R. Okamoto, Proceedings Symposium on Irradiation Phase Stability, Pittsburg, Pennsylvania (1980).
(Si 57) R. H. Silsbee, *J. Appl. Phys.* **28**:1246 (1957).
(Si 68) P. Sigmund, G. P. Scheidler, and G. Roth, BNL 50083 (C-52):374 (1978).
(Si 74) P. Sigmund, *Appl. Phys. Lett.* **25**:169 (1974).
(Sp 74) J. A. Sprague, J. E. Westmoreland, F. A. Smidt Jr., and P. R. Malinberg, *J. Nucl. Mater.* **54**:286 (1974).
(Sp 79) J. A. Spitznagel, F. W. Wiffen, and F. V. Nolfi, *J. Nucl. Mater.* **85**, **86**:629 (1979).
(St 81a) A. Y. Stathopoulos, Harwell Report AERE-R 9784, *Phil. Mag. A* **44**:285 (1981).
(St 81b) A. Y. Stathopoulos, C. A. English, B. L. Eyre, and P. B. Hirsch, Harwell Report, AERE-R 9785, *Phil. Mag. A* **44**:309 (1981).
(Te 78) A. Tenenbaum, *Phil. Mag.* **37**:731 (1978).
(Th 69a) M. W. Thompson, *Defects and Radiation Damage in Metals*, Cambridge University Press, Cambridge (1969).
(Th 69b) L. E. Thomas, T. Schober, and R. W. Balluffi, *Radiat. Eff.* **1**:257 (1969).
(Th 78) D. A. Thompson and R. S. Walker, *Radiat. Eff.* **30**:91 (1978).
(To 72a) I. M. Torrens and M. T. Robinson, in Proceedings of International Conference on Radiation-Induced Voids in Metals, Albany, New York (1972), p. 734.
(To 72b) I. M. Torrens and M. T. Robinson, in *Proceedings of Conference on Interatomic Potentials and Simulation of Lattice Defects*, Plenum Press, New York (1972), p. 423.
(Ve 70) J. A. Venables, in *Atomic Collision Phenomena in Solids*, North-Holland, Amsterdam (1970), p. 132.
(Wa 78) R. S. Walker and D. A. Thompson, *Radiat. Eff.* **37**:113 (1978).
(We 70) J. E. Westmoreland and P. Sigmund, *Radiat. Eff.* **6**:189 (1970).
(We 78) C-Y-Wei and D. N. Seidman, *Phil. Mag. A* **37**:257 (1978).
(We 81a) C-Y-Wei, M. I. Current, and D. N. Seidman, Cornell University Materials Science Center Report No. 4235, *Phil. Mag. A* **44**:459 (1981).
(We 81b) C-Y-Wei and D. N. Seidman, Cornell University Materials Science Center, Report No 4088, *Phil. Mag. A* **43**: 1419 (1981).
(Wi 70) K. B. Winterbon, P. Sigmund, and J. B. Sanders, *K. Dan. Vid. Selsk. Mat. Fys. Medd* **37**:No. 14 (1970).
(Wi 71) M. M. Wilson, *Phil. Mag.* **24**:1023 (1971).
(Wi 72) M. M. Wilson and P. B. Hirsch, *Phil. Mag.* **25**:983 (1972).
(Wi 76a) M. Wilkens, Proceedings of International Conference on Fundamental Aspects of Radiation Damage in Metals, USERDA Conf. 751006-Pi, (M. T. Robinson and F. Young, eds.), (1976), p. 98.
(Wi 76b) T. M. Williams and B. L. Eyre, Proceedings 6th European Congress on Electron Microscopy, Jerusalem, Israel (1976), p. 550.
(Wi 77) T. M. Williams, D. R. Arkell, and B. L. Eyre, *J. Nucl. Mater.* **68**:69 (1977).
(Wi 79a) T. M. Williams and D. R. Arkell, *J. Nucl. Mater.* **80**:78 (1979).
(Wi 79b) T. M. Williams, *J. Nucl. Mater.* **78**:28 (1979).
(Wi 80) T. M. Williams, *J. Nucl. Mater.* **88**:217 (1980).

9

WILLIAM A. LANFORD is Professor of Physics at the State University of New York in Albany. He was educated at the University of Rochester, receiving the Ph.D. in 1971. Following two years as an Assistant Professor at the Michigan State University, he joined the Yale faculty as Assistant and later Associate Professor and remained at Yale until 1979, when he took up his present position.

Analysis with Heavy Ions

WILLIAM A. LANFORD

1. Introduction

In the past decade, nondestructive analysis using heavy-ion beams has become increasingly important both in fundamental physics and in technology. For example, see (An 80) and (Du 79). Perhaps the most obvious example is the important contribution Rutherford backscattering spectrometry has made in the development of the planar thin-film technology that has made possible the present revolution in microelectronics. Indeed, many of the major corporations in this industry now have their own accelerators used exclusively for analysis. While it is still too early to judge the full impact in other fields, there are strong indications that analysis using heavy ions will make equally important contributions to such diverse areas as fundamental surface science, energy, corrosion, and material science.

The principal techniques presently using heavy ions for nondestructive analysis are nuclear reaction analysis, Rutherford backscattering spectrometry, and nuclear recoil analysis. All of these methods rely on purely nuclear processes and, as such, are insensitive to chemical bonding in the material being studied. As a result, these methods are inherently easily made quantitative. Most importantly, these methods are insensitive to poorly understood atomic or solid state matrix effects which make techniques based on sputtering or Auger or x-ray emission so difficult to quantify.

A second important characteristic of these heavy-ion analytical methods is their ability to measure the concentration vs. depth (concentration profiles) with high sensitivity and good depth resolution.

There are studies of analytic methods based on essentially all radiations emitted by solids under ion bombardment, including x rays and Auger

WILLIAM A. LANFORD • Department of Physics, State University of New York/Albany, 1400 Washington Avenue, Albany, New York 12222.

electrons as well as nuclei or nuclear reaction products. Heavy-ion-induced x rays or Auger electrons are not very useful as analytic methods because of all the complicated processes which can contribute, such as quasimolecular x-ray emission, resulting in complex spectra. In general, heavy-ion-induced emissions have few advantages over mthods such as proton-induced x-ray emission, which is highly developed and widely used for studies of air and water pollution (Ca 80).

2. Nuclear Reaction Analysis of Hydrogen in Solids

Historically, the study of the effects of hydrogen on materials has presented some of the most difficult analytical problems. The presence of hydrogen can have dramatic effect on the physical, chemical, and electrical properties of many materials. The hydrogen-induced embrittlement of steel (Be 74), the role of hydrogen ions in mediating water-based chemical reactions including corrosion (La 79a), the influence of hydrogen on the critical temperature of superconductors (La 78a), and the dramatic effect of hydrogen on the semiconducting properties of amorphous silicon solar cell materials (Br 77) are a few examples. Unfortunately, hydrogen is "invisible" in most traditional analytic methods such as Auger or x-ray spectroscopies, neutron activation, or Rutherford backscattering spectrometry. Hence, because of this lack of effective analytic methods, many of the problems involving hydrogen in materials have not been fully studied and are incompletely understood.

This situation was abruptly changed in the 1970s with development and application of quantitative, largely nondestructive techniques for measuring hydrogen concentration vs. depth in any solid using nuclear reactions (Co 72, Le 73, Ad 74, Le 76, La 76). One of the first important general conclusions to come from nuclear reaction analysis was that hydrogen is a very common contaminant, probably the most universally present contaminant of all the elements. This observation has further increased the interest in and the importance of using heavy-ion-induced nuclear reactions as probes of hydrogen in solids.

Table 1 lists the nuclear reactions which have been used as probes for hydrogen (^{1}H) in solids. There are also a number of reactions used to study deuterium, but these will not be discussed here. As can be seen in this table, a variety of heavy-ion beams and associated nuclear reactions have been used and many more could be used. It is perhaps worth noting that while material scientists and metallurgists have long recognized the great need for a method of quantitatively measuring hydrogen concentration vs. depth in solids, and that while nuclear physicists have also long known that one could use nuclear reactions and scattering to probe for hydrogen, the

Table 1. Nuclear Reactions Used to Measure Hydrogen in Solids

Reaction	Signal detected	References
$^{1}H + {}^{19}F \rightarrow {}^{4}He + {}^{16}O + \gamma$ ray	γ ray	(Le 73)
$^{1}H + {}^{15}N \rightarrow {}^{4}He + {}^{12}C + \gamma$ ray	γ ray	(La 76)
$^{1}H + {}^{7}Li \rightarrow {}^{4}He + {}^{4}He + \gamma$ ray	γ ray	(Ad 74)
$^{1}H + {}^{11}B \rightarrow 3{}^{4}He$	^{4}He	(Li 74)
$^{1}H + {}^{1}H \rightarrow {}^{1}H + {}^{1}H$	^{1}H (coincidence)	(Co 72)
$^{1}H + {}^{35}Cl \rightarrow {}^{1}H + {}^{35}Cl$	^{1}H	(Le 76)
$^{1}H + {}^{4}He \rightarrow {}^{1}H + {}^{4}He$	^{1}H	(Do 79)

communication between these two scientific disciplines was so poor that the application of nuclear techniques for hydrogen analysis was slow in being put to use. The first paper that uses a high-energy nuclear physics technique to measure hydrogen content of solids is that by B. Cohen and collaborators at the University of Pittsburgh (Co 72), whose motivation resulted in part from discussions with members of the U.S. steel research staff. These workers utilized *p–p* elastic scattering (in coincidence) to measure hydrogen concentration vs. depth in metal foils. They clearly demonstrated the power of this technique, but, apparently because of the lack of interdisciplinary communication, this technique was not widely used to solve hydrogen-related problems.

Indeed, it seem that it was largely nuclear physicists wanting to solve other fundamental physics problems that led to implementation of nuclear techniques. For example, T. Tombrello became interested in the possibility of learning about the history of the solar wind by measuring the hydrogen implanted by the solar wind into moon rocks. To do this he needed a suitable technique for measuring hydrogen concentration profiles. Hence, he and D. Leich developed the $^{19}F + {}^{1}H \rightarrow {}^{16}O + {}^{4}He + \gamma$ ray resonant reaction as a probe for measuring hydrogen profiles (Le 73). W. A. Lanford became interested in the potential problems caused by hydrogen bombardment of first walls of fusion reactors and hydrogen impurities on the surfaces of ultracold neutron "bottles," and, along with colleagues at IBM and Yale, he developed the ^{15}N hydrogen profiling technique (La 76), which then became widely used in many hydrogen-related technology problems. Because of our own experience, we will discuss the ^{15}N hydrogen profiling technique in detail and then give examples of the use of this and some of the other reactions in solving particular problems.

The ^{15}N hydrogen profiling technique (La 76) makes use of the resonances in the $^{15}N + {}^{1}H \rightarrow {}^{12}C + {}^{4}He + 4.43$ MeV gamma-ray reaction. Because of its importance to astrophysical processes, the cross section for this reaction has been measured in great detail most recently by C. Rolfs and W. Rodney (Ro 74) at CalTech. This cross section is given in Figure 1

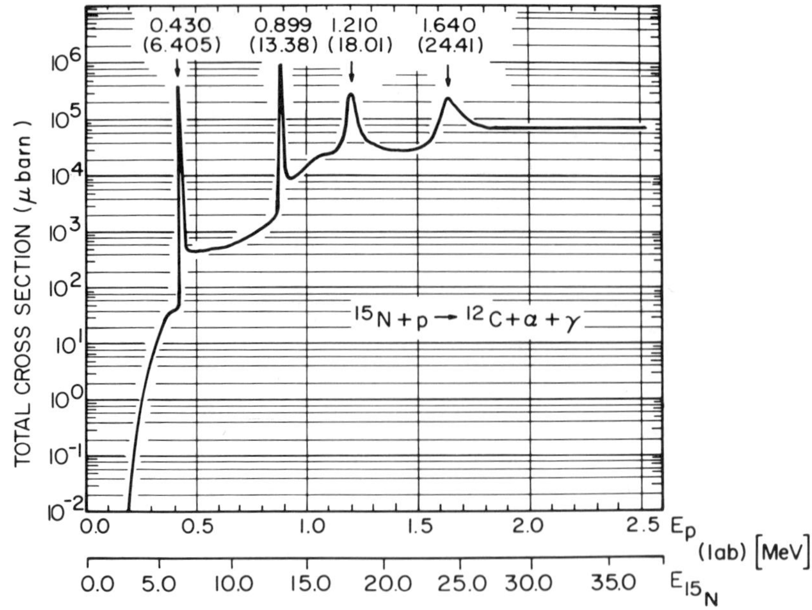

Figure 1. The cross section for the $^{15}N + {}^{1}H \rightarrow {}^{12}C + {}^{4}He$ + 4.43-MeV gamma ray. The cross section is given as a function of energy in the laboratory frame of reference both for proton beams (E_p) and nitrogen beams (E_N). From (La 81a) after (Ro 74).

(La 81a). As seen in Figure 1, there is a narrow isolated resonance at a proton energy of 430 keV (lab) with a peak cross section of 450 mb. In the use of this reaction as a probe for hydrogen, the sample being investigated is bombardment with ^{15}N at or above the resonance energy. In the laboratory system in which ^{15}N is accelerated, the 430-keV resonance energy is 6.405 MeV and its width $\Gamma = 6$ keV.

The ^{15}N hydrogen profiling technique is shown schematically in Figure 2. Essentially the same method is used in all resonant nuclear reaction profiling techniques, such as those using ^{19}F and ^{7}Li listed in Table 1. If a sample is bombarded with ^{15}N ions at the resonance energy (E_{res}), the yield of characteristic gamma rays is proportional to the hydrogen concentration at the surface of the sample. If the ^{15}N beam is above the resonance energy, the yield of gamma rays from reactions with hydrogen on the surface is neglible because the off resonance cross section is three orders of magnitude smaller than the peak cross section. However, as the ^{15}N ions lose energy penetrating the sample, they reach the resonance energy at some depth. The measured gamma-ray yield is now proportional to the hydrogen content at this depth. Hence, by measuring the yield of characteristic gamma rays vs. ^{15}N beam energy, the concentration of hydrogen vs. depth is determined.

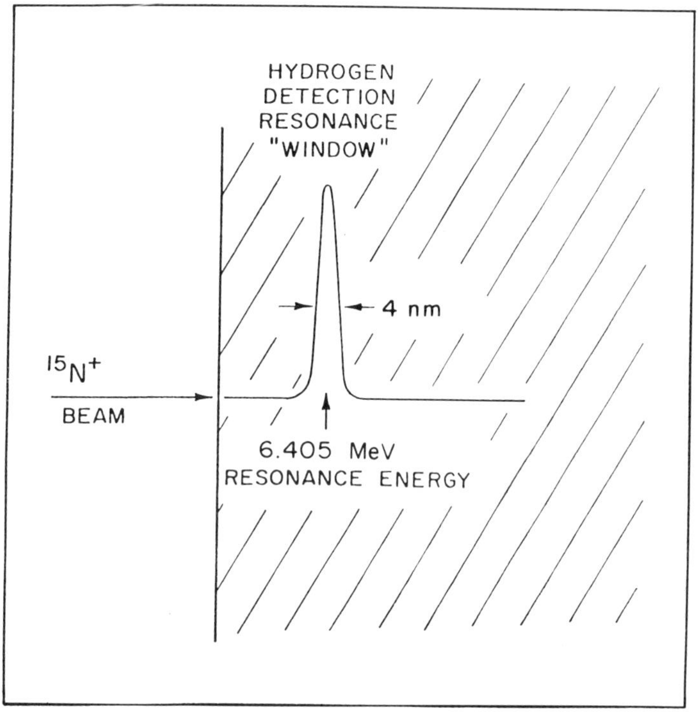

Figure 2. A schematic representation of the ^{15}N hydrogen profiling technique. Because the reaction is a resonance reaction, the gamma-ray yield is a measure of the amount of hydrogen within a "hydrogen detection window" the depth of which is determined by the beam energy. From (La 76).

Figure 3 shows the experimental chamber used at SUNY/Albany for measuring hydrogen concentration profiles. The samples to be analyzed are mounted on the outer radius of a wheel, allowing for rapid sample changing by simply rotating this wheel. Directly behind the sample is a 3-in. by 3-in. NaI scintillation counter for detecting the 4.43-MeV characteristic gamma rays. The whole analysis chamber is electrically isolated so that the number of incident ^{15}N ions can be determined by integrating the beam current. Because of the seemingly universal presence of hydrogen, it is necessary that all beam-defining aperatures be far away from the NaI detector in order not to introduce background from reactions with hydrogen on the slits. A typical accelerator vacuum of 10^{-6} Torr seems completely satisfactory for these measurements, except in cases where hydrogen adsorption on surfaces is being studied, when a vacuum of the order of 10^{-10} Torr may be needed.

Shown in Figure 4 are data for the hydrogen profile of silicon implanted with 10^{16} H/cm^2 at 40 KeV (La 80). There are a number of interesting

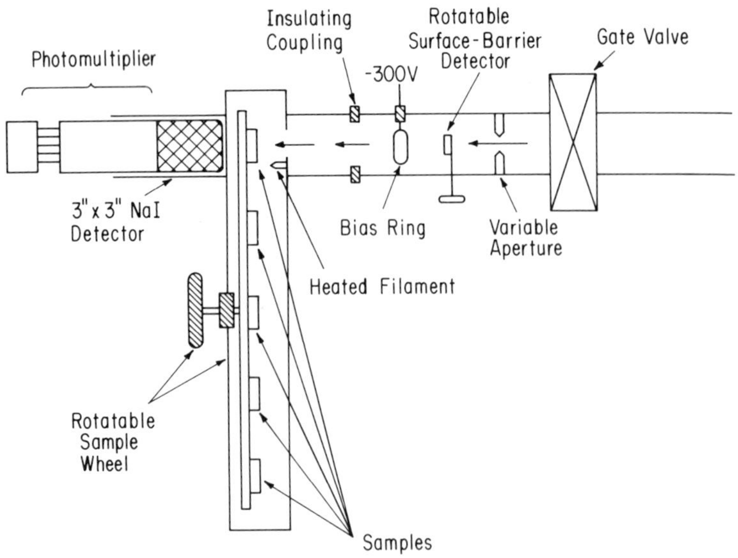

Figure 3. A schematic representation of the nuclear reaction and Rutherford backscattering analysis chamber used at SUNY/Albany.

features shown in Figure 4. First, the method is quantitative. The integrated area under the profile is 0.97×10^{16} H/cm^2 compared to the implanted dose of 1.0×10^{16} H/cm^2. Second the method is both rapid and has good sensitivity. Using a ^{15}N beam of 100 nA, each data point requires 100 s, and the background count rate in this time, which is the result of cosmic rays, corresponds to about 0.2×10^{20} H/cm^3 or about 0.4 parts per thousand (atomic). Third, the method has good depth resolution. Near the surface of this sample, the depth resolution is given simply by the ratio of the resonance width (Γ) to the rate of energy loss of the ^{15}N ions in the solid, assuming the energy spread of the ^{15}N beam from the accelerator can be neglected. This resolution depends somewhat on the material but is typically a few nanometers. Deeper into the sample, the straggle in the ^{15}N energy loss (ΔE_s) becomes important. We can use the formula of Bohr (Bo 48) to obtain a rough estimate for this:

$$\Delta E_s = 2.35(4\pi e^4 Z_1^2 Z_2 N x)^{1/2} \tag{2.1}$$

where Z_1 and Z_2 are the atomic numbers of the projectile and target atoms, respectively, N is the density of atoms in the target, and x is the depth into the target. For ^{15}N in silicon, this becomes $\Delta E_s \simeq 75$ keV/(μm)$^{1/2}$. Using the fact that the rate of energy loss of 6.4 MeV ^{15}N in silicon is 1.50 MeV/μm, the straggle given by the Bohr formula implies that the depth resolution becomes $\simeq$46 nm at a depth of 1 μm.

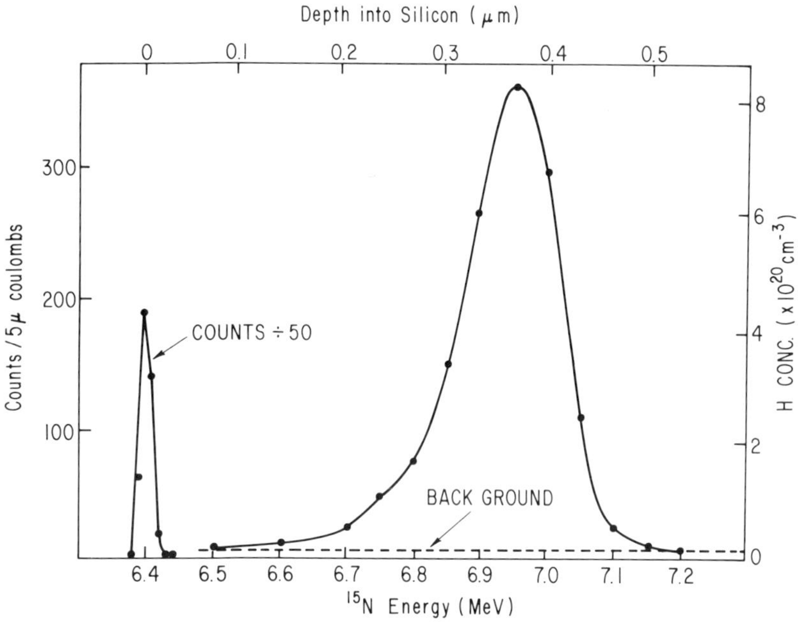

Figure 4. A typical ^{15}N hydrogen profile. This shows the yield of characteristic gamma rays vs. beam energy of a sample of silicon ion implanted with 10^{16} H/cm^2 at 40 KeV. From (La 80).

The conversion of the raw data (i.e., counts vs. E_{beam}) into hydrogen concentration vs. depth is very simple. It happens that dE/dx for ^{15}N is constant to within a few percent for energies between 6.4 and 9 MeV. Hence, the depth of hydrogen analysis (x) is given by $x = (E - E_{\text{res}})/(dE/dx)$. The concentration at this depth is given by

$$\rho_H = K(dE/dx)(\text{yield of } \gamma\text{-ray counts}) \tag{2.2}$$

where K is a constant reflecting the $^{15}\text{N} + {}^{1}\text{H}$ reaction cross section and gamma-ray detector efficiency. This constant can be calculated from the known reaction cross section and a calculation of the detector efficiency, or it can be evaluated for a given experimental setup by examining a sample of known hydrogen content. All investigators to date have principally relied on the latter approach using various ion implanted materials, plastics, minerals, and metal hydrides as standards (Zi 78, La 80, Be 80). Indeed, the sample used to obtain the data shown in Figure 4 could be used as such a standard by equating the measured area under the hydrogen distribution with the known implant dose.

2.1. Hydrogen Profiles of Lunar Material and the History of the Solar Wind

The Apollo moon probes in the early 1970s made lunar material available for laboratory study for the first time. T. A. Tombrello at California Institute of Technology realized that these materials may contain a valuable record of the effect of the solar wind on the exposed surface of the moon. In the simplest case, perhaps the silicate minerals on the moon's surface would retain the hydrogen "ion implanted" by the solar wind over the more than three billion years of exposure. One might hope that by measuring the hydrogen concentration vs. depth in the exposed surfaces, a record of both the integrated solar wind intensity and its energy spectrum could be extracted.

It was the possibility of discovering such fundamental information about our solar system which motivated Tombrello and his co-workers to develop and utilize the $^{19}F + {}^{1}H \rightarrow {}^{16}O + He + \gamma$ ray resonant reaction (Le 73) as a means of measuring hydrogen concentration vs. depth. Figure 5 (To 80, Le 74) shows some of their data. Indeed, one sees hydrogen penetrating into these samples several thousand angstroms. The precise origin of the hydrogen in these samples is not clear. While this hydrogen

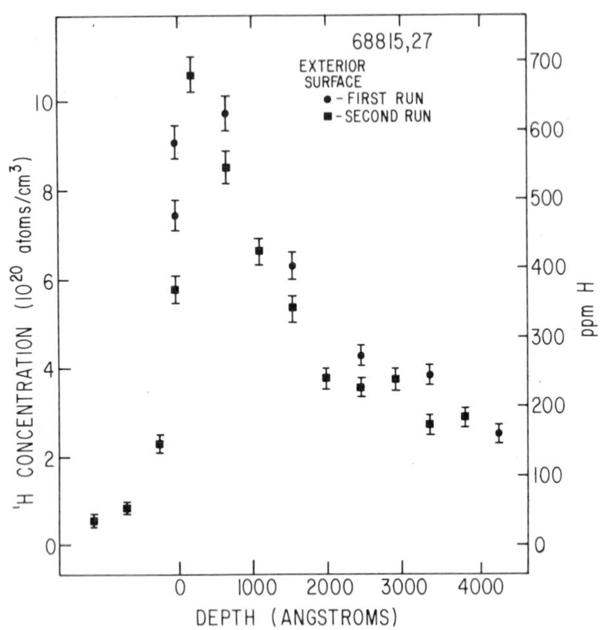

Figure 5. A hydrogen profile of the surface of some lunar material measured using the ^{19}F hydrogen profiling technique. After figures in (Le 74) and (To 80).

may be from the impact of the solar wind, it extends farther into the sample than expected. In addition, one expects the solar wind not only to deposit hydrogen but also to sputter the surface, to create damage, and to ion beam-mix the surface material with that deeper in the solid. All of these can be expected of affect the concentration profile resulting from the implant of the solar wind (To 80). Since none of these effects are presently well understood, the final conclusions about the history of the solar wind recorded in the surface of moon rocks are not yet available. However, as is so often the case, this one investigation is leading to the study of other fundamentally important ion solid interaction effects, such as radiation enhanced diffusion, defect trapping of hydrogen, and preferential sputtering.

2.2. Hydrogen Surface Contaminations and the Containment of Ultracold Neutrons

A second area where nuclear reaction analysis of hydrogen has contributed to a problem of fundamental interest in physics is in the study of the containment of ultracold neutrons in material "bottles" (Go 79). While one normally thinks of neutrons as very penetrating particles, at sufficiently low energy they should undergo total reflection of many solid surfaces. This is the same effect which is used in making reactor neutron guide tubes where neutrons are totally reflected for glancing angles of incidence, below some critical angle (Fe 46). At sufficiently low energies, typically of order 10^{-7} eV for materials such as carbon or copper, this critical angle goes to 90°. As such, ultracold neutrons can be stored in material bottles for long periods of time. This neutron containment leads to the possibility of substantial improvements of such fundamentally important experiments as the search for the electric dipole moment of the neutron (a test of time reversal invariance) and the measurement of the free neutron β-decay lifetime (important in evaluating the weak interaction coupling constants).

While the first experiments with ultracold neutrons confirmed the expectation that they could be confined for tens and even hundreds of seconds in material bottles (Lu 68, St 69), there was a large disagreement between the measured storage times and the storage times predicted based on expected loss mechanisms such as neutron β-decay, neutron capture, or inelastic scattering from near surface nuclei. For a bottle wall material such as graphite or boron-free glass, the measured neutron loss rates per reflection from the wall were two orders of magnitude larger than theoretical prediction (La 77a). One possible explanation was that the wall surface did not have the same composition as the bulk, but rather had large contaminations of elements with large neutron reaction cross sections. Hydrogen was the most likely candidate.

W. Lanford and R. Golub made measurements of the hydrogen

contaminations in neutron bottle materials using the ^{15}N hydrogen profiling method, and they found enough hydrogen to explain the enhanced loss rate (La 77a). Following this work, P. LaMarche, W. Lanford, and R. Golub proceeded to investigate methods of making large area, hydrogen-free surfaces, again using the ^{15}N hydrogen profiling as their primary analytic tool (La 81a, La 81b). It is expected that these results will lead to improved sensitivity in experiments such as the search for the elastic dipole moment using bottled neutrons which is presently underway at the Institute Laue Langevin in Grenoble.

2.3. Hydrogen in Thin Film High Technology Materials

Thin film materials have become increasingly important both in fundamental materials research and in such applied technologies as microelectronics and computer design. Thin film technologies have advantages over more traditional methods because (1) smaller amounts of precious materials are needed, (2) smaller devices and, hence, higher packing density can be made, and (3) some interesting materials can be made only in thin films. In recent years it has been realized that hydrogen plays important roles, often initially introduced as an impurity, in the chemical and electrical properties of a number of thin film materials (Br 77, La 78a, La 78b). In general, nuclear reaction analysis is the only known method for quantitatively measuring the hydrogen content of thin films (Zi 78, La 80). Below we will briefly discuss the role of hydrogen in a thin film superconductor, semiconductor, and insulator.

2.3.1. Nb_3Ge Superconductor

Nb_3Ge is the material with the highest known superconducting transition temperature, $Tc \simeq 23$ K. The high Tc phase of this material is metastable and can only be prepared in thin films using evaporation or sputtering techniques. It is of obvious interest to try to understand why Nb_3Ge has such a high Tc. It was recently realized by J. Rowell and coworkers at Bell Laboratory that this material can be systematically doped with hydrogen (Ro 77) with the consequence that the crystal lattice expands and Tc decreases. In order to extract detailed information about the transition to the superconducting state in this material, it was necessary to quantify this effect, i.e., to measure accurately the hydrogen doping level in the Nb_3Ge. ^{15}N hydrogen profiling measurements were carried out (La 78a). Some of these data are given in Figure 6. These results were helpful in evaluating various models which had been proposed to explain why Nb_3Ge had such a high Tc. Furthermore, an unanticipated result of these

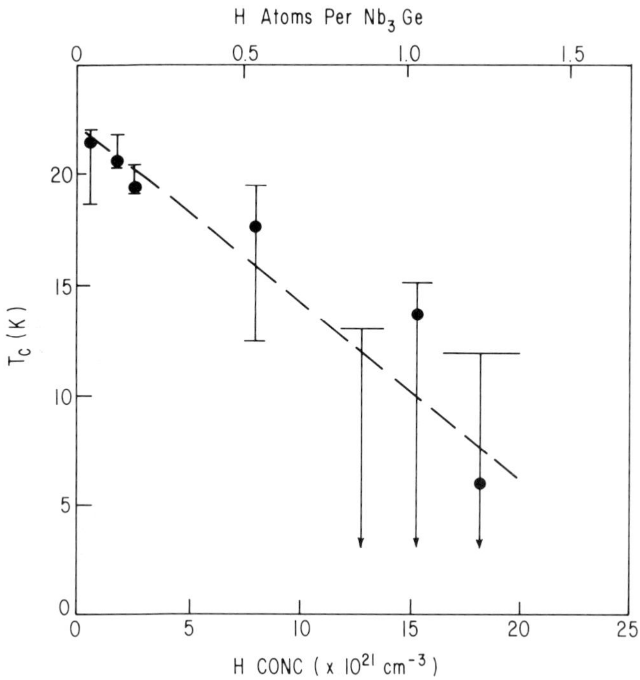

Figure 6. A plot showing the dependence of the superconducting critical temperature on hydrogen content for Nb_3Ge. The vertical lines through the data points represent the range of temperatures over which the transition to the superconducting state takes place. From (La 78).

measurements was the indication that hydrogen impurities may play an important role in the phase stabilization of the high *Tc* phase of Nb_3 Ge.

2.3.2. Amorphous Silicon Solar Cells

Recently W. Spear and P. LeComber discovered that amorphous silicon made by a glow discharge deposition technique from a silane (SiH_4) plasma was a good semiconductor (Sp 75). Because of the possibility of using this material to make large area, inexpensive solar cells, a worldwide study of this material was initiated immediately. The reason the observations of Spear and LeComber were so surprising is that previous research with amorphous silicon, made, for example, by evaporation of pure silicon, had shown this material to be a very poor semiconductor, with its electrical properties dominated by the "dangling" bonds. These dangling bonds would seem to be inherent in anything but fully fourfold coordinately single-crystal silicon.

While initially there was great confusion over the reason that plasma-deposited silicon was so much superior to evaporated silicon, it is now

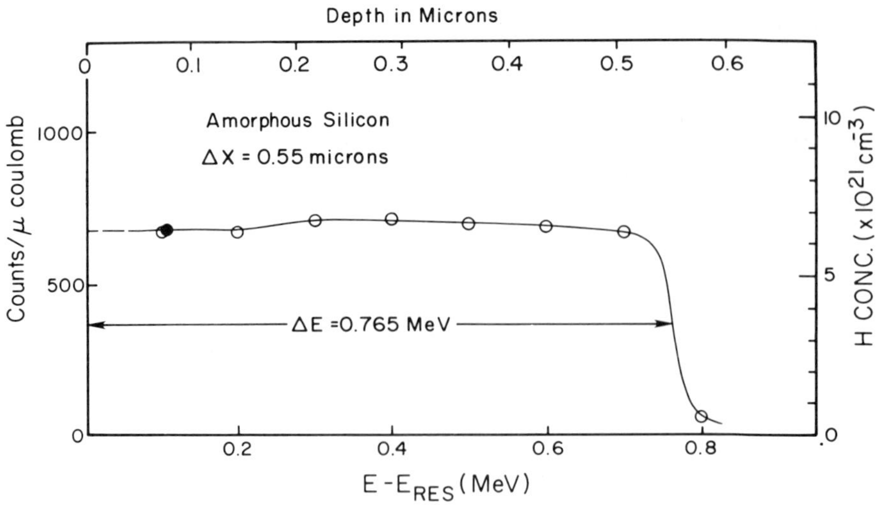

Figure 7. A plot of hydrogen content vs. depth for hydrogen in thin film amorphous silicon. The solid data point was measured after the complete profile had measured to test reproducibility.

generally recognized (Co 80) that impurity hydrogen introduced from the silane gas used to generate the plasma was being incorporated into the amorphous silicon and passifying the dangling silicon bonds. Much of the confusion over the role of hydrogen in this material arose from spurious results from attempts to measure the hydrogen content of these thin films using such traditional techniques as measuring the gas evolved during heating of the film to crystallization (La 80). Most of the groups working in this field now rely on nuclear reaction analysis as the primary method for quantitatively measuring the hydrogen content of this films (Co 80). Shown in Figure 7 are the first quantitive measurements of hydrogen in these materials showing that they are typically 10%–30% (atomic) hydrogen (Br 77).

While amorphous silicon solar cells are already commercially manufactured to power small hand-held calculators, they have an efficiency of about only 7% in sunlight. At present there is a great effort to increase this efficiency by tailoring the hydrogen and other impurity content with the hope of making inexpensive cells that are efficient enough for wide-scale application.

2.3.3. *Insulators for Encapsulation of Microelectronic Circuits*

Plasma-deposited silicon nitride is becoming widely used as an insulator for encapsulation of microelectronic circuits. This material is typically made by deposition from a SiH_4–NH^3 or SiH_4–N_2 plasma and may be expected to contain some hydrogen. Indeed, nuclear reaction analysis showed that this material typically contains as many hydrogen atoms as silicon or nitrogen atoms (La 78b, Pe 79). Furthermore, the hydrogen has important effects on the chemical durability and stress of the film. For example, shown in Figure 8 is a plot of etch rate vs. hydrogen content (Ch 80) showing that the incorporation of hydrogen can dramatically influence chemical reaction rate. In microelectronic device manufacturing, properties such as etch rate may need to be accurately controlled. Further, for use in devices which are sensitive to hydrogen impurities, it may be essential to control the amount and bonding of hydrogen in the silicon nitride if it is going to be used as an encapsulant. Again we see that nuclear reaction analysis is providing unique information which is of value both to the technological development of this material and to the understanding of the fundamental plasma chemistry used to make this material.

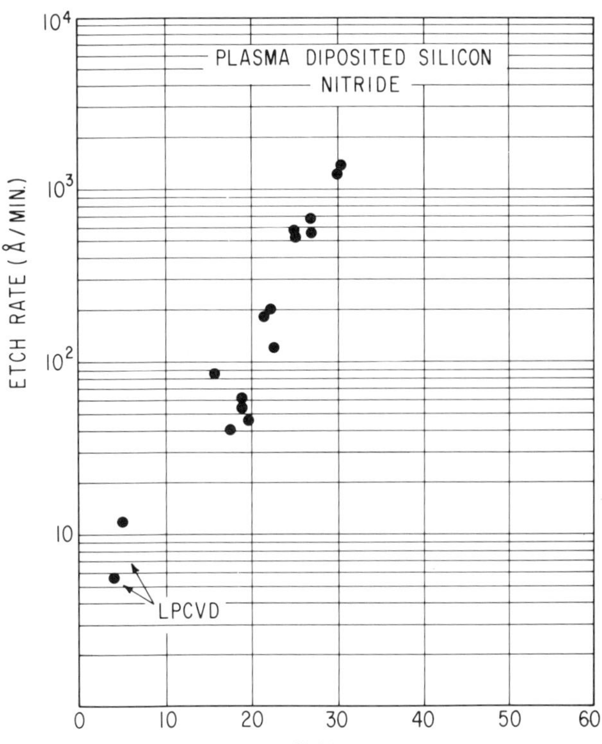

Figure 8. The etch-rate vs. hydrogen content for plasma deposited silicon nitride and two samples of low-pressure chemical vapor deposited (LPCVD) silicon nitride. Data from (Ch 80).

3. Rutherford Backscattering Spectrometry

Rutherford backscattering spectrometry experiments are probably the most common use of MeV ion accelerators today (Ch 78, An 80, Du 79). Many 1–5-MeV accelerators which were initially built to study nuclear physics are now used only for ion beam analysis studies, and a whole new generation of accelerators is being built exclusively for such purposes. Rutherford backscattering spectrometry, because of its reliable quantitive nature, because of the ease and simplicity of making measurements, and because of the wide range of interesting applications, is the most widely used of the MeV ion beam analysis techniques.

The history of the development and application of Rutherford scattering to nonnuclear problems shows some of the typical characteristics of interdisciplinary science. In this case, an analytic method was well understood by scientists in one discipline and a series of problems requiring such an analytic method was recognized by scientists in another discipline. It seems in retrospect that there was very slow application of the technique because of poor interdisciplinary communication. Rutherford scattering has always been and continues to be used by nuclear physicists to identify impurities in this targets. However, it was not until the mid to late 1960s that significant numbers of applications of Rutherford scattering were made outside nuclear science.

The mostly widely publicized early application of Rutherford backscattering spectrometry, one which was probably important in breaking

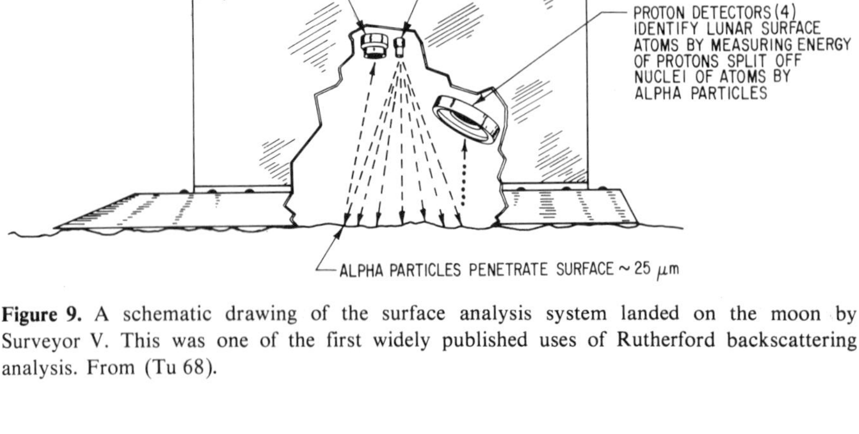

Figure 9. A schematic drawing of the surface analysis system landed on the moon by Surveyor V. This was one of the first widely published uses of Rutherford backscattering analysis. From (Tu 68).

down indisciplinary barriers, was the backscattering experiment carried on lunar probe Surveyor V, which made a soft landing on the mon in 1967 (Tu 68). This experiment was designed to measure the surface composition of the moon. Figure 9 schematically shows the experimental arrangement. However, it was in the development of the planar technology of the microelectronics industry that backscattering became widely recognized outside of nuclear physics as an excellent quantitive technique for measuring elemental distributions near surfaces.

3.1. Principles and Applications of Rutherford Backscattering Spectrometry

Backscattering spectrometry relies on the fact that, if a low-energy ion (such as He^+, Li^+, O^+) scatters from a target, there are precise relationships: (a) between the energy of the scattered particle and the mass of the scattering atom, and (b) between the probability for scattering and the atomic number (Ch 78) of the target atom. These relationships are

$$\frac{E_1}{E_p} = \left\{ \frac{[1-(M_1/M_2)^2 \sin^2 \Theta]^{1/2} + (M_1/M_2)\cos\Theta}{1+M_1/M_2} \right\}^2 \tag{3.1}$$

and

$$\frac{d\sigma}{d\Omega} = \left(\frac{Z_1 Z_2 e^2}{4E}\right)^2 \frac{4}{\sin\Theta} \frac{\{[1-(M_1/M_2)^2 \sin^2\Theta]^{1/2} + \cos\Theta\}^2}{[1-(M_1/M_2)^2 \sin^2\Theta]^{1/2}} \tag{3.2}$$

where E and E_0 are the backscattered energy and the initial energy, respectively; Θ is the laboratory scattering angle, and M_1, Z_1 and M_2, Z_2 refer to the projectile and target mass and Z, respectively. The differential cross section is given in the laboratory frame of reference. Hence, in the simplest case of backscattering from a thin target or a thin layer of heavy elements on the surface of a light mass substrate, measuring the energy and number of backscattered ions from the target uniquely determines the mass (from the energy of the backscattered ion) and the amount (from the number of backscattered ions) of the heavy elements present.

3.1.1. Thin Target Analysis: Metal Impurities in Tokamak Plasmas

One example of the application of Rutherford backscattering spectrometry in its simplest form is in the analysis of heavy impurities in plasmas of Tokamak fusion reactions (De 78). Tokamaks are toroidal fusion devices which induce large currents in their plasmas. These currents both heat and confine the plasma. While these devices show great promise for

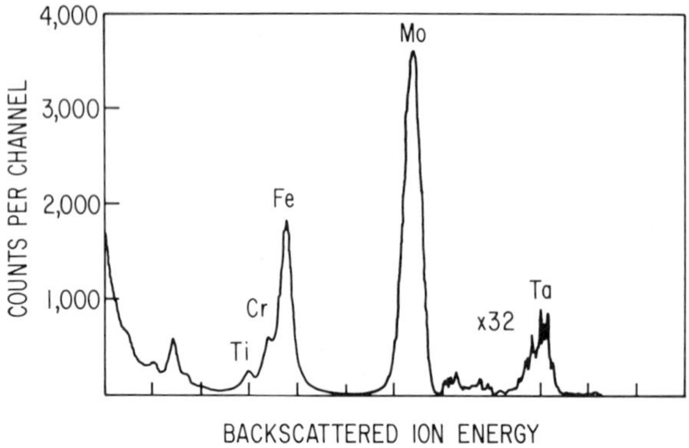

Figure 10. The Rutherford backscattering spectrum measured from a collector exposed to a Tokamak discharge. From (De 78).

exceeding the critical temperature–density requirements for fusion, the presence of heavy element impurities in the plasma causes serious problems. Most importantly, the presence of high-Z ions in the plasma provides a mechanism for the radiation of high-energy x rays from the plasma. For example, iron impurities can emit their characteristic 6.4-keV K x rays, which are high energy enough to have a high probability of escaping the plasma and, hence, of lowering the plasma temperature. The corresponding transitions in the H and He of the plasma are of eV energies and would be reabsorbed by the plasma.

One technique for studying this impurity problem was demonstrated by G. Dearnaley *et al.* (De 78), who used Rutherford backscattering to analyze the elements deposited on carbon collectors exposed to the Tokamak discharge. In order to have high sensitivity, they used 3.5-MeV nitrogen ions, which are too heavy to backscatter from the carbon collectors. Some of these results are shown in Figure 10. As seen in Figure 10, a variety of heavy metal impurities are observed. By scanning the carbon collectors through the tokamak, they can deduce the radial distribution of these elements, and, by exposing different parts of the collectors at different times, they can record the buildup of these impurities as a function of time during a single discharge. For this latter application, it is important to have high sensitivity, which in this case was about 10^{11} atoms/cm^2 (10^{-4} monolayers).

3.2.1. Thick Target Analysis: Glass Surfaces

In the case of backscattering from a thick target, it is necessary to take into account the effect of the energy loss both before and after the scattering event. One of the most obvious differences to be expected in going from thin to thick target backscattering is that for thick targets one no longer has

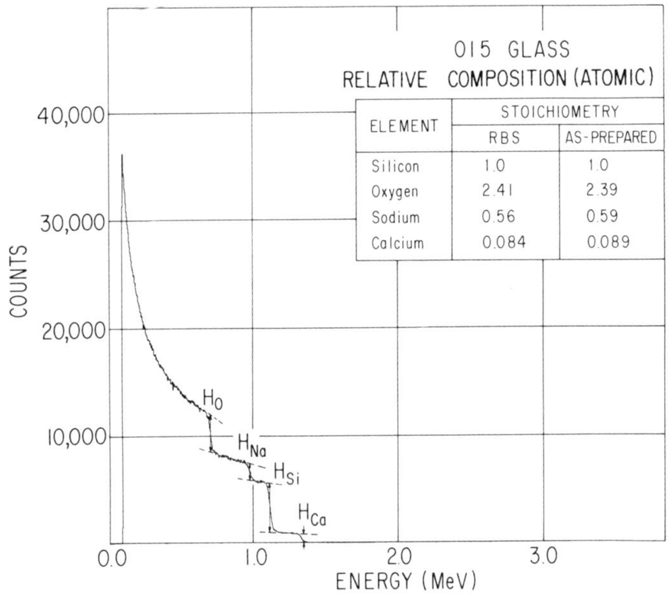

Figure 11. The Rutherford backscattering spectrum recorded from a common soda-lime glass. The measurement of the step heights can be used to accurately determine the composition of the glass.

discrete peaks for each element (such as in Figure 10) but rather steps. For each element on the surface of a sample there will be a corresponding step (H_A) in the energy spectrum of backscattered particles at the kinematic energy E_1, given by equation (3.1). This is illustrated in Figure 11, which shows a backscatter spectrum for 2-MeV He ion incident on a typical metal oxide glass (Me 81). This glass contains four elements (O, Na, Si, Ca) and for each element there is a corresponding step in the backscatter yield. By combining the cross sections [equation (3.2)] and the kinematics [equation (3.1)] with the rate of energy loss of He in this material, the measured step heights for all elements present can be used to determine accurately the stoichiometry of this glass. For the case of determining the ratio of concentration of two elements (say Ca to Si in this glass) from the ratio of step heights, the expression is (Ch 78, Ma 77)

$$\frac{N_{\mathrm{Ca}}}{N_{\mathrm{Si}}} = \frac{H_{\mathrm{Ca}}}{H_{\mathrm{Si}}} \frac{(d\sigma/d\Omega)_{\mathrm{Si}}}{(d\sigma/d\Omega)_{\mathrm{Ca}}} \frac{\varepsilon_{\mathrm{Ca}}}{\varepsilon_{\mathrm{Si}}} \tag{3.3}$$

Here

$$\varepsilon_A = \text{stopping factor}$$

$$= \frac{E_1}{E_0}\,\varepsilon(E_0) + \frac{1}{\cos\Theta}\,\varepsilon(E_1)$$

where E_1/E_0 is given by equation (3.1), Θ is the scattering angle, and ε is the stopping cross section (energy loss/atom). As indicated in Figure 11, Rutherford backscattering can given rapid, quantitive analysis of bulk compositions.

It is not usual to use backscattering to perform simple bulk analyses, such as that illustrated above, because there are other means for analyzing bulk composition. Such analyses are usually incidental to the use of backscattering to obtain other information. For example, the results in Figure 11 were obtained as part of a study of the kinetics of the chemical reactions between water and glass. Exposure of glasses to water results in changes in the surface composition which can be easily measured using backscattering but which are otherwise difficult or impossible to obtain. Shown in Figure 12 are enlargements of the regions of the backscatter spectra corresponding to the Na edge (Me 81). For these data, the only difference between the samples is the length of time they were exposed to water at 90°C. As can be seen in Figure 12, the sodium is rapidly becoming depleted from the surface region. Although not shown, careful analysis of the complete backscatter spectra along with nuclear reaction analysis for hydrogen profiles show that essentially the only change in the glass is the diffusion of sodium out of the glass and its replacement with hydrogen (La 79). Studies of the surface durability of glass are important to such diverse fields as glass science (chemical durability, fracture strength), energy [consolidation of radioactive wastes (La 79b)], archaeology [dating (La 77b, La 78c)], and physical chemistry (ion transport in glasses).

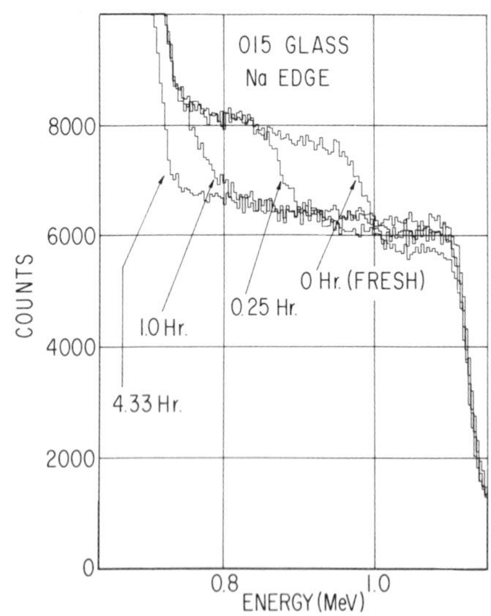

Figure 12. An enlargement of the region of the sodium edge in the Rurtherford backscattering spectrum of the glass used in Figure 11. The different glasses correspond to glasses exposed to water for varying lengths of time. From (Me 81).

3.2. Analysis of Ion Implanted Samples

It was in the development of thin film planar microelectronic devices where Rutherford backscattering first became established as one of the principal analytic methods. Part of the reason for its great success there was the fact that usually a great deal was known about the samples in terms of the elements present and very specific information—such as depth distributions—was needed. For example, if ion implantation of As is used to dope silicon, how far in does the As go? Where does the As reside in the Si crystal lattice? What happens when the sample is annealed? Shown in Figure 13 are some data illustrating how backscattering (Si 75) and channeling (Ha 72) can be used to answer these questions. The illustration on the left in Figure 13 shows the backscattering spectrum from a silicon crystal implanted with As at 250 keV. The energy and width of the As backscatter peak given the mean depth and spread in depth of the As implant. The illustration on the right in Figure 13 shows the As backscatter yield near a channeling direction. (See Chapter 5 of this volume.) The presence of a dip in the yield at this and other channeling directions indicates that the As is located substitutionally in the Si crystal lattice. Because Rutherford backscattering combined, in some cases, with channeling and blocking techniques provided rapid and quantitative answers to key questions, it played a central role in the development of such common devices as the microcomputers or hand-held calculators which use ion implantation in their manufacture.

3.3. Thin Film Analysis: Ion-Beam-Mixing Induced Silicide Formation

Recently it has been realized that one can induce mixing between thin films or between a thin film and a substrate using the recoil cascade setup as an ion bombards a surface (Ma 80). This process is of great fundamental importance in the general study of particle–solid interactions. It is also becoming of great technological importance because it may allow the formation of new and useful compounds which could be used in components of microelectronic cricuits or other devices. Refractory metal silicides are a case in point. WSi_2, $TaSi_2$, and $MoSi_2$ have become increasingly important because of their applications in integrated circuits.

The formation of tungsten silicide by ion implantation of As ion at 130 keV is illustrated in Figure 14 (La 81c). The upper part of this figure shows the backscatter spectrum of a sample of silicon with ~200 Å of W evaporated on its surface. The high-energy peak results from the W and the lower-energy "step" results from the thick Si substrate.

The lower part of Figure 14 shows what happens as a result on

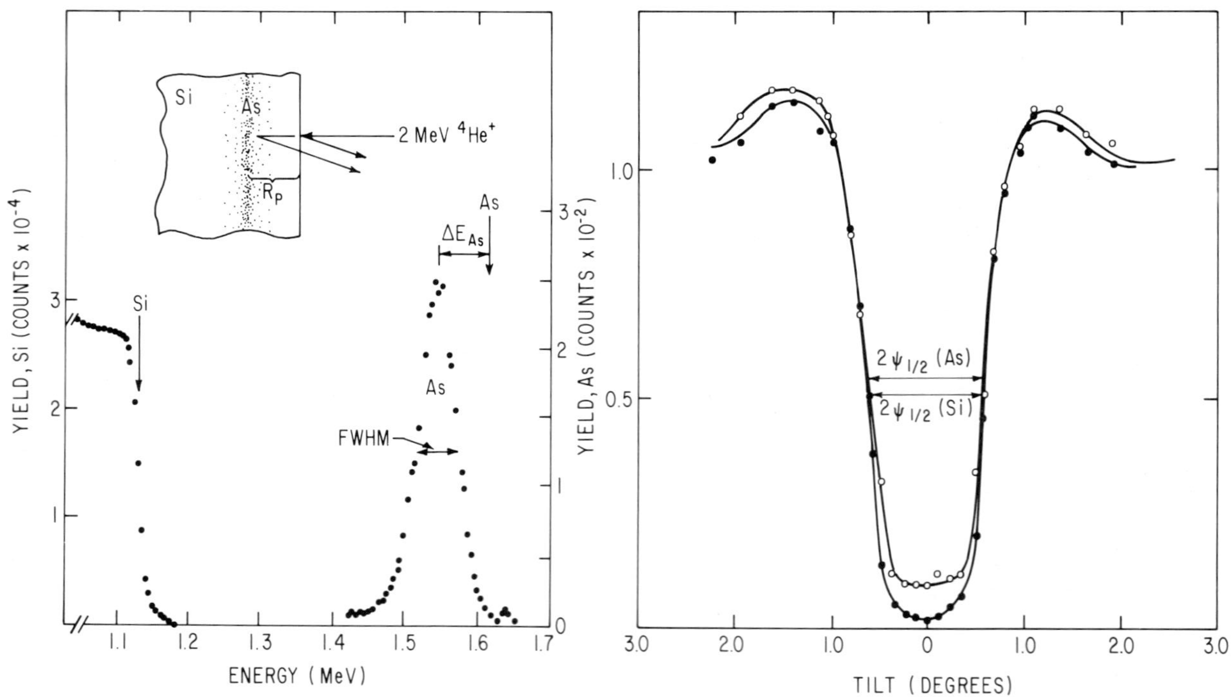

Figure 13. Left: The Rutherford backscattering spectrum record from a sample of silicon ion implanted with As. From (Ch 8), p. 138. Right: The yield of backscattering from As ion implanted into silicon vs. tilt angle from a channeling direction. The dip in yield in the channeling direction shows the As is subsitutional. From (Ch 78), p. 268.

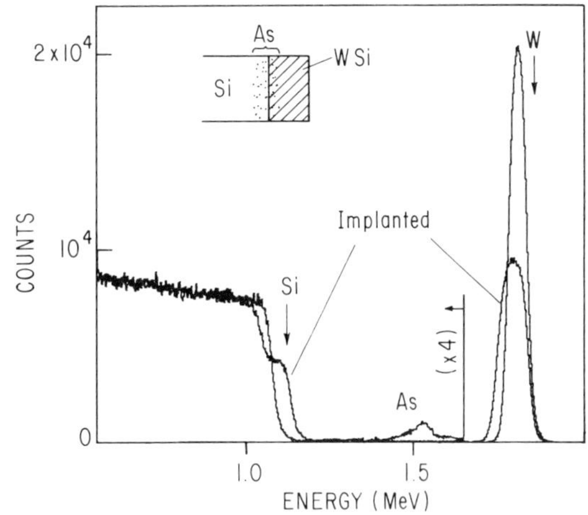

Figure 14. The Rutherford backscattering spectra from two samples with 20 nm of W evaporated onto silicon. One was implanted with 10^{15} As/cm^2 which ion-beam-mixed the W and Si to form a tungsten silicide. From (La 81c).

implanting 10^{16} As ions/cm^2 at 130 keV. Both these spectra were recorded with the same number of 2-MeV He ions incident on the sample. Notice that the W peak has decreased in intensity (in counts/channel) more than a factor of 2 and is correspondingly broader; the silicon step now has a ledge; and there is a peak as a result of the implanted As. Analysis of this spectrum shows that the As ions have intermixed the As and silicon to form a homogeneous tungsten silicide, slightly rich in Si. This follows from both the decrease in the intensity of the W peak (the W is "diluted" with Si) and the growth of a silicon step beginning at the surface and continuing to the pure silicon substrate. The ultimate goal in a device application would be to use the As beam both to form the silicide and to introduce a dopant; the observation of the As peak in the backscatter spectrum gives the depth distribution of this dopant and is a useful bonus in this study.

3.4. Analysis of Surfaces with Heavy-Ion Beams

The study of surfaces and surface structure has become increasingly important both to the fundamental understanding of material properties and to technology. Surfaces play dominant roles in such diverse phenomena as catalysis, chemical durability, and fracture where the surface indicates the initial development of cracks which subsequently result in fracture. Heavy-ion beams provide a number of unique analytic capabilities for studying surfaces; this application of heavy-ion beams is answering many long-standing questions as well as discovering a number of unanticipated phenomena.

One class of questions has to do with the geometrical position of the surface layer of atoms in a crystal relative to the underlying crystal lattice. If a single crystal is cleaved, do the atoms on the newly created surface stay in the same site or do they "relax," i.e., is the distance between the surface layer and the underlying layer the same as that between layers in the bulk? Another possibility is for surface atoms to "reconstruct" by moving laterally, e.g., to pair up to make use of bonds which otherwise would "dangle" in free space. What is the effect of the presence of impurities on the position of surface atoms?

To illustrate the power of these heavy-ion techniques and the richness of this scientific field, the role of hydrogen as a "surface terminator" will be described. Shown in Figure 15 are schematic representations of a crystal. In the upper part, the crystal is assumed to be unreconstructed, i.e., the atoms in the surface layer are in the same location they would be in if they were in the bulk. In the lower part, the crystal is assumed to have a reconstructed surface.

If backscattering yields are measured from these crystals in a chan-

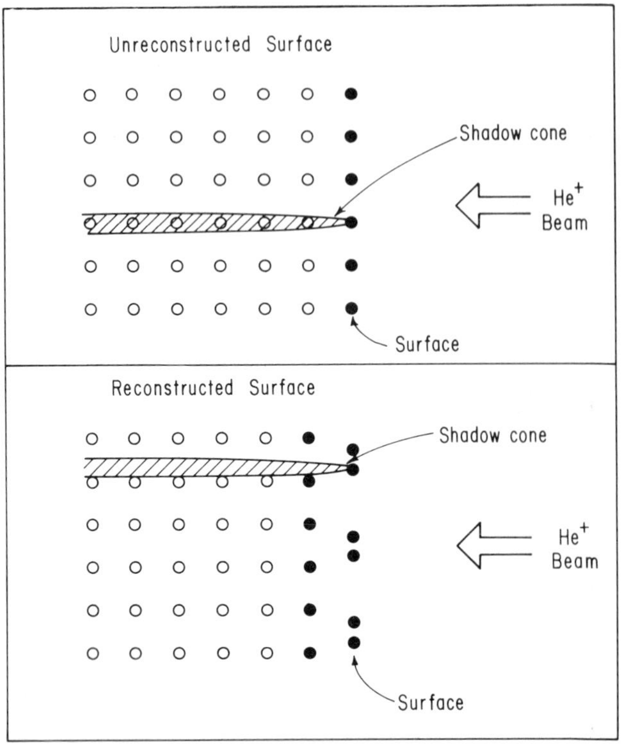

Figure 15. Schematic representations of crystal surfaces with unreconstructed surface (top) and reconstructed surfaces (bottom). The surface atoms casting shadow cones are shown as black dots.

neling direction, one can not only distinguish these two possibilities but also make quantitative statements about the extent of the reconstruction. Because channeling is described in Chapter 5 of this volume, we will not go into detail here except to point out that the surface atoms cast a "shadow cone" whose radius at the second layer of atoms is given by

$$R = 2\left(\frac{Z_1 Z_2 e^2 d}{E}\right)^{1/2}$$

where Z_1 and Z_2 are the atomic number of the projectile and target atoms, respectively, d is the distance to the second layer, and E is the incident energy (Ch 78). This shadow cone is illustrated schematically in Figure 15. In a channeling geometry, the incident ion beam never penetrates inside this shadow cone because of the focusing effect of the Coulomb repulsion between the ion and the crystal lattice. This is the same effect that keeps the inident ions trapped in the channel.

Hence, in a backscattering measurement, as the crystal is rotated to the channeling direction, the yield of backscattered ion drops from that typical of a randomly oriented thick target to that of the layer surface atoms (shown as black circles in Figure 15), all of the other atoms being within the shadow cone of surface atoms. The yield of backscattered ions, then, directly measures the number of unshadowed surface atoms. However, this number is sensitive to the position of the surface atoms relative to the underlying lattice. For example, for the simplified case illustrated in Figure 15, the backscatter yield from the reconstructed surface is just twice that from the unreconstructed surface.

Shown in Figure 16 are the backscatter spectra from reconstructed and unreconstructed W(001) surfaces measured by I. Stensgaard *et al.* at Bell Laboratories (St 79). As can be seen, the yield from the reconstructed surface is almost twice as large as that from the unreconstructed surface, showing directly that reconstruction involves lateral movement of the surface atoms.

You may wonder how both reconstructed and unreconstructed surfaces of the same material can be prepared. In this case and in others, the clean surface is reconstructed, presumably in order to satisfy as many dangling surface bonds as possible by pairing up surface atoms. If one exposes such a clean surface to a small amount of hydrogen, the surface unreconstructs, with hydrogen attaching to the dangling surface bonds (St 79, Na 81). In order to quantitatively evaluate this model, it is necessary to be able to measure the hydrogen adsorbed on such surfaces. Again, nuclear reaction analysis (such as ^{15}N hydrogen profiling) provides the analytic method needed to confirm this model. Indeed, such measurements by Stensgaard (St 79) did demonstrate the hydrogen uptake in the transition from the

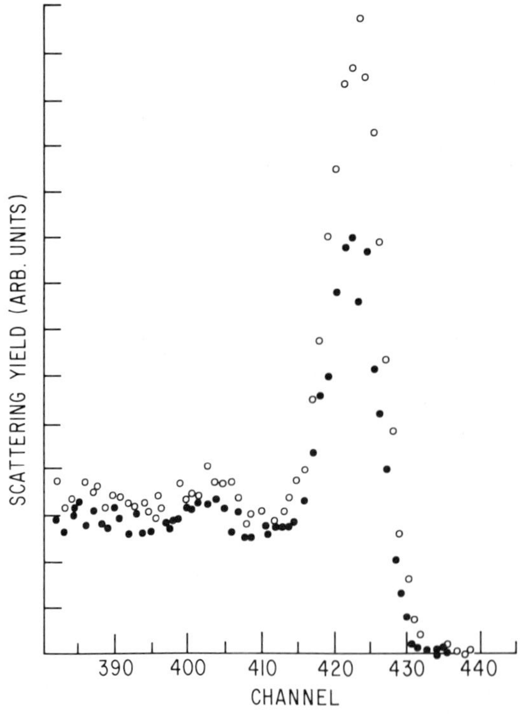

Figure 16. The Rutherford backscattering yield for 2-MeV He from a reconstructed (open circles) and unreconstructed (black dots) W surfaces masured in a channeling direction. From (St 79).

reconstructed to unreconstructed state in W, as did measurements by T. Narusawa *et al.* (NA 81) at SUNY/Albany for the silicon case.

Hence, the use of heavy-ion beams for backscattering and nuclear reaction analysis is beginning to suggest that hydrogen may commonly play a role as a "surface terminator," the surface being terminated by attaching hydrogen to surface bonds which otherwise would dangle in free space. This surface effect is no doubt related to the universal presence of hydrogen found in the ultracold neutron study discussed above (La 77a, La 81a, La 81b) and may have implications in such technologically important phenomena as hydrogen-induced fracture of steel.

4. Nuclear Recoil Analysis

While nuclear reaction analysis continues to be the most widely used method for profiling light elements, especially hydrogen and helium, a method based on the detection of high-energy recoiling nuclei has been described and shows great promise (Le 76, Do 79). The technique is very much like Rutherford backscattering except instead of detecting the backscattered ion, the recoiling target atom is observed. The motive for developing nuclear recoil analysis is the hope of developing a method with the generality of application of Rutherford backscattering but one applicable to light elements where kinematics and small cross sections make backscattering measurements impossible.

The basic experimental arrangement is as shown in Figure 17 (Do 79). A heavy-ion beam is incident on the sample at a glancing angle of incidence ($\sim 15°$), and a detector is placed at a corresponding forward angle ($\sim 15°$ to the surface). Neglecting energy losses in the target, if an elastic scattering event occurs when a target atom recoils into this detector, its energy is kinematically determined to be

$$E_{\text{recoil}} = E_{\text{incidence}} \frac{4M_1 M_2}{(M_1 + M_2)^2} \cos^2 \Theta \tag{4.1}$$

where M_1 and M_2 are the masses of the beam and recoiling atom, respectively, and Θ is the scattering angle. See Figure 17. Because of the energy loss of the incident and recoiling ions in penetrating the target, the energy of the detected ions is less than E_{recoil} by an amount dependent only on the depth at which the scattering took place. Hence, by measuring the energy spectrum of recoiling atoms, a depth profile is obtained.

This technique can be easily made quantitative if the incident ion and its energy are chosen such that it is always below the Coulomb barrier for elements in the target. Then the Rutherford cross section can be used, which in terms of Θ_{lab} (Figure 17) is

$$\frac{d\sigma}{d\Omega} = \left[\frac{Z_1 Z_2 e^2 (M_1 + M_2)}{2M_1 E}\right]^2 \frac{1}{\cos^3 \Theta} \tag{4.2}$$

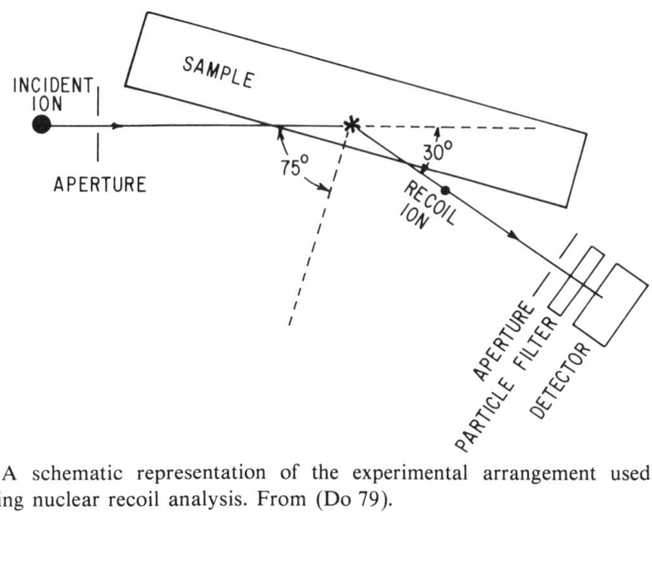

Figure 17. A schematic representation of the experimental arrangement used to analyze surfaces using nuclear recoil analysis. From (Do 79).

where Z_1 and Z_2 are the atomic numbers of the incident and recoiling nuclei and E is the energy of the incident ion.

Because of the simplicity of working below the Coulomb barrier, the group at Montreal which did much of the pioneering work in developing this method (Le 76, Tc 77) typically used 30-MeV ^{35}Cl beams. With such a beam in the geometry shown in Figure 17, one obtains information not only about the H and He content of the sample but also about the C and O content. One of the great advantages of this method is that, potentially, profiles for all of the light elements present can be recorded *simultaneously.* Because of the common presence of light element impurties of H, C, and O, and because of the difficulty of quantitative analysis for all of these elements, nuclear recoil analysis has the potential of making great contributions. An example (Te 77) of the energy spectrum recorded by B. Terreault *et al.* at Montreal for a complex target is given in Figure 18. The target was a thin film of copper coated on both sides with LiOH and implanted with 1-keV He. Because this was a thin film, the detector could be placed at $\theta = 0$. As seen in Figure 18, recoils are observed for ^6Li, ^7Li, and ^1H from both sides and from the He implant. A large carbon contamination was also observed.

One of the experimental difficulties with this technique is the fact that the number of recoiling ions is always much less than the number of elastically scattered beam particles. At some sacrifice in depth resolution, this problem is usually avoided by covering the detector with a thin absorber which stops the beam particle but passes the lighter recoiling nuclei. In the

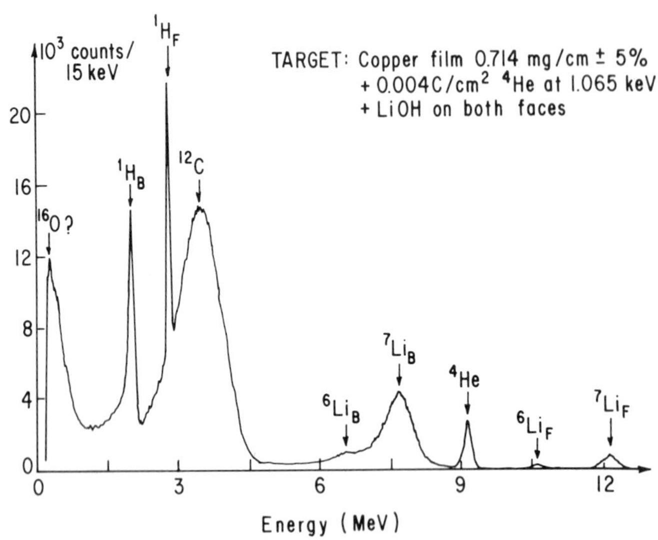

Figure 18. A spectrum of the recoiling particles measured from a thin copper sample coated with lithium and implanted with He. From (Le 76).

cases of thin films where a transmission geometry is possible, often the film itself can be used to stop the beam particle.

The use of 30–40-MeV mass 30–40 projectile for nuclear recoil analysis has not yet been widely applied to problems in physics or technology. However, B. Doyle and P. Peercy (Do 79) have used 2.4-MeV He and 12-MeV ^{12}C beams for recoil analysis of hydrogen in a variety of materials such as then film chemical vapor deposition silicon nitride.

In the future, with the development of ultrahigh vacuum chambers attached to heavy-ion accelerators, heavy-ion-induced recoil analysis is likely to become an important tool for the study of surfaces and the interaction between surfaces and the common light element contaminants: hydrogen, carbon, and oxygen. The use of heavy-ion counter telescopes in order to have redundant identification of the recoiling ions will also become more common.

Acknowledgments

The author gratefully acknowledges helpful discussions with many of his colleagues especially Paul LaMarche and Chandra Burman and the financial support of the Alfred P. Sloan Foundation and the Army Research Office. The author would also like to thank those researchers who kindly provided figures for illustration.

References

(Ad 74) P. N. Adler, E. A. Kamykowski, and G. M. Padwar, *Hydrogen in Metals*, American Society for Metals, Cleveland, Ohio (1974), p. 623.

(An 80) H. H. Andersen, J. Bottiger, and H. Knudsen, eds., *Proceedings of Fourth International Conference of Ion Beam Analysis*, North-Holland, Amsterdam (1980).

(Be 74) I. M. Bernstein, R. C. Dermott, and A. W. Thompson, eds., *Hydrogen in Metals*, American Society for Metals, Cleveland, Ohio (1974).

(Be 80) R. E. Benenson, L. C. Feldman, and B. G. Bagley, *Nucl. Instrum. Methods* **168**:547 (1980).

(Bo 48) N. Bohr, *K. Dan. Vidensk. Selsk. Mat.-Fys. Medd.* **18**:10 (1948).

(Br 77) M. H. Brodsky, M. A. Frisch, J. F. Ziegler, and W. A. Lanford, *Appl. Phys. Lett.* **30**:561 (1977).

(Ca 80) T. Cahill, Proceedings Pan-American Conference on New Uses of Accelerators, Mexico City, October 27–31, 1980.

(Ch 78) W. K. Chu, J. W. Mayer, and M. A. Nicolet, *Backscattering Spectrometry*, Academic Press, New York (1978).

(Ch 80) R. Chow, W. Lanford, and Wang Ke-Ming, Proc. Meeting of Electrochemical Society, Hollywood, Florida, October 5–10, 1980.

(Co 72) B. L. Cohen, C. L. Fink, and J. D. Degan, *J. Appl. Phys.* **43**:19 (1972).

(Co 80) T. Coatts, L. L. Kazmerski, and S. Wagner, eds., *Proc. Photovoltaic Material and Device Measurement Workshop*, San Diego, January 3–4, 1980, published in *Solar Cells* **2**:185 (1980).

(De 78) G. Dearnaley, G. M. McCracken, J. F. Turner, and J. Vince, *Nucl. Instrum. Methods* **149**:253 (1978).

(Do 79) B. L. Doyle and P. S. Peercy, *Appl. Phys. Lett.* **34**:811 (1979).

(Du 79) J. L. Duggan, I. L. Morgan, and J. A. Martin, eds., Proceedings of the Fifth Conference on the Applications of Small Accelerators in Research and Industry, *IEEE Trans. Nucl. Sci.* **26**:989 (1979).

(Fe 46) E. Fermi and W. N. Zinn, *Phys. Rev.* **70**:103 (1947).

(Go 79) R. Golub and J. M. Pendleburg, *Rep. Progr. Phys.* **42**:439 (1979).

(Ha 72) J. Haskell, E. Rimini, and J. W. Mayer, *J. Appl. Phys.* **43**:3425 (1972).

(La 76) W. A. Lanford, H. P. Trautvetter, J. Ziegler, and J. Keller, *Appl. Phys. Lett.* **28**:566 (1976).

(La 77a) W. A. Lanford and R. Golub, *Phys. Rev. Lett.* **59**:1509 (1977).

(La 77b) W. A. Lanford, *Science* **196**:975 (1977).

(La 78a) W. A. Lanford, P. H. Schmidt, J. M. Rowell, J. M. Poate, R. C. Dynes, and P. D. Dernier, *Appl. Phys. Lett.* **32**:339 (1978).

(La 78b) W. A. Lanford and M. J. Rand, *J. Appl. Phys.* **49**:2473 (1978).

(La 78c) T. Laursen and W. A. Lanford, *Nature* **276**:153 (1978).

(La 79a) W. A. Lanford, K. Davis, P. LaMarche, T. Laursen, R. Groleau, and R. H. Doremus, *J. Non-Crystalline Solids* **33**:249 (1979).

(La 79b) W. A. Lanford, *IEEE Trans. Nucl. Sci.* **26**:1795 (1979).

(La 80) W. A. Lanford, *Solar Cells* **2**:351 (1980).

(La 81a) P. LaMarche, W. A. Lanford, and R. Golub, *Nucl. Instrum. Methods* **189**:533 (1981).

(La 81b) P. LaMarche, Ph.D. Thesis, Yale University (1981).

(La 81c) W. A. Lanford, C. Burman, and R. Chow, private communication (1981).

(Le 73) D. A. Leich and T. A. Tombrello, *Nucl. Instrum. Methods* **108**:67 (1973).

(Le 74) D. A. Leich, R. H. Foldberg, D. S. Burnett, and T. A. Tombrello, Proceedings of Fifth Lunar Science Conference, Vol. 2 (1974), p. 1869.

(Le 76) J. L'Ecuyer, C. Brassard, C. Cardinal, J. Chabbal, L. Deschenes, J. P. Labrie, B. Terreault, J. G. Martel, and R. St.-Jacques, *J. Appl. Phys.* **47**:381 (1976).

(Li 74) E. Ligeon and A. Guivarc'h, *Radiat. Eff.* **22**:101 (1974).

(Lu 68) V. I. Lushikev, Y. N. Pokoftlousky, A. V. Strelkov, and F. L. Shapiro, JINR, Dubna, Preprint P3-4127 (1968).

(Ma 77) J. W. Mayer and E. Rimini, eds., *Ion Beam Handbook for Material Analysis*, Academic Press, New York (1977).

(Ma 80) J. W. Mayer, B. Y. Tsaur, S. S. Lau, and L. S. Hung, in *Proceeding of the Second International Conference on Ion Beam Modification of Materials*, Albany, July 14–18, 1980, North-Holland, Amsterdam (1980).

(Me 81) Y. Mehrotra, Ph.D. Thesis, R.P.I. (1981); and R. H. Doremus, Y. Mehrotra, W. A. Lanford, and C. Burman, *J. Mater. Sci.* **18**:612 (1983).

(Na 81) T. Narusawa, K. Kinoshita, W. M. Gibson, and L. Feldman, SUNY/Albany PSI Institute preprint (1981).

(Pe 79) P. S. Peercy, H. J. Stein, B. L. Doyle, and V. A. Wells, Proc. 7th International Conference on Chemical Vapor Deposition, Los Angeles, October 14–19 (1979).

(Ro 74) C. Rolfs and W. S. Rodney, *Nucl. Phys.* **A235**:450 (1974).

(Ro 77) J. M. Rowell, P. H. Schmidt, E. G. Spencer, P. D. Dernier, and D. C. Joy, *IEEE Trans. Magn.* **MAG-13**:644 (1977).

(Si 75) T. W. Sigmon, W. K. Chu, H. Muller, and J. W. Mayer, *J. Appl. Phys.* **5**:347 (1975).

(Sp 75) W. E. Spear and P. G. LeComber, *Solid State Commun.* **17**:1193 (1975).

(St 69) A. Steyerl, *Phys. Lett.* **29B**:33 (1969).

(St 79) I. Stensgaard, L. C. Feldman, and P. J. Silverman, *Phys. Rev. Lett.* **42**:247 (1979).

(Te 77) B. Terreault, J. G. Martel, R. G. St. Jacques, and J. L'Ecuyer, *J. Vac. Sci. Technol.* **14**:492 (1977).

(To 80) T. A. Tombrello, *Nucl. Instrum. Methods* **168**:459 (1980).

(Tu 68) A. L. Turkevich, in Surveyor Project Final Report, Nat. Aeronaut. and Space Administration Tech. Rep. 32-1265 (1968), p. 303.

(Zi 78) J. F. Ziegler, C. P. Wu, P. Williams, C. W. White, B. Terreault, B. M. U. Scherzer, R. L. Schulte, E. J. Schneid, C. W. Magee, E. Ligeon, J. L'Ecuyer, W. A. Lanford, F. J. Kuchne, E. A. Kamykowski, W. O. Hofer, A. Guivarch, C. H. Filleux, V. R. Deline, C. A. Evans, B. L. Cohen, C. J. Clark, W. K. Chu, C. Brassard, R. S. Blewer, R. Behrisch, B. R. Appleton, and D. D. Allred, *Nucl. Instrum. Methods* **149**:19 (1978).

10

RICHARD C. ARNOLD is currently on the staff of the Max Planck Institut für Quantenoptik in Garching. Educated at the University of Chicago, he received his Ph.D. in Elementary Particle Physics in 1962. From 1962 through 1965, he held a position as Research Assistant and Lecturer at the University of California at Los Angeles and then joined the staff of the High Energy Physics Group at the Argonne National Laboratory. He was Leader of the Theoretical Particle Physics Group from 1970 through 1975. Beginning with a sabbatical year at the Rutherford Laboratory in 1975–1976, he worked on the development of a heavy-ion-induced fusion power program and continued this work on his return to Argonne. In 1982 he joined the Max Planck Institut in Garching.

Heavy-Ion-Induced Fusion Power

RICHARD C. ARNOLD

1. Introduction

1.1. Inertial Confinement Fusion

Generation of useful power from nuclear fusion reactions is generally regarded as a realizable goal. There are basically two paths to this end, differing in their fundamental methods of confining the fusion fuel during thermonuclear burn.

Magnetic confinement fusion (MCF) requires a low-density plasma to be contained in a suitable magnetic field configuration while essentially steady-state thermonuclear reaction conditions are maintained; fuel densities are of order 10^{14} ions/cm^3. Inertial confinement fusion (ICF) utilizes instead a fast, nonequilibrium propagating burn wave in highly compressed fusion fuel; densities are of order 10^{24} ions/cm^3. Heavy-ion fusion (HIF) is a method for achieving ICF.

To produce sufficiently high fuel densities and an ignition temperature of order 5 keV per ion at the center of the fuel, a solid mass of fuel about a millimeter in radius needs to be compressed in a few nanoseconds (10^{-9} s) to densities of order 10^3 times that of normal solid hydrogen. This compression is achieved through the heat deposited by intense beams striking an external region of the fuel-containing pellet.[1] Instantaneous power of order 10^{14} W must be deposited in a small volume of order 10^{-2} cm^3. The heat thus generated is used to compress the fuel core, and ignite the burn through shock implosion. For power production, fuel pellets would be imploded at a rate of a few per second (see Figure 1).

RICHARD C. ARNOLD • Argonne National Laboratory, Argonne, Illinois 60439. Current address: Max Planck Institut für Quantenoptik, D-8046 Garching, Federal Republic of Germany.

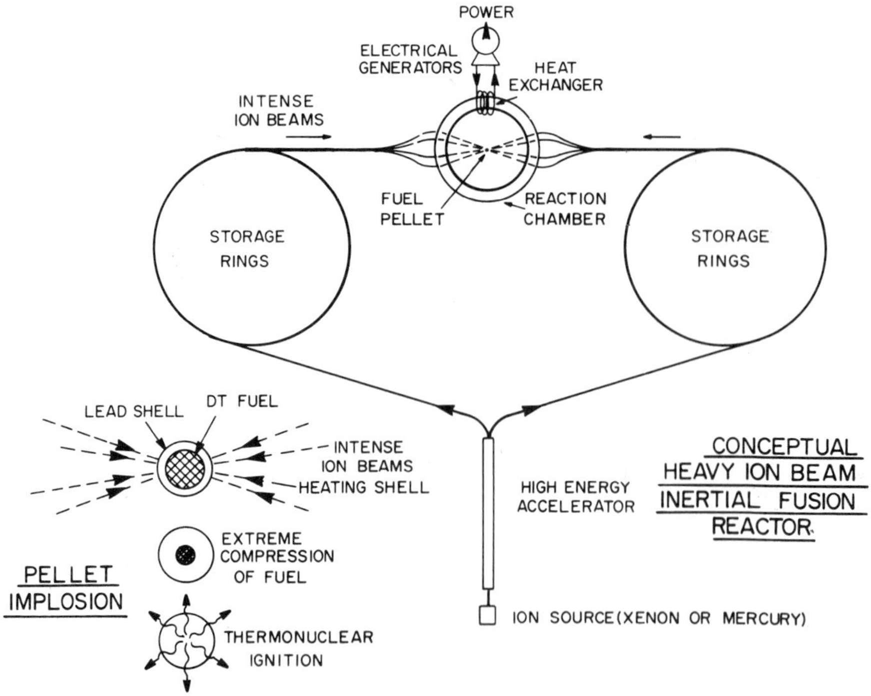

Figure 1. Schematic drawing of an HIF power reactor system with rf linac driver.

The instrument producing the beams is called the driver. There are several possibilities in principle for the technology of drivers: lasers, pulsed power charged particle sources, and heavy-ion drivers have been widely studied.[2] Each has advantages and disadvantages for ICF research.

A few driver types are expected to be efficient enough for commercial power production. The heavy-ion driver is regarded by many as the leading candidate for such an application if a conservative path is desired, as it is not only efficient and of high repetition rate, but the technology requires little development.

From a commercial perspective, ICF presents several advantages over MCF:

1. The reactor design can be simplified in important ways because the highly sensitive part of the system (the driver) can be physically isolated from the region of fusion power generation (the reactor chamber).
2. The basic reaction physics is similar in many respects to thermonuclear weapons physics; this lends confidence to feasibility studies of fusion burn and its effects on the reactor chamber.

3. In the case of HIF, the highest technology driver components are similar in design to that of existing research accelerators, so that little extrapolation is required for feasibility studies compared to MCF.

1.2. Driver Technologies

Lasers using Nd-doped glass, CO_2 gas, and a few other media have been used in ICF research. Most of the current data on the implosions of ICF targets have come from glass laser experiments with pulse energies of 10 kJ or less. That is about a factor of 100 smaller than the pulse energy which will be required for energy production, according to current estimates based on computer simulation of the plasma hydrodynamics in the pellets.

Charged-particle beam drivers (using either light or heavy ions) are expected to offer great advantages in electrical efficiency compared to lasers. For ion beams, energy deposition in the target is expected to be reliably estimated with classical concepts. These factors motivated the development of both pulsed-power light-ion ICF drivers and heavy-ion accelerator drivers[3] (see Figure 2).

Light-ion drivers use capacitor accumulation of the electrical input energy, followed by stages of power flow conditioning to increase the beam power on target to reach thermonuclear burn conditions in the target. Capacitors and other passive storage devices can supply at best a few megavolts of potential; thus the beam currents must reach several megamperes to deliver the instantaneous pulse power required. Such high currents require in turn neutralization of their space charge, and collective plasma phenomena are intrinsic in the transport and focusing of the beams. Experiments are currently under way to determine the technical limitations of this approach.[4]

High-energy beams of heavy ions can deposit the same power in comparable volumes of target material using 10^3 times more kinetic energy, as their range is short compared to light ions. This enables 10^3 times less beam current to be used, and it is possible to avoid collective plasma phenomena in the final stage. In estimating the behavior of the driver systems suitable for a power reactor, it is a definite advantage to be able to avoid difficult theoretical questions concerning beam transport to the target.

The energy accumulation in a heavy-ion driver is primarily in the kinetic energy of the ion beam, which receives a series of small impulses over a time during which the beam is accelerated and stored (in the rf systems). This method of accumulation is qualitatively different from the methods used in lasers or pulsed power accelerators, and is expected to be reliably estimated through experience with many high-energy research accelerators.

In this article we will discuss the heavy-ion driver approach to ICF.

Although no demonstration heavy-ion drivers have yet been constructed, there is high confidence among the community of accelerator designers that such a driver is feasible. This confidence has been expressed through the large workshops which were held in the US in 1976,[5] 1977,[6] 1978,[7] and 1979,[8] and other technical reviews.[9]

1.3. Power Production Requirements

In order that ICF power production be commercially feasible, an ICF reactor must produce much more electrical energy than is required by the driver to compress and ignite the fuel. This requirement constrains the electrical efficiency of the driver and the thermonuclear gain of the pellet to be high.

Let G be the target gain, defined as the ratio of thermonuclear fusion energy released when a pellet ignites to the driver beam energy incident upon the pellet. Let η_e be the electrical efficiency of the driver, the ratio of beam pulse energy to electrical energy required to accumulate that pulse. Let η_c be the conversion efficiency of thermonuclear burn energy into electrical energy, including heat-exchanger losses as well as Carnot efficiency of the electrical generators in the power plant.

If W is the beam energy per pulse, then the output electrical energy per pellet is $=\eta_c W$. Let the recirculating power fraction in the power plant be denoted by $r = W/\eta_e$. Then we have

$$r = (\eta_e \eta_c G)^{-1}$$

For commercial power plants we must have a small value of r; otherwise the operating cost will be excessive, since the recirculated power is not available for sale. Conventional planning criteria specify r to be equal to (or less than) 0.15.[9]

For $\eta_c = 0.3$ (a reasonable thermal efficiency) and $\eta_e = 0.4$ (possible for ion drivers), with $r \leqq 0.15$ we obtain the constraint

$$G \geqq (r\eta_e \eta_c)^{-1} \cong 50$$

Such a large gain requires a high burn efficiency. Detailed theoretical estimates for several types of HIF targets indicate that a driving energy of order 5 MJ would be required to achieve such a gain.

Overall power plant systems studies based on heavy-ion drivers have been carried out; capital costs are estimated to be reasonable, especially if a single accelerator can drive several reactor chambers making up a relatively large energy generating complex. Smaller systems may also be feasible, however; there has been little work to date on optimization of target–driver combinations which might provide lower driver costs. We will further examine cost scaling in a later section.

Thermonuclear mixtures containing tritium are easiest to ignite, but require breeding the tritium, e.g., from lithium. Other fuel cycles are possible, including pure deuterium burn (DD reaction) or the pB reaction. These are more technologically demanding, requiring higher ignition temperatures. However, neither of these reactions require nuclear fuel processing, and both use highly abundant raw materials. In addition, the pB reaction produces three alpha-particles instead of neutrons. This would allow more compact shielding and greatly reduced structure activation. Thus an ultimate fusion power scenario might include burning only D or B, although no specific economically feasible reactor has yet been designed for these fuels.

Electrical power production is not the only possible commercial use of ICF. Three other possibilities have been studied to some extent: synthetic fuel production, fissile fuel breeding, and vehicle propulsion.

Synthetic fuel processes, such as hydrogen production from water or methane production from coal, can be driven with any intense source of energy. ICF could provide such a source if a suitable reactor chamber blanket is engineered. This possibility has been studied in the MCF context, but most of the basic concepts can also be applied to ICF.

Fissile fuels to supply conventional nuclear power plants can be bred with fusion neutrons if a chamber blanket containing fertile material is used. Again, studies have been done primarily for MCF.

Direct impulse propulsion of very large spacecraft might be feasible with ICF. This concept was proposed some time ago using small weapons-type explosives (Project Orion); some studies of the ICF version have subsequently been published. The use of fusion power for smaller vehicles appears impractical at this writing because the driver is of such large scale, but future technology advances could change that picture.

2. Heavy-Ion Drivers

2.1. Basic Concepts

Two disparate approaches to the design of high-energy heavy-ion drivers have been pursued to date. The first (chronologically) uses circular storage rings loaded by rf linear accelerators and/or synchrotrons. It is modeled after conventional designs used in high-energy physics research. The second, less highly developed, uses a linear induction accelerator for single-pass acceleration of a single bunch of ions. In this discussion we will treat the rf approach in more detail, as it is closer to existing accelerator technology.[7]

In each case, ions of mass number A equal to or greater than 100 are generated in a singly charged state by an intense ion source. Acceleration by dc potentials is used to achieve a kinetic energy around 1 MV. Then, typically along with stripping to a somewhat higher charge state (2–4), a repetitive (rf or pulsed cavity) accelerator structure is used to increase the ion energy to 10 GV or more.

Radiofrequency designs accelerate a few tenths of an ampere of current and use storage rings at the full energy to accumulate circulating currents of order 20–30 A in each ring. When fully loaded, the circulating beams are extracted, compressed in time to a pulse length of order 10 ns, and focused on a target spot with radius of order 1 mm. A conceptual reactor scheme is shown in Figure 1.

The linear induction design uses about 10^3 ferromagnetic induction modules, each giving one longitudinal accelerating and/or bunching impulse to the single bunch of ions passing through. The final pulse length is adjusted to reach the value required by the target (again of order 10 ns) and focused on the target spot (of order 1 mm).

Radiofrequency driver technology would be based on adapting high-energy accelerator designs. The technical issues which arise have been thoroughly discussed in the workshops. The following key areas, which have a large influence on driver feasibility, were deemed worthy of further research and development efforts:

1. Ballistic transport physics in the reactor chamber;
2. Final focusing aberrations;
3. Bunching limitations;
4. Storage ring tolerances and beam loss mechanisms;
5. Low-velocity acceleration and preservation of beam quality in the accelerator.

The beam quality in a pulse is characterized by the longitudinal momentum spread, $\Delta p/p$, and the transverse emittances ε_x and ε_y which are a measure of the volume in four-dimensional transverse phase space occupied by the beam pulse.

Each of the five areas above influences in some way the number of storage rings and beamlines necessary for a given target requirement, and hence is important in estimating the cost and complexity of the driver. The state of the art at this writing appears adequate to ensure that a driver is quite feasible, both technically and economically, for delivering 5-MJ pulses of 10–20-GeV ions of $A \geqq 200$ onto a spot of radius 2 mm. Conceptual designs will be outlined in the next section.

2.2. Reference Designs

Detailed prototype design concepts for rf linac, synchrotron, and linear induction ion drivers for power reactors were presented at the 1978 HIF workshop. Other reference designs have been proposed since then, taking advantage of community criticisms and refined design considerations.

In addition, designs and proposals for smaller accelerator systems for research have been presented since 1979 by Argonne and Berkeley groups. Balance between cost and capability is difficult to define for research systems which would require $50M or less for construction. Some target experiments appear feasible in that regime, but the same cost scaling which is favorable for large pulse energy is definitely unfavorable for small energies. It is this circumstance which may shift the emphasis in HIF planning to emphasize modification of existing or planned accelerators to allow beam and target experiments, rather than building specialized small ICF drivers. We will review here design concepts for multimegajoule ion drivers.

A heavy-ion driver system of each type can be conceptually divided into several stages to facilitate discussion of the technical issues:

1. Ion source and preaccelerator;
2. Low-velocity linac;
3. Main accelerator system;
4. Bunching/accumulation stage;
5. Beam transport to reaction chamber;
6. Final focus lens arrays.

We will sketch three reference driver designs spanning the range of options for rf accelerators, and one linear induction design example.

The rf accelerator system options discussed and shown in Figure 2 are

A. RF linac with accumulator ring;
B. Single-pulse synchrotron with rf linac injector;
C. Single-pulse synchrotron with rapid cycling synchrotron injector.

Each of these examples is a design of a system to accelerate Hg ions and delivers 5 MJ in a 25-ns pulse to a target spot of radius 2 mm, with a specific energy deposition of 20 MJ/g. The minimum pulse repetition rate is 2 s for case C, of order 1 s for case B, and of order 0.1 s for case A. Because of the superior repetition rate and electrical efficiency of a linac, case A is a plausible candidate for a power plant driver. The other two circular accelerators should be more appropriate to a high-gain research facility.

The parameters given are not necessarily accurately determined, but rather serve as typical values for such driver designs. These design parameters have not been constrained to eliminate the longitudinal

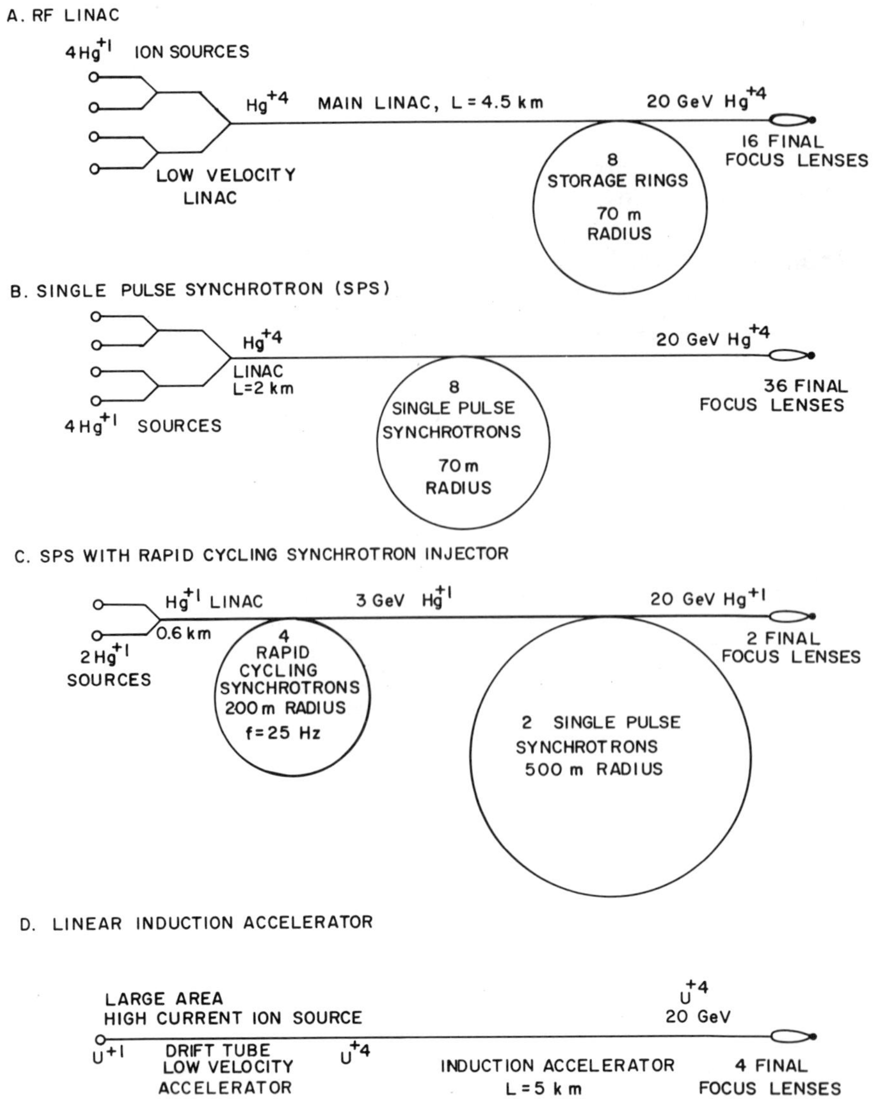

Figure 2. Reference designs for HIF drivers, with parameters.

microwave instability (see Section 2.3). If damping of that instability is not feasible, more rings and beamlines may be necessary.

The driver architecture and parameters are shown in Figure 2, which is not to scale. A linear induction scenario, D, is also shown for comparison.

Analysis at Argonne[(10)] of the comparative capital costs of the rf drivers showed that C was least expensive, and A was the most expensive. However, the synchrotron systems (B and C) would not be feasible if the ion–ion charge-changing cross sections were pessimistically estimated using geometric ion sizes. Measurement of such cross sections would be essential before further substantial effort on these systems would be warranted.

2.3. Feasibility Constraints

There is widespread agreement[5–7] that accelerator systems can be designed which meet the presently understood requirements for driving high gain targets. However, with limited resources for research and lack of unanimity on priorities, costs are an important issue. To some extent it is possible to trade off cost issues against conservative design principles if the margin for error were to be better known in all parts of the system. Several technical uncertainties lead to constraints in design which could have a substantial influence on the cost or technical margin for error in operation of such a driver. The areas of these uncertainties include: focusing constraints, beam transport in a gas-filled chamber, current limits on beam transport in beamlines, storage ring instability control (rf system) or longitudinal beam bunch control (linear induction system), and ion–ion charge-changing cross sections (in synchrotron systems). In each case, there exists knowledge adequate for a conservative design (at least with an rf linac), but optimizing the driver may require additional information.

2.3.1. Final Focus

Aberrations are important in focusing the ion beams on small target spots of order 0.1 cm^2 area or less, with magnetic quadrupole lens arrays. These aberrations can depend both on beam size (geometric aberrations) and on longitudinal momentum spread (chromatic aberrations). Both types increase with beam emittance, and can be reduced at small ε. However, very small emittances of beams in the accelerator system are difficult to maintain and may lead to costly drivers. Accurate estimates of the spot sizes achievable with given ε and $\Delta p/p$ are therefore valuable in the planning of a system.

Arrays of correction elements (sextupole and octupole magnets) can be placed in the beamlines to decrease the aberrations. Analysis of the feasibility of such correction arrays is required in each driver design if optimum ε is to be obtained; but this process is not yet well systematized, and a comprehensive method is needed.

2.3.2. Transport in Gas

If the reaction chamber contains a gas pressure of 10^{-4} Torr or higher, calculations show that plasma effects on the ballistic beam transport may be important. Such a gas may arise from a variety of sources; pellet debris, wall or protective layer evaporation, or an external source of gas fill. The latter has been proposed as one mechanism for partially protecting the inner wall of the reaction vessel against x-ray damage from the pellet burn, and also for partially charge-neutralizing the ion beam to reduce space-charge defocusing effects.

Reactor concepts with chamber pressures below 10^{-4} Torr have been investigated; both HIBALL and the Westinghouse HIF study conservatively adopted such a scheme. The introduction of a gas fill could, however, increase the range of reactor and driver design options and perhaps lead to a more reliable and/or less costly system. There is therefore a need for reliable prediction of ion beam propagation in gases.

2.3.3. Beam Line Transport

The number of beamlines and their cost depends in many cases upon the maximum current which can be transported in each without excessive emittance growth. In a perfectly matched periodic line, the transverse emittance does not grow. However, space charge effects prevent perfect matching uniformly through the volume in the beamlines. In addition, instabilities can be shown to develop at sufficiently high currents. Some of the linear problems have been investigated theoretically, and conservative estimates have been achieved which constrain the beam currents. Experimental verification of these results is needed. In addition, experimental investigation in parameter regimes in which strong emittance growth is predicted would be valuable to find out if the expected emittance growths are realistic or whether the theory is too conservative.

2.3.4. Storage Ring Instabilities

The rf reference designs discussed earlier, as well as those in the 1979 workshop, have been checked for possibly troublesome instabilities affecting the accumulated beam of ions in the storage rings. Most possibilities have been ruled out on theoretical grounds. However, one instability appears to be a strong constraint on the current which can be accumulated in a ring. This is the "longitudinal microwave instability," which arises from high-frequency beam–wall interactions and has a disruptive effect on the longitudinal momentum spread in a bunch. The threshold and growth rate for this instability allow it to appear in HIF circular machines.[(8)] The effect has been seen experimentally in proton machines.

This instability can be avoided in principle by using a larger number of storage rings, each with a beam of smaller emittance and lower current compared to a system with a minimal number of rings. A larger number of beamlines would be expected also. Such a solution would lead to a somewhat more expensive driver; however, the cost differential is not well established at present.

Experiments on this instability can be carried out on some intense circular proton machines, such as the SNS at the UK Rutherford Laboratory. The linear theory should be verified and possible damping mechanisms (intrinsic or extrinsic) should be explored.

2.3.5. Longitudinal Control in Induction Accelerators

A linear induction accelerator for high-energy heavy ions would be composed of a large number (10^3 to 10^4) of induction modules, each driven with a shaped electrical impulse to accelerate and/or shorten the beam pulse. If the beam pulse is not to suffer dilution in longitudinal phase space the phasing and pulse shape of each impulse must be precisely controlled and precisely induced into the ion bunch. There is as yet no operating experience with more than a few modules, and there only with relativistic electrons which do not require such precise pulse control. Some theoretical studies have been done, but more are required before it is clear just how easy or difficult the design of the induction linac driver will be. Beam experiments in this area could be carried out with a proton linear induction accelerator if available.

2.3.6. Ion–Ion Cross Sections

If a synchrotron is used as the main accelerator there is a need for maximum storage time in the rings, since slower cycle times lead to less complication in preserving the longitudinal beam quality during acceleration compared to rapid cycling systems. Storage time is limited both by collisions with background gas and by collisions between ions circulating in the beam. The former can be alleviated by using a higher vacuum (state of the art in proton storage rings), but the latter is intrinsic with the beam parameters, and must be reduced by choosing a species of ion which has a low cross section and if necessary by reducing the circulating currents.

Theoretical estimates of reasonable reliability are available for the high-velocity background gas collisions. However, the cross sections for ion–ion charge-changing collisions at the modest center-of-mass energies encountered in HIF drivers are very difficult to estimate theoretically.[11] Various models give quite different estimates. Conservative estimates, based on the geometric sizes of the ions, are adequate for rf linac or linear induction devices. However, those cross sections are often too large for interesting synchrotron possibilities. For research machines (as distinct from reactors) high charge states may be preferable, as we indicate in the next section. Thus experiments on charge-changing cross sections for ions such as Xe^{+8} would be valuable. Existing data cover a few singly charged ions only.[11] The relevant collision energies are around one keV per nucleon.

2.4. Costs

One of the attractive features about heavy-ion fusion is the applicability of existing technology in deriving construction and operating cost estimates for a large driver which will be necessary for a power plant. Research machines for high-energy physics have been operating for a decade which have the size and scale appropriate for HIF reactors.

Scaling studies show a nonlinear relation between attainable pulse energy and driver cost, in which a great economy of scale is apparent at large energies. One of the basic reasons for the nonlinear cost scaling is the decrease in space-charge effects at high kinetic energies, for a given amount of pulse energy. This effect can be qualitatively illustrated by a simplified analytic cost scaling analysis for rf linac drivers. A similar analysis can be carried out for synchrotron drivers, with qualitatively similar results.

The choice of charge state to be used in the ions can be varied to minimize the overall cost of the driver. We adopt a cost function consisting of the sum of linac and storage ring terms: $C = C_L + C_R$.

The cost of a high-energy rf linac accelerator depends strongly on the final energy, but more weakly upon the current to be accelerated, provided the structure costs are larger than the power supply costs. This circumstance holds over a limited range of design parameters where instantaneous linac power requirements are not too great. It is a useful simplifying assumption for the purposes of our discussion. We will model the linac cost therefore by

$$C_L = aV$$

where V is the final kinetic energy of the ions divided by their final charge state, and a is considered constant over the range of driver parameter variations considered.

The storage rings are assumed now to be responsible for the bulk of the remainder of the driver cost. This has been found to be the case in several reference designs.

The cost of a storage ring appropriate for HIF scales roughly as the product of magnet aperture area and average magnetic rigidity; the latter is proportional to the ion momentum p divided by the ion charge state q. This scaling holds only at large apertures.[12] We will further simplify by assuming the aperture area scales proportionately to the emittance of the beam (assumed equal in the two transverse planes), ε. This is valid when the period of the betatron oscillations is held constant as the size of the ring is varied. We thus model the cost of the storage rings by

$$C_R = bN_R\, p\varepsilon/q$$

where N_R is the number of storage rings required, and b is constant over the parameter range to be considered.

The relation between N_R and other parameters is determined by choosing N_R to be the minimum which allows a given total circulating ion energy W to be stored, consistent with constraints on beam intensity in each ring arising from space charge defocusing. If N_i is the number of ions stored in one ring, the space charge constraint is conventionally estimated as

$$N_i = 1 \times 10^{13}\, A \in B_F \beta^2\gamma^3/q^2$$

where A is the atomic mass of the ion, q its charge state, ε the emittance of the circulating beam, B_F the longitudinal bunching factor (fraction of ring circumference occupied by circulating ion bunches) during accumulation and storage, $\beta = v/c$, and $\gamma = 1/(1-\beta^2)^{1/2}$, where v is the velocity of the ion and c the velocity of light.

We will consider in this discussion only ions in the nonrelativistic regime (v/c less than about 1/3), so that nonrelativistic kinematics are adequate. Then we can use the approximations

$$A\beta^2\gamma^3 \cong 2T$$

where T is the kinetic energy of the ions; and (in suitable units)

$$p \cong (2AT)^{1/2}$$

Then if we consider B_F to be held constant over the parameter variations considered,

$$N_R = W/N_i T \propto Wq^2/T^2\varepsilon$$

The total driver cost is then modeled by

$$C = C_L + C_R = aT/q + bWA^{1/2}q/T^{3/2}$$

where the constant b has been redefined.

Note the stored beam emittance does not appear explicitly here. For small ring apertures, the ring cost does not decrease with emittance and the driver costs for a given W increase as the emittance decreases.

We consider T and W as fixed by target requirements, in the following. We find the cost-minimizing charge state for a given T and W by setting $\partial C/\partial q = 0$, which gives

$$q_{\min} \propto T^{5/4} W^{-1/2} A^{-1/4}$$

Substituting this result into the cost formula we obtain

$$C_{\min} \propto T^{-1/4} A^{-1/4} W^{1/2}$$

This simple model suffices to illustrate several features found consistently in detailed rf system designs:

1. Cost per megajoule decreases as W increases.
2. There is an optimum charge state, which decreases as W increases.
3. High kinetic ion energies T, if allowed by target design, can enable costs of the driver to be reduced; if there is no cost penalty for reducing emittance, T can be increased to the limit of feasibility, thus lowering C.

Exploitation of the latter feature requires that target and driver parameters be mutually constrained. However, only a small range of target parameters have been systematically explored to date.

Synchrotron accelerators may provide an additional way of reducing the costs of producing high-kinetic-energy beams. More work on these aspects of HIF should be useful in reducing driver costs, especially for research facilities preliminary to full-scale reactor drivers.

3. Targets

3.1. Principles

Compression and ignition of the fuel in an ICF target entails first a conversion of beam energy into thermal energy in an outer portion of the target, followed by compression of the core of fusionable fuel. The former process can be highly efficient with ion-beam drivers, but the latter has an efficiency of typically 1/10. Fusion power generation requires an overall target gain of at least of order 10^2. Thus the fusion yield of the fuel core must be at least 10^3 times the energy delivered into the fuel for compression and ignition.

For homogeneous cores the highest gain is obtained with a DT fuel mixture, with equal numbers of deuterium and tritium ions, which undergo the $D + T \rightarrow n + \alpha$ reaction. If DT (at any density) was uniformly heated to a temperature of 10 keV, as would be necessary for a burn to proceed faster than explosive disassembly in a uniform core, then the fusion gain of the fuel core would be roughly 500; the burn efficiency is roughly 30% under these conditions, and the output energy is that carried by the 14-MeV neutrons. This highly idealized model would have a gain which was not quite adequate for power production. Imperfections and additional losses would further decrease the yield.

Thus to obtain sufficient gain it will be necessary to produce a thermonuclear burn which in partially self-igniting.[13] Only a central portion of the fuel core should reach 5–10 keV temperatures; the early burn process at the center should then heat the remainder of the fuel to ignition temperatures. Computer simulations and simple analytic models for yields of 10^3 MJ or less show that the majority of the energy delivered to the fuel should be used to compress, rather than to heat, the fuel, to achieve gains of 10^2 or greater.[14,15]

The compression criterion for a burn to propagate with high efficiency is

$$\int_0^\infty \rho \, dr \gg \lambda_\alpha$$

where ρ is the final fuel density, r is the radial coordinate measured outward from the center, and λ_α is the range of the 3.5-MeV α particles generated in the fusion burn. These α particles redeposit their kinetic energy in the remainder of the fuel and will be responsible for the self-igniting aspect of the burn, when the above criterion is satisfied.

The goal of target theory for commercial energy is thus to design a pellet which achieves the compression criterion above and heats a small central region of the compressed fuel to ignition temperature (about 5 keV) while minimizing beam power and fuel mass.

To achieve such a state of the fuel with high compression will require a high degree of spherical symmetry in the implosion process. Illumination symmetry is an important consideration. Hydrodynamic (e.g., Raylleigh–Taylor) instabilities may also arise during the compression. These questions have received a fair amount of theoretical study, but many issues will remain unresolved until ICF experiments at high compressions can be carried out with adequate diagnostics. At this writing, volume compressions of DT to around 100 times solid density have been reached by 10-kJ laser experiments, with poor statistics.

3.2. Designs

One of the simplest spherical target concepts using an ion beam driver is shown in Figure 3. A metal shell approximately 1 mm inner radius is filled with DT gas or liquid. The thickness of the shell is determined to be slightly greater than the range of the ions. The beams thus penetrate only partially through the shell. Their kinetic energy heats the deposition region to temperatures between 100 and 400 eV. The inner unheated shell layer, called the pusher layer, is compressed inwards by the reaction from the rapid expansion of the outer part of the shell. The pusher in turn compresses the DT fuel and serves as a tamper after burn begins. Maximum compression is

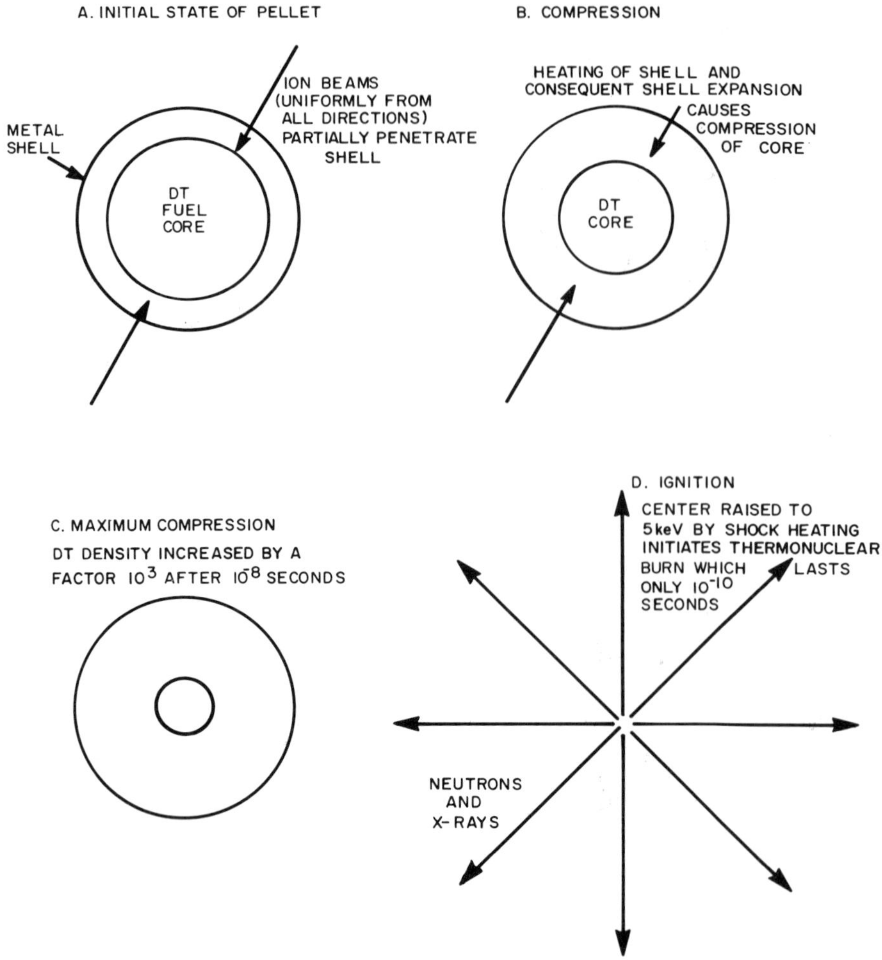

Figure 3. Simple HIF spherical target design.

reached after approximately 10 ns. If the fuel has undergone an appropriate compression history it will be heated to ignition temperature (about 5 keV) at maximum compression, and a self-igniting burn begins. The burn is complete in less than 0.1 ns. With such a simple target, it is not known how to arrange an efficient compression and ignition sequence; the calculated gains are not high enough for power production.

Calculations of the performance of such targets were first published by Clauser in 1975.[16] Those calculations showed that fusion gains of order 10 could be obtained with a pulse shape constant in time, both in power and particle kinetic energy. A more detailed study was later carried out by Gula and Magelssen,[17] and gains of order 30 were calculated with constant 3-MJ pulses of 10-GeV Hg ions.

Higher gains for such simple targets have been calculated[18] by Magelssen using pulses of ions of two different kinetic energies; a low-energy and low-power compression pulse of duration about 10^{-7}s, followed by a final high-energy, high-power pulse. The compression pulse, penetrating about halfway through the shell, compacts the inner part of the metallic shell; a dense pusher layer is created, and the fuel is partially compressed. The final high-energy pulse penetrates partially into that pusher layer, causes some additional fuel compression, and shock-heats the center of the fuel to ignition temperatures. Gains of 175 have been calculated, using roughly 1 MJ of 5-GeV Hg ions at 10 TW for compression and 2 MJ of 50-GeV Hg ions at 500 TW for the final pulse. Drivers to produce such ion beams appear roughly as feasible as single energy ion beam drivers. Separate storage rings would be used in rf systems.

Irradiation symmetry for such targets is a detailed issue, involving ion ranges. Preliminary studies[19] indicate that for thick enough pusher layers, more easily produced in the dual-energy approach, adequate compression uniformity may be obtained with 20 or more ion beams. Thin pushers might require many times more beams.

Multiple shell targets are another approach to obtaining higher gain. Instead of multiple energy beams, the mechanical target construction can be adapted to produce a more efficient compression–ignition sequence. Calculations of such targets have been published by Bangerter and collaborators, and gains exceeding 100 are predicted when driven with appropriate ion beams.[20]

Another possible target geometry is an imploding cylinder. The operation would be similar to that of the simple spherical shell target, but shell heating would be accomplished by long-range small-diameter ion beams which penetrate the shell parallel to the axis. Driver design in this case would be oriented toward high-kinetic-energy ions with small emittance beams, compared to spherical targets. A typical example considered by the author is a driver producing 5 MJ of 500-Ge Hg ions focused to 100 μm diameter needle beams. To maintain adequate implosion symmetry the beams can be scanned in a circle by electromagnetic deflectors with a frequency of several GHz. Computation of target performance indicates this geometry can achieve fusion gains comparable to that of simple spherical targets.

Thus, at the time of this writing, there are several opportunities for high gain targets which are still relatively simple in design. The actual target performance could be predicted with reasonable accuracy if the relevant physics data bases for intense ion beam interactions are able to include the equations of state of various shell materials in regions of 50 to 500 eV temperatures, at both low and high densities, and data on energy deposition and energy partition in shell materials at temperatures from 50 to 500 eV at normal or lower densities.

Some information of this kind can be obtained when any ion beams of sufficient power density become available. An example of possible research in this area is provided by proton beams which in principle could be focused to very tiny spot sizes (10–100 μm radius). Beams of sufficiently high intensity and high quality (low volume in phase space) might be obtained from existing high-energy physics accelerators such as those at Fermilab or CERN. This possibility has been explored[21] and preliminary results are encouraging. Although not cheap, such experiments would be possible without construction of completely new accelerators. Cylindrical geometry targets could be useful for implosion experiments using existing accelerators if they were modified to accelerate fully stripped heavy ions.

Physics information about fuel behavior, e.g., the equation of state, relevant to compression and shock heating may be obtained from other sources such as laser and light ion ICF experiments, and study of weapons physics.

Other target designs using concepts or data currently restricted by classification have been studied at Livermore and Los Alamos. Some of these include the recently declassified concept of radiation coupling.[22] These targets cannot be discussed at the present time in the open literature. However, independent research efforts may show the way to utilization of such concepts.[23]

4. Reactors

4.1. General Considerations

In an ICF power plant, the energy released by pellet explosions must be contained and transformed into useful forms. For DT fuel, about 80% of the output energy is carried by 14-MeV neutrons which escape from the pellet fireball. These can be absorbed in a "blanket" of material about one meter thick around the target chamber. The blanket material heats up, and a heat exchanger communicates this thermal energy to an electrical generation system, for example a water boiler and steam turbine generator.

Tritium for the fuel pellets can be bred in the blanket structure if lithium is introduced, either as an integral part of the blanket (e.g., liquid Li) or in a circulation system which is separate from the bulk of the blanket material.

Design of such reactor systems has received a considerable amount of attention. Two detailed feasibility studies have been carried out specifically for heavy-ion drivers: HIBALL,[24] a collaboration between the University of Wisconsin and several groups in the Federal Republic of Germany, and a study by Westinghouse[25] funded by USDOE. Both studies reach optimistic conclusions on reactor feasibility. The Westinghouse study stresses maintenance and material areas, while HIBALL addresses also driver and target design questions in some detail.

The lifetime of the inner (first) wall of the reactor chamber is an important factor in these feasibility studies. X-ray bursts from the fusion burn can heat this surface to sublimation temperatures. There may then be excessive erosion and/or shock damage to the wall structure. Pellet yields for a power reactor may be limited by these considerations. Excessively large reaction chambers are expensive and difficult to maintain, and driver costs increase if the pellet requires more driving power, or if the lens-to-target distance is increased, all features of a reactor which has no wall protection scheme.

One method of protection of the inner wall against x-ray ablation damage is to use a low-pressure fill of gas such as neon, which can absorb part of the x rays and slow down the thermal energy transport to the wall. This gives time for conduction of heat away from the inner surface so that ablation temperatures are not reached. However, shock waves would be generated in the gas, and these might damage the reactor walls if the gas density is excessive. In addition, plasma phenomena could disrupt the beam of heavy ions. Theoretical studies have indicated that a gas pressure of order 1 Torr might be useful.

Another approach to wall survivability is through careful matching of materials and pellet yields. The Westinghouse study proposed using the same high-sublimation metal (Ta) for first wall and pellet shell, and also limiting the pellet yield to just avoid heating the first wall to sublimation temperatures. Thus not only was shock damage reduced, but wall erosion did not take place; instead, the pellet debris depositing on the reactor walls thickened them with time. A specially tailored pellet was fired on occasion to clean off the excess.

4.2. Examples

HIBALL was a conceptual design study of a heavy-ion beam driven fusion reactor used as an electrical power plant. Its parameters were chosen to represent a design which would be plausible from both technical and economic viewpoints. Parameters given here are rounded from values contained in a preliminary report written in 1981.[24]

The net electrical power output was 4 GW, obtained from 8 GW of fusion power release. An rf linac accelerator was designed, accelerating Bi^{+2} to 10 GeV with 30% electrical efficiency. The beam pulse energy on target was 5 MJ, delivered in 50 ns. The pellet repetition rate was 20 Hz, divided into four reaction chambersoperating at 5 Hz each. Target gain was 80; target shells were mainly PbLi with a Pb coating. 20 beams were used in each reactor chamber; each carried 2.5 kA of current.

In the reactor walls, PbLi compound was used as a breeder and coolant, which reached a maximum temperature of 500°C. The tritium inventory was 1 kg. A first wall lifetime of 20 yr was expected. The first wall was made of ferritic steel and was 7 m from the target. It was protected from x-rays, ions, and neutrons by a dense array of porous SiC tubes containing a flowing PbLi, placed 2 m in front of the first wall. This array requires replacement every 2 yr.

Preliminary economic analysis yielded a capital cost of $1800 (in 1981) per kWe, and a busbar cost of electricity of 41 mills per kWh. These costs were considered competitive with earlier magnetic fusion plant estimates by the Wisconsin group.

A picture of a scale model of another HIF power plant concept is shown in Figure 4.

The Westinghouse HIF study,[25] contemporaneous with the preliminary HIBALL work, also used rf linac technology, with 30% electrical efficiency. A 2-MJ driver pulse energy and a pellet gain of 175 were assumed, a more optimistic pellet performance than the HIBALL study. Design of such a pellet was not attempted. One reactor chamber was used; the repetition rate was 10 Hz; and approximately 1300 MWe was generated. The reactor cavity was 10 m in radius. Tantalum was used as a coating on the first wall, and also as a pellet shell material. Liquid lithium at 400°C was circulated in tubes to form a tritium breeding blanket behind the first wall. A sodium heat exchange circuit carried thermal energy to generating systems.

Special attention was paid to maintenance procedures and down-time requirements from a utility viewpoint, and environmental considerations. The study concluded that HIF reactors showed great potential as a commercially feasible source of power from fusion.

5. Prospects

5.1. Secrecy Problems

The classification[26] of a substantial portion of ICF target research (theory as well as experiment) has the consequence that technical

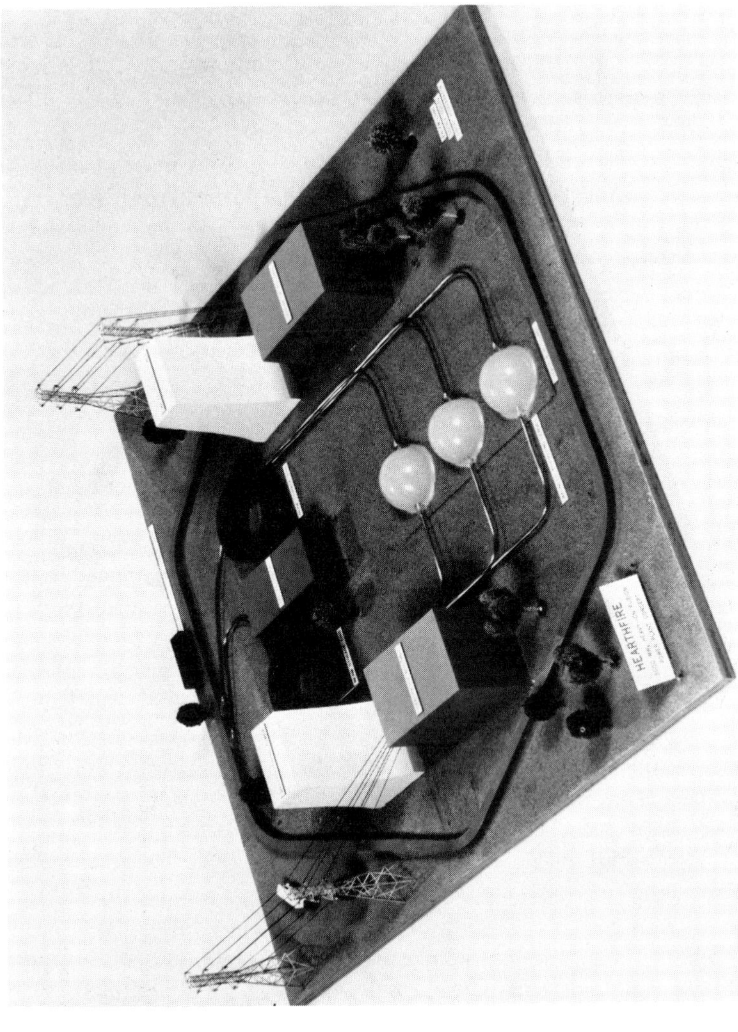

Figure 4. Scale model of an rf linac driven HIF power plant. This conceptual design delivers 3000 MWe and utilizes a 2.5-GV linear Hg accelerator, coiled twice around the site periphery. The hexagonal accelerator building is approximately 200 m on each side.

assessments of this work only draw on expertise from within the nuclear weapons community. Two unfortunate aspects of ICF research thereby arise.

Theoretical physicists cannot contribute critical judgement and innovative suggestions as a worldwide community, and thus the institutional processes so vital to scientific progress are denied in this field.

Competition for governmental program funds in this area is confined to those programs directed and administered through the weapons laboratories. This effectively prevents the development of ICF programs which are not oriented toward weapons physics since their scope is greater than private industry can support.

The reasons for tight classification restrictions on ICF remain controversial. From a purely technical viewpoint, declassification of some key concepts which took place during 1980 was considered by some experts to be a constructive step. However, there has been no subsequent declassification followup; rather, governmental restictions (at least in the US) have been tightened in many areas of sensitive technology.

The declassified material referred to here is reproduced as part of Appendix A. The release was apparently forced by legal activities in conjunction with the case of the *Progressive* magazine's article on thermonuclear weapons.[27]

5.2. Development Programs

Ideally, research and development programs should be carried out to eliminate the uncertainties described in Section 2. However, as of this writing (1982) the USDOE is not supporting research and development on any civilian power applications of ICF, including HIF. Some development work was done on HIF, primarily at Argonne and Berkeley, in 1977–1981.

An ion source, 1.5 MV preaccelerator, and several rf acceleration cavities were in operation at Argonne in 1981 at the time funding was stopped and the group disbanded. The preaccelerator and ion source are shown in Figure 5. Acceleration of 30 mA of Xe^+ with beam quality adequate for a full-scale driver had been demonstrated. This segment of the acceleration system was considered as the most questionable for rf linac development, and the Argonne program demonstrated its feasibility.[28]

At Berkeley, a Cs^+ surface ionization source adequate for beginning development of the linear induction approach was demonstrated. Some hardware development in addition was carried out at a low level. Theoretical work, e.g., on beam transport, is continuing.

At Los Alamos some accelerator development took place to adapt the low-velocity rf quadrupole concept to HIF. USDOE chose Los Alamos to lead the research in any applications to weapons development arising from HIF.

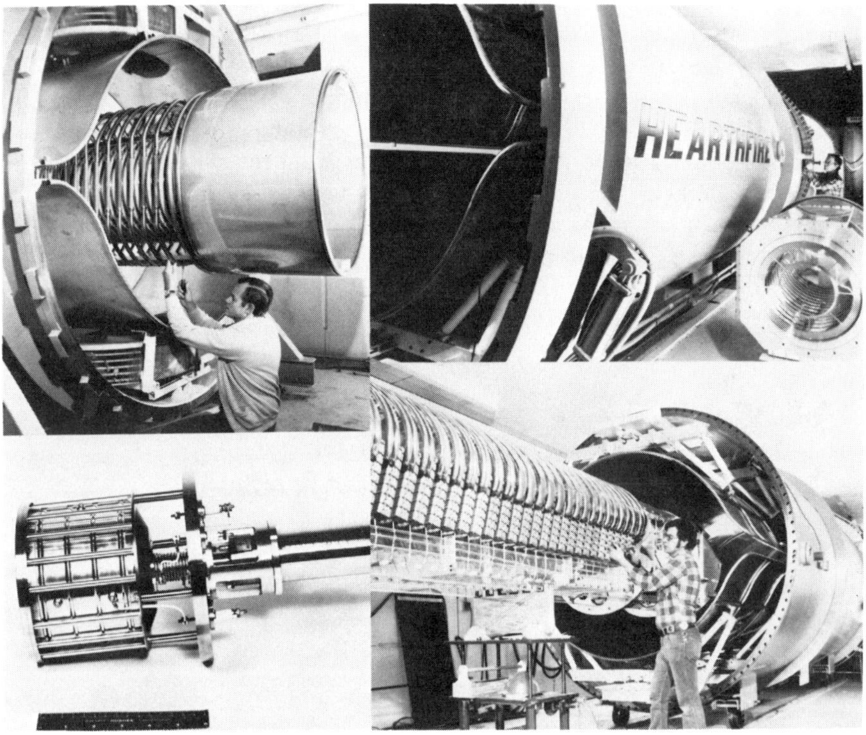

Figure 5. The Argonne heavy-ion preaccelerator and ion source. The modified Dynamitron high-voltage equipment shown here was obtained largely from a surplus NASA laboratory accelerator in 1976 and subsequently modified at Argonne for high-current beams. Clockwise, starting at upper left, is shown the high-gradient accelerating column, the pressurized enclosure, the 1.5-MV power supply, and the Xe ion source. The last item was developed and manufactured by Hughes Research Laboratories.

In Europe, two design study programs for civilian power applications of HIF are funded, one in the UK and the other in the Federal Republic of Germany. A small amount of FRG funding was shared with the University of Wisconsin in the HIBALL study. In the GSI laboratory at Darmstadt, some relevant studies of low-velocity heavy-ion linacs are being carried out. The UK studies have not resulted in published work to date, although informal reports on driver and target concepts have been circulated.

A small study program on HIF driver design exists also in Japan.[8]

The effectiveness of HIF research programs in the future will depend critically upon political decisions such as the US priorities for developing fusion power compared to weapons research.[29] A four-year proposal for HIF technology development first presented by Argonne in 1979 (and at one time approved by USDOE) is shown in Figure 6. At a cost of \$30M to \$40M (1980 dollars), demonstration of acceleration, storage, and targetting of Xe^{+8} at the 1–10 kJ level was to be carried out. Some target physics experiments could also be done. This proposal has been described at several conferences. Berkeley has also proposed a "test-bed" of similar scale for demonstration of key areas of linear induction technology for heavy ions. No other groups have put forth concrete facility proposals as of this writing.

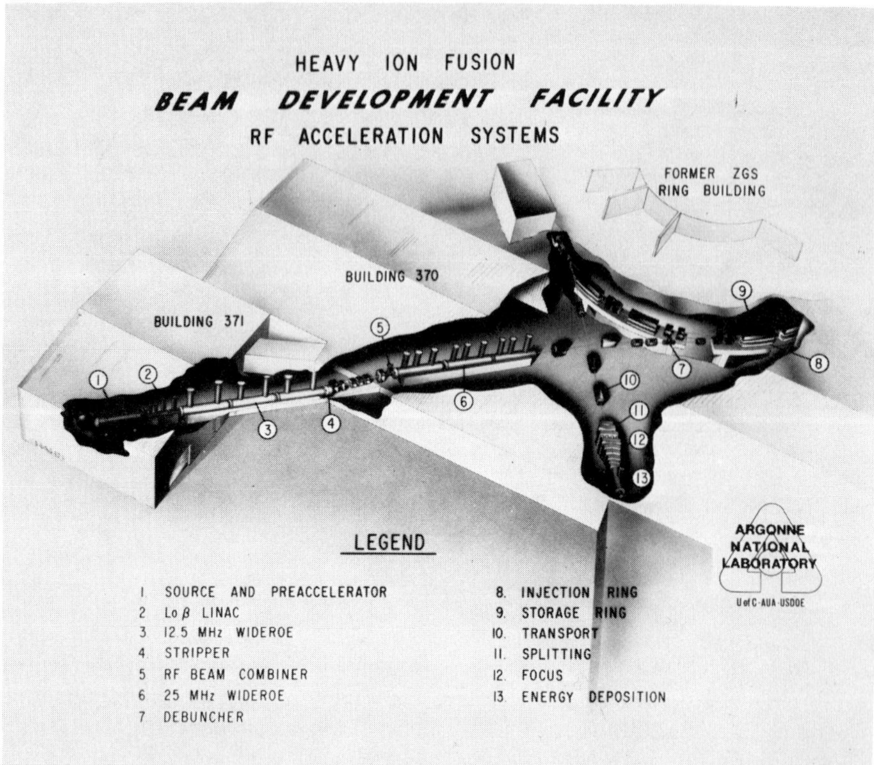

Figure 6. Artist's conception of the proposed heavy-ion Beam Development Facility (BDF) at Argonne. This facility was designed to provide a demonstration of acceleration, storage, and targeting of high-current, high-quality xenon beams. Synchrotron acceleration of the ions to 10 GeV was planned, to provide sufficient beam power for studies of ion energy deposition in dense plasma targets at 50 eV temperatures. The source, preaccelerator, and low-velocity linac structures were constructed by 1981.

5.3. Overview

Heavy-ion-induced nuclear inertial fusion is a promising technology for electrical power generation. In common with other nuclear fusion reactor concepts, HIF would provide an inexhaustible, fail-safe energy supply, ultimately using raw material for fuel which is available in all regions of the world in sea water. Research and development required for the driver is of reasonable scope. The fuel cycle and reactor development would be comparable to any other method for fusion power. Target design will require detailed experiments, but at the present time the available information from all sources (including classified data) leads to gain estimates which would suffice for power production with driver energies of a few megajoules.

Demonstrations of HIF technology is thus not obstructed for technical reasons. An investment within the capability of any highly industrialized nation would be required. Sooner or later it will be done, depending on the political priorities and the effectiveness of secrecy constraints.

The Energy Research Advisory Board of USDOE held a review of ICF under the guidance of Dr. John Foster. Excerpts from his report to ERAB are reproduced in Appendix B. The implication emerges clearly that driver development is closely connected with a choice of goals. Power production as a goal would favor emphasis on heavy-ion drivers, while lasers are adequate for military applications of ICF.

Note Added in Proof

For HIF developments since 1981, see Reference 30.

Appendix A: Statement of USDOE on Declassification Actions Issued on September 4, 1980

On December 27, 1979, the responsible DOE official, Duane C. Sewell, Assistant Secretary of Defense Programs, declassified certain general principles of operation of thermonuclear weapons which had been compromised by publication of the Hansen letter in the *Madison Press Connection* and the Morland article in the *Progressive*. Declassification was carefully limited to the minimum required. Most aspects of the design and fabrication of nuclear weapons are still classified and protected as restricted data under the Atomic Energy Act.

Specifically, the following general statement was declassified:

> In thermonuclear weapons, radiation from a fission explosive can be contained and used to transfer energy to compress and ignite a physically separate component containing thermonuclear fuel.

The fact that certain ICF targets rely on similar physical principles was also declassified:

In some ICF targets, radiation from the conversion of the focused energy (e.g., laser or particle beam) can be contained and used to compress and ignite a physically separate component containing thermonuclear fuel.

Appendix B: Excerpts from the Public Transcript of the May 3, 1979 Meeting of the USDOE Energy Research Advisory Board; Testimony of Dr. John Foster, Reporting on the Conclusions of the ICF Review he Chaired in 1969. (pp. 146 ff.)

...let me start by discussing the objective... this question of objective is a difficult one for the community... When it was first thought of, it was... in terms of a scientific activity in pursuit of a scientific objective. Can one make, in the laboratory, deuterium (and) tritium burn to the point of an explosive? Exciting question.

A few years ago, when it was decided to make an enormous increase... in the funding of this kind of a notion, it was because, at least in the minds of the people in (DOE) headquarters and in the minds of people in the Congress, there was a different objective. There was the hope, or the expectation, or the promise that this kind of an approach could lead to power for the United States—power in the year 2020, 30, 40, 50, whatever. And it was the realization of the importance of having large amounts of energy from some sources in that period that led to the interest and the additional funding of this program. But that implies automatically a change in the objective, from one of science interest and scientific demonstration to one of solution of a national program. So, the objective now is to determine whether or not one can put such a concept to work on the U.S. grid, and that entails not just questions about whether or not you can deliver energy into some small spot and make it go off, but whether or not you could live with the total picture.

Given this objective... the question then is... what kind of strategy should one use, and there are a variety of strategems. There is only one strategy if you go to one laboratory or one individual to find out. But the trouble is that the government has to live with the whole. ...I think one of the basic questions in the strategy (is) ...how much risk do you want to run... If you really want to give yourself the highest chance of realizing the optimum out of this kind of concept, you can't get it by a single thread approach, without running the risk of a dead period for a decade or so, while you regroup and try some other (driver) approach.

...So, let me say just in conclusion what we would recommend. We would recommend that the Department (of Energy) not go ahead with NOVA as was requested by the (Livermore) laboratory, but recognizing that the program... needs a driver in the 100-kilojoule range, we would recommend that the best way to do it is to take the first piece of NOVA and build that, without prejudicing the question of whether you want to build any more of it.

CHAIRMAN BUCHSBAUM: Johnny, isn't this really a major point that the technology that underlies NOVA is known to be ultimately inadequate to do the job and, probably, therefore, will require a different technology?

FOSTER: Yes. And by a different machine.

BUCHSBAUM: Certainly by a different machine, certainly not glass, which is intrinsically inefficient. Have you addressed the question of how much money you want to pour into a technology which you know to be a dead-end technology?
FOSTER: We probably ought to be a bit cautious about dead-end technology. From the weapons point of view at say a megajoule, using a pulse once or twice a day—or once every few minutes, the glass is perfectly acceptable from the point of view of performance and economics.
BUCHSBAUM: I am talking about energy.
FOSTER: It is from the power application point that its efficiency is not adequate.
...
MR. FUBINI: Is it certain that the laser is a better solution than...
FOSTER: No. But let me say one thing. We did, as the group of individuals that I mentioned—we did ask ourselves... can we see the way through to the end from the systems studies that have been done? The answer was "yes," in the sense that if we were to take a high-energy, high-Z accelerator (for ions) and deliver a few megajoules in the time... indicated..., the physics of such machines seems to be reasonably enhanced so that those who have to study it would be willing to bet that they can do it. That machine seems to be compatible with the chamber which houses the pellet and in the chambers that we looked at we saw means not only to bring the beam in but to drop the pellets down and get the power out. Not that there weren't some enormously complicated difficulties, but for each, at least that we could identify, we could see a way around or several ways around. We came to the conclusion, therefore, that there were no showstoppers. There was no technological problem which we could identify which we couldn't find some kind of plausible, practical solution to. Now, that is kind of exciting.
...
MR. TOM COCHRAN: When you were talking about no showstoppers, you were talking about heavy ions.
FOSTER: That would be the first candidate if you wanted to make a conservative approach.

References

1. C. M. Stickley, *Phys. Today* **May**, 50 (1978).
2. R. C. Arnold, *Nature* **276**:19 (1978).
3. R. C. Arnold and G. Yonas, article on particle-beam fusion in: *McGraw-Hill Encyclopedia of Energy*, 2nd Edition, McGraw-Hill, New York (1981), p. 508.
4. G. Yonas, *Sci. Am.* **239**:40 (1978).
5. Final Report of the ERDA Summer Study of Heavy Ions for Inertial Fusion, Berkeley report LBL-5543 (December 1976).
6. Proceedings of the Heavy-Ion Fusion Workshop held at Brookhaven National Laboratory, Brookhaven report BNL-50769 (February 1978).
7. Proceedings of the Heavy-Ion Fusion Workshop held at Argonne National Laboratory, Argonne report ANL-79-41 (1979).
8. Proceedings of the 1979 Workshop on Accelerator Issues in HIF, Berkeley report (1980).
9. K. A. Brueckner, ed., Electric Power Research Institute report (1980).
10. R. Martin, R. Arnold, and R. Burke, in Proceedings of 1980 ANS Topical Conference on Fusion Technology at King of Prussia, Pennsylvania (1980).
11. Y.-K. Kim and G. L. Magelssen, eds., Report of the Workshop in Atomic and Molecular Data Needs for HIF, ANL report ANL-80-17 (December 1979).

12. R. Burke *et. al.*, *IEEE Trans. Nucl. Sci.* Vol. NS-24, p. 996, (1977).
13. K. A. Brueckner and S. B. Jorna, *Rev. Mod. Phys.* **46**:325 (1974).
14. R. E. Kidder, *Nucl. Fusion* **16**:405 (1976); **19**:223 (1979).
15. S. E. Bodner, Critical Elements of High Gain Laser Fusion, NRL Memo Report 4453, Naval Research Laboratory, Washington, D. C. (January 1981).
16. M. Clauser, *Phys. Rev. Lett.* **35**:848 (1975).
17. G. L. Magelssen and W. Gula, *Phys. Fluids* **25**:898 (1982).
18. G. L. Magelssen, in Proceedings IEEE International Conference on Plasma Science, Santa Fe, New Mexico, May 18–20 (1981).
19. R. Sacks, R. Arnold, and G. Magelssen, *Nucl. Fusion* **22**:1421 (1982).
20. R. Bangerter, J. Mark, and A. Thiessen, *Phys. Lett* **88A**:225 (1982).
21. R. C. Arnold *et al.*, *Nucl. Instrum. Methods* **199**:557 (1982).
22. R. E. Kidder, *Laser Focus* **Jan.**, 65–66 (1982).
23. F. Winterberger, *Nature* **286**:364 (1980).
24. HIBALL: A Conceptual Heavy-Ion Beam Driven Reactor Study; University of Wisconsin and Institute for Nuclear Research at Karlsrühe [Univ. of Wisc. report UWFDM-450 (June 1981)].
25. E. W. Sucov, ed., ICF Central Station Electric Power Generating Plant: Final Report for March 1, 1979 to September 30, 1980; Westinghouse Fusion Power Systems report WFPS-TME-81-001 (February 27, 1981).
26. *Physics Today*, Vol. 33, #8, p. 77 (August 1980).
27. A. DeVolpi, G. E. Marsh, T. A. Postol, and G. S. Stanford, *Born Secret: The H-Bomb, the Progressive Case and National Security*, Pergamon Press, New York (1981).
28. J. M. Watson *et al.*, in Proceedings of the 1980 International Conference on Thermonuclear Fusion Research, at Berchtesgaden, Federal Republic of Germany (1980).
29. *Nature* **281**:414 (1979).
30. R. Bock, I. Hoffmann, and R. C. Arnold, in *Nucl. Sci. Applications*, Vol. 2 (in press, 1985).

Index